AN INTRODUCTION TO

ELECTRICAL MACHINES AND TRANSFORMERS

AN INTRODUCTION TO

ELECTRICAL MACHINES AND TRANSFORMERS

SECOND EDITION

George McPherson

University of Missouri–Rolla

Robert D. Laramore

Purdue University

WILEY

John Wiley & Sons

New York ▪ Chichester ▪ Brisbane ▪ Toronto ▪ Singapore

*Dedicated to the coming
of our Father's Kingdom*

Library of Congress Cataloging in Publication Data:

McPherson, George, 1921-
    An introduction to electrical machines and transformers
    George McPherson, Robert D. Laramore. — 2nd ed.
      p.   cm.

    Bibliography: p.
    Includes index.
    ISBN 0-471-63529-4
    1. Electric machinery.   2. Electric transformers.
    I. Laramore, Robert D.   II. Title.

TK2182.M32 1990   89-33699
621.31'042—dc20   CIP

Printed in the United States of America

10 9 8 7 6 5 4 3 2 1

PREFACE

Much of the preface to the first edition of this book applies to the second edition as well. The changes found in the present edition reflect the suggestions of people who have used the first edition as a course text. They usually wanted certain additional topics included, and we have endeavored to satisfy those needs. However, the material in the first edition filled to overflowing the standard three-semester hour course in electrical machinery. As a result, for students and teachers who will be involved in a standard course, the new material enhances the value of the book as a reference rather than as a text. Even so, we believe that the book has been improved as a standard text by the use of better notation, by clearing up some obscurities, and by the substantial revision of the transformer chapter.

The major contribution of the new material to the book, as a text, is that it allows the design of different courses. An emphasis on small machinery is possible, for example, by using the sections on PM alternators and brushless dc machines in Chapter 2, PMDC machines in Chapter 5, single-phase induction motors in Chapter 6, and special motors in Chapter 7.

The new Chapter 8 presents the "energy conversion" approach to the calculation of magnetically developed forces. It may be used for a machinery course by using the text material in the following order: Chapter 3, "Transformers," first 10 sections; Chapter 8, "Forces and Torques," first 6 sections; Chapter 2, "Synchronous Machines," first 11 sections; Chapter 4, "Asynchronous Machines," first 12 sections; Chapter 5, "DC Machines," Sections 5.1 to 5.8, 5.10, and 5.14.; and finally the remainder of Chapter 8. A similar plan has been used quite successfully in a senior elective machinery course at the University of Missouri–Rolla.

We express our appreciation to those who helped us in the development of this new edition, particularly Dr. Arthur Miles, now at South Dakota State University, and Dr. Elias Strangas, now at Michigan State University. The section on the isolated asynchronous generator is largely based on the thesis of Professor H. Ouazene of INELEC, Boumerdes, Algeria, and we are grateful to Professor Shi, Chuanli, of the North China Institute of Electric Power, Baoding, Hebi, China, for providing us with two of his published volumes of problems, based in part on the first edition of this book. We are appreciative of the assistance given by the reviewers of this book: Professor Harit Majmudar, Worcester Polytechnical Institute; Professor Allen Drake, University of New Hampshire; Professor Richard Hoft, University of Missouri; Professor Arturo Miles, University of Missouri; and Professor Frederic L. Zeigler, Lawrence Institute of Technology. We are especially grateful for the secretarial assistance provided by Janice Spurgeon, Krista Fester, and Connie Knowles. Additional thanks are due to Kathy Collins for her work on the Instructor's Manual.

GEORGE MCPHERSON
ROBERT D. LARAMORE

PREFACE TO THE FIRST EDITION

I wrote this book because I needed one like it. It has developed out of a quarter century of teaching electrical machinery. More directly, it is the product of nine year's struggle to teach a *meaningful* course in the subject in only three semester hours. Such an objective may well be impossible of attainment, but this book, I hope, will make it more nearly possible.

The book contains sufficient material for two courses and some practical information that may be more useful as reference material. The early sections of each chapter form the basis for a three-semester hour, junior-level, electrical engineering course in machines and transformers. The remaining sections may be used for a second or elective course.

The book has several features designed to conserve class time. First, the figures and text material are quite detailed, and I have attempted to answer questions as they might arise in the mind of the reader. Thus the lecture may be used to confirm what the student has learned, rather than to break new ground. The so-called "energy conversion" approach to torque has been abandoned because it requires much class time in developing concepts of use primarily to the designer. Instead, torque is taught qualitatively on the basis of attraction and repulsion of magnetic poles and quantitatively on the basis of conservation of power in the steady state. Finally, each device is discussed in terms of magnetic fields first, voltages second. This sequence makes it unnecessary to backtrack in order to explain the origin of the flux that generated the voltage.

Each chapter follows the same general plan. The early sections deal with essentials—information that should be a part of the knowledge of every electrical engineer. The middle parts provide a deeper insight into the characteristics of the devices considered, while the final sections deal with useful information of a more specialized nature. This organization of the material may seem awkward and somewhat illogical at first. I believe, however, that this arrangement, together with other features of the book, will enable an instructor to plan a course (or a set of two) that will provide students with considerable insight into the operation and characteristics of electrical machines and transformers. In addition to chapter design, the special features of this book include flexibility, rigorous development of the circuit model of each device, and the occasional insertion of advanced material marked "For Further Study" in places where it should occur logically.

Class time devoted to the logical development of each circuit model is well spent.

Today's student is impatient and uncomfortable with cookbook approaches to engineering. A clear understanding of the physical basis of a model allows him or her to use it more intelligently in the solution of problems. For this reason, the development of each model has been made as rigorous and truthful as possible, considering that the book is written on an introductory level. Bli and Blv are not mentioned because they are inappropriate for application to conductors shielded from the air-gap flux by iron teeth.

From the standpoint of course planning, I have tried to make the book extremely flexible. Curricula in which prerequisite courses are weak in magnetic and/or three-phase circuits are accommodated by the inclusion of appendixes on these subjects. The internal organization of each chapter contributes to this flexibility. If an instructor feels that synchronous machines are more important than transformers, he or she may take the class deeply into Chapter 2 and dabble only into the early sections of Chapter 3. The book is designed for hopping, skipping and jumping by making the later sections of each chapter more or less independent of each other. This permits the instructor to be eclectic in presenting the theory and applications of each device. The chapters themselves need not be used in sequence, although synchronous machines should be studied before the others. Even this priority may be avoided (at some cost in clarity) by including Section 1.7 as part of the introduction to the course.

Essential to what I consider to be an introductory course would be the first 6 sections of Chapter 1, plus Section 1.8; the first 15 sections of Chapter 2; the first 11 sections of Chapter 3; the first 12 sections of Chapter 4, and the first 9 sections of Chapter 5. These can be taught in less than three semester hours, leaving time for adding more meat to the course here and there.

Perhaps most important, I have tried to make each subject understandable to undergraduate students. Having used the book in loose-leaf form for several semesters, it does appear that my students find machinery less puzzling. Those advanced sections "For Further Study" need not be mentioned in an elementary course, but they will be read and appreciated by the more interested students. They are also useful in the development of a second course.

So many people have contributed to this book's evolution that I cannot begin to acknowledge them all. In the selection of content for this book I have been profoundly influenced by Professor Robert H. Nau, who has been my partner in teaching our undergraduate machinery course at the University of Missouri-Rolla. I particularly thank Dr. J. R. Betten and Dr. J. Derald Morgan, Electrical Engineering Department Chairmen who encouraged me in my work. I am grateful to Mrs. Michael McKinzie and Miss Laurie Armstrong, who worked overtime to type the final manuscript revisions, and to student draftsmen Ward Silver and Dave Tordoff. But the production of this book would have been utterly impossible without the help of Mrs. Michael Krone, who typed most of the first draft. She was somehow able to convert, swiftly and accurately, hundreds of pages of my miserable handwriting into typing that was clear and beautiful.

GEORGE MCPHERSON

ABOUT THE AUTHORS

GEORGE MCPHERSON was born at Westfield, New York, July 16, 1921. His parents were Ohio natives and soon returned to that state. He grew up and received his education in Ohio. He graduated from the Ohio State University with a B.E.E. degree in the middle of World War II, and went to work for the Naval Research Laboratory. Near the end of the war he was drafted into the Navy as a Chief Specialist. At the end of the war he returned to Ohio State to earn a B.Sc. in education and a M.Sc. in electrical engineering. He came out of the war a pacifist and is now a member of the Religious Society of Friends (Quakers).

After one year of teaching high school science and math, he joined the Battelle Memorial Institute in Columbus, Ohio, doing four years of research in vibratory oil well drilling techniques. Having decided to make electrical engineering education his life work, he joined the faculty of the University of Kentucky as assistant professor. He moved to the University of Missouri–Rolla (then the Missouri School of Mines and Metallurgy) in 1956, and retired from UMR as Professor Emeritus in 1985. After three years of retirement he has returned to UMR to teach electrical machinery part time.

Professor McPherson is a member of Sigma Xi and is a recipient of the University of Missouri's Thomas Jefferson Award. In 1984 he was commissioned a Kentucky Colonel by Governor Martha Layne Collins. He has been elected a Fellow of IEEE and has received the Education Award of the St. Louis section of that organization. The Small Motor Manufacturers Association gave him their 1989 Hall of Fame Award. He is a Registered Professional Engineer in the state of Ohio.

ROBERT D. LARAMORE was born July 15, 1952 at St. Louis, Missouri, where he continued to live throughout childhood. His primary and secondary education were received at the Normandy school district. He left the St. Louis area to attend the University of Missouri–Rolla, from which he graduated with B.S.E.E. and M.S.E.E. degrees in 1973 and 1975, respectively. While in college he met and married Carol Candler. Together, they have been blessed with four children.

Upon graduation, he took a position with Missouri Engineering Corporation, where he worked for five years as a project engineer. In 1980 he returned to UMR's Electrical Engineering department to teach undergraduate courses and coordinate the laboratory program. While there, he played an active role in the design, construction, and development of the Emerson Electric Company Machines and Drives Laboratory. In

1988 he moved to Lafayette, Indiana to join the faculty of Purdue University as assistant professor in the Electrical Engineering Technology department.

Mr. Laramore is a member of Tau Beta Pi and Phi Kappa Phi. He was included in Outstanding Young Men of America—1985. He is a member of IEEE and is a Registered Professional Engineer in the state of Missouri.

CONTENTS

1 WHAT MACHINES AND TRANSFORMERS HAVE IN COMMON

1

1.1 Purposes of Electrical Machines and Transformers, 1

1.2 Mechanical Torque and Power, 3

1.3 The Role of the Magnetic Field, 5

1.4 Types of Machines, 6

1.5 Machine Terminology, 7

1.6 Magnetic Poles in Machines, 13

1.7 Torque Development in the Three Basic Machines, 17

1.8 Efficiency, 18

1.9 Loss Phenomena, 20

1.10 Maximum Efficiency, 26

1.11 The Nameplate, 27

1.12 Regulation, 28

Problems, 30

2 SYNCHRONOUS MACHINES

35

2.1 Synchronous Machine Characteristics, 35

2.2 Construction of a Synchronous Machine, 36

2.3 Cylindrical-Rotor Theory Versus Salient-Pole Theory, 47

2.4 The Magnetomotive Force Field of the Rotor, 47

2.5 The Rotating MMF of the Stator Armature Winding, 50

2.6 Air Gap Magnetic Field and the Flux per Pole, 59

2.7 Voltage Induced in the Armature Winding of a Synchronous Machine, 64

2.8 Leakage Impedance and Equivalent Circuit, 78

2.9 Circuit Model of the Synchronous Machine, 82

2.10 Relative Magnitudes of Synchronous Machine Impedances: The Per-Unit System, 89

2.11 Power and Torque as Related to Power Angle δ, 90

2.12 Damping and Synchronous Motor Starting, 93

2.13 How Field Current Changes Are Reflected in the Circuit Model, 95

2.14 Operation as a Motor, 96

2.15 Operation as a Synchronous Condenser, 102

2.16 The Synchronous Machine as a Generator, 104

2.17 Measuring x_d, 107

2.18 The Short-Circuit Ratio, 113

2.19 Calculating Excitation Requirements for Given Operating Conditions, 115

2.20 Permanent-Magnet Alternators, 123

2.21 Single-Phase Synchronous Alternators, 125

2.22 "Brushless dc" Motor, 129

2.23 The Capability Curve, 141

2.24 Short-Circuit Current Transients in Synchronous Alternators, 144

Problems, 148

3 TRANSFORMERS

155

3.1 Why Transformers Are Essential to Power Systems, 155

3.2 Types of Power System Transformers, 157

3.3 Elements of a Transformer, 158

3.4 Faraday's Law and Lenz's Law, 161

3.5 Magnetic Fluxes of a Transformer, 163

3.6 Transformer Inductances, 164

3.7 Real-World Transformers Are Nearly Ideal, 171

3.8 The Ideal Transformer, 176

3.9 Core Excitation, 180

3.10 The Steinmetz "Exact" Circuit Model of a Transformer, 186

3.11 The Steinmetz Approximate Transformer Model, 189

3.12 Relative Magnitudes of the Circuit Model Constants: The Per-Unit System, 193

3.13 How to Find the Primary Voltage Necessary to Produce a Desired Secondary Voltage, 198

3.14 Transformer Voltage Regulation, 200

3.15 Transforming Three-Phase, 204

3.16 Three-Phase Transformation with Two Transformers, 212

3.17 Computing Transformer Efficiency, 218

3.18 Measuring Transformer Quantities, 220

3.19 Measurements on Three-Phase Transformers, 223

3.20 Autotransformers, 227

3.21 Instrument Transformers, 231

Problems, 232

4 INDUCTION, OR ASYNCHRONOUS, MACHINES 239

4.1 History and Application of Induction Machines, 239
4.2 Construction of Induction Machines, 240
4.3 How the Induction Motor Works, 242
4.4 The Concept of Slip, 245
4.5 The Frequency of Rotor Voltages and Currents, 246
4.6 The Induction Motor Under Load, 246
4.7 Circuit Model of the Induction Machine, 247
4.8 Losses, Power Flow, and Efficiency of Induction Motors, 256
4.9 Air Gap Power: The Magic Quantity, 257
4.10 Separation of Mechanical Load from Rotor Copper Loss in the Circuit Model, 259
4.11 Performance Calculations Using the Circuit Model, 260
4.12 Torque/Speed Characteristics of Induction Machines, 264
4.13 Some Useful Exact and Approximate Relationships, 269
4.14 Control of Performance Characteristics by Rotor Design, 271
4.15 Tests to Determine Circuit Model Impedances, 276
4.16 Starting Induction Motors, 284
4.17 Speed Control of Induction Motors, 287
4.18 The Asynchronous Generator, 302
Problems, 315

5 DIRECT-CURRENT MACHINES 319

5.1 The Importance of dc Machines, 319
5.2 Construction of dc Machines, 319
5.3 How dc Machines Work, 322
5.4 The Generated Voltage of a dc Machine, 325
5.5 Circuit Model of a dc Machine, 329
5.6 Developed Torque, 331
5.7 Brushless dc Machines, 333
5.8 Field Excitation of Wound-Pole Machines, 334
5.9 Finding $K_a\phi$, 336
5.10 Speed Control of Shunt and Permanent-Magnet Motors, 342
5.11 Per-Unit Values of dc Machine Resistances and Losses, 348
5.12 Direct-Current Motor Starting and Dynamic Braking, 352
5.13 Armature Reaction and Commutation, 359
5.14 Characteristics of Series and Compound dc Motors, 368
5.15 Permanent Magnet dc (PMDC) Machines, 370
5.16 Self-Excited dc Generators, 377

5.17 Armature Windings for Commutator Machines, 380

5.18 Metadynes, 384

Problems, 390

6 SINGLE-PHASE MACHINES

395

6.1 Why Single-Phase Motors Are Different, 395

6.2 The Universal Motor, 395

6.3 Single-Phase Induction Motors, 396

6.4 The Split-Phase Motor, 401

6.5 Capacitor Motors, 402

6.6 The Shaded-Pole Induction Motor, 405

6.7 Repulsion Motors, 410

6.8 The Circuit Model of Single-Phase Induction Motors, 411

6.9 Using the Single-Phase Induction Motor Model, 416

Problems, 420

7 MACHINES FOR SPECIAL JOBS

423

7.1 Introduction, 423

7.2 Hysteresis Motor, 423

7.3 Synchronous Reluctance Motors, 427

7.4 Linear Motors, 429

7.5 Stepper Motors, 431

7.6 Switched Reluctance Motors, 438

7.7 Low-Inertia dc Motors, 439

7.8 Torque Motors, 441

7.9 Synchros, 443

Problems, 445

8 FORCES AND TORQUES IN ELECTROMAGNETIC SYSTEMS

447

8.1 The Energy Conversion Approach to Force or Torque Calculation, 447

8.2 Energy Stored in a Magnetic Field, 447

8.3 Forces or Torques in Linear Systems with One Electrical Input, 451

8.4 Linear Systems with Two Electrical Inputs, 457

8.5 Nonlinear Systems, 460

8.6 Torque in Machines with Cylindrical Air Gaps, 469

8.7 Torque of Cylindrical-Rotor Synchronous Machines, 474

8.8 Torque Relations in Three-Phase Induction Motors, 475

8.9 Torque in dc Machines, 477

Problems, 482

APPENDIX **A** **A REVIEW OF MAGNETIC CIRCUITS**   **489**

    A.1   Magnetic Concepts, 489
    A.2   Magnetic Circuits, 491
    A.3   Two Types of Magnetic-Circuit Problems, 493
    A.4   The Magnetization Curve, 499
    A.5   Superposition in Nonlinear Magnetic Circuits, 500
    A.6   Parallel Magnetic Circuits, 502
    Problems, 503

APPENDIX **B** **BALANCED THREE-PHASE CIRCUITS**   **505**

    B.1   Why Study Balanced Three-Phase Circuits?, 505
    B.2   Single-Phase Power, 505
    B.3   Description of Balanced, Three-Phase Circuits, 507
    B.4   Definition of a Phase, 507
    B.5   Relationships Between Line and Phase Voltages, 507
    B.6   Load Impedance Connections, 510
    B.7   Solving a Δ Load, 511
    B.8   Solving a Y Load, 514
    B.9   Three-Phase Problems Requiring Phasor Notation, 517
    Problems, 525

APPENDIX **C** **SALIENT-POLE THEORY OF SYNCHRONOUS MACHINES**   **529**

    C.1   Derivation of the d–q Model, 529
    C.2   Determining x_d and x_q, 534
    C.3   Procedure for Finding I_f, 536
    C.4   Power-δ Characteristics of a Salient-Pole Synchronous
           Machine, 536
    C.5   The Salient-Pole Machine as a Motor, 539
    Problems, 544

APPENDIX **D** **A SHORT BIBLIOGRAPHY**   **547**

    GLOSSARY OF SYMBOLS   **549**

    INDEX   **563**

1

WHAT MACHINES AND TRANSFORMERS HAVE IN COMMON

1.1 PURPOSES OF ELECTRICAL MACHINES AND TRANSFORMERS

Engineers call electric motors and generators "electrical machines." The reason for this more general term is that the same device may operate either as a motor or as a generator. Electrical machines convert electrical energy into mechanical energy or mechanical energy into electrical energy. When the conversion is from electrical to mechanical, the machine is called a *motor*. When it is being used to convert from mechanical energy to electrical energy, the machine is called a *generator*.

These two modes of operation are illustrated in Figure 1.1. In Figure 1.1*a*, a single-phase machine is shown operating as a motor to drive a washing machine. Electrical energy is supplied to the motor through a line cord. This energy is converted to the mechanical form by the motor, which supplies the energy to the washer's mechanism through a belt. Figure 1.1*b* shows an alternator in a truck. An alternator is a polyphase synchronous machine used as a generator. Mechanical energy is delivered to this machine from the truck's engine through a belt. The alternating-current output energy is rectified to charge the car's battery.

Figure 1.2 shows the essentials of the two modes of operation. In either mode, some energy is lost in the form of heat so that the input power has to be greater than the output.

A *transformer* is a device that transfers energy from one ac system to another. A transformer can accept energy at one voltage and deliver it at another voltage. This permits electrical energy to be generated at relatively low voltages and transmitted at high voltages and low currents, thus reducing line losses, and to be used at safe voltages. Because transformers operate on many of the same principles as do machines, transformers and machines are usually studied together.

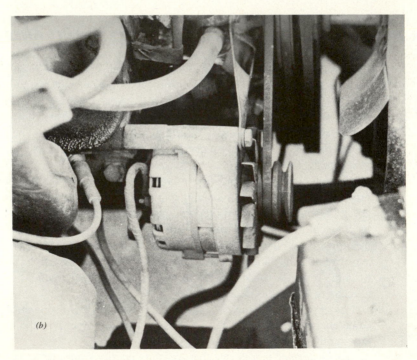

FIGURE 1.1 Two rotating machines, (*a*) Motor action. (*b*) Generator action.

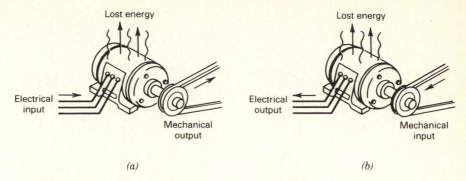

FIGURE 1.2 The two modes of machine operation. (*a*) Motor mode. (*b*) Generator mode.

1.2 MECHANICAL TORQUE AND POWER

The input to a generator and the output of a motor are mechanical in nature. It is well, then, to review the relationships between mechanical quantities, especially those having to do with rotation.

Isaac Newton defined mechanical work as the integral of force over the distance through which it acts:

$$W = \int f \, dx \quad \text{joules (J)} \tag{1.1}$$

In the S.I. (Système International) system of units, Equation 1.1 finds the work in joules (1 J = 1 W-s), if the force is in newtons and the distance x is in meters. The power, in watts, is the rate of doing work:

$$P = \frac{dW}{dt} = f \frac{dx}{dt} \quad \text{watts (W)} \tag{1.2}$$

since from Equation 1.1, $dW = f \, dx$. One watt equals one joule per second (1 W = 1 J/s).

Rotary work, power, and speed are easily visualized in terms of a hand-cranked ice-cream freezer, as shown in Figure 1.3. A hungry person has just applied a force of f newtons to the crank, causing it to move a differential distance dx meters (m). As a result, the crank arm rotated $d\theta$ radians (rad). We know from trigonometry that

$$dx = r \, d\theta \quad m \tag{1.3}$$

The work done in turning the crank is, by Equation 1.1

$$W = \int f r \, d\theta \quad J \tag{1.4}$$

Now in rotational mechanics, the tangential force times the radial distance at which it is applied, measured from the axis of rotation, is called the torque (τ):

$$\tau = fr \quad \text{newton-meters (N-m)} \tag{1.5}$$

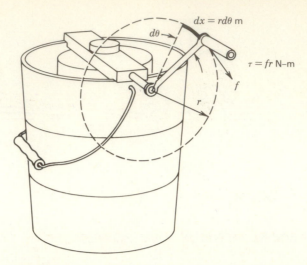

FIGURE 1.3 Ice-cream freezer with crank length of r meters.

Then Equation 1.4 may be written

$$W = \int \tau \, d\theta \qquad (1.6)$$

and $dW = \tau \, d\theta$. The mechanical power is given by

$$P = \frac{dW}{dt} = \tau \frac{d\theta}{dt} = \tau\omega \quad \text{W} \qquad (1.7)$$

where $\omega \equiv d\theta/dt$ is the *angular velocity* in radians per second. It follows, then, that torque may be calculated from the mechanical power if the speed is known:

$$\tau = \frac{P}{\omega} \quad \text{N-m} \qquad (1.8)$$

The so-called English system of units is still very widely used. It is well to be acquainted with the relationships between English and S.I. units. James Watt determined that the power capability of the average horse was 550 foot-pounds per second (ft-lb/s). One inch is 0.025400 m, and one pound is 4.448 N, with the following results:

$$\text{Torque in lb-ft} = 0.738 \cdot \text{torque in newton-meters} \qquad (1.9)$$

$$\text{Horsepower} = \frac{\text{watts}}{746} \qquad (1.10)$$

If rotational speed n is given in revolutions per minute, then

$$n = \frac{60\omega}{2\pi} \quad \text{rev/min} \qquad (1.11)$$

and

$$\tau = \frac{\text{horsepower} \times 5252}{n} \quad \text{lb-ft}$$

or

$$\tau = \frac{7.04 \times \text{power in watts}}{n} \quad \text{lb-ft} \tag{1.12}$$

1.3 THE ROLE OF THE MAGNETIC FIELD

The magnetic field in a machine forms the energy link between the electrical and mechanical systems. The magnetic field is produced by current flowing in coils of wire inside the machine, or by a combination of coils and permanent magnets. The magnetic field performs two functions:

1. Magnetic attraction and repulsion produces mechanical torque.
2. The magnetic field, by Faraday's law, induces voltages in the coils of wire.

Faraday's law states that the voltage e induced in a coil of wire by a changing magnetic flux is given by

$$|e| = \frac{dN\phi}{dt} \equiv \frac{d\lambda}{dt} \quad \text{V} \tag{1.13}$$

where N is the number of turns in a coil of wire and ϕ is the magnetic flux (units: webers, Wb) linking the coil. The term $\lambda \equiv N\phi$ is called the flux linkage of the coil.[1]

In actual operation, both magnetic phenomena are going on at the same time. The instantaneous power *input* to a *motor* is

$$p = vi \quad \text{W} \tag{1.14}$$

where v is the terminal voltage and i is the current, as shown in Figure 1.4a. If $v = 0$, no power can be absorbed from the line. The terminal voltage includes internal resistance and reactance voltage drops, but most of v is due to the Faraday's-law induced voltage e. It is this field-induced voltage that permits the motor to draw power from the line to be converted into mechanical power. While the magnetic field is inducing this voltage, it is also developing the mechanical output torque. Note that the induced voltage is in opposition to the current flow. It is often called the *countervoltage* or *counter-emf* for that reason.

Conditions for *generator* operation are shown in Figure 1.4b. Here the voltage induced by the field is in the same direction as the current and is called the *generated voltage*. What about the torque produced by the magnetic field when the machine is operating as a generator? In this mode of operation, the machine torque opposes that

---

[1]This definition of λ assumes that the same flux links all turns of the coil. The more general case is discussed in Chapter 3.

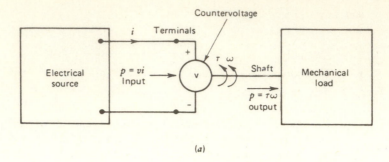

(a)

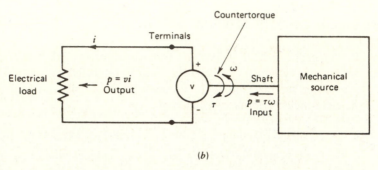

(b)

FIGURE 1.4 Terminal and shaft conditions for motor and generator operation of an electrical machine. (a) Motor action. (b) Generator action.

of the engine trying to drive the generator, and is called the *countertorque*. Since the mechanical input to the generator is, by Equation 1.7,

$$P_{\text{mech}} = \tau\omega \quad \text{W} \tag{1.15}$$

it is obvious that it would be impossible to have any mechanical input to the machine if $\tau = 0$ (no countertorque). In Equation 1.15, the torque is given in newton-meters, and the shaft speed ω is in radians per second.

Summing up, in *motor* action the magnetic field produces the output torque and, by inducing a countervoltage makes it possible for the machine to absorb power from the line to be converted into mechanical output. In *generator* action, the magnetic field induces the generated voltage and, by developing a countertorque, makes it possible for the machine to absorb power from the mechanical drive to be converted to electrical output.

1.4 TYPES OF MACHINES

Each machine has a stationary part and a moving part. The moving part is connected to the mechanical system and provides the means for mechanical input or output, depending on whether the machine is being used as a generator or a motor. The motion may be *linear,* as in the case of "linear motors," often used to convey parts in a manufacturing plant. The motion may be *reciprocating* or *vibrating*, as in the case of some electric razors. In most machines, however, the movable part *rotates*; and **this book is concerned primarily with rotating machines**.

There are three basic rotating machines:

THE THREE BASIC MACHINES

1. The polyphase synchronous machine
2. The polyphase asynchronous or "induction" machine
3. The dc machine

Once the operating principles of these three machines are understood, most other rotating and linear machines will be found to operate on the basis of the same principles. Examples of these other machines are single-phase induction motors, universal motors, and repulsion motors.

1.5 MACHINE TERMINOLOGY

Rotating machines have an outside (stationary) part, called the *stator*. The inner (rotating) part is called the *rotor*. The rotor is centered within the stator, so that the rotor axis is concentric with that of the stator (see Figure 1.5). The space between the outside of the rotor and the inside of the stator is called the *air gap*.

The rotor is mounted on a stiff rod (usually steel), called a *shaft*. The shaft is supported in bearings so that the rotor is free to turn. The shaft extends through one or both of the bearings to provide a means to connect the machine to the mechanical system (see Figure 1.6). As Figure 1.7 shows, this connection may be direct or, by means of a coupling, a pulley and belt or a gear. Note that the rotor is solidly fastened to the shaft so that the rotor and the shaft turn at the same speed. Therefore the terms "rotor speed," "shaft speed," or "machine speed" all mean the same thing and are used interchangeably.

The rotor and the stator each have three basic parts: the core, the windings, and the insulation. Thus it is proper to speak of "the rotor core," "the stator windings," "the rotor insulation system," and so on. The purpose of the rotor and stator cores is to conduct the magnetic field through the coils of the windings. The cores are almost always made of iron or steel.[2]

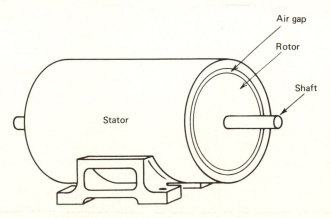

FIGURE 1.5 Basic parts of a rotating machine.

[2]Magnetic ceramic materials are sometimes used. The use of such ceramic materials is increasing as magnetic ceramic materials with improved properties are being developed.

End bell

Bearing housing

Shaft

Nameplate

Terminal box

FIGURE 1.6 Photograph of a single-phase induction motor. (Courtesy of Emerson Electric Company.)

The *windings* conduct electric currents that are the source of the magnetic field and provide closed loops in which voltages may be induced by the magnetic field, in accord with Faraday's law. If the current in a winding varies with the load on the machine, it is called *load current*. If the current in a winding merely provides a magnetic field and is independent of the load, it is called *magnetizing current*, or *exciting current*. A winding that carries *only* load current is called an *armature*. A winding that carries *only* magnetizing current is called a "field winding." The current in field windings is almost always dc. A winding which carries current that provides *both load and magnetizing* functions is called a *primary* and is usually the power-input winding. In such cases, the *power output* winding is called the *secondary*. Table 1.1 relates this terminology to the basic machines and transformers.

A winding that carries *load* current must handle all of the power being converted or transformed by the device; however, the *magnetizing* power requirement is relatively small. The steady-state power input to a field winding is only about $\frac{1}{2}$ to 2 percent of the rated power of the machine. This input power to dc field windings is all consumed as I^2R loss, except during the turn-on transient, lasting at most a second or two, during which energy is being stored in the magnetic field.

The winding conductors are almost always copper or aluminum. Depending on the amount of current to be carried, they may consist of coils of wire (Figure 1.8) or heavy bars. Currents of large amperage require conductors of large cross-sectional area. Each winding consists of several coils or bars in series or in series/parallel combinations. The ends of certain of the windings are brought out to terminals to permit easy connection to the electrical system.

In some machines, the windings of either the rotor or the stator may be placed around projecting magnetic pole pieces, called *salient poles* (salient means "prominent"

(b)

(a)

FIGURE 1.7 Ways to connect a motor to its mechanical load. (*a*) Fan load directly connected to motor shaft. (*b*) Drill press, belt driven by motor pulley. (Courtesy of Emerson Electric Company.)

FIGURE 1.7 Continued. (*c*) Motor connected to load by flexible coupling. (Courtesy of General Electric.) (*d*) Reduction gears connect motor to wheel of earthmover. (Courtesy of Emerson Electric Company.)

TABLE 1.1
Winding Terminology

Device	Winding Function	Winding Term	Location	Current Type
Synchronous machine	Input/output	Armature	Stator	ac
	Magnetizing	Field	Rotor	dc
dc machine	Input/output	Armature	Rotor	ac in winding dc at brushes
	Magnetizing	Field	Stator	dc
Induction machine	Input	Primary	Stator	ac
	Output	Secondary	Rotor	ac
Transformer	Input	Primary		ac
	Output	Secondary		ac

or "obvious"). Examples of this kind of core construction are shown in Figure 1.9. When a core has salient poles, the coils of the winding are wound around the waists of the pole pieces. These narrower parts of the salient poles are called the *pole cores*. The shaped ends of the poles are called *pole shoes*. Their purpose is to provide the correct flux density distribution in the air gap. Salient poles are used on the *stator* cores of *dc machines* and on the *rotor* cores of many *synchronous machines*, and in both cases they carry dc field windings. The mechanical weakness and air resistance of salient poles prohibit their use on rotors of large, high-speed synchronous generators designed to be driven by steam or gas turbines. Turboalternators, as these are called,

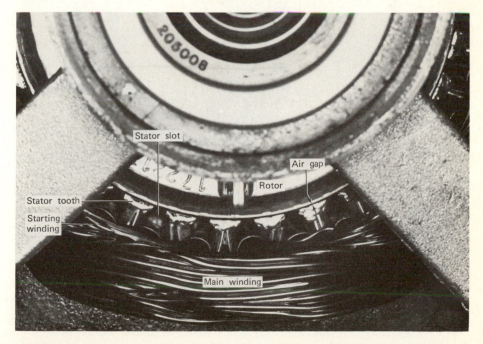

FIGURE 1.8 Closeup of a single-phase motor stator showing coils, slots, and teeth. (Courtesy of Emerson Electric Company.)

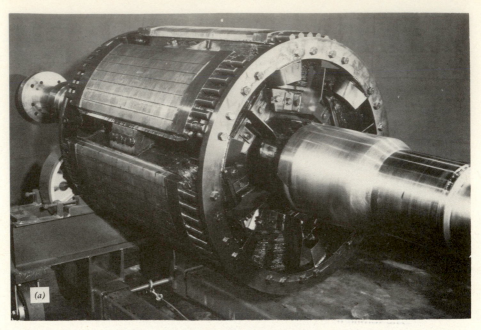

FIGURE 1.9a Salient-pole rotor for a synchronous machine. (Courtesy of Westinghouse Electric Corporation.)

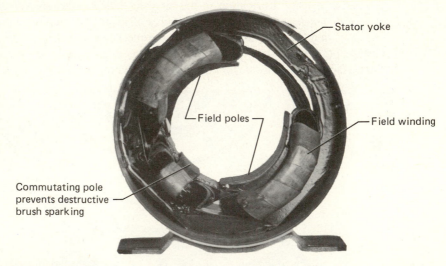

FIGURE 1.9b Stator of a two-pole dc machine, showing salient poles. (Courtesy of Baldor.)

always have cylindrical rotors with the field windings embedded in slots cut into the rotor surface.

Insulation systems prevent short circuits between turns of a given winding coil and insulate the windings from the core iron. The core is almost always grounded for safety's sake, and, in some large machines, the windings may be several thousand volts above ground in normal operation. Lighting strokes and switching surges in power systems can cause very high winding-to-ground voltages of short time duration, even in 115-V or 230-V motors. The insulation system must protect the machine against damage under all circumstances likely to be encountered.

From a thermal point of view, the insulation is the weakest part of the machine. High temperatures cause insulation to deteriorate, and I^2R losses in the winding conductors cause them to get hot. *Thus the ability of the insulation to withstand high temperatures sets the current limit for windings of a given cross section.* Insulating materials are rated according to the maximum permissible "hot-spot" temperature; that is, the temperature at the hottest point in the winding.

The insulation system consists of (1) the *conductor* or wire *insulation*, often natural or synthetic varnish; (2) the *coil insulation*, usually some kind of tape or several layers of tape; and (3) the *slot liner*. Examples are shown in Figures 1.8, 2.3, and 2.5.

When the winding consists of coils that are located in slots, the coils are held in place in the slots by *slot wedges*. The projections of the iron core between the slots are called *teeth*. DO NOT CONFUSE THESE WITH SALIENT POLES. Reference to Figures 1.8 and 1.9 will show that salient poles are relatively much larger than teeth. In fact, a machine having both its rotor and stator windings in slots is often called a "nonsalient pole machine."

1.6   MAGNETIC POLES IN MACHINES

The torque developed by electrical machines is the result of attraction and repulsion between magnetic poles on the rotor and stator. It is well known that magnetic poles of like polarity (two north or two south poles) will produce a mechanical force of repulsion, while two poles of unlike polarity (one north and one south pole) will produce an attractive force. In machines, the windings are so designed that they develop magnetic poles on the inside surface of the stator and the outside surface of the rotor. In other words, these poles appear on the inner and outer surfaces of the air gap. Repulsion and attraction between the rotor poles and those of the stator produce a torque on the rotor and a reaction torque against the stator. Figure 1.10 shows the rotor and stator magnetic poles for a machine having salient stator poles and a cylindrical rotor.

To produce a continuous unidirectional torque, the number of rotor poles must equal the number of stator poles. The number of poles on either the rotor or the stator

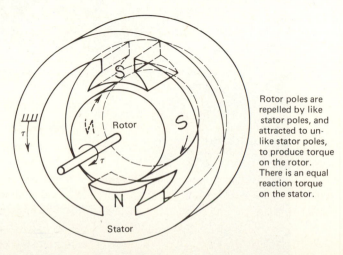

Rotor poles are repelled by like stator poles, and attracted to unlike stator poles, to produce torque on the rotor. There is an equal reaction torque on the stator.

FIGURE 1.10 Magnetic poles on the stator and rotor air gap surfaces of a two-pole, salient-pole electrical machine.

must be even (there must be as many south poles as north poles). If a machine has four stator poles, there must be four rotor poles, and the machine is said to have four poles. This number of poles will be given the symbol p. If the machine is in the motor mode, the torque on the rotor causes it to rotate. If the machine is operating as a generator, the torque developed is a countertorque, and the mechanical drive must overcome this torque and force the machine to rotate in the opposite direction.

to the countertorque

Magnetic Polarity

The direction of the magnetic field of a current-carrying coil of wire may be determined by Ampere's right-hand rule, as shown in Figure 1.11. If a coil is grasped by the right hand so that the fingers encircle the coil in the direction of current flow, the thumb will point in the *direction* of the magnetic flux field produced by the coil. If the coil is considered as an electromagnet, the thumb is directed toward the "north" end of the magnet (the top of the coil in Figure 1.11). We think of flux as streaming *out* of a north magnetic pole and flowing *in* at a south magnetic pole. The actual pattern of magnetic flux around such a coil is shown in Figure A.1 in Appendix A. The direction of flux flow provides one way of determining magnetic polarity. If flux is coming out of a pole, it is a north pole. If the flux is entering, it is a south pole.

When, as in Figure 2.12 (p. 46) or Figure 5.7a (p. 326) coils are wound around the iron cores of salient poles, it is not difficult to visualize the location or to determine the polarities of the poles. When the coils are embedded in slots cut into the surface of a cylindrical core, the location of the magnetic poles on the core surface, and the determination of their polarities, are not so obvious. Examples of such windings are

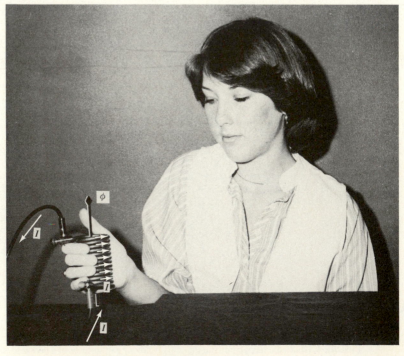

FIGURE 1.11 Ampere's right-hand rule.

FIGURE 1.12 Application of the right-hand rule to an embedded winding. (*a*) Stator of a two-pole, single-phase induction motor. (*b*) End turns cut away to show slot conductors. (Continued on p. 16.)

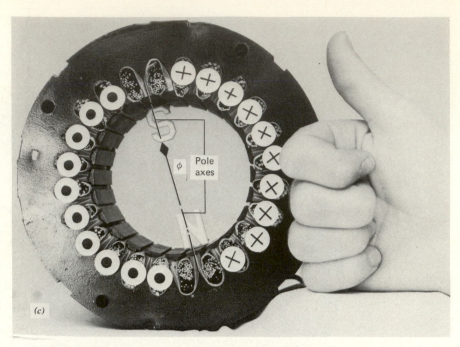

FIGURE 1.12 Continued. (*c*) Current sectors and magnetic poles at a given instant.

shown in Figures 2.4 (p. 38), 2.8 (p. 41), 2.10*a* (p. 43), 2.11*b* (p. 45), 5.3 (p. 322), and 5.6 (p. 325).

Application of the right-hand rule to embedded windings is shown in Figure 1.12. The winding is that of a two-pole stator for a single-phase induction motor. For Figure 1.12*b* the front-end turns of the coils have been sawed off so that the slot conductors may be seen. At a given instant, the directions of the currents in the slot conductors are shown in Figure 1.12*c*, where a dot indicates the point of the current arrow coming out of the paper, and a cross represents the tail of a current arrow directed into the paper. Note that a current pattern is developed which is arranged into sectors such that currents in a given sector are all in the same direction. The current is alternating, so that after a half cycle the slot conductor currents in both sectors will have reversed.

Consider a radial axis extending through the center of the machine to form a dividing line between these sectors. Current is circulating around that axis. At the instant of time illustrated by Figure 1.12, current is flowing inward through the slot conductors in the sector on the right and across through the end turns at the back of the core and coming outward through the conductors in the sector on the left. The circuit is completed through the end turns that have been cut away.

Application of the right-hand rule shows that the flux field developed is in the direction shown by the φ arrow. Flux is flowing upward along the axis. It is coming out of the slotted inside surface of the core at the lower end of the axis and entering the core at the top. The currents flowing in these windings have thus produced two magnetic poles on the inside surface of the core: a north pole at the division between the current sectors at the bottom and a south pole at the division between the sectors at the top. When the current reverses, so does the polarity of the magnetic poles.

1.7 TORQUE DEVELOPMENT IN THE THREE BASIC MACHINES

Figure 1.13 shows the magnetic pole locations in a dc machine. The commutator is a mechanical switch that changes connections to the conductors in the rotor slots so that the current pattern remains stationary, even as the rotor revolves. This keeps the rotor magnetic poles approximately halfway between those of the stator and produces a constant torque.

The *stator* windings of both polyphase synchronous machines and polyphase induction machines are alike. Nearly all polyphase motors are designed for three-phase operation, but, for simplicity, a two-phase stator is shown in Figure 1.14. (Three-phase stator windings are discussed in detail in Chapter 2.) A two-phase electrical system supplies two currents that are 90° out of time phase. These currents flow in two separate stator windings, the slot conductors of which are labeled a–a' and b–b' in the figure. The current patterns and magnetic pole positions are shown for two instants: when $\omega t_1 = 45°$ and when $\omega t_2 = 135°$. Note that the poles rotate 90° in

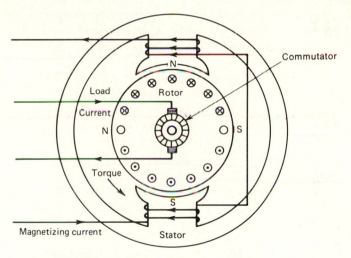

FIGURE 1.13 Magnetic poles of a dc machine.

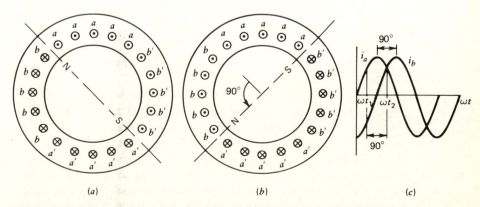

FIGURE 1.14 Rotating magnetic poles produced by a two-phase stator winding. (*a*) Time = t_1. (*b*) Time = t_2. (*c*) Winding currents.

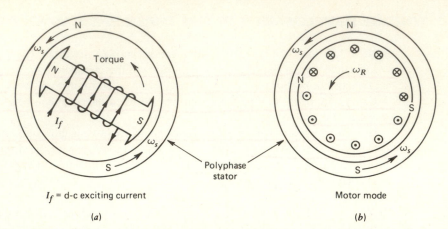

I_f = d-c exciting current

(a)

Motor mode

(b)

FIGURE 1.15 Torque in the basic polyphase machines. (*a*) Synchronous machine torque. (*b*) Induction machine torque.

space as the currents go through 90° of their cycle in time. It is easily shown that the stator poles rotate smoothly and maintain constant magnetomotive force. The rotational speed of these stator magnetic poles is called the *synchronous speed* of the machine, and is given the symbol ω_s.

Figure 1.15 shows how the magnetic poles of polyphase stators similar to that of Figure 1.14 interact with the rotors of synchronous and induction machines to produce torque. The rotor of a synchronous machine is supplied with direct current through slip rings. The current in the stator windings is all load current, hence the stator winding is called an *armature winding*. The rotor winding is called the *field*. Note that the rotor rotates at the same speed as the poles of the stator field, which is why the machine is said to be "synchronous."

The induction machine needs no electrical connections to its rotor windings. Their ends are merely shorted together. Voltages are induced into the rotor windings in accord with Faraday's law. That means that the rotor conductors must be in motion relative to the air gap flux; in other words, the rotor must turn at a speed different from that of the stator poles if torque is to be developed. Motor action is illustrated in Figure 1.15*b*. Here the rotor speed ω_R is less than the stator-pole speed ω_s. However the induced current pattern rotates at the stator-pole speed. This results in a constant angular displacement between rotor and stator magnetic poles, and a uniform torque.

In the generator mode, ω_R is greater than ω_s, with the result that the rotor current directions, the rotor pole polarities, and the torque direction are all reversed.

1.8 EFFICIENCY

The average input power to a machine or a transformer is always greater than the average output power, because some of the input energy is converted into heat by physical phenomena occurring naturally within the device. The student will immediately think of several of these phenomena, for example I^2R in the windings and friction in the bearings of machines. The energy escaping in the form of heat is considered "lost," and the difference between the input power and the output power is defined as the sum of the "losses" of the device.

Efficiency has just two definitions for an engineer. Let the symbol for efficiency be η.

Power efficiency:

$$\eta = \frac{\text{output power}}{\text{input power}} \equiv \frac{P_{\text{out}}}{P_{\text{in}}} \qquad (1.16)$$

Energy efficiency:

$$\eta = \frac{\text{output energy}}{\text{input energy}} \qquad (1.17)$$

By the **law of conservation of energy**, the input power P_{in} may be obtained from the output power P_{out} if the losses are known:

$$P_{\text{in}} - \text{losses} = P_{\text{out}}$$
$$P_{\text{in}} = P_{\text{out}} + \text{losses} \qquad (1.18)$$

Then Equation 1.16 may be written in two additional useful forms:

$$\eta = \frac{P_{\text{out}}}{P_{\text{out}} + \text{losses}} \qquad (1.19)$$

$$\eta = 1 - \frac{\text{losses}}{P_{\text{in}}} \qquad (1.20)$$

■ Example 1.1

A 5-horsepower (hp), 230-volt (V), three-phase induction motor draws 7.0 A from the line at 0.80 power factor when operating at half load. What is the efficiency of the machine at this load and how much power is lost as heat?

Solution.

$$P_{\text{in}} = \sqrt{3}\, V_L I_L \cos \theta$$
$$= \sqrt{3} \cdot 230 \cdot 7.0 \cdot 0.80 = 2230 \text{ W}$$
$$P_{\text{out}} = \text{horsepower} \times 746 \text{ W/hp}$$
$$= 2.5 \cdot 746 = 1865 \text{ W}$$
$$\eta = \frac{1865}{2230} = 0.836$$
$$\text{Losses} = P_{\text{in}} - P_{\text{out}} = 2230 - 1865 = 365 \text{ W} \qquad ■$$

■ Example 1.2

In Example 1.1, the input volts, amperes, and power factor are given, the losses are known to be 365 W, but the output power is not known. Calculate the efficiency and output power. (This is a realistic problem, because often the losses may be calculated from machine data, but no means are available to measure the mechanical output.)

Solution. As before,

$$P_{\text{in}} = 2230 \text{ W}$$

By Equation 1.20,

$$\eta = 1 - \frac{365}{2230}$$

$$= 1 - 0.164 = 0.836, \quad \text{or } 83.6\%$$

By conservation of energy,

$$P_{\text{out}} = 2230 - 365 = 1865 \text{ W}$$

$$\frac{1865}{746} = 2.50 \text{ hp} \qquad \blacksquare$$

1.9 LOSS PHENOMENA

Losses in machines and transformers may be summarized as follows:

Losses Occurring in both Machines and Transformers.

Electrical: copper loss $=$ total I^2R in all windings

Magnetic: core loss $=$ hysteresis loss $+$ eddy current loss

Mechanical: cooling and ventilating losses

Stray load loss: The total of miscellaneous load-dependent I^2R and core losses not otherwise accounted for

Additional Losses in Machines Only.

Electrical: brush drop loss $= V_{BD}I$

Mechanical: friction and windage

Hysteresis denotes a nonlinear phenomenon in which the response to a driving force in one direction is different from the response to a force in the opposite direction. Figure 1.16c is a typical magnetic *hysteresis loop* for iron. The area enclosed by the loop represents the energy lost per cubic meter of iron during one complete cycle of flux variation (joules per m^3 per cycle). To illustrate this point, suppose that the given loop is characteristic of the iron sample being excited by a coil of N turns in the circuit of Figure 1.16a. The "remanent flux density" (in teslas) is B_r. (Obviously, if B_r is large, the material would make a good permanent magnet, but the energy loss in a core made of such material would be large. A good core material should have a small B_r and a "skinny" hysteresis loop to keep the loss low.) Suppose that the iron has been magnetized previously in a reverse direction and is at $-B_r$, and that at $t = 0$, $v = 0$, and $i = 0$. At any time, the flux linkage of the coil is given by

$$\lambda = NBA \text{ Wb-turns} \qquad (1.21)$$

where A is the cross-sectional area of the sample and B is the instantaneous internal flux density. Let ℓ_{Fe} be the average length of the path traversed by the flux flowing

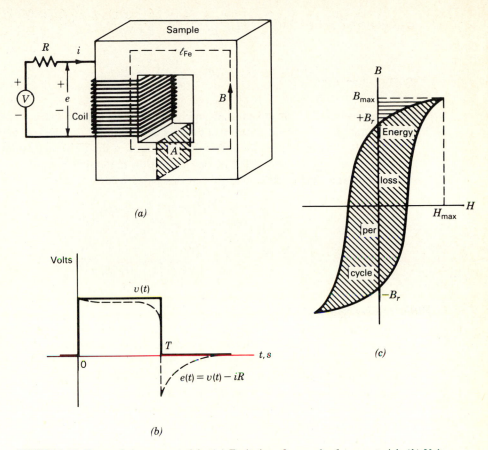

FIGURE 1.16 Hysteresis in core materials. (*a*) Excitation of a sample of core material. (*b*) Voltages in (*a*) as functions of time. (*c*) Hysteresis loop of sample.

in the iron sample. The magnetomotive force applied to the iron is Ni ampere-turns, and the average magnetic field intensity experienced by the iron is

$$H = \frac{Ni}{\ell_{Fe}}$$

or

$$i = \frac{H\ell_{Fe}}{N} \text{ A} \qquad (1.22)$$

Let R be the coil resistance in ohms and e be the Faraday's law voltage induced in the coil by flux variations:

$$e = \frac{d\lambda}{dt} = N\frac{d\phi}{dt} \text{ V} \qquad (1.23)$$

Integrating Equation 1.23 shows that a change in flux in the sample can be effected by applying a voltage pulse to the coil:

$$\frac{1}{N} \int e \, dt = \phi = BA \text{ Wb} \tag{1.24}$$

$$\Delta B = \frac{1}{NA} \int_0^T e \, dt \tag{1.25}$$

Suppose a rectangular voltage pulse, as in Figure 1.16b, is applied to the coil terminals for the duration of T seconds, such that the flux density in the sample is changed from $-B_r$ to $+B_{max}$ by this pulse. Note that the amplitude of the pulse will be larger than the induced voltage e, as a result of the iR drop in the coil resistance. The energy absorbed by iron during the application of this pulse will be

$$W = \int_0^T ei \, dt = \int_0^T \left(NA\frac{dB}{dt}\right) \frac{H\ell_{Fe}}{N} \, dt$$

$$= A\ell_{Fe} \int_{-B_r}^{B_{max}} H \, dB \text{ J} \tag{1.26}$$

The integral in Equation 1.26 is the area between the right side of the hysteresis curve and the B axis, including the part at the top with horizontal shading. After $v(t)$ goes to zero at $t = T$ seconds, $e = -iR$. The current then decays to zero as B drops from B_{max} to $+B_r$, and the energy represented by the horizontally shaded area is dissipated in the coil resistance. Note that the quantity $A\ell_{Fe}$ is the volume of the iron sample, so that the area of the right half of the hysteresis loop is energy *per unit volume* required to reverse the magnetization from $-B_r$ to $+B_r$. The area of the left half of the loop represents the joules required to remagnetize the core to $-B_r$. *Thus the area enclosed by the loop is the hysteresis energy loss per unit volume per cycle*, and the power loss in watts will be the joules lost per cycle times the number of cycles per second (Hz).

When the applied voltage is a sinusoidal function of time, there is a definite relationship between the rms induced voltage and B_{max}, the peak flux density in the core. In Equation 1.25, let $e = \sqrt{2}E \cos 2\pi ft$, where E is the rms value of the induced voltage and f is the frequency in Hz. Then

$$B = \frac{1}{NA} \int_0^T \sqrt{2}E \cos 2\pi ft \, dt$$

$$= \frac{\sqrt{2}E}{NA2\pi f} \sin 2\pi fT \tag{1.27}$$

The maximum value of $\sin 2\pi fT$ is 1.0, so

$$B_{max} = \left(\frac{1}{\sqrt{2}\pi A}\right)\left(\frac{E}{Nf}\right) \tag{1.28}$$

Thus for sinusoidal voltage, the peak flux density is proportional to the rms induced voltage and is inversely proportional to the number of coil turns and the frequency. Usually the induced voltage is nearly equal to the applied voltage, and Equation 1.28 is approximately correct when V, the rms value of the applied voltage, is used instead of E. Note that for constant frequency the peak flux density and the peak flux ($A \cdot B_{max}$) depend on the volts per turn.

Steinmetz showed experimentally that the area of a hysteresis loop is given by γB_{max}^x, where γ is a constant, and x is "Steinmetz's exponent," which is usually slightly less than 2.0. The hysteresis loss in joules per cycle is thus given by

$$W_h = (\text{core volume}) \cdot \gamma B_{max}^x \text{ J/cycle} \qquad (1.29)$$

Hysteresis loss in watts may be obtained by multiplying energy lost per cycle times the frequency. Let P_h be the symbol for the hysteresis loss in a core. Then from equation 1.27,

$$P_h = (\text{core volume}) \cdot f \gamma B_{max}^x \text{ W} \qquad (1.30)$$

The hysteresis loss is thus proportional to:

The volume of the core

The frequency

The peak flux density to the power of Steinmetz's exponent x

Eddy current loss is the result of electric currents in iron magnetic cores which circulate in closed paths like eddies in a creek. To induce voltages in the windings of a machine or transformer, the magnetic flux threading the windings must be alternating, and this requires alternating flux to flow in the core structure. Alternating Lenz's-law voltages are induced in the iron in directions to oppose the flux alternations, and these voltages drive the eddy currents. The eddy currents produce I^2R losses and heating in the iron, and their effects are to be minimized.

In very early machines, eddy current losses were impractically large. The solution to the problem has been to insert insulating material in the core to break up the eddy current paths, but to do it in such a way as to maintain a high permeance to the flow of flux. One common method involves building the core of thin sheets of iron alloy, called "laminations." The sheets are insulated from each other by iron oxide or special coatings. A close look at Figures 1.9a and 1.12a will show that the core structures are stacks of thin laminations, and the laminated nature of motor and transformer cores is evident by referring to Figures 2.2 and 3.4. In a machine or a transformer, the coils of the windings are placed around the cores in such a way that the magnetic flux is directed through the iron laminations and is not required to cross through the insulation separating them.

Elaborate computer programs involving finite-element methods ("FEM") have been developed to trace flux paths through electromagnetic devices and to determine core losses. However, there are very good approximate methods which tell much about the factors affecting eddy current loss.[3] These methods show that eddy current loss is proportional to:

The volume (or weight) of the core

The square of the thickness of the laminations

The square of the peak magnetic flux density (assuming sinusoidal flux variation)

The electrical conductivity of the magnetic material of the core

---

[3]See, for example, L. Matsch and J. Morgan, *Electromagnetic and Electromechanical Machines*, 3rd ed. (New York: Harper & Row, 1986), pp. 46–52.

Thus a core should be made of thin laminations of magnetic material of good permeability and low conductivity. Laminations that are too thin present structural problems. A typical lamination thickness for a 60-Hz transformer would be about 0.4 mm.

Let P_e be the eddy current loss in watts and P_{h+e} the total core loss:

$$P_e = (\text{core volume})(K_e f^2 t^2 B_{max}^2) \tag{1.31}$$

where K_e is a constant, and

$$P_{h+e} = (\text{core volume})(\gamma f B_{max}^x + K_e f^2 t^2 B_{max}^2) \tag{1.32}$$

Brush drop loss occurs when large amounts of current must be conducted to moving parts through sliding electrical contacts, particularly in dc machines. The stationary parts of these sliding contacts are called "brushes." Brushes are usually shaped blocks of especially developed carbon compounds that are held against the rotating part of the contact under spring pressure. The voltage drop across such sliding contacts is essentially independent of current and is called the "brush drop" V_{BD}. The resulting loss is simply

$$\text{Brush drop loss} = V_{BD}I \tag{1.33}$$

Windage loss is the result of friction between the moving parts of a machine and the air inside the machine. Most machines require some means of circulating air around the core and windings to keep the temperature within safe limits. This may be done by having some simple fan blades attached to the shaft inside the machine housing.

Very large machines require external cooling systems, and some are filled with hydrogen to reduce windage loss and to take advantage of the better cooling qualities of hydrogen. Motors that must operate at speeds too low to make shaft-mounted fans effective also require external cooling. The power required to operate such external cooling systems is considered part of the losses attributed to the machine and is sometimes called **ventilation loss**.

■ Example 1.3

Two cores, roughly similar in shape to the one shown in Figure 1.16a, are made up of laminations of silicon steel having a hysteresis coefficient γ of 500 and a Steinmetz exponent of 1.9. Core A has a cross-sectional area of 0.0025 m² and a mean flux path length (ℓ_{Fe}) of 0.500 m. Core B has a cross-sectional area three times that of A, and its $\ell_{Fe} = 0.866$ m. Each core is wound with 100 turns of wire, and both coils are connected in parallel to a 60-Hz sinusoidal voltage source of 80.0 V, rms. Find the hysteresis loss in each core.

Solution. Equation 1.27 shows that the peak flux produced by either coil depends only on the volts per turn and the frequency. In either case,

$$\phi_{max} = \frac{\sqrt{2} \cdot 80}{100 \cdot 2\pi 60} = 0.00300 \text{ Wb}$$

Neglecting any leakage, the peak flux density in core A will be

$$B_{max} = \frac{0.003}{0.0025} = 1.200 \text{ T}$$

Since core B has three times the cross-sectional area of A, its peak internal flux density will be one third as much, or 0.400 T. The volume of core A is its area times its ℓ_{Fe}, or

$$\text{Vol}_A = 0.0025 \cdot 0.5 = 0.00125 \text{ m}^3$$

Then for core A,

$$P_h = 0.00124 \cdot f\gamma B_{max}^x = 0.00125 \cdot 60 \cdot 500 \cdot 1.2^{1.9}$$
$$= 53.0 \text{ W}$$

Similarly, for core B,

$$P_h = 0.0065 \cdot 60 \cdot 500 \cdot 0.4^{1.9} = 34.2 \text{ W}$$

Note that the total flux depends *only* on the volts per turn and the frequency, and that although core B has over 5 times (actually $3 \cdot \sqrt{3}$ times) the volume of A, its hysteresis loss is less because the larger area results in a lower flux density. ∎

■ Example 1.4

A single-phase transformer has three windings on its core. Two of its windings are not connected to anything (terminals open), and the third winding, which has 100 turns, is connected to a 240-V, 60-Hz source through a wattmeter. The wattmeter reads 400. W, which may be considered to be the total core loss since under these conditions the I^2R loss in the winding is quite small. The frequency is then reduced to 30 Hz, and the voltage is simultaneously lowered to keep the flux density the same as it was at 60 Hz. The wattmeter now reads 169. W. What should the voltage be at 30 Hz? What is the eddy current loss at 60 Hz? What is the hysteresis loss at 60 Hz?

Solution. Note in Equation 1.27 that the core flux density is proportional to the applied voltage and inversely proportional to the frequency. If the frequency is reduced to half its original value without changing the voltage, the flux density would double. Therefore the voltage should also be cut to half its value at 60 Hz:

For constant B_{max}:

$$V_2 = \frac{f_2}{f_1} V_1 = \frac{30 \cdot 240}{60} = 120 \text{ V}$$

Since in this example the core volume, γ, B_{max}, and the lamination thickness are all constants, Equation 1.32 may be written in terms of two constants:

$$P_{h+e} = C_h f + C_e f^2$$

The trick in solving the second part of the problem is to divide this equation through by the frequency:

$$\frac{P_{h+e}}{f} = C_h + C_e f$$

The simultaneous equations for the two frequencies are

$$\frac{400}{60} = 6.67 = C_h + 60C_e$$

$$5.63 = C_h + 30C_e$$

from which $C_h = 4.59$ W/Hz and $C_e = 0.0347$ W/Hz^2. Then at 60 Hz, $P_h = C_h \cdot 60 = 275$ W and $P_e = C_e \cdot 3600 = 125$ W. ■

1.10 MAXIMUM EFFICIENCY

It is often desirable to design a machine or transformer to have maximum efficiency at some particular load. When the line voltage is approximately constant, the "load" and the current are proportional. When the voltage and frequency are both constant, the core losses are nearly constant. When the speed is approximately constant, the friction, windage, and ventilation losses are practically constant.

There are many practical situations in which the line voltage, line frequency, and shaft speed may be considered constant for the purpose of efficiency calculations. The losses may then be divided into three categories:

$$\left.\begin{array}{l}\text{Constant losses: friction + windage + ventilation + core losses}\\ \text{(+ field copper losses in some machines)}\end{array}\right\} = P_k$$

Losses proportional to (line current)^2: copper losses = I^2R

Losses proportional to line current: brush drop loss = $V_{BD}I$

Under these circumstances Equation 1.19 may be written, for a device with electrical output,

$$\eta = \frac{CVI}{CVI + P_k + I^2R + V_{BD}I} \tag{1.34}$$

where C is a constant. For example, in a three-phase machine, $C = \sqrt{3} \cdot$ (power factor). The efficiency will be a maximum or minimum when

$$\frac{\partial \eta}{\partial I} = 0$$

or when

$$\frac{(CVI + P_k + I^2R + V_{BD}I) \cdot CV - CVI \cdot (CV + 2IR + V_{BD})}{(CVI + P_k + I^2R + V_{BD}I)^2} = 0$$

or

$$CVI + P_k + I^2R + V_{BD}I - (CVI + 2I^2R + V_{BD}I) = 0$$

Note that the terms CVI and $V_{BD}I$ cancel. *This means that losses proportional to* I *to the first power do not affect the determination of the load at which maximum efficiency occurs.* The final result is

$$I^2R = P_k \qquad (1.35)$$

Thus **maximum efficiency occurs when those losses proportional to the square of the load current are equal to the constant losses of the device**. (This relationship is sometimes *incorrectly* stated as "maximum efficiency occurs when the variable losses equal the constant losses.")

A distribution transformer has only copper and core losses. In this case

$$P_k = \text{core losses}$$

Example 1.5

A 10-kVA distribution transformer has a 240-V secondary winding. The equivalent internal resistance of this transformer, referred to that winding, is 0.048Ω. The core loss of this transformer is 75 W. At what kVA load will this transformer operate at maximum efficiency? Let I_2 be the current in the secondary winding.

Solution. From Equation 1.35:

$$P_k = I^2R$$

$$75 \text{ W} = I_2^2 \times 0.048$$

$$I_2 = \sqrt{75/0.048} = 39.5 \text{ A}$$

Then at this current, the output volt-amperes of the transformer would be

$$|S| = V_2I_2 = 240 \cdot 39.5 = 9480 \text{ VA, or 95\% of rated load} \qquad ■$$

1.11   THE NAMEPLATE

Each machine or transformer carries a *nameplate,* usually of metal or plastic. The nameplate gives the following information.

1. What kind of a device (transformer, dc motor, etc.) it is
2. Name of manufacturer
3. Rated voltages and frequency
4. Rated currents and volt-amperes
5. Rated speed and horsepower (if a machine)

The nameplate may give additional useful information, such as power factor, internal impedance, connection diagram, and so on.

Faraday's law shows us that, by integration of Equation 1.13,

$$\phi = \frac{1}{N} \int e \, dt \tag{1.36}$$

This implies that the magnetic field flux will depend on the terminal voltage or, more completely, on the volts per turn. The designer adjusts the flux level in the machine by deciding how many turns to have in each winding, given the rated operating voltage and frequency. The name plate gives the voltage for which the machine or transformer was designed: the "rated" voltage. If this voltage is exceeded, the flux will be excessive. The core will be more highly saturated and the core losses will be higher than normal. The machine will run hotter as a result. *The core losses are approximately proportional to the square of the terminal voltage, since Steinmetz's exponent is nearly 2.0.*

Obviously, *the copper losses are proportional to the square of the winding current.* The heat resulting from I^2R loss is applied directly to the insulation. Since heat deteriorates insulating materials, the current cannot be allowed to be so high that it shortens the life of the insulation seriously. It is this heat limit that determines the current rating of the device. By inscribing voltage and current ratings on the nameplate of the machine, the manufacturer is saying, in effect: "If you do not operate this machine above the stated voltage and current, this machine will not get too hot and will give long and satisfactory service. If you operate this machine at much below the rated voltage, the magnetic field will be weak, and you will not be able to get rated output without excessive current."

Nothing says that the device *must* operate at rated voltage, current, speed, or what have you. DO NOT ASSUME IN ANY PROBLEM IN THIS BOOK OR IN THE PRACTICE OF ENGINEERING THAT THE DEVICE YOU ARE STUDYING IS OPERATING AT THE RATED VALUES GIVEN ON THE NAMEPLATE, unless you KNOW it to be so. Be very careful about this.

1.12 REGULATION

To engineers, the word "regulation" means the variation of the output speed or voltage of a device as the load on that device is increased from zero to some specified fraction of full load.

Regulation may be expressed as follows:

$$\text{Speed regulation} = \frac{\text{no-load speed} - \text{speed under load}}{\text{speed under load}} \tag{1.37}$$

$$\text{Voltage regulation} = \frac{\text{no-load voltage} - \text{voltage under load}}{\text{voltage under load}} \tag{1.38}$$

In many devices, distribution transformers, for example, a constant output voltage is desirable. This would correspond to zero voltage regulation. It is often preferable that the speed of a motor remain essentially constant; that is, zero speed regulation is ideal in many motor applications. In other words, a *big* regulation is most often considered to be a *bad* regulation. The regulation is, then, a "figure of merit" for a given device. For some applications, however, large speed or voltage regulations are desirable or even necessary.

In most cases, the engineer is primarily interested in the *full-load regulation*, that is, the variation in speed or voltage when full load is applied to a machine or transformer that was previously operating without load. When the word "regulation" is used without qualification, one may assume that "full-load regulation" is what is meant. It is usually assumed that rated speed or voltage occurs at full load. Then *for full-load regulation*, the expressions of Equations 1.37 and 1.38 become

$$\text{Speed regulation} = \frac{\text{no-load speed} - \text{speed at rated load}}{\text{speed at rated load}} \tag{1.39}$$

$$\text{Voltage regulation} = \frac{\text{no-load voltage} - \text{voltage at rated load}}{\text{voltage at rated load}} \tag{1.40}$$

where "voltage (or speed) at rated load" is usually taken to be the nameplate value.

■ **Example 1.6**

A 5-hp, 208-V, three-phase, 60-Hz induction motor runs at 1746.0 rev/min when delivering 5 hp, and the input voltage and frequency are at the rated values given. With the mechanical load completely disconnected from the shaft, the motor's speed is found to be 1799.5 rev/min, with the input voltage and frequency held at 208 V and 60 Hz. What is the full-load speed regulation of this motor?

Solution.

$$\text{Regulation} = \frac{1799.5 - 1746.0}{1746.0} = 0.0306, \text{ or } 3.06\%$$

Note that, to calculate the regulation of the motor itself, it was necessary to know that the input voltage and frequency were the same for both the full-load and no-load speed measurements. If the voltage and frequency were permitted to vary with motor load, the regulation measured would not be that of the motor alone, but that of the motor and the electrical systems combined. ■

■ **Example 1.7**

Two tests were made to determine the voltage regulation of a transformer to be used in a regulated power supply: one for unity power factor load and one for load of 0.7 power factor lagging. The transformer nameplate ratings are: 115-24 V, 48 VA. Note that rated secondary current is 48 ÷ 24 = 2.0 A. Test data are as follows:

Primary Volts	Secondary Volts	Secondary Amperes	Power Factor
115	24.0	2.00	1.00
115	27.3	0	
115	22.1	2.00	0.70 lag
115	27.3	0	

Solution. Full-load voltage regulation:
(a) For unity power factor,

$$\frac{V_{nl} - V_{fl}}{V_{fl}} = \frac{27.3 - 24.0}{24.0} = 0.137, \text{ or } 13.7\%$$

(b) Determining a meaningful value for the regulation from the data supplied at 0.7 power factor, lagging, requires some engineering judgement. To raise the primary voltage so that rated secondary voltage would be obtained at rated current would not be too realistic, considering that the line voltage in practice will not accommodate itself to fit a neat definition of regulation found in some book. (In large power systems, line voltages *are* adjusted to account for regulation in big power transformers.) Note in the statement below Equation 1.40 that the denominator in the regulation expression is usually taken to be the rated value. Note also that the secondary voltage varies from 13.8 percent above rating voltage at no load to 7.9 percent below rated voltage at rated current; that is, it is at rated voltage at some reasonable load. The most meaningful value for regulation in this case would place rated voltage, rather than the test value at rated current, in the denominator:

$$\text{Regulation} = \frac{27.3 - 22.1}{24.0} = 0.217, \text{ or } 21.7\%$$

This is a small transformer. Large transformers for power systems have regulations in the order of 3 to 5 percent. ■

PROBLEMS

1.1.   Define the following:
(a) Electrical machine. (b) Motor. (c) Generator. (d) Transformer.

1.2.   Define the following, as referred to electrical machines:
(a) Input power. (b) Output power. (c) Rotor. (d) Stator. (e) Shaft. (f) Air gap.

1.3.   Define the following:
(a) Load current. (b) Exciting current. (c) Armature. (d) Field winding. (e) Primary. (f) Secondary.

1.4.   It has been said that a human being develops his or her maximum horsepower in running up a flight of stairs. When this author was in high school, one of our best athletes, to provide an example for our physics class, developed $\frac{3}{4}$ hp in running up two floors. If he traveled a total vertical distance of 20 ft and weighed 185 lb, how long did it take Jim to run up the stairs?

1.5.   An ice-cream freezer has a crank 8 in. long. At a stage near the end of the freezing process, a tangential force of 5 lb must be applied to the handle to turn the crank. If the crank is turned at the rate of 1 revolution per second, what is the mechanical power input to the freezer in watts? In horsepower? How many joules of work are done in 1 minute? Where does this energy go?

1.6.   A 10-kW, single-phase generator is designed for emergency service. When it is operated at its rated speed, 3600 rev/min, and rated voltage, and is delivering 10 kW output to an electrical load, the machine has an efficiency of 85%. How many output horsepower must a gas engine develop at this speed in order to drive the generator? What torque must the engine produce?

1.7. A gallon of water weighs 8.33 lb. A small dam provides a head of 10 ft so that each gallon of water spilling over the dam has a potential energy of 83.3 ft-lb to drive a water turbine. If the turbine–generator combination converts this potential energy to electricity at an efficiency of 75%, how many gallons per minute must flow through the turbine to light a single 100-W bulb?

1.8. In Example 1.4, what are the hysteresis and eddy current losses of the transformer at 30 Hz, with 120 V applied to the winding?

1.9. (This is an exercise to learn the general effects of changing transformer dimensions. It is NOT a good way to design a new transformer.) A successful single-phase, two-winding transformer, when operated at rated voltages and frequency, has a core loss equal to 1% of the rated volt-amperes of the primary winding. A new, larger transformer is to be designed. Each linear dimension of the core of the new transformer is to be k times that of the original model. That is, the mean length of flux path, ℓ_{Fe}, will be k times that of the old transformer, and the cross-sectional area of the new core will be k^2 times that of the old. The peak flux density in the new core is to be the same, the windings are to have the same voltage ratings, and the transformer is to operate at the same frequency. (a) How will the volume of the new core compare with that of the original? (b) How will the core losses compare? (c) If it is desired that the core loss be 1% of the volt-ampere rating of the new transformer, how will its rating compare with the original? (d) By what factor must the current ratings of the windings be increased to accommodate the new volt-ampere ratings? (e) In order to carry this additional current, by what factor must be cross-sectional area of the wire in each winding be increased? (f) How will the core loss of the new transformer compare with that of the original? (g) By what factor must the number of turns in each winding be changed? (h) By what factor must the resistances of the windings be changed if the original transformer has its maximum efficiency at full load and the new transformer is also to have maximum efficiency at full load?

1.10. There are means available for measuring the core loss in a rotor core of a dc machine under no-load conditions. Consider a two-pole machine such as that in Figure 1.13. Under no load, there will be essentially no rotor current, and hence no magnetic poles as shown on the rotor surface. The core loss will be the result of rotation of the rotor core iron through the field produced by the stator poles, and the frequency of flux density reversal will be proportional to the speed. In a dc machine, it is easy to keep the no-load peak flux density in the rotor iron constant by keeping the strength of the stator poles constant. If the rotor core loss is found to be 200 W at 1200 rev/min and 80 W at 600 rev/min, what are the hysteresis and eddy current losses in the rotor core at 1200 rev/min?

1.11. Make a list of power losses of machines and transformers, grouped into: (a) Electrical losses. (b) Core losses. (c) Mechanical losses.

1.12. A transformer is operating with a load (output) of 50,000 VA at 0.800 power factor. This transformer has losses of 1200 W.
(a) What is the input power? (b) What is the efficiency of the transformer?

1.13. A single-phase, two-winding transformer has a core loss of 1000 W. The secondary (output) voltage is 460 V and the output power is 48,000 W at 0.800 power factor, lagging. Under these conditions the current in the primary winding is 13.06 A. The secondary winding resistance is 0.0200Ω and the primary winding resistance is 2.00 Ω.

(a) What is the secondary current?
(b) What is the complex impedance of the load connected to the secondary terminals?
(c) What are the I^2R losses in the primary and secondary windings?
(d) What is the input power?
(e) What is the transformer efficiency?

1.14. A 10-hp, three-phase induction motor draws 26.0 A from a 208-V, 60-Hz line at full load. Calculate the internal power loss and efficiency if the power factor is 0.880.

1.15. A $1/3$-hp, single-phase motor draws 3.90 A from a 115-V line at rated load, with a power factor of 0.850, lagging. At half load, the motor losses are 75 W and the input current is 2.56 A.

(a) Calculate the efficiency of the motor at full load.
(b) At what power factor and efficiency does the motor operate at half load?

1.16. A single-phase transformer has a core loss of 5600 W. When the secondary current is 250 A, the copper loss is 6400 W. At what secondary current will this transformer have maximum efficiency?

1.17. If, at maximum efficiency, the secondary voltage of the transformer of Problem 1.16 is 2300 V and the secondary current lags this voltage by 25°, what are the input power and efficiency?

1.18. A dc motor is operating under the following conditions. Output power 6.70 hp, armature current 25.0 A, brush drop 2.0 V, armature winding resistance 0.80 Ω, field current 0.20 A, field winding resistance 1000 Ω, total of friction, windage, and core losses 700 W.

(a) Calculate the armature and field-winding copper losses.
(b) Calculate the input power.
(c) Calculate the efficiency.
(d) At what value of armature current would the efficiency of this motor be maximum, assuming the field current is constant at 0.20 A?

1.19. The voltage induced in the primary winding of a transformer is very nearly equal to the voltage of the electrical source. A certain transformer has a primary of 115 turns and is connected to a 60-Hz sinusoidal ac voltage source of 230 rms V.

(a) Find the peak core flux, assuming any flux component due to the constant of integration has decayed to zero.

(b) What will be the peak flux if the number of primary turns is reduced to 90?

1.20. The speed of a synchronous motor is the same at full load as it is at no load. What is its speed regulation?

1.21. What is the voltage regulation of a synchronous alternator if, when tripped off the line by a circuit breaker, its terminal voltage rises from 17 kV to 23 kV, assuming that it was supplying rated load before the trip? (The regulation of synchronous generators is so poor that they nearly always require automatic voltage regulators.)

2

SYNCHRONOUS MACHINES

2.1 SYNCHRONOUS MACHINE CHARACTERISTICS

Synchronous machines are called "synchronous" because their speed is directly related to the line frequency:

$$\omega_s = \frac{\omega_e}{p/2} = \frac{2\pi f}{p/2} \text{ rad/s}$$

or (2.1)

$$n_s = \frac{120f}{p} \text{ rev/min}$$

where $p/2$ is the number of pairs of magnetic poles designed into each machine. In this relationship, ω_s is the angular *velocity* of the shaft and ω_e is the angular *frequency* of the electrical system. Thus when two or more synchronous machines are connected to the same ac line, they will all run in synchronism because they all are operating at the same line frequency, and ω_s is called the *synchronous speed* of a given machine. Of course, if one machine has 2 poles and another has 14, the 14-pole machine will run at precisely $\frac{1}{7}$ the speed of the 2-pole machine.

Like other electrical machines, synchronous machines may be operated either as motors or as generators. A polyphase synchronous machine operated as a generator is called an *alternator*. The largest electrical machines in the world are synchronous alternators. Some are rated as much as 1.7 billion watts (1700 MW). Although designed to operate as generators, even these large machines can operate as motors and sometimes do "motor" under abnormal system conditions. For pumped-storage stations, synchronous machines with ratings in the order of 50 MW are designed for both motor and generator operation; and many smaller polyphase synchronous machines (5 to 8000 hp) are designed primarily for motor applications.

Since electrical energy can be transported most economically by three-phase transmission systems, nearly all synchronous generators larger than, say, 10 kVA, and most industrial motors, are designed for three-phase operation. For this reason three-phase machines are the basis of this discussion.

2.2 CONSTRUCTION OF A SYNCHRONOUS MACHINE

Figure 2.1*a* shows a synchronous machine designed for use as an auxilliary alternator to provide 75 kVA of 400-Hz, three-phase power (90 KVA peak) in a Boeing 767 aircraft. It is driven at a controlled speed of 12,000 rev/min by a gas turbine separate from the main engines. Figure 2.1*b* shows the essential parts of a similar machine for a Fokker F-100, rated at 30 kVA. In both devices, there are really three machines in one. At the rear is an "exciter" generator to supply dc to the field windings of the four salient poles that form the rotor of the main machine. At the shaft end is a permanent-magnet (PM) emergency alternator capable of supplying enough power to keep the most vital systems of the plane operating should the main alternator fail. (The PM alternator for the 767 is rated at 0.8 kVA.) Its 12-pole Alnico PM rotor will be seen just behind the roller bearing in Figure 2.1*a*.

The 75-kVA machine is 13.6 in. long (including the shaft) and 9.2 in. in diameter

(a)

(b)

FIGURE 2.1 Three-phase, 400-Hz synchronous machines disassembled. (a) 75-kVA, 12,000 rev/min aircraft alternator. (b) Stators and rotor of a 30-kVA aircraft alternator. (Courtesy of Sunstrand Corporation, Electric Power Division.)

FIGURE 2.2 Assembly of the laminated stator core of a synchronous machine. (Courtesy of Westinghouse Electric Corporation.)

FIGURE 2.3 Partially wound stator core. (Courtesy of Westinghouse Electric Corporation.)

and weighs 62.5 lb. The construction of the main alternator is similar to that of much larger synchronous machines. The stator winding is the source of voltage and electrical power when the machine is operating as a generator, and it is the input winding when operating as a motor. It is thus the *armature* of the machine. The coils on the salient poles of the rotor are connected in series to form the dc *field* windings. *Armature* and *field* are defined on p. 8.

The stator core consists of a stack of slotted, ring-shaped laminations (see Figure 2.2). When these laminations are stacked and bolted together, a cylindrical core results, with axial slots on its inner surface. These slots run the length of the stack and provide for the insertion of the coils of the armature winding. Figure 2.3 shows a partially wound stator core. A completed stator is shown in Figure 2.4.

A typical coil for a three-phase stator winding is shown in Figure 2.5. A coil is said to have two "ends" and two "sides." The sides of the coils are placed in the slots of the stator core, with the ends appearing at each end of the core, as shown in Figure 2.6. Most three-phase windings are "double layer"; that is, *two* coil sides are placed in each slot. If a given coil has one of its sides in the bottom of a slot, its other side will be found in the top of a slot, the top being that position in the slot nearest the air gap. Since there are two coil sides per slot and each coil has two sides, *the number of coils in a double-layer winding is equal to the number of slots.*

A coil may have one or more turns. The number of turns per coil will be given the symbol N_c. The number of conductors in each coil side is obviously equal to N_c, so the number of slot *conductors* provided by each coil is $2N_c$. The peculiar kink in each coil end provides mechanical clearance between adjacent coils.

FIGURE 2.4 Completed stator for a synchronous machine. (Courtesy of Westinghouse Electric Corporation.)

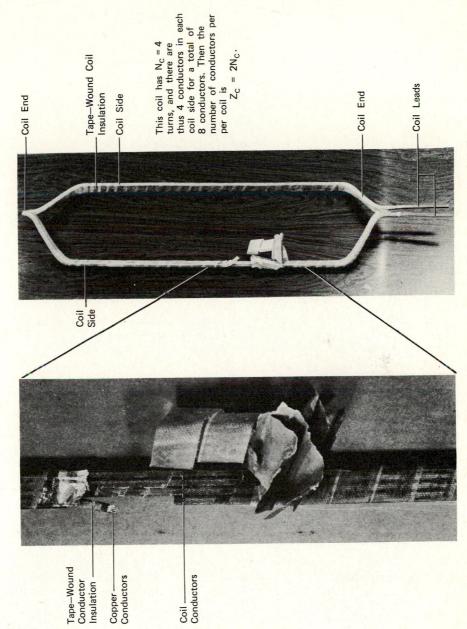

Coil End

Tape—Wound Coil
Insulation

Coil Side

This coil has $N_C = 4$
turns, and there are
thus 4 conductors in each
coil side for a total of
8 conductors. Then the
number of conductors per
per coil is

$$Z_C = 2N_C.$$

Coil End

Coil Leads

Coil
Side

Tape—Wound
Conductor
Insulation

Copper
Conductors

Coil
Conductors

FIGURE 2.5 Typical coil for a synchronous or induction machine stator winding.

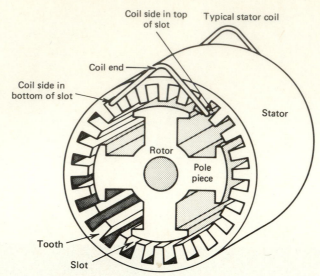

FIGURE 2.6 Core configuration of a four-pole synchronous machine having a salient-pole rotor.

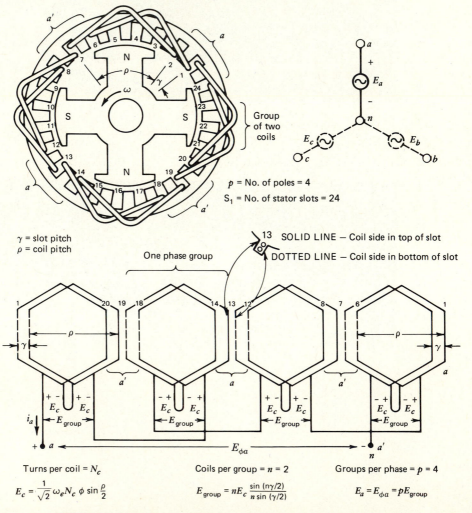

p = No. of poles = 4

S_1 = No. of stator slots = 24

γ = slot pitch
ρ = coil pitch

SOLID LINE — Coil side in top of slot

DOTTED LINE — Coil side in bottom of slot

One phase group

Turns per coil = N_c

$$E_c = \frac{1}{\sqrt{2}}\,\omega_e N_c\,\phi \sin\frac{\rho}{2}$$

Coils per group = n = 2

$$E_{\text{group}} = nE_c\,\frac{\sin\,(n\gamma/2)}{n \sin\,(\gamma/2)}$$

Groups per phase = p = 4

$$E_a = E_{\phi a} = pE_{\text{group}}$$

FIGURE 2.7 Some features of one phase winding of a three-phase machine.

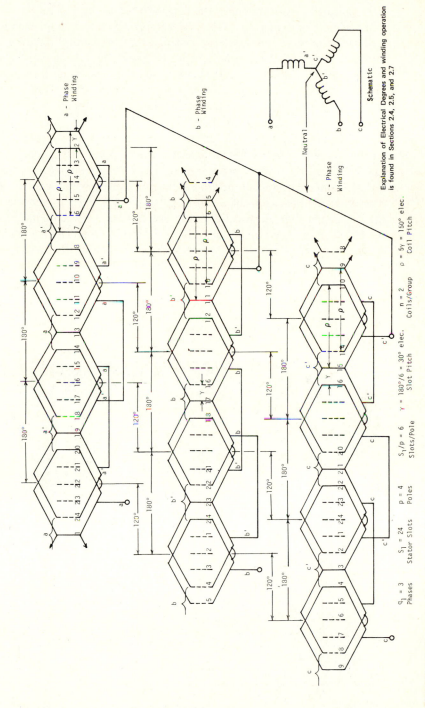

FIGURE 2.8 Four-pole, three-phase winding having two coils per group showing 180 electrical degree placement between groups in one phase and 120° displacement between windings. Y-connection illustrated.

The coils of the armature winding in the stator are connected to form three independent *phase windings*. Figure 2.7 shows one phase winding for the 24-slot stator of Figure 2.6. Note that the coils composing the winding are so placed in the slots as to form equally spaced groups. The number of groups in each phase winding is equal to the number of rotor poles. When current flows in the winding, each group produces a magnetic pole having a polarity dependent on the current direction and a magnetomotive force (MMF) proportional to the amperage. The groups of a given phase winding are connected so that when one group presents a north magnetic pole to the air gap, the groups on either side of it produce south poles. Thus for a given instantaneous current direction, the magnetic polarity of the groups is alternately N, S, N, S, and so on, as viewed from the air gap.

The layout of a complete four-pole, three-phase winding for a 24-slot stator is shown in Figure 2.8. Coil sides located in the bottoms of slots are shown as broken lines. The *angular distance between two adjacent rotor poles is defined as 180 electrical degrees*, and the centers of the groups of a given phase winding are also 180 electrical degrees apart. The width of a coil is called the *coil pitch* ρ, and may be in inches, meters, degrees, or electrical degrees. The distance between adjacent slots is called the *slot pitch* γ. The number of coils in one phase group (n) is equal to the number of slots per pole per phase. If S_1 is the number of slots in the stator core and q_1 is the number of phases, then the number of coils per group is given by

$$n = \frac{S_1}{q_1 p}$$

FIGURE 2.9 Slotting a large synchronous alternator rotor. (Courtesy of the General Electric Company.)

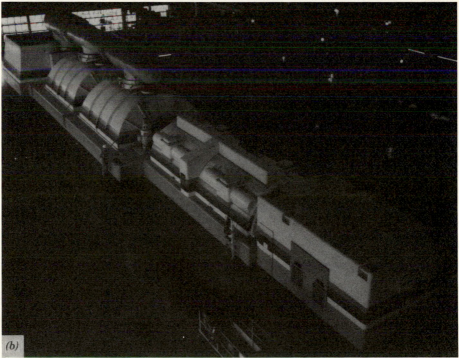

FIGURE 2.10 Typical large cylindrical-rotor synchronous machine. (*a*) Cylindrical rotor being assembled into a stator for testing. (*b*) A 618 MW steam turbine-generator unit showing the three major components: the collector house containing equipment for supplying field current to the rotor, the synchronous alternator itself, and the two-stage steam turbine at the far end. (Courtesy of General Electric Company.)

When balanced three-phase currents flow in the windings, magnetic poles are produced on the inner surface of the stator. As the currents alternate in time sequence (120° apart in time), these magnetic poles move smoothly from group to group along the stator surface to produce a rotating magnetic field. The speed of rotation is given by Equation 2.1. The number of poles in this rotating field is equal to the number of coil groups in any one phase winding.

Synchronous machine *rotors* are simply rotating electromagnets or permanent magnets built to have as many poles as are produced by the stator winding. Electromagnetic rotor poles are magnetized by dc currents flowing in the field coils surrounding each pole. The magnetic field produced by the rotor poles locks in with the rotating stator field, so that the shaft and the stator field rotate in synchronism.

Rotors of synchronous machines are of two types—*salient pole* and *cylindrical*. The machine of Figure 2.1 has salient poles, as do most synchronous machines. However salient poles are too weak mechanically and develop too much wind resistance and noise to be used in large, high-speed alternators driven by steam or gas turbines. For these big machines, the rotor must be a solid, cylindrical steel forging to provide the necessary strength. Axial slots are cut in the surface of the cylinder (see Figure 2.9) to accommodate the field windings. A large cylindrical-rotor machine is shown in Figure 2.10, together with a completed cylindrical rotor. Most cylindrical rotors have only two poles (3600 rev/min at 60 Hz). A few have four poles (1800 rev/min at 60 Hz). Figures 2.11a and b show drawings of the stator of Figure 2.7 with the two types of rotor inserted. Both rotors have four poles in this figure.

A rotor for a synchronous machine having 24 salient poles is shown in Figure 2.12. Applying Equation 2.1, this rotor would be for a 300 rev/min machine at 60 Hz. What is the 60-Hz speed for the rotor of Figure 1.9a?

Sixty-hertz alternators designed to be driven by diesel engines often have 14 poles, because these engines can be designed for high efficiency at the corresponding speed. Alternators for use with hydroturbines often have more than 100 poles. Hydroalternators always have salient poles.

Since the rotor poles have constant polarity their windings must be supplied with direct current. This current may be provided by an external dc generator or by a rectifier. In this case the leads from the field winding are connected to insulated rings mounted concentrically on the shaft. Stationary contacts called brushes ride on these slip rings to carry current to the rotating field windings from the dc supply. The brushes are made of a carbon compound to provide a good contact with low mechanical friction. An external dc generator used to provide field current is called an *exciter*.

An alternative method for supplying dc current to the rotor windings is to mount the armature of a relatively small exciter alternator on the shaft of the synchronous machine. The field of this alternator is stationary. Diodes installed in the rotor rectify the output of the alternator to provide dc for the rotor field coils. No mechanical sliding contact is required—the coupling between the exciter alternator's armature and field is magnetic. This means of supplying field current to the rotor coils is called *brushless excitation* and is being used on most new synchronous machines. The armature of the exciter alternator for the field of the aircraft alternator of Figure 2.1 may be seen behind the main rotor poles. Figure 2.13 shows the armature and rectifying diodes of the brushless exciter for a 25-kW, three-phase synchronous machine. The rectifier output leads will be connected to the coils on the salient field poles before the unit is assembled.

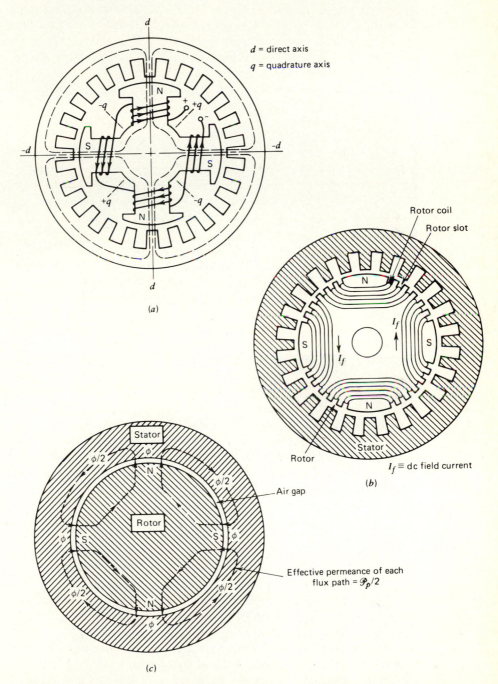

FIGURE 2.11 Magnetic geometry of synchronous machines. (*a*) Four-pole synchronous machine with salient-pole rotor. (*b*) Four-pole synchronous machine with cylindrical rotor having three coils per pole. (*c*) Magnetic geometry assumed for cylindrical rotor theory.

FIGURE 2.12 Salient-pole rotor for a synchronous motor. (Courtesy of Westinghouse Electric Corporation.)

FIGURE 2.13 Rotor of a 25-kVA 3-phase synchronous machine featuring brushless excitation. Armature of the exciter alternator is in the foreground, together with the six-diode field rectifier bridge assembly. (Photograph taken at the plant of Kato Engineering/Reliance Electric, Mankato, Minnesota.)

Magnetic Axes of the Rotor

The axis of symmetry of the north magnetic poles of the rotor is called the *direct*, or *d axis*. That of the south magnetic poles is the *negative d axis*. The axis of symmetry halfway between adjacent north and south poles is called the *quadrature axis*, or *q axis*. The *q* axis lagging the north pole is considered the positive *q* axis, as in Figure 2.11*a*. The quadrature axis is so named because it is 90 *electrical* degrees (one quarter cycle) away from the direct axis.

2.3 CYLINDRICAL-ROTOR THEORY VERSUS SALIENT-POLE THEORY

Cylindrical-rotor synchronous machines are much easier to analyze than those having salient-pole rotors. The reason is that cylindrical-rotor machines have a relatively uniform air gap. It can be assumed that a radial MMF field will produce the same air gap flux regardless of its angular direction. This cannot be said of a salient-pole machine, because the air gap is much greater between the poles (i.e., along the quadrature axis) than it is at the centers of the poles (i.e., on the direct axis).

Fortunately, *cylindrical-rotor theory is fairly accurate in predicting the steady-state performance of salient-pole machines*, as well as that of machines having cylindrical rotors. As a result, the extra complications of salient-pole theory are required only when an unusually high degree of accuracy is needed, or when problems involving transients or power system stability are being considered. In this chapter, then, the primary emphasis is on the simpler and largely adequate cylindrical-rotor theory.

2.4 THE MAGNETOMOTIVE FORCE FIELD OF THE ROTOR

Compare Figure 2.11*b* with Figure 2.14. Figure 2.14 is a developed cross-sectional view of the machine with the cylindrical rotor, stator slots neglected. ("Developed" means that a drawing of a circular device has been mentally slit along a radius and flattened out.) The ampere turns per coil and the slot spacing are designed to produce a MMF function $\mathcal{F}_2(\theta_2)$ on the rotor surface which is, as nearly as possible, sinusoidal. The subscript "2" designates rotor quantities, while "1" as a subscript indicates stator quantities. The angle θ_2 is measured from an axis fixed on the rotor. A sinusoidal MMF applied to an uniform air gap will result in a sinusoidal flux density distribution. A rotating flux density field having such a distribution along the rotor surface will induce sinusoidal voltages in the stator coils, a very desirable result.

Although the rotor MMF $\mathcal{F}_2(\theta_2)$ is drawn as a stepped function, leakage and iron saturation, together with the fact that the rotor slots have a finite width, tend to round off the corners of the steps. The result is that the flux density distribution becomes very nearly sinusoidal, and little error is introduced in taking the rotor MMF as being equal to the fundamental component of the Fourier series for the actual MMF waveform. In cylindrical-rotor theory, it is usually assumed that

$$\mathcal{F}_2(\theta_2) = F_2 \cos\left(\frac{p}{2}\,\theta_2\right) \qquad (2.2)$$

where F_2 is the amplitude of the fundamental term in the Fourier series representing $\mathcal{F}_2(\theta_2)$. All other harmonics are neglected.

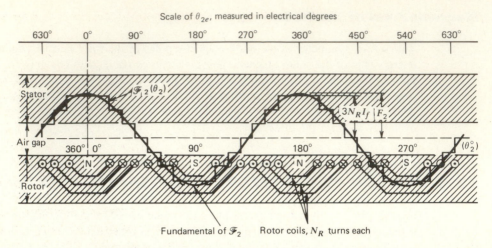

Scale of θ_{2e}, measured in electrical degrees

FIGURE 2.14 MMF of a cylindrical rotor of four poles ($p = 4$).

The *effective turns per pole of the field winding*, N_f, is defined by the relationship

$$F_2 \overset{\triangle}{=} N_f I_f \tag{2.3}$$

where I_f is the dc exciting current.

The concept of electrical degrees is very useful in the study of machines. If, in Equation 2.2, a new angular measure is defined such that

$$\theta_{2e} \overset{\triangle}{=} \frac{p}{2}\,\theta_2 \tag{2.4}$$

where θ_{2e} is in *electrical* degrees or radians, then Equation 2.2 becomes

$$\mathscr{F}_2(\theta_{2e}) = F_2 \cos\theta_{2e} \tag{2.2a}$$

The advantage of this notation is that expressions written in terms of electrical angles apply to machines having any number of poles. To make a clear distinction, θ_2 is said to be in "mechanical" degrees or radians.

FOR FURTHER STUDY OF CYLINDRICAL ROTORS

FINDING F_2 AND N_f

The stepped waveform of Figure 2.13 shows the magnetomotive force developed by the excited field winding of a four-pole cylindrical rotor. The fundamental component of the Fourier series of this complex waveform may be determined easily by taking advantage of symmetry, and of the fact that the function is the sum of the MMF contributions of the individual coils.

By symmetry, the coils of the winding may be considered in pairs. Consider a typical pair of coils of equal pitch ρ_a centered about adjacent north and south poles. The pitch angle ρ_a is in electrical degrees. These two coils each have N_a turns and are carrying the dc field current I_f, amperes. The magnetomotive force developed by this coil pair is a rectangular function of θ_{2e}, as shown in Figure 2.14. This

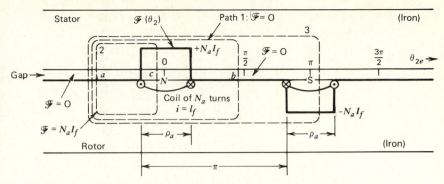

FIGURE 2.15 Magnetomotive force of a typical coil pair in a cylindrical-rotor winding.

rectangular wave may be found by the application of *Ampère's law in circuital form,* which states: "The magnetomotive force $\mathscr{F}$, found by integrating H around any closed path, is simply equal to the net current enclosed by the path."

Figure 2.15 shows a "developed," or flattened-out, section of the rotor, stator, and air gap. For the purpose of the analysis, we assume that permeability of the iron of the rotor and stator core is infinite, so that all of the rotor MMF is applied to the air gap. Note that path 1 encloses zero net current, since the total amperes going into one coil side is equal to the amperes coming out of the other side. Thus the air gap MMF provided by the two coils at points a and b is zero. However path 2 encloses the current in one coil side, $N_a I_f$, so the $\mathscr{F}$ around path 2 is $N_a I_f$. Since the gap MMF at a is zero and there is no drop in the infinitely permeable iron, the MMF at c must be $N_a I_f$. The direction of $\mathscr{F}$ around the path is determined by applying the right-hand rule to the enclosed current. In this case, $N_a I_f$ is directed out of the rotor at point c. By extending the width of path 2, we will see that the air gap MMF is constant over the coil pitch ρ_a. Similarly, path 3 shows that the air gap MMF between the slots containing the second coil is $-N_a I_f$.

If θ_{2e} is measured from the center of a north rotor pole, the MMF of one coil pair may now be written as

$$\mathscr{F}_a = \begin{cases} N_a I_f, & -\dfrac{\rho_a}{2} \le \theta_{2e} \le \dfrac{\rho_a}{2} \\[2mm] 0, \dfrac{\rho_a}{2} \le \theta_{2e} \le \left(\pi - \dfrac{\rho_a}{2}\right), \left(\pi + \dfrac{\rho_a}{2}\right) \le \theta_{2e} \le \left(2\pi - \dfrac{\rho_a}{2}\right) \\[2mm] -N_a I_f, \left(\pi - \dfrac{\rho_a}{2}\right) \le \theta_{2e} \le \left(\pi + \dfrac{\rho_a}{2}\right) \end{cases}$$

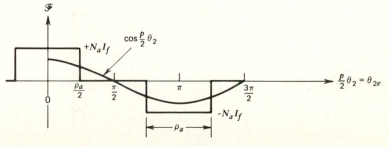

FIGURE 2.16 Components of the integrand for amplitude of Fourier fundamental.

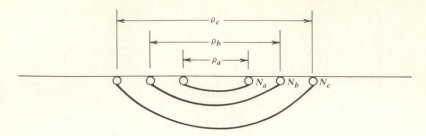

FIGURE 2.17 Typical set of three field coils.

By symmetry, the Fourier series for this function contains only cosine terms. The Fourier formula for the amplitude of the fundamental component is, then,

$$A_{1a} = \frac{1}{\pi} \int_0^{2\pi} \mathcal{F}_a(\theta_{2e}) \cos\theta_{2e} d\theta_{2e}$$

Note in Figure 2.16 that each time $\mathcal{F}_a(\theta_{2e})$ changes sign, $\cos\theta_{2e}$ also changes sign. The integral thus has four equal components, and may be written as

$$A_{1a} = \frac{4}{\pi} \int_0^{\rho_a/2} N_a I_f \cos\theta_{2e} d\theta_{2e} = \frac{4}{\pi} N_a I_f \sin\frac{\rho_a}{2}$$

Since all of the coils of each rotor pole are concentric, as in Figure 2.17, the fundamental components of their individual MMF waves will all be in phase. The amplitude of the fundamental MMF for the entire rotor will thus be the sum of the fundamental amplitudes associated with each coil. This peak value of the fundamental rotor MMF is called the *rotor magnetomotive force per pole* and will be given the symbol F_2:

$$F_2 = A_{1a} + A_{1b} + A_{1c} + \cdots$$

or

$$F_2 = \frac{4}{\pi} I_f \left(N_a \sin\frac{\rho_a}{2} + N_b \sin\frac{\rho_b}{2} + N_c \sin\frac{\rho_c}{2} + \cdots \right)$$

The effective field turns per pole, N_f, is defined by Equation 2.3. Comparing this with the expression for F_2 leads to the following expression for N_f in the case of a cylindrical rotor:

$$N_f = \frac{4}{\pi} \left(N_a \sin\frac{\rho_a}{2} + N_b \sin\frac{\rho_b}{2} + \cdots \right)$$

2.5 THE ROTATING MMF OF THE STATOR ARMATURE WINDING

The MMF of the *a*-phase winding of Figure 2.7 is shown as a function of both time and θ_1 in Figure 2.18. The subscript "1" is used to designate stator quantities. Scales of θ_1 in both mechanical and electrical radians are shown for comparison. Note that

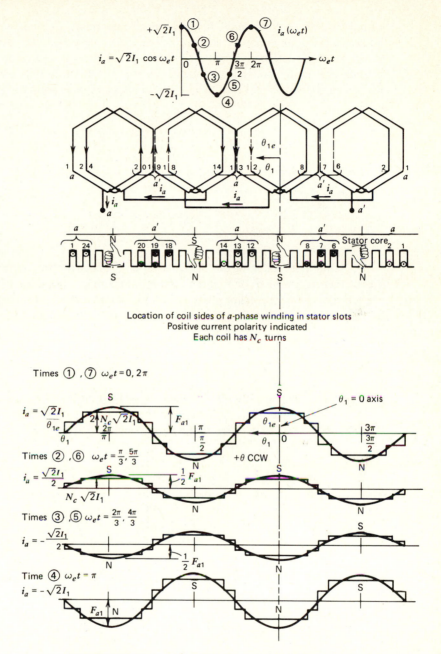

FIGURE 2.18 The breathing field of one phase winding.

positive current is taken to be flowing into the a' terminal of the winding and out of the a (unprimed) terminal. Considering the voltage polarity (a positive, a' negative), the current direction corresponds to operation in the generator mode. The side of each coil that is electrically closer to the a terminal is designated as an a side, while that closer to the a' terminal is marked a'. Note that positive current is directed inward in the a' sides and outward in the a sides.

Various instants in the a-phase current cycle are indicated by 1, 2, 3, and so on. The MMF of the a-phase winding is shown for seven points in the cycle. This kind of field is called a "breathing" field. It expands, contracts, expands with the opposite polarity, and again contracts, and so on; *but it does not move*. The assumed positive direction of magnetomotive forces and fluxes is radially outward from the rotor. Thus a south pole on the stator is considered positive as is a north pole on the rotor. The fundamental component of the field of the *a*-phase winding may be written

$$\mathscr{F}_a = (F_{a1} \cos \omega_e t) \cos \frac{p}{2} \theta_1 \text{ ampere-turns} \tag{2.5}$$

where θ_1 is mechanical degrees and ω_e is the angular frequency ($2\pi f$) of the phase-winding current. As in the case of the rotor, Equation 2.5 may be written in terms of electrical degrees:

$$\mathscr{F}_a = (F_{a1} \cos \omega_e t) \cos \theta_{1e} \tag{2.6}$$

F_{a1} is the peak value of the fundamental component of the Fourier series for the stepped waveform of $\mathscr{F}_a(\theta_1)$, at the instant of maximum current in the winding. The rms current in any one stator phase is I_1. Thus $\sqrt{2} I_1$ is the peak value of the current in the *a*-phase winding. By the methods of pages 56 to 59 inclusive, it is shown that

$$F_{a1} = \left(\frac{4}{\pi} n N_c k_w \right) \sqrt{2} I_1 \tag{2.7}$$

where *n* is the number of coils per group, N_c is the number of turns in each coil, and k_w is a "winding factor" calculated from the coil pitch ρ and the slot pitch γ. Since the units of F_{a1} are ampere-turns, this expression for F_{a1} may be simplified by letting

$$F_{a1} \stackrel{\triangle}{=} N_g I_1 \tag{2.8}$$

where N_g is the effective turns per phase group:

$$N_g = \frac{4\sqrt{2}}{\pi} n N_c k_w \tag{2.9}$$

As shown in Figures 2.8 and 2.19, the coil groups of the three phase windings are displaced from each other by 120 electrical degrees. Each phase winding produces a breathing field; however when balanced three-phase currents flow in the three phase windings, the three fields perform their stationary oscillations 120° out of phase with each other in time. The *combined* field of the the phase windings moves along the inside surface of the stator core at synchronous speed, given by Equation 2.1. This can be seen in two ways. First, mathematically, the three phase-winding fields may be written in terms of electrical degrees:

$$\mathscr{F}_a = (F_{a1} \cos \omega_e t) \cos \theta_{1e} \tag{2.10}$$

$$\mathscr{F}_b = [F_{a1} \cos (\omega_e t - 120°)] \cos (\theta_{1e} - 120°) \tag{2.11}$$

$$\mathscr{F}_c = [F_{a1} \cos (\omega_e t - 240°)] \cos (\theta_{1e} - 240°) \tag{2.12}$$

The total stator field is then

$$\mathscr{F}_1 = \mathscr{F}_a + \mathscr{F}_b + \mathscr{F}_c \tag{2.13}$$

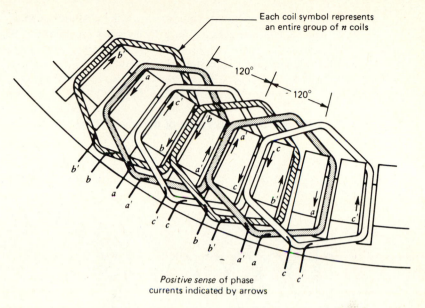

Each coil symbol represents
an entire group of *n* coils

120°

120°

Positive sense of phase
currents indicated by arrows

FIGURE 2.19 Sequence of phase coil groups in a section of the armature winding of a synchronous machine.

Applying $\cos a \cos b = \frac{1}{2}[\cos (a + b) + \cos (a - b)]$:

$$\mathcal{F}_1 = F_{a1} \cdot \tfrac{1}{2}[3 \cos (\omega_e t - \theta_{1e}) + \cos (\omega_e t + \theta_{1e})$$

$$+ \cos (\omega_e t + \theta_{1e} - 240°) + \cos (\omega_e t + \theta_{1e} - 480°)] \qquad (2.14)$$

The last three terms in the brackets sum to zero. (This may be seen easily by considering them as three phasors of equal magnitude at angles $0°$, $-240°$, and $-480°$. It will be observed that they add to zero.) Then

$$\mathcal{F}_1 = \tfrac{3}{2} F_{a1} \cos (\omega_e t - \theta_{1e})$$

$$= \tfrac{3}{2} F_{a1} \cos \left(\omega_e t - \frac{p}{2} \theta_1 \right) \qquad (2.15)$$

$$= \tfrac{3}{2} N_g I_1 \cos \left(\omega_e t - \frac{p}{2} \theta_1 \right)$$

This equation shows that:

1. The amplitude of the stator field is constant, to be defined as F_1:

$$F_1 = |\mathcal{F}_1| = \tfrac{3}{2} F_{a1} = \tfrac{3}{2} N_g I_1 \qquad (2.16)$$

Note that the MMF per pole of the stator field is $^3/_2$ the peak MMF per pole of one phase winding.

2. The field as *p* poles. This is evident by letting $t = 0$ and allowing θ_1 to vary from 0 to 360°.

3. The field travels with a speed of ω_s rad/s when θ_1 is measured in mechanical radians.

4. When i_a is at its maximum value ($t = 0$), the peak of the MMF coincides with the a-phase axis, since Equation 2.15 becomes identically $\frac{3}{2}$ of Equation 2.10.

To further simplify the expression for F_1, define

$$N_{I1} \overset{\triangle}{=} \tfrac{3}{2}N_g$$

as the effective current turns per pole of the stator. Then

$$F_1 = N_{I1}I_1 \tag{2.17}$$

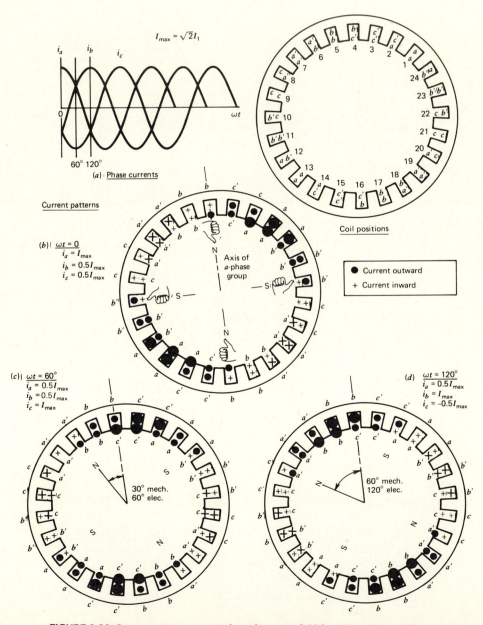

FIGURE 2.20 Stator current patterns and rotating stator field for a three phase winding.

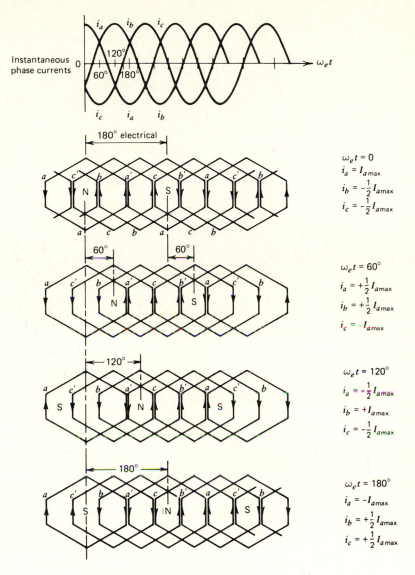

NOTE: Phase groups are viewed looking outward from rotor

FIGURE 2.21 Moving field developed by balanced three-phase currents in phase-winding coil groups displaced 120° electrical from each other.

where I_1 is the rms stator phase current under balanced conditions, and

$$N_{I1} = \tfrac{3}{2} \frac{4\sqrt{2}}{\pi} n N_c k_w \text{ (see pp. 56–59)} \qquad (2.18)$$

A second, qualitative approach to an understanding of the rotating stator field is through examination of the current patterns in the winding. Figure 2.20 shows the winding of Figures 2.7 and 2.8 in cross section. At a, the three-phase currents are shown as functions of time. Figures 2.20b, c, and d show the current patterns for $\omega t = 0$, $\omega t = 60°$, and $\omega t = 120°$, respectively. Dots and crosses in two sizes show the current magnitudes and directions. Note that the currents fall into sectors, all currents in each sector being in the same direction. Equal and opposite currents in a slot contribute zero MMF. Note also that these current sectors rotate with time.

Applying the right-hand rule on the axes separating the current sectors shows the existence of the stator magnetic poles, four in each case, on these axes. Observe that the stator poles fall on the axes of phase a when the current i_a is at its maximum. Note also that the poles rotate at an angular velocity of $\omega_s = \omega_e/2$, where $(p/2) = 2$.

The current pattern in plan view is shown in Figure 2.21. In this figure, each coil symbol represents a phase group, as in Figure 2.19. Note that the field moves one electrical degree for each degree of the current cycle. Also note that the field moves one pole pitch in $\frac{1}{2}$ current cycle. This is necessary because in one half cycle all currents become exactly the negatives of the values they had at the beginning of the half cycle. That means that every north pole of the winding is a south pole one half cycle later. The speed of the field is thus one pole pitch in $\frac{1}{2}$ cycle. The angular distance traveled is $2\pi/p$ mechanical radians in $1/2f$ s. Then

$$\omega_s = \frac{2\pi/p}{(1/2f)} = \frac{4\pi f}{p} = \frac{\omega_e}{p/2} \qquad (2.19)$$

This figure also demonstrates that the poles of the rotating stator field are centered on the coil groups of a given phase at the instant the current is a maximum in that phase.

FOR FURTHER STUDY

CALCULATION OF THE MMF OF THE ARMATURE $\vec{F}_1$

Consider only the first coil in each phase group in two adjacent pole windings of the armature winding of a synchronous machine, shown in Figure 2.22. (The case illustrated is a double-layer, three-phase winding of three slots per pole, the coil pitch being seven slots.) The plot of the MMF applied to the magnetic circuit by this coil pair, $\mathcal{F}(\theta_e)$, is shown by the solid line, as derived by Ampère's law. The broken line is the fundamental component of the Fourier series for $\mathcal{F}(\theta_e)$. The angle θ_e is in electrical degrees measured from the axis of the first coil. The amplitude of the fundamental is found as follows:

$$F_{1c} = \frac{1}{\pi} \int_0^{2\pi} \mathcal{F}(\theta_e) \cos \theta_e \, d\theta_e = 4 \left(\frac{1}{\pi} \int_0^{p/2} N_c i_1 \cos \theta_e \, d\theta_e \right)$$

The quantity in the parenthesis has the same value for each quarter period because $\mathcal{F}(\theta_e)$ in this case is symmetrical with $\cos \theta_e$ and always has the same sign as $\cos \theta_e$. The result of the integration is

$$F_{1c} = \frac{4}{\pi} N_c i_1 \sin \frac{\rho}{2}$$

where $\sin (\rho/2)$ will be recognized as the pitch factor involved in voltage calculations (see pp. 68–69). Then

$$F_{1c} = \frac{4}{\pi} k_p N_c i_1$$

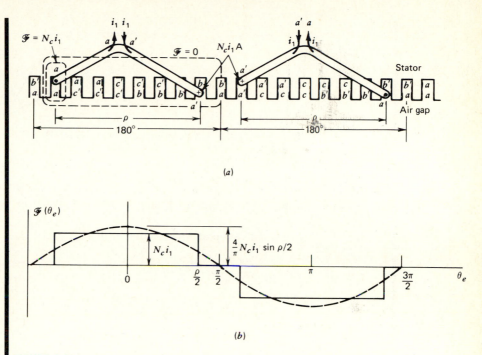

(a)

(b)

FIGURE 2.22 The magnetomotive force pattern of the first coils in two adjacent phase groups. (a) Placement of a pair of coils in adjacent phase groups. (b) MMF pattern of the two coils.

Expressing the fundamental in terms of space angle θ_e,

$$F_{1c}(\theta_e) = \frac{4}{\pi} k_p N_c i_1 \cos \theta_e$$

The fundamental MMFs of all the coils in one phase group are illustrated in Figure 2.23 for the case of three coils per group. It is evident that sinusoidal functions of space angle may be resolved in the same manner as phasors are as shown in Figure 2.24.

By analogy with Figure 2.33, it is seen that

$$F_{1group} = nF_{1c} \frac{\sin(n\gamma/2)}{n \sin(\gamma/2)} \equiv nF_{1c}k_d$$

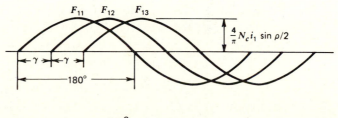

$n = 3$
$\gamma =$ slot pitch $= \dfrac{180}{nq_1}$

FIGURE 2.23 Fundamental components of the MMFs of the coils of one phase group.

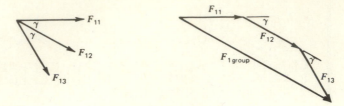

FIGURE 2.24 Phasor addition of coil MMFs.

where n is the number of coils per group (three coils per group illustrated in Figures 2.23 and 2.24) and k_d is the same distribution factor employed in voltage calculations.

If $i_1 = \sqrt{2}I_1 \cos \omega_e t$, the peak value, in time, of $F_{1\text{group}}$ is

$$F_{1m\text{group}} = nF_{1\text{max}}k_d$$

$$= n\frac{4}{\pi} k_p N_c \sqrt{2}I_1 k_d$$

$$= \frac{4}{\pi} nN_c k_w \sqrt{2}I_1 \equiv N_g I_1$$

This is the peak value of the fundamental component of the breathing field of Figure 2.18. The quantity nN_c is the total number of series turns in one phase group, since there are n coils of N_c turns in each group. There are as many groups as there are poles in the complete phase winding. Therefore

$$N_\phi = pnN_c$$

or

$$nN_c = \frac{N_\phi}{p}$$

Then

$$F_{a1} = \frac{4}{\pi} \frac{N_\phi}{p} k_w \sqrt{2}I_1$$

It was pointed out in Equation 2.16 that the amplitude of the rotating armature field (all three phases working together with balanced, three-phase currents flowing in them) is given by

$$F_1 = \tfrac{3}{2}F_{a1}$$

In general, it can be shown for a q_1-phase machine that

$$F_1 = \frac{q_1}{2} F_{a1}$$

Then

$$F_1 = \frac{3}{2} \frac{4}{\pi} \frac{N_\phi}{p} k_w \sqrt{2} I_1, \text{ or}$$

$$= \frac{q_1}{2} \frac{4}{\pi} \frac{N_\phi}{p} k_w \sqrt{2} I_1$$

Define

$$N_{I1} \triangleq \frac{q_1}{2} \frac{4}{\pi} \frac{N_\phi}{p} k_w \sqrt{2}$$

Then the amplitude of the stator rotating MMF field may be written simply as

$$F_1 = N_{I1} I_1$$

2.6 AIR GAP MAGNETIC FIELD AND FLUX PER POLE

The basic operation of ac machines depends only on the fundamental components of the rotor and stator fields. The harmonics cause some undesirable side effects, but these may be minimized by design. The fundamentals are, by definition, sinusoids. It has been pointed out that the rotor and stator fields travel at the same speed. They must maintain the same angular relationship with each other in the steady state, although there are transient readjustments in position between the two fields when the torque changes or when the field current is adjusted.

Since the rotor and stator MMFs are considered to be synchronous sinusoids, their sum is a sinusoid. They are synchronous in the sense that they rotate at the same angular velocity and thus maintain fixed angular relationships. Their resolution in space may be carried out as if they are two-dimensional vectors. The same techniques are used as in dealing with phasors, which represent quantities that are both sinusoidal and synchronous, that is, they have the same angular frequency. The resolution of the two MMFs may thus be written

$$\vec{R} = \vec{F}_1 + \vec{F}_2 \tag{2.20}$$

where $\vec{R}$ is the total MMF applied to the magnetic circuit of the machine. $\vec{R}$ will be called the "resultant field."

Figure 2.25 shows a developed, cross-sectional view of a synchronous machine. In this example, the stator armature winding has three coils per group and the coils are short-pitched two slots. The positions of the rotor and stator MMF fundamental waves are shown at the instant the a-phase current is at its maximum, and, as a result, the stator MMF poles are just passing centers of the a-phase groups as they move at synchronous speed. *Note:* In this figure, positive θ_1 is to the left. The rotor poles are shown leading the stator MMF field by an angle β. This angle will be constant in the steady state. Plots of the rotor and stator MMFs are also shown in the figure. When these two are added point by point along the air gap, the waveform of the resultant $\vec{R}$ is obtained.

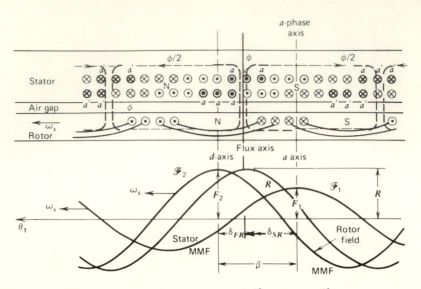

FIGURE 2.25 Machine flux produced by resultant of rotor $(\vec{F}_2)$ and stator $(\vec{F}_1)$ MMF fields shown at the instant of maximum current in phase a.

The magnetic circuit sees not $\vec{F}_1$ alone, not $\vec{F}_2$ alone, but $\vec{F}_1 + \vec{F}_2$. In other words, the flux in the machine is determined by $\vec{R}$. Figure 2.25 also shows the path of this flux, ϕ, as it flows across the air gap and through the rotor and stator iron. The peak flux density in the air gap coincides with the peak of $\vec{R}$.

Figure 2.25 illustrates the following definitions:

$$\beta = \text{angle by which } \vec{F}_2 \text{ leads } \vec{F}_1$$

$$= \text{angle by which } d \text{ axis leads } \vec{F}_1$$

$$\delta_{FR} = \text{angle between } \vec{F}_2 \text{ and } \vec{R}$$

$$\delta_{SR} = \text{angle between } \vec{F}_1 \text{ and } \vec{R}$$

$$F_1 = |\vec{F}_1| = N_{I1}I_1$$

$$F_2 = |\vec{F}_2| = N_f I_f$$

$$R = |\vec{R}|$$

Figure 2.26 is a schematic, two-pole representation of the situation in Figure 2.25. (The poles are shown as salient simply to sharpen the mental impression as to their relative location in space, even though cylindrical-rotor theory is being considered.) Such a representation contains a great deal of information in compact form and is quite useful in analyzing the operation of synchronous machines. Note the coincidence of the vector $\vec{F}_2$ with the d axis of the rotor and that the flux axis coincides with the vector $\vec{R}$. Note also the vector resolution of $\vec{F}_1$ and $\vec{F}_2$ to obtain $\vec{R}$. The same resolution is shown in Figure 2.27.

A computer-generated flux plot for a six-pole, salient-pole synchronous machine is shown in Figure 2.28. The figure demonstrates the displacement of the axis of peak air gap flux density (the R axis) from the field pole axis as a result of the flow of

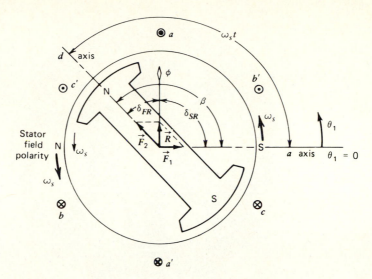

FIGURE 2.26 Schematic, two-pole representation of Figure 2.25.

armature currents in the stator windings. Compare this figure with the approximate drawing of the flux pattern of a cylindrical-rotor machine in Figure 2.32.

If the MMF $\vec{R}$ distributed along the air gap is considered to be a sinusoidal function, the air-gap flux density will also vary sinusoidally. Not quite sinusoidally, however, because the iron carrying the peak flux density will be more saturated than that experiencing lesser flux density. The result is that an instantaneous plot of flux density along the air gap would be a slightly flattened wave. This flattening results in some third harmonic voltage being induced in the coils, but it can be shown by circuit theory that these voltages do not appear at the terminals of the machines when the three windings are connected in either Δ or Y. So the flux density can be considered sinusoidal when plotted as a function of θ_{2e}.

The flux per pole is calculated as follows, if the peak flux density is known. See Figure 2.29. In this figure, angular distance along the air gap is measured in electrical degrees from the instantaneous location of peak $\vec{R}$. Note that both $\vec{R}$ and B waves are traveling around the air gap at speed ω_s, given by Equation 2.1. The positive direction flux density B is radially outward from the rotor surface.

The differential area dA is illustrated in Figure 2.30 and is given by

$$dA = Lr\, d\theta \tag{2.21}$$

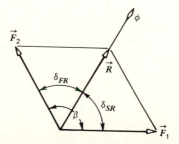

FIGURE 2.27 Resolution of MMF vectors in a synchronous machine.

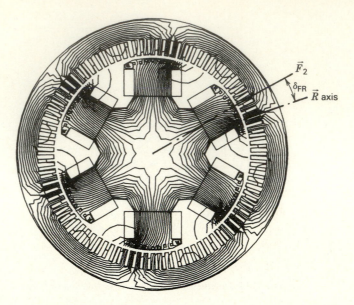

FIGURE 2.28 Computer flux plot for a six-pole synchronous machine. (Courtesy of Prof. Elias Strangas, Michigan State University.)

where θ is in mechanical degrees. In terms of θ_e, measured in electrical degrees

$$dA = Lr d\left(\frac{2\theta_e}{p}\right) = \frac{2Lr}{p} d\theta_e \tag{2.22}$$

from Equation 2.4. Now in general, flux is related to flux density by

$$\phi = \iint B \cdot dA \tag{2.23}$$

The flux density may be expressed as

$$B(\theta_e) = B_{\max} \cos \theta_e \tag{2.24}$$

The flux per pole of the air gap magnetic field is given by[1]

$$\phi = \frac{2Lr}{p} \int_{-\pi/2}^{\pi/2} B_{\max} \cos \theta_e \, d\theta_e$$

$$= \frac{2Lr B_{\max}}{p} \sin \theta_e \Big|_{-\pi/2}^{\pi/2}$$

or

$$\phi = \frac{4Lr B_{\max}}{p} \tag{2.25}$$

[1]Alternative method for ϕ: since $B(\theta_e) = B_{\max} \cos \theta_e$, the average value over one pole pitch is $B_{\text{ave}} = 2B_{\max}/\pi$. The area of the rotor surface per pole is $A_{\text{pole}} = 2\pi r L/p$. Then $\phi = B_{\text{ave}} A_{\text{pole}} = 4Lr B_{\max}/p$.

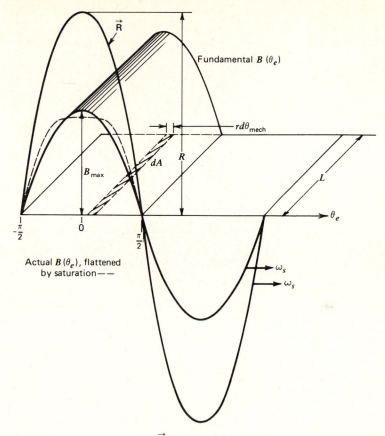

FIGURE 2.29 Relationships between $\vec{R}$, air gap flux density, and differential air gap area.

■ Example 2.1

The inside diameter of the stator of a small, three phase, six-pole synchronous machine is 0.300 m, and the length of the stator stack is 0.250 m. If the air gap flux density is sinusoidally distributed at the inside surface of the stator and has a peak value of 0.96 T, find the flux per pole.

Solution

$$\phi = \frac{4LrBl_{max}}{p} = \frac{4 \cdot 0.250 \cdot 0.150 \cdot 0.96}{6} = \underline{0.024 \text{ Wb}}$$

■

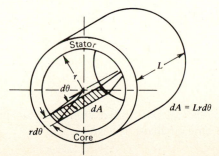

FIGURE 2.30 Definitions of machine dimensions as related to the differential area perpendicular to the radial air gap flux.

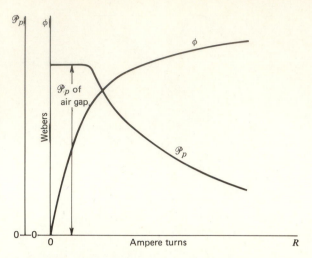

FIGURE 2.31 Effect of iron saturation on $\mathscr{P}_p$.

A useful concept in relating the air gap flux per pole to the sinusoidally distributed MMF causing it is that of *effective permeance per pole* of the magnetic circuit of the machine. The symbol for this quantity is $\mathscr{P}_p$ defined by

$$\phi = \mathscr{P}_p R \tag{2.26}$$

Since the value of ϕ depends on the degree of iron saturation, $\mathscr{P}_p$ is not a constant, but becomes less as saturation increases. See Figure 2.31.

2.7 VOLTAGE INDUCED IN THE ARMATURE WINDING OF A SYNCHRONOUS MACHINE

Recall from Chapter 1 that the magnetic field, in addition to developing torque or countertorque, induces voltage in the windings of a machine. This is the generated voltage of a generator or the countervoltage of a motor. The process by which this voltage is induced gives additional insight into how the synchronous machine operates.

Figure 2.32 shows a typical stator coil being linked by the flux produced by one pole of the resultant MMF, $\vec{R}$.

The angular distance between pole centers is called the pole pitch. In other words, the pole pitch is $2\pi/p$ radians, where p is the number of equally spaced poles on the rotor. Note also that one pole pitch corresponds to one half cycle of the flux density wave. One half cycle of a flux density wave occupies π electrical radians, or 180 electrical degrees. Thus the pole pitch is *always π electrical* radians.

If the flux density distribution in the air gap due to each pole of the air gap flux is sinusoidal, at the *instant* shown in Figure 2.32 the air gap flux density may be expressed as follows:

$$B = B_{\max} \cos\frac{p}{2}\,\theta = B_{\max} \cos\theta_e \tag{2.27}$$

where θ_e and θ are measured from the coil axis. In Equation 2.27, θ is in mechanical radians or degrees and θ_e is in electrical radians or degrees.

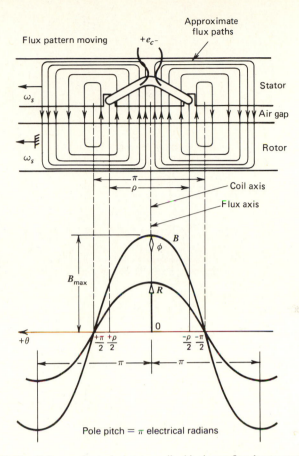

FIGURE 2.32 Flux linkages of a typical stator coil with air gap flux due to resultant MMF.

The total flux through the coil may be calculated on the basis that

$$\phi_c = \iint B \cdot dA_{\text{coil}} \tag{2.28}$$

As before, the differential area through which the flux passes can be seen from Figure 2.30 to be

$$dA = Lrd\theta = \frac{2Lr}{p} d\theta_e$$

Now the flux pattern is actually rotating with an angular velocity ω_s mechanical radians per second. The flux density wave may be written

$$B(\theta, t) = B_{\max} \cos\frac{p}{2}(\theta - \omega_s t)$$

$$= B_{\max} \cos(\theta_e - \omega_e t) \tag{2.29}$$

where ω_e is the electrical angular *frequency* and is related to the mechanical angular *velocity* by

$$\omega_e \stackrel{\Delta}{=} \frac{p}{2}\omega_s \qquad (2.30)$$

Then

$$\phi_c = \int_{-\rho/2}^{+\rho/2} [B_{max} \cos(\theta_e - \omega_e t)]\frac{2Lr}{p} \, d\theta_e \qquad (2.31)$$

But $\cos(x - y) = \cos x \cos y + \sin x \sin y$, so

$$\phi_c = B_{max} Lr \int_{-\rho/2}^{+\rho/2} (\cos \theta_e \cos \omega_e t + \sin \theta_e \sin \omega_e t)\frac{2}{p} \, d\theta_e \qquad (2.32)$$

Now from Equation 2.25, $4B_{max} Lr/p = \phi$, the flux per pole; then

$$\phi_c = \frac{\phi}{2}\left[\sin \theta_e \cos \omega_e t - \cos \theta_e \sin \omega_e t \right]_{-\rho/2}^{+\rho/2}$$

$$= \phi \cos \omega_e t \sin \frac{\rho}{2} \qquad (2.33)$$

As would be expected, the maximum coil flux is dependent on the coil pitch ρ. Equation 2.33 contains what is called the *pitch* factor k_p, which is defined by

$$k_p \stackrel{\Delta}{=} \sin \frac{\rho}{2} \qquad (2.34)$$

where ρ is in electrical degrees or radians.

The coil voltage may now be obtained by Faraday's law. If the coil has N_c turns,

$$\lambda = N_c\phi_c = N_ck_p\phi \cos \omega_e t \qquad (2.35)$$

With the assumed polarity of e_c indicated in Figure 2.32, the application of Lenz's and Faraday's laws gives the instantaneous coil voltage:

$$e_c = -\frac{d\lambda_c}{dt}$$

$$= \omega_e N_c k_p\phi \sin \omega_e t \qquad (2.36)$$

The frequency of this ac voltage is evident from the fact that

$$\omega_e = 2\pi f$$

The peak coil voltage is

$$E_{c\ max} = \omega_e N_c k_p\phi \qquad (2.37)$$

and its rms value is

$$E_c = \frac{E_{c\ max}}{\sqrt{2}} = \frac{\omega_e N_c k_p\phi}{\sqrt{2}}$$

or

$$E_c = \frac{2\pi f N_c k_p \phi}{\sqrt{2}}$$

$$= \sqrt{2}\pi f k_p \phi$$

$$\cong 4.44 f N_c k_p \phi \tag{2.38}$$

■ Example 2.2

The machine of Example 2.1 has 36 slots in its stator; that is, there are 6 slots per pole. Since each pole corresponds to 180 electrical degrees, there are 180°/6, or 30 electrical degrees per slot. If each coil has a span of 5 slots, what are the coil pitch ρ and the pitch factor k_p?

Solution

$$\rho = 5 \cdot 30 = 150° \qquad k_p = \sin\frac{\rho}{2} = \sin 75° = 0.9659$$

What is the maximum flux linkage with each coil if the coils have two turns?

Solution

From Equation 2.35, $\lambda_{c\ max} = N_c k_p \phi$. From Example 2.1 $\phi = 0.024$ Wb/pole. Then

$$\lambda_{c\ max} = 2 \cdot (\sin 75°) \cdot 0.024 = 0.046 \quad \text{Wb turns}$$

If the machine is running at 1000 rev/min, what rms voltage is induced in each coil?

Solution

From Equation 2.1,

$$f = \frac{pn_s}{120} = \frac{6 \cdot 1000}{120} = \frac{1000}{20} = 50 \text{ Hz}$$

Then from Equation 2.38,

$$E_c = \sqrt{2}\pi f N_c k_p \phi = \sqrt{2}\pi \cdot 50 \cdot 0.046 = 10.3 \text{ V} \qquad ■$$

The Voltage Induced in One Phase of a Three-Phase Winding

Figure 2.7 shows one third of the stator winding of a three-phase machine, that is, those coils associated with one phase. This is designated as phase a in the figure. The relative polarities of the coil sides and group and phase terminals are indicated by "primed" and "unprimed" phase symbols. Thus a' designates the same polarity as the neutral, while the absence of the "prime" indicates the same polarity as line terminal *a*. Figure 2.20 shows the location of the coil sides in the slots. Note that both the bottom and top layers of coil sides follow the same sequence counterclockwise: a, c', b, a', c, b'. Note also that a given slot may contain coil sides of two different phase belts. This is the result of using coils having a pitch less than 180 electrical degrees. The winding is described as follows:

1. Three phase ($q_1 = 3$), Y connected
2. Four pole ($p = 4$)
3. Double layer (there are two coil sides per slot)
4. Lap (each coil laps over the next)
5. Six slots per pole, $S_1/p = 6$, and $\gamma = 180/6 = 30°$
6. Two slots per pole per phase ($n = 2$)
7. Coil pitch five slots [$\rho = (\frac{5}{6})180° = 150°$]
8. Two coils per group (n = 2)
9. Each phase series connected

Note that the number of groups in each phase winding is equal to p, the number of poles. The following definitions are implied:

$q_1 \triangleq$ number of stator phases

$n \triangleq$ slots/pole/phase $\equiv$ coils per group in a double-layer winding[2]

$\gamma \triangleq$ slot pitch in electrical degrees $= 180/q_1 n = 180 \div$ slots per pole

$S_1 \triangleq$ number of stator slots, and $n = S_1/pq_1$

Most four-pole windings would have more than two coils per phase group, so Figure 2.7 is a somewhat simplified picture, but one that demonstrates the basic principles involved in calculating phase voltage. Note that the individual coil voltages in each group will be out of phase with each other by γ electrical degrees. The phase difference is due to the fact that the coils experience the rotating air gap flux in time sequence. Since they are separated from each other by one slot pitch, their voltages are out of phase by the slot pitch γ.

Also observe that the alternate group voltages are exactly 180° out of phase, because when one group is under the influence of a north pole, the next group is being acted upon by a south pole. However the groups are interconnected in such a way that all of the group voltages add in phase. Since there are p groups, the phase voltage is given by

$$E_\phi = pE_{\text{group}} \tag{2.39}$$

If there are n coils per group, then there are n coil voltages to be added to obtain the group voltage. Each coil voltage will be out of phase with the next by the slot pitch γ. Figure 2.33 shows how the group voltage is calculated. The rms value of the phasor sum of the n coil voltages is

$$|E_{\text{group}}| = nE_c k_d = n\frac{\omega_e N_c \phi}{\sqrt{2}} k_p k_d = \sqrt{2}\pi f n N_c \phi k_p k_d \tag{2.40}$$

---

[2]Each coil has two sides, and in a double-layer winding there are two coil sides in each slot. Thus the number of coils equals the number of slots. Then if there are n slots per pole per phase, each pole group has n coils.

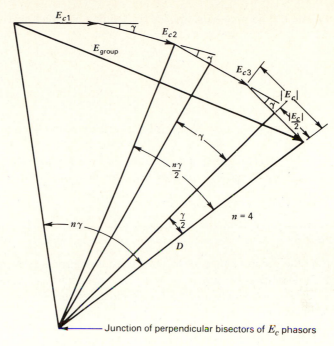

FIGURE 2.33 Calculation of E_{group} and distribution factor k_d for a group of four coils.

where k_d, the "distribution factor," is given by

$$k_d = \frac{\sin(n\gamma/2)}{n\,\sin(\gamma/2)} \qquad (2.41)$$

Finally, since the phase winding is composed of p groups in series,

$$|E_\phi| = |pE_{\text{group}}| = \omega_e \frac{pnN_c\phi}{\sqrt{2}}\,k_p k_d = \sqrt{2}\pi fpnN_c\phi k_p k_d \qquad (2.42)$$

Note that pnN_c is the total number of turns in the phase winding. Let the total series turns per phase be given the symbol N_ϕ:

$$N_\phi \triangleq pnN_c \qquad (2.43)$$

Then

$$|E_\phi| = \sqrt{2}\pi N_\phi f\phi k_p k_d \qquad (2.44)$$

The product $k_p k_d$ is often called the "winding factor" k_w:

$$|E_\phi| = \sqrt{2}\pi N_\phi f\phi k_w = \frac{\omega_e N_\phi \phi}{\sqrt{2}}\,k_w \qquad (2.45)$$

Equation 2.45 may be simplified by defining N_{E1} such that

$$|E_\phi| = N_{E1}\omega_e\phi, \text{ where}$$

$$N_{E1} \triangleq \frac{N_\phi k_w}{\sqrt{2}}$$

(2.46)

The quantity N_{E1} may be termed the "effective voltage turns per phase of the stator winding."

When ϕ is the flux per pole due to $\vec{R}$, E_ϕ is the induced phase voltage under load. When $F_1 = 0$ and ϕ is the flux per *rotor* pole, E_ϕ is the open-circuit phase voltage.

■ **Example 2.3**

Find the group voltage in the three-phase synchronous machine of Examples 2.1 and 2.2. Assuming the groups of one phase winding to be connected in series, find the induced phase winding E_ϕ. If the armature winding is Y-connected, find E_ϕ on a line-to-line basis.

Solution. The number of coils in a phase group is equal to the number of slots per pole per phase:

$$n = \frac{S_1}{q_1 p} = \frac{36}{3 \cdot 6} = 2 \text{ coils per group}$$

From Example 2.1, $\gamma = 30°$; then

$$k_d = \frac{\sin(n\gamma/2)}{n \sin(\gamma/2)} = \frac{\sin(2 \cdot 30°/2)}{2 \sin 15°} = \frac{0.500}{0.518} = 0.966$$

From Example 2.2, $E_c = 10.3$ V. Then by Equation 2.40,

$$|E_{\text{group}}| = nE_c k_d = 2 \cdot 10.3 \cdot 0.966 = 19.9 \text{ V}$$

The number of groups in a phase winding is equal to the number of rotor poles. Then by Equation 2.42,

$$|E_\phi| = p|E_{\text{group}}| = 6 \cdot 19.9 = 119 \text{ V}$$

For a Y-connected stator, the line-to-line induced voltage is

$$\sqrt{3}E_\phi = 207 \text{ V}$$

Alternatively, $N_\phi = nN_c p = 2 \cdot 2 \cdot 6 = 24$ turns. Then by Equation 2.44,

$$|E_\phi| = \sqrt{2}\pi N_\phi f \phi k_p k_d = \sqrt{2}\pi \cdot 24 \cdot 50 \cdot 0.024 \cdot 0.966^2$$
$$= 119 \text{ V}$$

■

Voltage of the Complete Stator Winding

A balanced set of three-phase voltages consists of three equal voltages of the same frequency that are mutually 120° out of phase. To generate balanced three phase, the stator winding of a three-phase machine must have three sets of coils, each set having the same N_ϕ. Each phase coil set must have a number of coil groups equal to the number of rotor poles. The voltages induced in the three phase windings, E_a, E_b, and E_c, must be 120° apart in *time*. This means that the coil groups that comprise the three windings must be displaced from each other in *space* by 120 *electrical* degrees. Figure 2.8 shows how this is accomplished in the simple winding illustrated.

When the three phase windings are connected in delta, the line-to-line terminal voltage is equal to the phase terminal voltage and the rated line current is $\sqrt{3}$ times the rated phase current. When the phases are connected in Y, the line voltage is $\sqrt{3}$ times the phase terminal voltage and the line and phase currents are equal. It is interesting to note that the same winding design that produces the desired rotating field of the synchronous machine also develops balanced three-phase voltages.

FOR FURTHER STUDY

THE REDUCTION OF HARMONICS IN VOLTAGE AND MMF WAVEFORMS

The Nature of Space Harmonics

That harmonics are present in the MMF waveforms of $F_1(\theta_{1e})$ and $F_2(\theta_{2e})$ is evident by the nonsinusoidal waveforms of Figures 2.14 and 2.18. These are "space harmonics," rather than time harmonics. Both F_1 and F_2 are stationary with respect to the rotor under steady-state conditions, so both may be expressed as functions of θ_{2e}. Their sum, point by point, is $R(\theta_{2e})$:

$$R(\theta_{2e}) = F_2(\theta_{2e}) + F_1(\theta_{2e})$$

Obviously $R(\theta_{2e})$ will also have harmonics, which will cause harmonics in $B(\theta_{2e})$, which in turn will result in harmonics in $E_\phi(t)$.

Consider the fundamental and fifth-harmonic waveforms of a MMF field of two poles, as illustrated in Figure 2.34. It is evident that the fifth harmonic has ten poles! By Equation 2.1 it travels at one-fifth the speed of the fundamental field. In the figure, if the fundamental is written $F_1 \sin \theta_{2e}$, then the fifth harmonic is $F_5 \sin 5 \theta_{2e}$. In other words, one electrical degree of the fundamental contains five electrical degrees of the fifth harmonic. For a harmonic order of ν,

$$\theta_\nu = \nu\theta_{2e}$$

where θ_{2e} is in fundamental electrical degrees.

The phase groups in the stator winding are spaced 120° apart, electrically, for the fundamental and $\nu \cdot 120°$ apart for the νth harmonic. For the fifth, the spacing is 600°, which is equivalent to 360° + 240°. Thus as far as the fifth harmonic is concerned, the phase sequence is opposite to the fundamental, and the $5p$-pole field

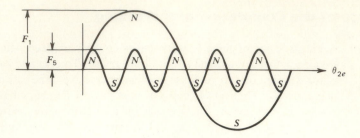

FIGURE 2.34 Fundamental and fifth-harmonic components of a nonsinusoidal MMF field.

rotates in a direction opposite to that of the main field. The seventh harmonic has seven times as many poles as the fundamental. However it rotates in the same direction as the fundamental field ($7 \times 120 = 840 = 2 \times 360 + 120°$).

Suppressing Harmonics in Machines

The symmetry of the phase groups precludes any even-order harmonics. Of the odd harmonics, only certain ones cause trouble in a three-phase machine. Consider the third harmonic. If the fundamentals of the voltages of the three stator phase windings are given by

$$E_{1max} \sin \omega t$$

$$E_{1max} \sin(\omega t - 120°)$$

$$E_{1max} \sin(\omega t + 120°)$$

then the three third-harmonic voltages will be given by

$$E_{3max} \sin(3\omega t + \alpha)$$

$$E_{3max} \sin(3\omega t - 360 + \alpha)$$

$$E_{3max} \sin(3\omega t + 360° + \alpha)$$

Note that the third-harmonic voltages of the three phases are all equal in magnitude and are in phase. If the machine is connected in Y, as shown in Figure 2.35, then

$$E_{3ab} = E_{3a} - E_{3b} = 0$$

$$E_{3bc} = E_{3b} - E_{3c} = 0$$

$$E_{3ca} = E_{3c} - E_{3a} = 0$$

Thus even if third-harmonic voltages are generated, they will not appear at the terminals.

Consider a delta connection, shown in Figure 2.36. Let the internal impedance of each phase of the machine to third-harmonic currents be designated Z_3. Let the

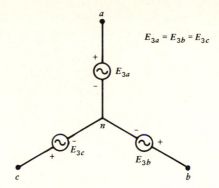

FIGURE 2.35 Third-harmonic voltages of a Y-connected machine.

three equal third-harmonic phase voltages each be E_3 V. A third-harmonic current will circulate in the phase windings:

$$I_3 = \frac{3E_3}{3Z_3} = \frac{E_3}{Z_3}$$

The third-harmonic voltages appearing at the terminals will be

$$E_{3ab} = E_3 - I_3Z_3 = E_3 - \frac{E_3}{Z_3}Z_3 = 0$$

$$E_{3bc} = E_3 - I_3Z_3 = 0$$

$$E_{3ca} = E_3 - I_3Z_3 = 0$$

Thus, regardless of whether the machine is connected in Y or delta, no third-harmonic voltage will appear at the line terminals. The same is true for harmonic orders that are multiples of 3, the number of phases, such as the ninth, fifteen, twenty-first, and so on. (Even harmonics do not exist and are thus not included in the list.) These are called the *triplen* harmonics. An important reason for connecting machines in Y is to avoid the circulating third-harmonic currents present in delta connections. In smaller machines there are often other, more compelling reasons for connecting the stator windings in delta.

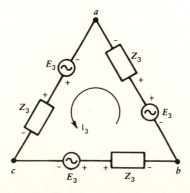

FIGURE 2.36 Third-harmonic voltages in a Δ-connected machine.

The pitch and distribution factors will be different for each harmonic, since the angles for the vth harmonic are v times as great as for the fundamental:

$$k_{pv} = \sin\frac{v\rho}{2}$$

$$k_{dv} = \frac{\sin(vn\gamma/2)}{n\sin(v\gamma/2)}$$

Since even harmonics do not exist and the triplen harmonics are naturally suppressed, only certain harmonics are troublesome for a symmetrical, three-phase machine:

$$v = 5, 7, 11, 13, 17, 19, 23, 25, 29, \text{etc.}$$

It is the nature of the Fourier series that the order of the harmonic occurs in the *denominator* of the expression for its amplitude. Thus the higher-order harmonics are quite small. The most troublesome harmonics are the so-called *belt harmonics* ("belt" is another name for a phase group of coils):

$$v_{\text{belt}} = 2q_1 \pm 1$$

where q_1 is the number of phases in the stator winding. Then for a three-phase machine,

$$v_{\text{belt}} = 6 \pm 1 = 5, 7$$

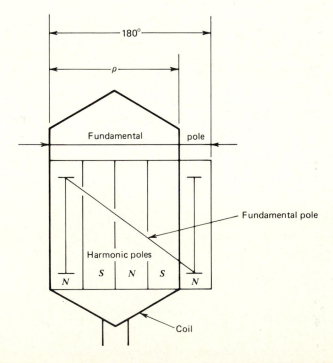

FIGURE 2.37 Short-pitched coil acted upon by the fundamental and fifth-harmonic fields.

Consider a coil under the influence of the field of Figure 2.34, consisting of a fundamental and fifth-space harmonic. The situation in plan view is shown in Figure 2.37. Note that if the coil pitch is four fifths of 180° (said to be "$\frac{4}{5}$ pitch"), the net fifth-harmonic flux enclosed is zero and no fifth-harmonic voltage can be induced in it:

$$k_p = \sin\frac{144°}{2} = 0.951$$

$$k_{p5} = \sin\frac{5 \cdot 144}{2} = 0$$

Similarly, the seventh harmonic may be completely eliminated by a coil of $\frac{6}{7}$ pitch, or 154.3°. Since both the fifth and seventh need to be suppressed, a good compromise is $\rho = 150°$, for which $k_{p5} = k_{p7} = 0.259$. The student should recognize that $\rho = 150°$ is not possible for all values of n (the number of coils per pole per phase), because the coil pitch must be an integral number of slots.

The fact that the coils in each group are distributed over several slots also reduces the harmonic content of both the voltage waveform and the $\mathcal{F}_1(\theta)$ waveform. Take for example a winding having three coils per group. Then for three phases there are $3n$, or 9 slots per pole, and $\gamma = 180/9 = 20°$. The distribution factors for the fundamental, fifth, and seventh are

$$k_d = \frac{\sin 30°}{3 \sin 10°} = 0.960$$

$$k_{d5} = \frac{\sin 150°}{3 \sin 50°} = 0.217$$

$$k_{d7} = \frac{\sin 210°}{3 \sin 70°} = -0.177$$

The negative sign indicates a reversal in phase of the seventh harmonic.

NOTE:

1. Since $k_{wv} = k_{pv}k_{dv}$, the suppression of the belt harmonics and by pitch and distribution is quite effective.
2. Since the pitch and distribution factors apply to both voltage and MMF functions, both waveforms are improved by the same measures.

Slot Harmonics

Certain higher-order harmonics cannot be suppressed by short pitching and distribution. These are harmonics of the order

$$\nu_{slot} = \frac{2MS_1}{p} \pm 1$$

where M is any integer, S_1 is the number of stator slots, and p is the number of poles. For example, in a 54-slot, six-pole machine, the two lowest slot harmonics are $(2 \cdot 54/6) \pm 1 = 17, 19$. If you substitute the above expression for ν into the

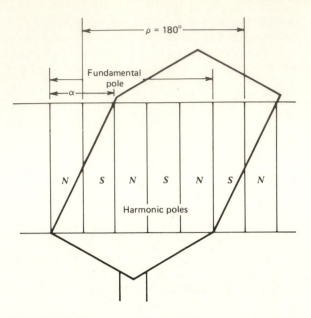

FIGURE 2.38 Effect of skewed coil on fifth-harmonic flux linkage.

expressions for k_p and k_d, you will find that the values obtained will be the same as those for the fundamental. In other words, *slot harmonics cannot be reduced by short pitch or distribution without reducing the fundamental a like amount.*

Slot harmonics are troublesome in induction motors and in large hydroalternators. In hydroalternators with many poles, the slot harmonic orders are relatively low. There are two means by which their effects may be effectively eliminated: skewing and fractional-slot windings. For a study of fractional-slot windings, refer to *Alternating-Current Machines,* by Liwschitz-Garik and Whipple (see Appendix D for reference). Skewing is discussed briefly here. Figure 2.37 in modified form is shown in Figure 2.38. In this case, the coil is full-pitched. However the slots in the stator are no longer parallel to the shaft, but are skewed two harmonic-pole pitches. The net harmonic flux is again zero, since the coil encloses the flux of $2\frac{1}{2}$ north and $2\frac{1}{2}$ south harmonic poles. If the harmonic flux density of the νth harmonic of the air gap flux density is given by

$$B_\nu(\theta,t) = B_{m\nu} \cos \nu(\theta_e - \omega_e t)$$

it can be shown by integration that the total harmonic flux linking a skewed coil is given by

$$\phi_{c\nu} = \frac{8 B_{m\nu} L r}{\nu^2 p \alpha} k_{p\nu} \sin \frac{\nu\alpha}{2} \cos \nu\omega_e t$$

where α is the skew angle in fundamental electrical degrees. The harmonic flux linkage is given by

$$\lambda_{c\nu} = N_c \phi_{c\nu}$$

and the coil voltage harmonic is given by

$$e_{cv} = -\frac{d\lambda_{cv}}{dt}$$

Its rms value is found to be

$$E_{\alpha v} = \frac{8 N_c B_{mv} L r \omega_e}{\sqrt{2} v p \alpha} k_{pv} \sin \frac{v\alpha}{2}$$

By a similar process, the vth-harmonic voltage induced in an unskewed coil ($\alpha = 0$) is found to be

$$E_{0v} = \frac{4 N_c B_{mv} L r \omega_e}{\sqrt{2} p} k_{pv}$$

The ratio of the voltage induced into a skewed coil to that of the same coil, unskewed, is called the *skew factor* k_s:

$$k_{sv} = \frac{E_{\alpha v}}{E_{0v}} = \frac{2 \sin(v\alpha/2)}{v\alpha}$$

where α must be in fundamental electrical radians. It is impractical to eliminate the fifth harmonic by this method because the amount of skew is mechanically excessive. However the slot harmonics are always of high order and may be "skewed out" with *relative* ease.

Rotor slots are usually skewed in induction motors and sometimes in dc machines. In larger hydroalternators, fractional-slot stator windings are usually employed. It is possible to skew the salient rotor poles instead of the stator slots to obtain reduction of slot harmonics. Some hydroalternators of European manufacture have been built with skewed poles.

The winding factor for a winding in skewed slots is, for the fundamental,

$$k_w = k_p k_d k_s$$

$$= \sin \frac{\rho}{2} \cdot \frac{\sin(n\gamma/2)}{n \sin(\gamma/2)} \cdot \frac{2 \sin(\alpha/2)}{\alpha}$$

and for the vth harmonic

$$k_{wv} = \sin \frac{v\rho}{2} \cdot \frac{\sin(nv\gamma/2)}{n \sin(v\gamma/2)} \cdot \frac{2 \sin(v\alpha/2)}{v\alpha}$$

Summary

Even-order harmonics are eliminated by symmetry.
Triplen harmonics do not appear at the machine terminals by circuit analysis.

Belt harmonics are greatly reduced by short-pitched coils and distributed coil groups.

Slot harmonics are reduced or eliminated by skew or fractional slot windings.

The result is that machines develop good voltage waveforms and their MMF fields are nearly sinusoidal functions of space angle.

2.8 LEAKAGE IMPEDANCE AND EQUIVALENT CIRCUIT

There are two kinds of magnetic fields that play a part in the way machines and transformers operate. First is the mutual field that links the windings of both rotor and stator or, in the case of transformers, the primary and secondary windings. This is the field that is involved in the energy conversion process. The second kind includes those fields which link only one winding.[3] These fields do not transfer energy from the input to the output of the device. They are called "leakage fields."

Leakage fields in machine are often segregated into three categories, for design purposes:

Slot leakage
End leakage
Differential leakage

These are illustrated in Figure 2.39.

Slot leakage includes those fluxes that link the coils of a winding by crossing the slots in which the coils are imbedded. Since some slots contain coil sides of more than one phase, the slot leakage for those slots includes mutual flux between the two phases. While most of the path of the slot leakage is in iron, the higher permeability of the iron makes the air path across the slot the major contributor to the total reluctance of the slot leakage paths. Thus the slot leakage is nearly a linear function of the winding current.

End leakage includes all the leakage fluxes around the coil ends. Since this leakage is in air, it is directly proportional to winding current.

Differential leakage is due to fluxes produced by a given winding that cross the air gap but do not link the other winding, at least in such a way as to be involved in the energy conversion process. They are due largely to the space harmonics that were neglected when it was assumed that the rotor and stator MMFs are sinusoidal functions of angle. As in the case of slot leakage, the major part of the reluctance of the differential leakage paths is due to air. So it can be said that the total leakage flux linkage of a given winding is very nearly proportional to the current in that winding.

When the current in a winding is ac, the leakage fluxes expand, collapse, and reverse in proportion to the current. Since they link the coils of the winding, they induce a voltage in the winding. By Faraday's law,

$$e_l = \frac{d\lambda_l}{dt} \tag{2.47}$$

---

[3]A field may link two stator windings, such as two phase windings, and still be considered leakage if it does not link the rotor winding.

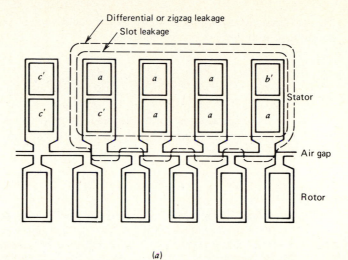

(a)

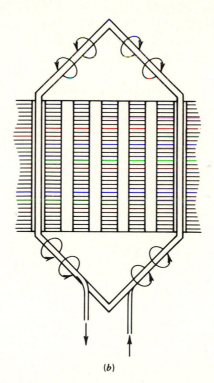

(b)

FIGURE 2.39 Leakage fields of a three-phase winding. (a) Leakage flux paths for one stator phase group. (b) End leakage flux of a typical coil.

where e_l^* is the leakage-induced voltage in a given winding and λ_l is the total leakage flux linkage with that winding. When the winding current is sinusoidal,

$$\lambda_l = K_l I_{\max} \sin \omega_e t \tag{2.48}$$

and by Equation 2.47,

$$e_l = K_l I_{\max} \omega_e \cos \omega_e t \tag{2.49}$$

The leakage-induced voltage is shown by this equation to be proportional to the current and to lead it by 90°. In phasor notation,

$$E_l = j\omega_e K_l \frac{I_{max}}{\sqrt{2}} = j\omega_e K_l I_{rms} \qquad (2.50)$$

We see by this equation that K_l, the constant of proportionality between the current and the leakage flux linkage, has the properties of inductance and is, in fact, called the "leakage inductance" L_l:

$$K_l \equiv L_l \qquad (2.51)$$

Then $\omega_e K_l \equiv \omega_e L_l$ is called the "leakage reactance" of the winding:

$$x_1 = \omega_e L_l \qquad (2.52)$$

This discussion of the leakage reactance per phase, x_1, has assumed balanced operation of the machine with the rms phase currents equal to each other and 120° apart in phase. Since leakage flux is, in part, mutual between the phase windings, the leakage reactance of the individual phase windings will vary under unbalanced conditions.

Each winding also has some resistance. At the time of this writing efforts are underway to develop an alternator with superconducting field windings operated at liquid helium temperatures. The field windings in these machines have zero resistance. However the armature windings of these "superconducting" generators still have resistance. The resistance of one phase winding of the armature will be assigned by the symbol r_1.

Each phase winding of the armature of a synchronous machine may be treated as a voltage source (the voltage induced by the air gap field, ϕ) having an internal resistance r_1 and an internal reactance x_1. The Thevenin equivalent circuit for the synchronous machine is shown in Figure 2.40. It is assumed that all three phase windings are subject to the same amount of leakage and have the same resistance.

Since the same things happen to all three phases in time sequence in the balanced, steady-state condition, much time and effort can be saved by considering only one

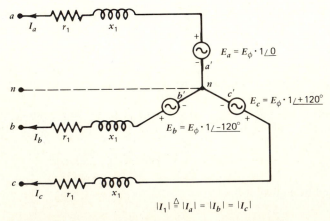

FIGURE 2.40 Thevenin equivalent of a synchronous machine.

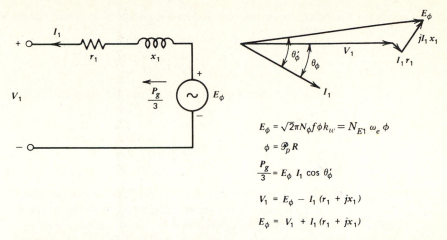

$$E_\phi = \sqrt{2}\pi N_\phi f \phi k_w = N_{E1}\,\omega_e\,\phi$$

$$\phi = \mathcal{P}_p R$$

$$\frac{P_g}{3} = E_\phi\,I_1\,\cos\theta_\phi'$$

$$V_1 = E_\phi - I_1\,(r_1 + jx_1)$$

$$E_\phi = V_1 + I_1\,(r_1 + jx_1)$$

FIGURE 2.41 Phase equivalent circuit of a three-phase synchronous machine, generator mode.

phase of the machine. It is a relatively simple matter to set down conditions in the other phases after one phase has been analyzed. The equivalent circuit for one phase is shown in Figure 2.41. The impedance $r_1 + jx_1$ is called the "leakage impedance" of the machine.

Power and Torque Implications

Figure 2.41 models only the electrical aspect of the synchronous machine, although the mechanical torque is implied. It has been pointed out that, in the steady state, the rotor winding has a fixed angular relationship to the flux ϕ. The magnitude of ϕ is constant since F_1 and F_2 are constant. (F_1 is constant by Equation 2.8 when balanced, three-phase currents flow in the stator windings.) Thus $d\lambda/dt = 0$ for the rotor windings. The rotor current is limited by the winding resistance R_f. Then in the steady state,

$$V_f = I_f R_f \quad \text{V, dc} \tag{2.53}$$

and the electrical input power to the rotor is

$$P_f = V_f I_f = I_f^2 R_f \quad \text{W} \tag{2.54}$$

In other words the *electrical* input power to the rotor is all converted to heat.

Most of the *mechanical* input to the rotor is converted to electrical energy. A small part is dissipated as friction, windage, and as stator core losses.[4] The power converted in each phase is $|E_\phi|\,|I_1|\,\cos\theta'_\phi$, where $\cos\theta'_\phi$ is the powerfactor angle between E_ϕ and I_1. The total power crossing the air gap will be called the "air gap power" P_g.

---

[4]The core losses appear as if they were mechanical losses. Eddy circuits in the stator iron set up magnetic poles that interact with the rotor poles to produce a drag on the rotor. A similar retarding torque is developed by hysteresis. As the flux field sweeps through the stator iron, the iron at any one spot is magnetized alternately in one direction and then the opposite. This phenomenon results in hysteresis loss and at the same time induces a rotating set of magnetic poles on the stator surface which pull backward on the rotor. Thus in overcoming these forces the mechanical power source that turns the rotor is required to supply the stator core losses.

This is the power being converted from the mechanical to the electrical form. Then for all three phases,

$$P_g = 3 |E_\phi| |I_1| \cos \theta'_\phi \qquad (2.55)$$

If losses are neglected, the torque required to drive the machine is the converted power divided by ω

$$\tau_d = \frac{P_g}{\omega} \qquad (2.56)$$

This is called the "developed torque." It is the basic output torque of a motor or the countertorque of a generator. (Generator quantities are shown positive in Figures 2.40 and 2.41.) The total torque required to drive an alternator is then

$$\tau = \frac{P_g + \text{friction} + \text{windage} + \text{core loss}}{\omega} \qquad (2.57)$$

For a synchronous machine, the shaft speed ω in Equations 2.56 and 2.57 is the synchronous speed ω_s given by Equation 2.1.

2.9 CIRCUIT MODEL OF THE SYNCHRONOUS MACHINE

The phasor diagram for the equivalent circuit and the vector diagram for the resolution of magnetic MMF fields are both shown in Figure 2.42. It would be most useful to combine the two. As a start, a relationship between $| E_\phi |$ and R is found by combining Equations 2.26, 2.45, and 2.46:

$$| E_\phi | = \sqrt{2}\pi N_\phi f k_w \phi$$
$$= \sqrt{2}\pi N_\phi f k_w \mathscr{P}_p R = \omega_e N_{E1} \mathscr{P}_p R \qquad (2.58)$$

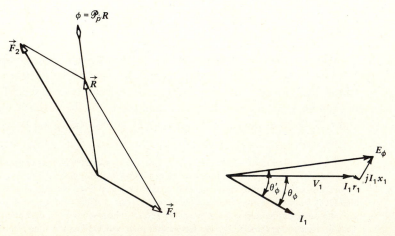

FIGURE 2.42 Vector and phasor diagrams of the synchronous machine.

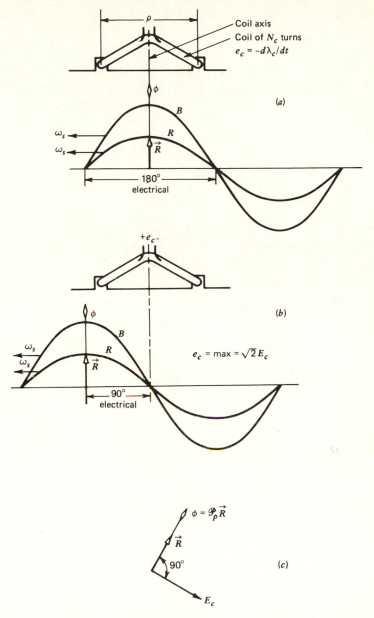

FIGURE 2.43 Relationship between vector $\vec{R}$ and phasor coil voltage defined. (a) Position of $\vec{R}$ relative to a stator coil for $\lambda_c = $ max. $\lambda_c = N_c\phi \sin(\rho/2)$. ($b$) Position of $\vec{R}$ relative to a stator coil for $(-d\lambda_{coil}/dt) = $ max and $e_c = +E_c$ max. (c) Phasor/vector relationship between $\vec{R}$ and E_c.

Let a factor m be defined[5] that includes all of the constants of Equation 2.45, together with the effective permeance per pole, $\mathscr{P}_p$ (which varies with saturation):

$$m \stackrel{\Delta}{=} \sqrt{2}\pi N_\phi fk_w \, \mathscr{P}_p = \omega_e N_{E1}\mathscr{P}_p \qquad (2.59)$$

[5]The factor m includes (1) $\mathscr{P}_p$, which represents the nonlinear relationship between the flux per pole, ϕ, and the peak value of the resultant MMF, R; (2) the effective voltage turns per phase, N_{E1}; and (3) ω_e, the electrical angular frequency.

Then

$$|E_\phi| = mR \qquad (2.60)$$

Now examine Figure 2.43. If the time at which $\vec{R}$ is aligned with a given coil axis is associated with the *position* of vector $\vec{R}$, maximum voltage induced by the flux produced by $\vec{R}$ occurs at a time 90° later in the electrical cycle. But during this interval the vector $\vec{R}$ has rotated 90 electrical degrees in space. Thus the coil voltage may be considered a phasor lagging the $\vec{R}$ vector by 90°. Figure 2.44 shows the same relationship schematically for the phase-induced voltage E_ϕ. Now that a time/space relationship between E_ϕ and $\vec{R}$ has been defined, the magnitude and phase information may be combined:

$$E_\phi = m\vec{R} \cdot 1\underline{|-90°} = -jm\vec{R} \qquad (2.61)$$

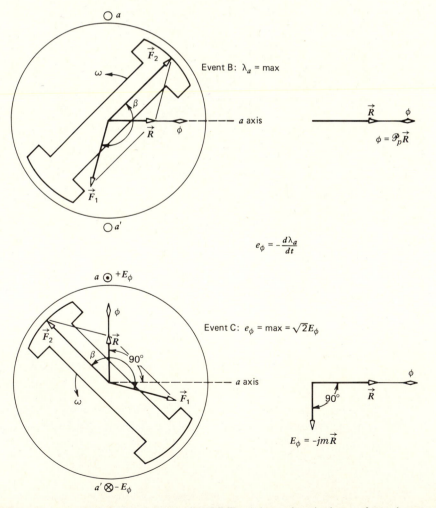

FIGURE 2.44 Phasor relationship between resultant MMF and phase voltage in phase *a* of a synchronous machine.

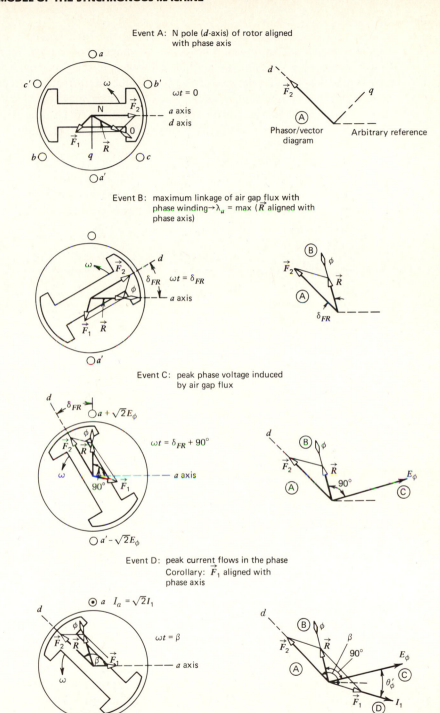

FIGURE 2.45 Sequence of events as they affect one phase during each cycle and are related to the phasor/vector diagram of a synchronous machine.

A phasor diagram is really a calendar giving the time sequence of events and including magnitude information as well. The diagram of the synchronous rotating MMF vectors of a synchronous machine tell not only their magnitudes and relative angular positions but also the time sequence in which they affect a given phase winding in the stator. It is enlightening to study the sequence of events in each cycle of operation of a synchronous machine as they affect one phase winding on the stator. This will permit building a combination **phasor/vector diagram,** which in turn will lead to a **circuit model** for the synchronous machine.

Consider the events depicted in Figure 2.45. Time and angles will be measured from the instant the center of the north pole of the rotor passes the axis of the phase under consideration—phase a in the figure. If there are more than two poles, this event may be described as the instant the north rotor poles are aligned with the centers of the a-phase coil groups. This event will be called event A. Other ways of describing event A are:

$\vec{F}_2$ aligned with a axis

Peak value of $F_2(\theta)$ passing centers of coil groups

d axis of rotor aligned with a axis

$\omega t = 0$ by assumption

(Note that for more than two poles, angular rotation would be measured in electrical degrees: $(p/2)\omega t = \omega_e t$.)

Figure 2.45 also shows the development of the phasor/vector diagram. Event A is denoted by vector $\vec{F}_2$ at some arbitrary angle. The reason for this is that it is often desirable in calculations to choose some other phasor as a reference, usually the terminal voltage V_1 or the phase current I_1.

The second event of interest, event B, is the instant when the maximum amount of the air gap flux ϕ links the phase winding under consideration. This event is illustrated by Figure 2.32, if the single coil in that figure is taken to represent a phase group whose center coincides with the axis of the coil shown. The term "air gap flux" refers to the flux mutual to the rotor and stator windings, excluding leakage fluxes. This is the flux of Equation 2.26. The peak density of this flux coincides with $\vec{R}$. When the $\vec{R}$ vector is centered on the phase axis, the linkage of the phase winding with the air gap flux is maximum and the instantaneous value of the induced phase voltage is zero. Between event A and event B, the rotor has rotated through an angle equal to the angle between $\vec{F}_2$ and $\vec{R}$ (i.e., δ_{FR} electrical degrees).

The next event of interest is the instant of peak induced phase voltage, because this locates the E_ϕ phasor. Call this event C. This has been shown (Figures 2.43 and 2.44 and Equation 2.61) to occur 90 electrical degrees of rotor rotation after event B.

The last event to be considered occurs at the instant the current in the phase is at its positive maximum. Call this event D. It has been pointed out that, at this instant, the poles of the rotating stator field $\vec{F}_1$ are instantaneously aligned with the centers of the coil groups of the phase winding. In other words, $\vec{F}_1$ is aligned with the phase axis. Event D, then, occurs θ'_ϕ degrees after event C, since the angle between the E_ϕ phasor and the I_1 phasor represents the time between the occurrence of the peak values of these two quantities. The angle between $\vec{F}_2$ and $\vec{F}_1$ has been defined as β. Since $\vec{F}_2$ was aligned with the a axis at the time of event A, and $\vec{F}_1$ is on the same axis as event D, the rotor has rotated β degrees at the instant of event D. These considerations allow the vector and phasor diagrams to be combined, with the $\vec{F}_1$ vector lying along the I_1 phasor. The combined phasor/vector diagram, based on cylindrical rotor theory, is shown in Figure 2.46.

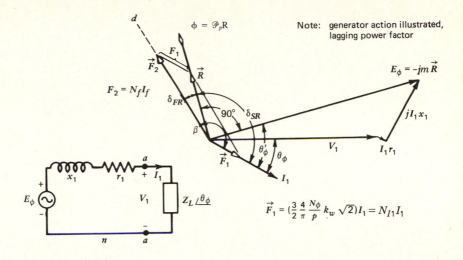

$$\text{KVL: } E_\phi = V_1 + I_1(r_1 + jx_1)$$

FIGURE 2.46 Combined phasor/vector diagram for one phase of a cylindrical-rotor synchronous machine.

It is now possible to include the MMF quantities in an expanded circuit model of the synchronous machine. Note that the resolution of MMF vectors is linear. The magnetic nonlinearity of the iron does not come into play until *after* the resultant has been found and ϕ and E_ϕ are to be calculated. The trick in modeling the resolution of the linear resolution of $\vec{F}_1$ and $\vec{F}_2$ to obtain $\vec{R}$, is to represent them as voltages to be added as phasors. Equation 2.61 establishes a relationship between E_ϕ and $\vec{R}$. The same relationship is now applied to obtain fictitious phasor voltages *representing* $\vec{F}_2$ and $\vec{F}_1$ in the model:

$$\text{Let } E_{ar} \overset{\Delta}{=} -jm\vec{F}_1 \tag{2.62}$$

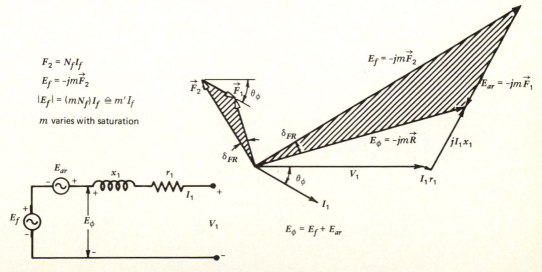

FIGURE 2.47 Intermediate model of a synchronous machine.

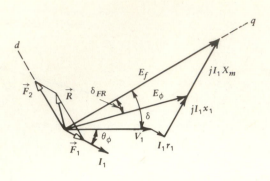

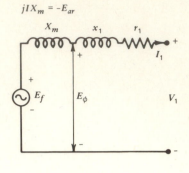

$$E_\phi = V_1 + I_1 (r_1 + jx_1)$$

$$E_f = V_1 + I_1 [r_1 + j(x_1 + X_m)]$$

(a)

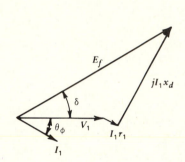

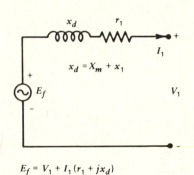

$$E_f = V_1 + I_1 (r_1 + jx_d)$$

Note: r_1 is about 1/2% to 2% of x_d and may be
          neglected except for efficiency calculations.

(b)

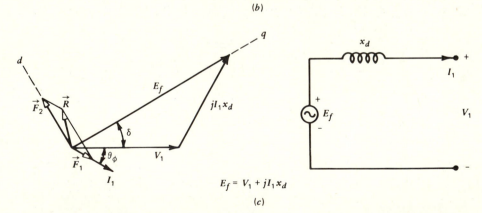

$$E_f = V_1 + jI_1 x_d$$

(c)

FIGURE 2.48 Final model of the cylindrical-rotor synchronous machine. (*a*) Complete initial model. (*b*) Combination of X_m and x_1 to form x_d, the *synchronous reactance*. (*c*) Model with winding resistance r_1 neglected.

and

$$E_f \overset{\Delta}{=} - jm\vec{F}_2 \qquad\qquad (2.63)$$

Then

$$-jm(\vec{F}_2 + \vec{F}_1) = - jm\vec{R} = E_\phi$$

or

$$E_f + E_{ar} = E_\phi \tag{2.64}$$

The resulting phasor diagram and circuit model are shown in Figure (2.47). Note that E_ϕ has been replaced by the *sum* of two *fictitious* voltages E_f and E_{ar}. The subscript f is a reminder that E_f *represents* the effect of the field MMF, $\vec{F}_2$, while the subscript ar indicates that E_{ar} *represents* the armature reaction vector $\vec{F}_1$.

This model may be simplified. Note that by Equation 2.59,

$$-E_{ar} = +jmF_1 = j(\omega_e N_{E1} \mathscr{P}_p)N_{I1}I_1 = j\,\omega_e(N_{E1}N_{I1}\mathscr{P}_p)I_1 \tag{2.65}$$

The quantity $(N_{E1}N_{I1}\mathscr{P}_p)$ is an inductance (units: henries). Let $\omega_e(N_{E1}N_{I1}\mathscr{P}_p)$ be defined as the *magnetizing reactance X_m*, also called the *reactance of armature reaction*. Thus $-E_{ar}$ has the properties of a reactance drop resulting from the effects of the armature current I_1. When $-E_{ar}$ is replaced by the reactance voltage drop $j\omega_e X_m I_1$, the circuit model becomes that of Figure 2.48a. Furthermore, since $-E_{ar}$ is an extension of the jI_1x_1 drop, let a new reactance x_d be defined, equal to the sum of the effects of leakage flux and armature reaction:

$$jI_1x_1 + jI_1X_m \overset{\Delta}{=} jI_1x_d$$
$$x_d \overset{\Delta}{=} x_1 + X_m \tag{2.66}$$

The resulting model, shown in three forms in Figure 2.48, is amazingly simple, considering all the complex phenomena it represents. It has been stated that x_1 is the leakage reactance of one phase of the armature winding. The sum of x_1 and X_m, x_d, has a very fancy name, the *direct-axis synchronous reactance*, or *synchronous reactance* for short.

2.10 RELATIVE MAGNITUDES OF SYNCHRONOUS MACHINE IMPEDANCES: THE PER-UNIT SYSTEM

The rated phase voltage is called the *base phase voltage $V_{\phi B}$* of the machine. Rated line voltage is called the *base line voltage V_{LB}*. For Y connection, $V_{LB} = \sqrt{3}V_{\phi B}$. The rated phase current is called the *base current $I_{\phi B}$*. The *base impedance* is defined by

$$Z_B \overset{\Delta}{=} \frac{V_{\phi B}}{I_{\phi B}} \tag{2.67}$$

A load having an impedance Z_B, when connected across one phase winding will draw rated current at rated voltage. When an actual quantity, say, the actual value of I_1, is divided by the corresponding base quantity, the result is called the *per unit* value of the quantity:

$$I_{1\text{pu}} = \frac{I_1}{I_{\phi B}}$$

$$V_{1\text{pu}} = \frac{V_1}{V_{\phi B}} \tag{2.68}$$

$$x_{d\text{pu}} = \frac{x_d}{Z_B}$$

On this basis, r_1 varies from about 0.02 per unit in small machines to 0.005 in large ones. On the other hand, x_{dpu} is about 1.0 per unit. In other words, r_1 is usually negligible compared to x_d, as indicated in Figure 2.46c. This large value of x_d means that the $I_1 x_d$ drop at full load is about equal to rated terminal voltage! If

$$|I_1| = \text{rated} = I_{\phi B}$$

and

$$\frac{x_d}{Z_B} \cong 1.0$$

then in per-unit,

$$\frac{|I_1 x_d|}{V_{\phi B}} \cong \frac{I_{\phi B} Z_B}{I_{\phi B} Z_B} = 1.0$$

and

$$|I_1 x_d| \cong V_{\phi B}$$

Of the two components making up x_d, the leakage reactance x_1 is about 0.20 per unit, leaving X_m at about 0.80 per unit.

2.11 POWER AND TORQUE AS RELATED TO POWER ANGLE δ

When the phase winding resistance is neglected, the circuit model of a synchronous machine is that of Figure 2.48c. This is one of a class of circuits in which the voltages at two ends of a purely inductive transmission line are known—Figure 2.49. Can the power transmitted be found if only E_s and E_R are known? The input power is equal to the output power in this circuit, since no power is dissipated in a pure inductance:

$$P_{\text{in}} = P_{\text{out}} = P = |E_R|\,|I|\,\cos\theta_\phi \qquad (2.69)$$

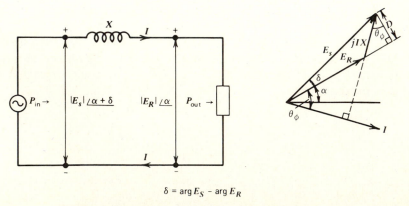

$$\delta = \arg E_S - \arg E_R$$

FIGURE 2.49 Lossless, purely inductive transmission line.

If a line D is drawn from the end of the E_s phasor, perpendicular to an extension of E_R, the angle it forms with the jIX phasor will be the power factor θ_ϕ. This is true because θ_ϕ is the angle between E_R and I, jIX is perpendicular to I, and D was made perpendicular to E_R. By inspection of Figure 2.49,

$$D = |E_s| \sin \delta = |I| X \cos \theta_\phi \qquad (2.70)$$

Then

$$|E_R \| E_s| \sin \delta = |E_R \| I| X \cos \theta_\phi = XP \qquad (2.71)$$

or

$$P = \frac{|E_R \| E_s|}{X} \sin \delta \qquad (2.72)$$

Applying this relationship to the synchronous machine, $E_s \to E_f$, $E_r \to V_1$, the power **per phase** is

$$P_\phi = \frac{|V_1| \, |E_f|}{x_d} \sin \delta \qquad (2.73)$$

and the total power of a three-phase machine is

$$P = 3 \frac{|V_1 \| E_f|}{x_d} \sin \delta \qquad (2.74)$$

Comparing Figure 2.48c with Figure 2.49, it is evident that this is power flowing *out* of the machine, (i.e., generated power).

A plot of Equation 2.74 is given in Figure 2.50, assuming $|E_f|$ and $|V_1|$ are constant. This figure illustrates two important points:

1. There is a definite upper limit to the power capability of a synchronous machine.
2. When δ becomes negative, power flow reverses. Power then flows into the electrical terminals, indicating that the machine is acting as a *motor:*

$$+ \, \delta \to \text{generator mode}$$

$$- \, \delta \to \text{motor mode}$$

The developed torque is given by

$$\tau_d = \frac{P}{\omega_s} = \frac{3}{\omega_s} \frac{|V_1 \| E_f|}{x_d} \sin \delta \text{ N-m} \qquad (2.75)$$

where ω_s is the synchronous speed given by Equation 2.1, in mechanical radians per second. For this reason, δ is also called the torque angle. Note that the torque reverses direction (sign) when the machine goes from generator operation to motor operation. Equation 2.75 was derived on the assumption that positive power flow is out of the terminals (i.e., generation). Thus positive torque in this equation is a *counter*torque:

$+ \ \tau_d$ implies τ_d is opposite to ω_s

$- \ \tau_d$ implies τ_d is in the same direction as ω_s

The angle δ is between an *actual* voltage V_1 and a *fictitious* voltage E_f. It therefore has little physical significance. Reference to Figures 2.48a and 2.49 shows that the power is also given by

$$P = 3 \frac{|E_\phi|\,|E_f|}{X_m} \sin \delta_{FR} \qquad (2.76)$$

and the torque by

$$\tau_d = \frac{3}{\omega_s} \frac{|E_\phi|\,|E_f|}{X_m} \sin \delta_{FR} \qquad (2.77)$$

Now δ_{FR} is the angle between two fictitious voltages, but it is also the angle between two real entities, the rotor poles $\vec{F_2}$ and the air-gap flux axis ϕ. Torque is produced as the rotor magnetic poles to try to align themselves with the flux, as illustrated by Figure 2.51. When the machine is operated as a generator, the prime mover drives the poles ahead of the flux. The poles are pulled backward to produce a positive countertorque. When operating as a motor, the mechanical load pulls the rotor poles back from the flux. The torque then is developed in the same direction as ω_s (negative τ_d) to drive the load. With no load on the shaft, the rotor poles are aligned with the flux, except for the small deflection required to overcome friction, windage, and core loss torques.

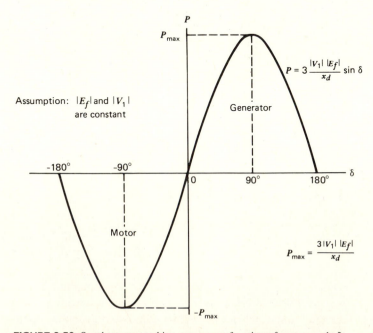

FIGURE 2.50 Synchronous machine power as a function of power angle δ.

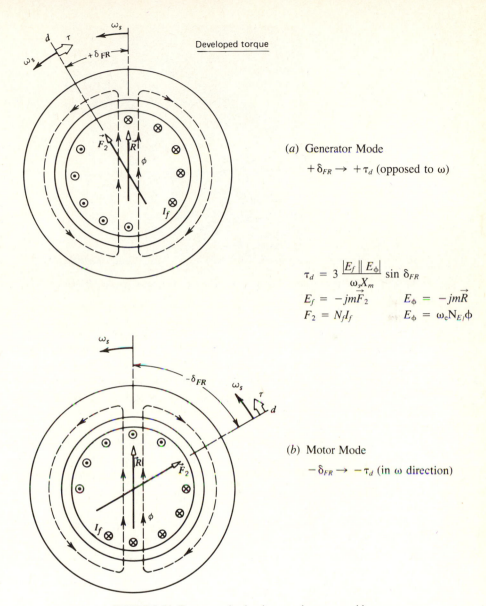

(a) Generator Mode

$$+\delta_{FR} \rightarrow +\tau_d \text{ (opposed to } \omega)$$

$$\tau_d = 3 \frac{|E_f||E_\phi|}{\omega_s X_m} \sin \delta_{FR}$$

$$E_f = -jm\vec{F}_2 \qquad E_\phi = -jm\vec{R}$$

$$F_2 = N_f I_f \qquad E_\phi = \omega_e N_{Ef}\phi$$

(b) Motor Mode

$$-\delta_{FR} \rightarrow -\tau_d \text{ (in } \omega \text{ direction)}$$

FIGURE 2.51 Torque production in a synchronous machine.

2.12 DAMPING AND SYNCHRONOUS MOTOR STARTING

The means of torque production just described requires that the rotor be turning at the same speed as the stator field (i.e., synchronous speed). If the rotor is going at some other speed, the rotating stator-field poles will be moving past the rotor poles—first attracting, then repelling them. The average torque under these conditions is zero. As a result, a "pure" synchronous machine will not start. A prevalent practice years ago was to bring the machines up to synchronous speed by means of a dc motor before connecting the machine to the three-phase line. Today the most widely used means of starting is a winding consisting of heavy copper bars installed in slots in the pole

FIGURE 2.52 Salient-pole assembly for a synchronous machine showing damper winding. (Courtesy of Westinghouse Electric Corporation.)

faces. These bars are all shorted together at both ends of the rotor. Currents induced in the bars by the rotating air-gap field interact with the field to produce a torque in the direction of field rotation. In other words, the machine is started as an induction motor, the bars in the pole-face slots forming a sort of squirrel-cage rotor. These bars and their shorting end rings are visible in Figure 2.12. A single pole with the bars installed is shown in Figure 2.52.

Induction motor action will bring the machine to nearly synchronous speed. At synchronous speed, there is no relative motion between the poles of the air-gap field and the pole-face bars. No current is induced in the bars at synchronous speed, and no torque would be produced by them. However, the top speed developed on induction motor action is so near synchronous speed that the rotor "falls into step" when the dc field current is turned on.

The field winding terminals are often shorted through a resistor during starting, until time to excite the field. This has two advantages. First, it protects the slip-ring insulation from the high ac voltages induced in the field during starting. Second, the circulating ac current in the shorted field windings contributes a small additional amount of acceleration torque.

The heavy bar "windings" in the pole faces serve an additional purpose. When there is a sudden change in load, δ_{FR} changes to adjust to the new torque requirement. The magnetic attraction between the air gap flux and the rotor MMF poles has a springlike quality and the rotor has a considerable moment of inertia. Consequently, any change in load results in an oscillatory motion superimposed on the normal, synchronous rotation of the shaft. This motion is called *hunting*. In a salient-pole machine there is very little to damp out this oscillation, unless the shorted pole face bars are present. Changes in δ_{FR} causes these bars to move relative to ϕ, with the result that currents are induced that circulate through the bars and end connections. The field set up by these currents produces a torque that opposes any change in δ_{FR}, and the angular oscillation is quickly damped out. For this reason the pole face bars are called *damper bars* and the winding consisting of these heavy bars and their end connections is called the *damper winding,* or by their French name, the *amortisseur* winding. (Roughly translated, this means "killer winding.")

Because an intolerable amount of hunting is present in the absence of damper windings, they are installed in nearly *all* synchronous machines, *generators as well as motors,* having *salient* poles. *Cylindrical* rotor cores cannot be laminated and still maintain the required mechanical strength. Eddy currents therefore can circulate freely in the pole faces. Since the flux through the pole faces is constant under steady-state conditions, these eddy currents exist only during transients. The interaction of the eddy currents with the air gap flux provides the necessary damping action in cylindrical-rotor synchronous machines.

2.13 HOW FIELD CURRENT CHANGES ARE REFLECTED IN THE CIRCUIT MODEL

By Equation 2.63, the magnitude of E_f is mF_2 and, by Equation 2.3, $F_2 = N_f I_f$. Then

$$|E_f| = mN_f I_f = m' I_f \qquad (2.78)$$

where

$$m' \overset{\Delta}{=} mN_f \qquad (2.79)$$

So the field current determines E_f in the model.

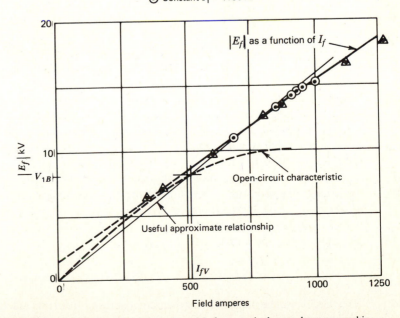

E_f as a function of field current for a 75–MW, 13,200–V (line-to-line) alternator

▲ Constant power, 75 MW
⊙ Constant $I_1 = 4100$ A

$|E_f|$ as a function of I_f

Open-circuit characteristic

Useful approximate relationship

I_{fV}

Field amperes

FIGURE 2.53 Relationship between E_f and I_f for a particular synchronous machine.

Now m is determined by the relationship between R and $|E_\phi|$ (Equations 2.60 and 2.67), which involves $\mathcal{P}_p$. Thus the value of m varies with the degree of saturation. The machine is usually operated above the knee of the saturation curve, Figure 2.37, so the coefficient m usually decreases slightly as I_f increases. However, $|E_f|$ always changes in the same direction as I_f and is *roughly* proportion to I_f.

Figure 2.53 is a plot of $|E_f|$ as a function of I_f for a cylindrical-rotor machine rated at 93.75 MVA at 0.8 power factor, under two load conditions. Note the relationship is quite linear over a wide range. However $|E_f|$ is not quite proportional to I_f, since an extrapolation of the linear region results in an intercept that is not at the origin. Under *no load* ($I_1 = 0$) the plot of $|E_f|$ as a function of I_f is the open-circuit characteristic. For accurate calculation, the "saturated synchronous reactance method" (p. 115) should be used for finding $|E_f|$ and I_f. For many practical calculations, however, it is often assumed that $|E_f|$ varies along a straight line drawn from the origin through rated voltage on the open-circuit characteristic. The slope of this curve is defined as m_a in Section 2.19. For many approximate calculations, this value may be used for m' in Equation 2.78, thus assuming a proportional relationship between $|E_f|$ and I_f.

2.14 OPERATION AS A MOTOR

Figure 2.54a shows the phasor/vector diagram and circuit model of the synchronous machine in the generator mode, r_1 neglected, with I_1 lagging V_1. The two important KVL (Kirchhoff's Voltage Law) equations for the circuit are

$$E_\phi = V_1 + jI_1x_1 \qquad (2.80)$$

$$E_f = V_1 + jI_1x_d \qquad (2.81)$$

The power output is given by

$$P = 3\,|V_1|\,|I_1|\,\cos\theta_\phi = 3\,\frac{|V_1|\,|E_f|}{x_d}\,\sin\delta \qquad (2.82)$$

In the motor mode, δ is negative, as is δ_{FR}. Motor operation is illustrated by Figures 2.54b and c. Note that motor operation takes I_1, into the left half plane. This must be so, since $P = 3\,|V_1|\,|I_1|\,\cos\theta_\phi$ reverses direction, flowing into the electrical terminals instead of out of them, implying that $\cos\theta_\phi$ is negative.

In calculating power, engineers are usually more comfortable with the voltage and current in the same half plane. Motor power is more readily visualized as flowing into the terminals and motor current as flowing into the assumed positive terminal. This more facile point of view may be achieved by redefining the motor current as the negative of the generator current or by redefining the voltages as the negatives of their generator model phasors. Many authors take the second alternative. It has the advantage of keeping $\vec{F}_1$ and the stator current in phase. This author, being lazy by nature, would rather reverse one quantity than three. Therefore a new current will be defined to describe motor operation:

$$I_{1m} \equiv -I_1 \qquad (2.83)$$

The model and the phasor/vector diagram that result from this redefinition of the stator-phase current are shown in Figure 2.54c. Note that the $\vec{F}_1$ vector is diametrically

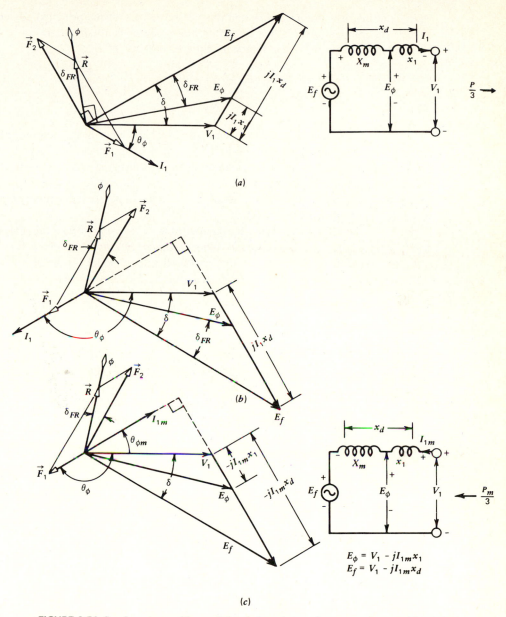

FIGURE 2.54 Synchronous machine model and phasor/vector diagram, motor quantities defined. (*a*) Generator mode: $\delta > 0$, $\delta_{FR} > 0$, $\cos\theta_\phi > 0$. (*b*) Motor mode: $\delta < 0$, $\delta_{FR} < 0$, $\cos\theta_\phi < 0$. (*c*) Motor mode: $I_{1m} \triangleq - I_1$, $\theta_{\phi m} \triangleq \pi + \theta_\phi$, $P_m \triangleq - P = 3|V_1||I_{1m}|\cos\theta_{\phi m}$.

opposed to the I_{1m} phasor. Positive power is now directed *into* the electrical terminals and is given by

$$P_m = 3 \, |V_1| \, |I_m| \, \cos\theta_{\phi m} \qquad (2.84)$$

$$= -3 \, \frac{|V_1| \, |E_f|}{x_d} \, \sin\delta$$

where $\theta_{\phi m}$ is the angle between V_1 and I_{1m}. Note that I_{1m} is *leading* V_1. It will soon be evident that it is not *necessary* that this be so, but one of the principal advantages of the synchronous motor is that *a leading power factor is possible*.

The KVL equations for the motor model are, in terms of I_{1m},

$$E_\phi = V_1 - jI_{1m}x_1 \qquad (2.85)$$

$$E_f = V_1 - jI_{1m}x_d \qquad (2.86)$$

Once a synchronous motor has been started and the rotor has synchronized with the air gap field, the speed depends only on the line frequency. The power output depends entirely on the mechanical load on the shaft. Since the speed does not change as the field current is varied, changing I_f does not affect the output power. The internal losses in a synchronous machine are small, so the ac input power is essentially constant when the mechanical load is constant.

What are the effects of varying the field current? First, *a high field current means stronger rotor poles and thus a greater torque required to pull the machine out of step* (i.e., to desynchronize it). This is demonstrated in the model by the fact that $|E_f|$ increases when I_f increases. Then from Equation 2.75,

$$\tau_{max} = \frac{3}{\omega_s}\frac{|V_1|}{x_d}|E_f| \qquad (2.87)$$

Second, I_f affects the power factor of the current drawn from the three-phase ac line. This is best understood in a situation in which the mechanical load is constant. Neglecting losses,

$$P_{mech} = -\tau_d\omega_s = \text{a constant} = P_k \qquad (2.88)$$

$$P_k = -3\frac{|V_1||E_f|}{x_d}\sin\delta = 3|V_1||I_{1m}|\cos\theta_{\phi m}$$

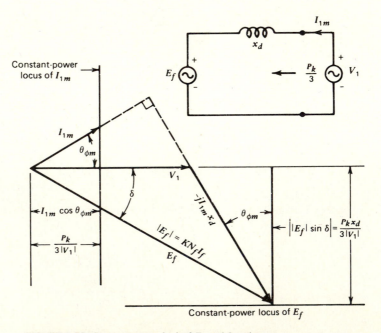

FIGURE 2.55 Constant-power loci of E_f and I_{1m} phasors.

Then

$$|E_f| \sin \delta = \frac{-P_k x_d}{3 \, |V_1|} \qquad \text{(constant)} \qquad (2.89)$$

$$|I_{1m}| \cos \theta_{\phi m} = \frac{P_k}{3 \, |V_1|} \qquad \text{(constant)} \qquad (2.90)$$

It should be remarked at this point that when analyzing a motor alone, the terminal voltage ($|V_1|$ in this case) is assumed to be constant in order to be fair to the motor. Equations 2.89 and 2.90 allow constant-power loci for the E_f and I_{1m} phasors to be drawn on the phasor diagram in Figure 2.55. When I_f is varied slowly enough to avoid

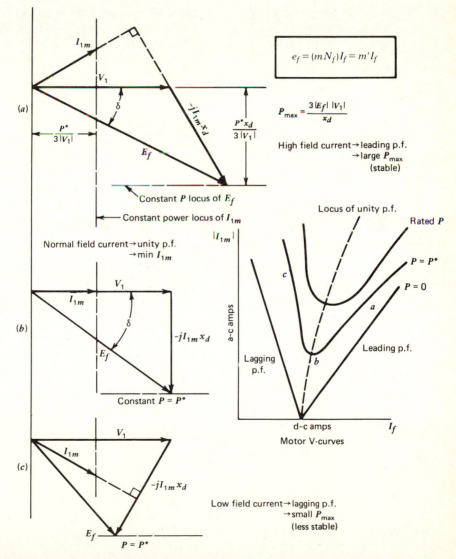

FIGURE 2.56 Effect of field current on power factor and line current   synchronous motor—the V curves. (*a*) Overexcited. (*b*) Normal excitation. (*c*) Underexcited.

hunting, E_f varies in magnitude and the tip of the E_f phasor moves along the constant-power locus so that $|E_f| \sin \delta$ remains constant. As a result of this restricted variation of E_f, the voltage drop across the synchronous reactance also varies:

$$I_{1m}(jx_d) = V_1 - E_f, \text{ or} \qquad (2.91)$$

$$I_{1m} = \frac{V_1 - E_f}{jx_d}$$

Circuit theory shows that the I_{1m} phasor must always remain perpendicular to the $-jI_{1m}x_d$ drop as the tip of the current phasor moves along its locus. Figure 2.56 shows the effect of varying I_f on the power factor and on $|I_{1m}|$.

The plots of $|I_{1m}|$ as a function of I_f for constant power are called the *V curves* of the machine. The curve connecting the minima of the curves for various power levels is called a *compounding curve*. Observe that minimum line current always corresponds to unity power factor. This is a useful bit of knowledge when adjusting the field current. Increasing the I_f beyond the level for minimum I_{1m} results in a leading power factor. Decreasing the field current below that for minimum line current results in lagging power factor.

■ Example 2.4

A 2300-V, three-phase, synchronous motor driving a pump is provided with a line ammeter and a field rheostat. When the rheostat is adjusted to that the ac line current is a minimum, the ammeter reads 8.8 A. Approximately what horsepower is being delivered to the pump? How should the rheostat be adjusted so that the motor is operating at 0.8 power factor, leading? How many kVARs is the motor supplying to the system at 0.8 power factor, leading?

Solution. At minimum line current, the power factor is unity. The power drawn from the line is thus

$$P = \sqrt{3} \, V_L I_L = \sqrt{3} \cdot 2300 \cdot 8.8 = 35 \text{ kW}$$

Neglecting losses, HP $\cong 35,000/746 = 47$ hp. The ac power is practically independent of field current. Then at 0.8 power factor,

$$|S| = \frac{P}{\text{p.f.}} = \frac{35,000}{0.8} = \frac{(\sqrt{3} \cdot 2300) \cdot 8.8}{0.8} = (\sqrt{3} \cdot 2300) \cdot 11.0 \text{ VA}$$

Thus the line current should be 11.0 A. To make sure the power factor is *leading*, the dc field current should be *increased* until the ac ammeter reads 11.0 A. This is accomplished by decreasing the field rheostat resistance. The kVARs supplied by the motor are given by

$$Q = |S| \sin \theta_{\phi m} = |S| \sin \cos^{-1}(\text{p.f.})$$

$$= \frac{35 \text{ kW}}{0.8} \sin \cos^{-1} 0.8$$

$$= \frac{35}{0.8} \cdot 0.6 = 26.25 \quad \text{kVAR} \qquad ■$$

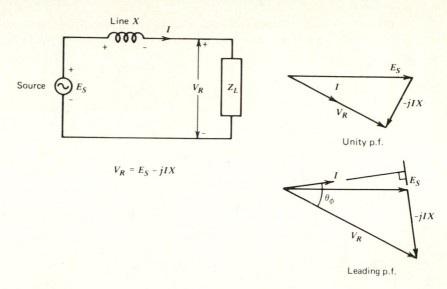

FIGURE 2.57 How leading current increases voltage at the receiving end of a transmission line.

Synchronous motors have many advantages when a constant speed is desirable. Typical applications are to drive pumps, fans, and dc generators. They can be operated at unity power factor, thus minimizing the line current. In sizes above 2000 hp, synchronous motors cost less than induction motors having static capacitors for power factor correction. A major advantage is that, by overexcitation, they can be caused to draw leading current and thus correct the overall power factor of a plant having other equipment drawing lagging current.

Synchronous motors have a natural tendency to maintain constant voltage at their terminals when supplied by lines having inductive reactance. It will be seen from the phasor diagrams of Figure 2.56 that a decrease in $|V_1|$ will cause the current to become more leading. Figure 2.57 shows how a more leading current will tend to cause the terminal voltage to rise.

If the terminal voltages supplied to a synchronous motor are unbalanced, the air gap flux will vary as it passes the different phase groups. The damper winding will have Lenz's law currents flowing in them which will tend to make flux constant. This improves the voltage balance.

It can be seen that a synchronous motor is a very good neighbor. It can be used to correct a lagging power factor, and it helps keep line voltages constant and balanced. It is more expensive than other kinds in small sizes. Its big disadvantage is the fact that direct current must be supplied to the rotor winding. (Permanent-magnet rotors do not permit control of the power factor.) Also, starting equipment is more complicated than for an induction motor.

Example 2.5

A manufacturing plant presents an electrical load to the power system of 5000 kW at 0.800 power factor, lagging. It has been decided to replace a 500-hp induction motor that drives a pump. This motor operates at an efficiency of 96 percent and a power factor of 0.900, lagging. If a synchronous motor is purchased as the replacement, which is capable of operating at 0.800 power factor, leading, what will be the new plant power factor? Assume the synchronous motor to have the same efficiency. What percent decrease in line current will result from the improved power factor?

Solution

1. Original complex power requirement of the plant:

$$S_p = 5000 + j5000 \tan \cos^{-1} 0.8 = 5000 + j3750 \quad \text{kVA}$$

2. Complex power of the induction motor:

$$\text{(a)} \quad P_{in} = \frac{P_{out}}{\eta} = \frac{0.746 \cdot 500}{0.96} = 389 \text{ kW}$$

$$\text{(b)} \quad S_m = 389 + j389 \tan \cos^{-1} 0.9 = 389 + j188 \quad \text{kVA}$$

3. Complex power of the synchronous motor:

$$S_s = 389 - j389 \tan \cos^{-1} 0.8 = 389 - j292 \quad \text{kVA}$$

4. New plant requirement:

$$S_p' = S_p - Q_m + Q_s$$

$$= 5000 + j3750 - j188 - j292$$

$$= 5000 + j3270$$

$$= 5974 \; \underline{|33.2°}$$

5. New power factor $= \cos 33.2° = 0.837$

6. Current reduction:

$$\text{Original current} = \frac{5000 \cdot 10^3/0.800}{\sqrt{3}V_L} = \frac{6250 \cdot 10^3}{\sqrt{3}V_L}$$

$$\text{New current} = \frac{5974 \cdot 10^3}{\sqrt{3}V_L}$$

$$\text{Percent reduction} = \frac{6250 - 5974}{6250} = 4.4\%$$

Note: No change in *power*. ■

2.15 OPERATION AS A SYNCHRONOUS CONDENSER

Condenser is an older word for capacitor. It has been shown that an overexcited synchronous motor draws leading power factor current from the line. When there is no mechanical load on the motor, the ac input power is only enough to supply the losses of the motor (friction, windage, core losses, and $3 |I_{1m}|^2 r_1$). These losses are

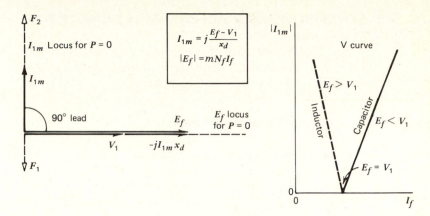

FIGURE 2.58 Synchronous condenser operation.

quite small, and the machine power factor is practically zero. The power angle $\delta \cong 0$ and $\theta_{\phi m} \cong \pm 90°$. Figure 2.58 shows this condition in terms of the phasor diagram and the V curves for $P = 0$. When I_f is adjusted so that E_f is greater than the terminal phase voltage, $\theta_{\phi m} \cong +90°$, the machine acts very much like a capacitor bank. Control of the field current provides smooth control of the leading VARs. A synchronous machine designed for this kind of service is called a *synchronous condenser*. Synchronous condensers are usually totally enclosed. The shaft does not extend outside the case of the machine. It is often found that, when a large number of VARs needs to be supplied to a power system, a synchronous condenser is more economical than ordinary capacitors.

■ Example 2.6

In the plant of Example 2.5, what should be the kVAR rating of a synchronous condenser to correct the original power factor to 0.900?

Solution

1. As before, the original complex power requirement of the plant is

$$S_p = 5000 + j3750 \quad \text{kVA}$$

2. At 0.9 power factor, the plant Q will be

$$Q_p' = 5000 \tan \cos^{-1} 0.9 = 2420 \quad \text{kVAR}$$

3. The difference must be supplied by the synchronous condenser:

$$Q_p + Q_s = Q_p'$$

$$3750 + Q_s = 2420$$

$$Q_s = -1330 \quad \text{kVAR}$$

■

2.16 THE SYNCHRONOUS MACHINE AS A GENERATOR

Polyphase synchronous machines operating in the generator mode are called alternators. The behavior of an alternator supplying an isolated electrical load is very much different from that of one connected to a large system.

Alternator Serving an Isolated Load

Figure 2.59 is a schematic diagram of a small, isolated power system, consisting of an engine, alternator, exciter generator, and three-phase ac load. There are many such small systems in the world. The load may consist of a combination of motors, lamps, and heaters. (Unfortunately, the load is usually unbalanced, but that problem will not be considered here.) The composite load is represented in Figure 2.59 as an equivalent balanced bank of three impedances connected in Y.

The operating characteristics of an isolated alternator are as follows:

1. The frequency depends entirely on the speed of the driving engine, waterwheel, or whatever the mechanical source may be. The driving mechanical source is called the *prime mover*. From Equation 2.1

$$f = \frac{p}{4\pi} \omega$$

A governor is required to keep the frequency constant. It is generally useless to try to operate an electric clock on such a system. Keeping time within a minute a day would require a governor accurate to 0.07 percent.

2. The power factor is the load power factor, and may vary as different electrically operated devices are switched on and off.

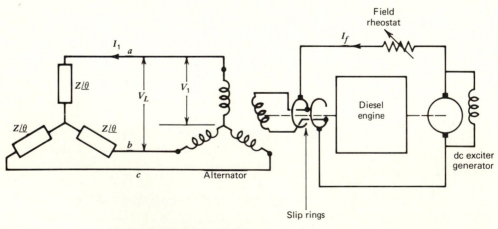

Characteristics of an Isolated Alternator
f depends on engine speed
V_L depends on speed, I_f, $|I_1|$, and θ
Power factor depends on load impedance

FIGURE 2.59 Isolated synchronous alternator.

3. The terminal voltage $V_1 = E_\phi - jI_1x_1$. It depends on

   (a) Speed: $|E_\phi| = (2\pi N_{E1}\mathcal{P}_p R)f$, $\quad f = \omega p/4\pi$

   (b) Field current: $\vec{R} = \vec{F}_2 + \vec{F}_1$, $\quad F_2 = N_f I_f$

   (c) Armature current: $F_1 = (4/\pi\ N_\phi/p\ k_w\ \sqrt{2})\ |I_1| \equiv N_{f1}\ |I_1|$

   (d) Power factor: Angle between $\vec{F}_2$ and $\vec{F}_1$ is β

$$\beta = 90° + \delta + \theta_\phi$$

jI_1x_1 is perpendicular to I_1

The speed is controlled by the governor. There is no control at the generating station over the armature current and power factor. The only control remaining is that of the field current. Since the effective internal impedance of the alternator (x_d) is so large, the voltage will swing wildly with load changes, unless I_f is controlled by an *automatic voltage regulator*.

4. The amount of power being generated is controlled by the load. As the load *watts* increase τ_d increases and the driving engine tends to slow down. This signals the governor to cause the engine to produce power in excess of the new requirement, so that the system speeds up. When the original speed is reached, the governor reduces the engine power to match the new load.

Alternator Connected to an Infinite Bus

In a generating station or substation, heavy bars of copper or aluminum connecting several circuits in parallel are called *bus bars*. Thus an alternator connected to a system is connected to some sort of "bus" consisting of three heavy phase conductors. If the system is imagined to be so large that it can absorb all the power an alternator can put out or supply all the power the machine would ever require as a motor, without any change in system frequency or voltage at the machine terminals, we say that the machine is connected to an *infinite bus*. This condition is very nearly realized in practice when a machine is operated as a part of a large power grid.

When connected to a large system an alternator has the following operating characteristics:

1. The speed is determined by the system frequency. The mechanical drive is forced to operate at a speed given by Equation 2.1.

2. The terminal voltage is the bus voltage.

3. The power factor is determined by the field current. This is illustrated by Figure 2.60. Note that overexcitation is required to supply *lagging* VARs to the system. That is one reason lagging VARs are expensive. Extra dc field power is required to generate them.

4. The amount of power generated is determined by the prime mover. The speed of prime mover is fixed, but its torque can be varied. This is usually accomplished by adjusting the characteristic of the governor so that it calls for more or less power at synchronous speed. The governor characteristic is often controlled by a small dc motor fed pulses from a central computer that is calculating the power requirement for each generator in the system. The effect of changing the governor characteristics on power output is illustrated in Figure 2.61.

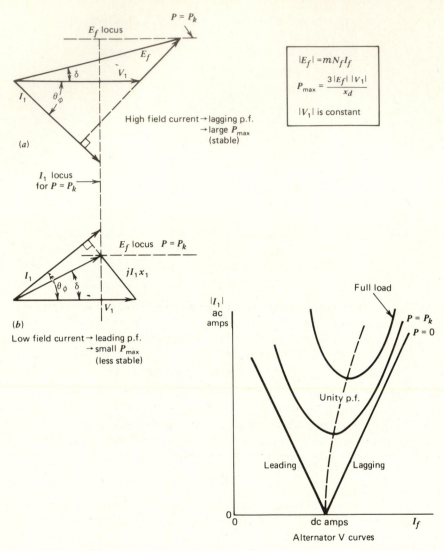

FIGURE 2.60 Effect of field current on the power factor of an alternator connected to an infinite bus. (*a*) Overexcited. (*b*) Underexcited.

When the torque of the prime mover is increased, the rotor poles are rotated farther ahead of the flux. The angle δ_{FR} is increased, δ increases, and the power increases.

Synchronizing

A large alternator cannot simply be connected across a bus of a large system without damage. Before closing the circuit breaker to connect the machine to the system, the three phase voltages must be very nearly equal to the system voltages and the two sets of voltages must match in phase. Under these conditions, the voltages across all of the three circuit breaker contacts will be zero before closing. This will not happen unless the phase sequence of the alternator and that of the system are the same. The phase sequence is carefully determined when the power plant is designed.

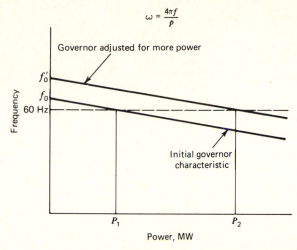

FIGURE 2.61 Power control by adjustment of governor of prime mover.

The sequence of events for synchronization is controlled by an *automatic synchronizer* as follows:

1. The machine is brought to nearly synchronous speed by the prime mover.
2. The field current is adjusted so that the terminal voltage of the machine matches the system bus voltage.
3. The voltage across the circuit breaker contacts will vary in magnitude as the alternator voltages vary in phase with the system voltages, due to the slight difference in frequency. At the instant all three contact voltages are zero, the circuit breaker is closed. There is now no power flow: $E_f = V_1$, $I_1 = 0$, and $\delta = 0$ in each phase.

Once the machine is synchronized, the system dispatcher now determines how much of the total system load is to be supplied by this particular generator, and the power output of the prime mover is increased until the desired generator output is obtained (δ increases). As the power output is rising, an automatic voltage regulator adjusts the field current to maintain the desired terminal voltage (E_f increases). As a result, the generator supplies its share of the systems VARs.

2.17 MEASURING x_d

If r_1 is neglected, x_d can be found by taking the ratio of the open-circuit phase voltage of the machine to the short-circuit current. This is easily seen from Figure 2.48c. The steady-state, short-circuit current is not too large, since the internal impedance x_d is nearly equal to Z_B. Thus if E_f is adjusted for rated voltage by adjusting the rotor current I_f, the short-circuit current would be about rated value. When the machine is first shorted, however, there are some heavy Lenz's law currents in the windings trying to prevent any change in air gap flux as $\vec{F}_1$ builds up. To avoid these high transient currents, I_f should be reduced to zero before the machine is shorted, and increased gradually.

The procedure in measuring x_d is to take data for a curve of open-circuit voltage

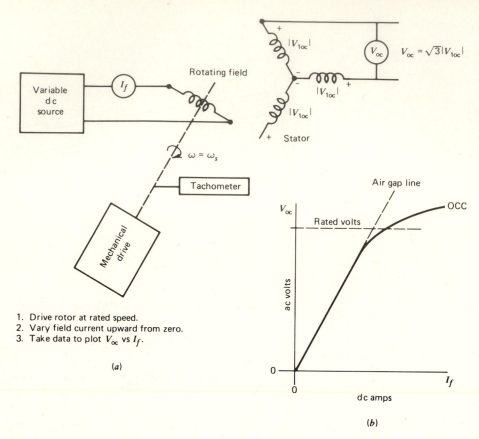

FIGURE 2.62 The open-circuit characteristic of a synchronous machine. (*a*) Schematic diagram. (*b*) Open-circuit characteristic.

as a function of I_f while the shaft is being driven at rated speed. This is called the *open-circuit characteristic* (OCC). Then the field current is reduced to zero and the machine is shorted through three ammeters. The field current is again increased to obtain a plot of short-circuit current as a function of I_f at rated speed. This is called the *short-circuit characteristic* (SCC). The ratio of open-circuit voltage to the short-circuit current measured at the same field current is an **approximate** value of x_d.

Figure 2.62*a* is a schematic diagram of the circuit used to determine the open-circuit characteristic. The shape of the curve is shown in Figure 2.62*b*. Note that the shape is the same as the plot of ϕ as a function of $|R|$ in Figure 2.31. This is true as a result of the following considerations. First, a glance at Figure 2.63 shows that, since the armature current is zero,

$$V_{1oc} = E_f = E_\phi \tag{2.92}$$

where

$$|V_{1oc}| = V_{oc} / \sqrt{3} \quad \text{(scalar)}$$

because all of the $I_1 x$ drops are zero. During this test, the frequency is kept at rated frequency by maintaining the speed at rated speed. Then by Equation 2.58, the open-circuit voltage is proportional to the flux per pole, ϕ:

$$V_{oc} = \sqrt{3}\,|E_\phi| \tag{2.93}$$

$$= \sqrt{3}(\omega_e N_{E1}) \cdot \phi$$

$$= \sqrt{3}(\omega_e N_{E1}) \cdot \mathscr{P}_p R_{oc}$$

Since the stator MMF, F_1, is zero when I_1 is zero, Equation 2.20 *for the open-circuit test* becomes

$$\vec{R}_{oc} = 0 + \vec{F}_2 \equiv \vec{F}_2$$

and

$$R_{oc} = F_2 = N_f I_f$$

$$I_f = \frac{R_{oc}}{N_f} \tag{2.94}$$

Combining Equations 2.93 and 2.94, the equation for the open-circuit characteristic is obtained, where $\mathscr{P}_p$ varies with I_f:

$$V_{oc} = \sqrt{3}(\omega_e N_{E1})\mathscr{P}_p I_f \tag{2.95}$$

Recall that the quantity $\omega_e N_{E1}\mathscr{P}_p$ has been defined (Equation 2.59) as m. Equation 2.94 may be written in terms of m:

$$V_{oc} = \sqrt{3}mN_f I_f \tag{2.96}$$

Let a $N_f m$ be replaced by m', as defined in Equation 2.79. Then

$$V_{oc} = \sqrt{3}m' I_f \tag{2.97}$$

Both m and m' represent a nonlinear relationship between V_{oc} and I_f, since they depend on $\mathscr{P}_p$, the permeance of the magnetic circuit of the machine.

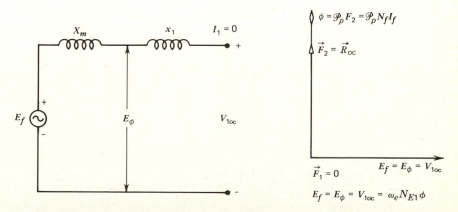

FIGURE 2.63 Circuit model and phasor/vector diagram for open-circuit conditions.

As a result of this analysis, two important things are to be noted about the open-circuit characteristic (OCC):

1. The OCC is a plot of ϕ as a function of R with changes of scale on both axes. This accounts for its shape.
2. The OCC is a plot of $\sqrt{3}\,|E_\phi|$ as a function of $(R/N_f) = I_f$.

The second statement implies that a value of m in Equation 2.78 may be found if $|\mathbf{E}_\phi|$ is known.

THE AIR GAP LINE

Unsaturated iron has a permeability several thousand times that of air. Therefore when I_f is small, $\mathscr{P}_p$ is limited almost entirely by the air gap. This explains why the initial part of the OCC is linear. An extension of this linear portion of the open-circuit characteristic is called the *air gap line,* as illustrated in Figures 2.31 and 2.62*b*.

The Short-Circuit Characteristic

The circuit for obtaining the short-circuit characteristic is shown in Figure 2.64, along with a typical curve. With the machine stopped, three ammeters are connected in Y

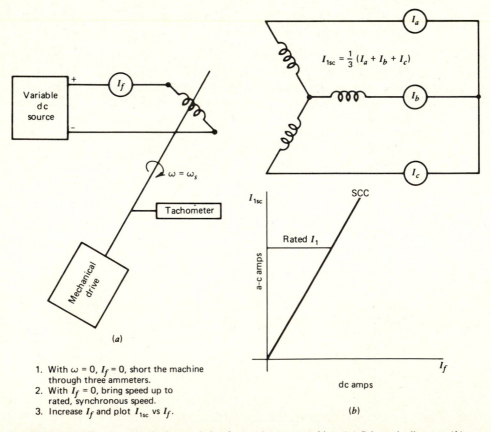

1. With $\omega = 0$, $I_f = 0$, short the machine through three ammeters.
2. With $I_f = 0$, bring speed up to rated, synchronous speed.
3. Increase I_f and plot I_{1sc} vs I_f.

FIGURE 2.64 The short-circuit characteristic of a synchronous machine. (*a*) Schematic diagram. (*b*) Short-circuit characteristic curve.

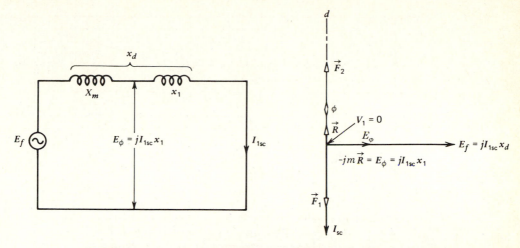

FIGURE 2.65 Circuit model and phasor/vector diagram under short-circuit conditions.

to its terminals. The machine is brought up to rated speed by means of the mechanical drive with $I_f = 0$. Then the field current is increased and the data are taken for a plot of short-circuit current (I_{1sc}) as a function of I_f. This is the short-circuit characteristic. If the ammeters are identical, all three currents should be equal. An unbalance of a few percent is tolerable. The average of the three ammeter readings is taken as I_{1sc}. Maximum short-circuit current should be limited to about 150 percent rated current during this test.

The circuit model of the synchronous machine is of great assistance in understanding the conditions inside the machine when it is shorted. Compare Figure 2.48c with Figure 2.65. The latter figure shows the phasor/vector diagram under short-circuit conditions. Since the internal impedance of the circuit model is a nearly pure inductive reactance, the short-circuit current I_{1sc} lags E_f by 90°. The armature reaction MMF $\vec{F}_1$ lies along the d axis and directly opposes the field MMF $\vec{F}_2$. The resultant MMF is thus quite small when the machine is shorted at its terminals. The evidence of this is the fact that E_ϕ is only large enough to overcome the leakage reactance drop $I_{1sc}x_1$. When the short-circuit current is at rated value, $|I_{1sc}| = I_B$, $|E_\phi|$ would only be about 0.2 per unit, or about 20 percent of its normal value. This means that ϕ is only about 20 percent of normal, so the magnetic circuit is unsaturated. As a result, the plot of $|I_{1sc}|$ as a function of I_f is a straight line.

The Approximate Saturated Synchronous Reactance

If the synchronous reactance is taken to be the ratio of open-circuit phase voltage to short-circuit phase current, both for the same value of I_f,

$$x_d = \frac{V_{oc}/\sqrt{3}}{I_{1sc}}\bigg|_{I_f = I_f^*} \quad \text{(approximate)} \quad (2.98)$$

where I_f^* is a chosen value of I_f, then it is obvious that the synchronous reactance varies with the degree of saturation of the magnetic circuit of the machine. This can be seen quite clearly in Figure 2.66. What value should be used? For *approximate*

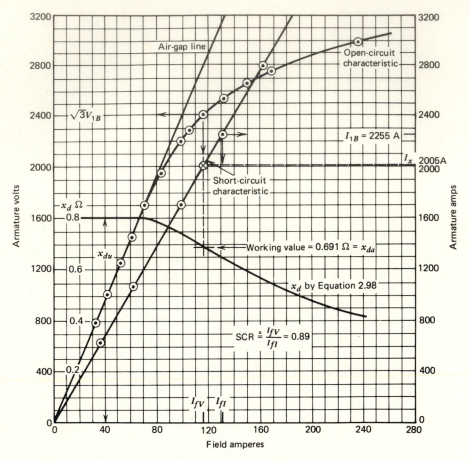

FIGURE 2.66 Characteristic curves of a 9375-kVA, two-pole, 60-Hz, 2400-V, three-phase synchronous machine.

calculations, the usual practice is to take the value corresponding to rated voltage on the open-circuit characteristic (V_{oc} = 2400 V, I_f^* = 116 A $\triangleq I_{fV}$ Figure 2.66). Then

$$x_{da} \triangleq \frac{V_{1B}}{I_x} = \frac{V_{LB}/\sqrt{3}}{I_x} \tag{2.99}$$

where I_x is the value of $|I_1|$ corresponding to $I_f = I_{fV}$ on the short-circuit characteristic. This value of x_d (i.e., x_{da} of Equation 2.99) will be called the *approximate saturated synchronous reactance*.

The student will recognize two inadequacies of this approach to finding x_d. First, it takes as constant a quantity that varies widely. Second, the method itself is suspect since it divides a voltage, V_{1oc}, measured when the machine is magnetically saturated, by a current, I_{1sc}, measured when the machine is unsaturated. The value of x_d thus obtained (x_{da}) however gives fair results when used to calculate excitation requirements. A more accurate way of calculating x_d under saturated conditions remains to be discussed.

Unsaturated Synchronous Reactance

When R is small, the machine is unsaturated and x_d has its maximum value, as indicated in Figure 2.66. The *unsaturated synchronous reactance* x_{du} is found as in Equation 2.98, except that the voltage is taken on the air gap line instead of on the open-circuit characteristics:

$$x_{du} = \frac{V_{a-g}}{\sqrt{3}I_{1sc}} \bigg|_{I_f = I_f^*} \qquad (2.100)$$

It does not matter that value of I_f^* is chosen to locate corresponding values of V_{a-g} and I_{1sc}. Since the SCC and the air gap line are both straight lines through the origin, x_{du} *is independent of I_f^* and is a constant of the machine.*

2.18 THE SHORT-CIRCUIT RATIO

The short-circuit ratio is a constant of a synchronous machine of more value to company executives than to engineers. In bragging about a new machine, the executive will look wise and say ". . . and the short-circuit ratio is . . ." such and such a number. The proper response to such a comment is "Golly!", "Really!", or some similar remark. The short-circuit ratio (SCR) has some physical significance, but not a great deal.

The short-circuit ratio is defined as the ratio of the field current that produces rated voltage on open-circuit (I_{fV}) to the field current that produces rated short-circuit current (I_{fI}):

$$\text{SCR} \overset{\Delta}{=} \frac{I_{fV}}{I_{fI}} \qquad (2.101)$$

In Figure 2.66, I_{fV} is 116 A, corresponding to 2400 V line to line on the open-circuit characteristic. Rated current for this machine is

$$I_{1B} = \frac{\text{rated volt amperes}}{\sqrt{3} \text{ rated } V_{L-L}}$$

$$= \frac{9375 \text{ kVA} \cdot 1000}{\sqrt{3} \cdot 2400} = 2255 \text{ A}$$

The field current corresponding to this current on the short-circuit characteristic is $I_{fI} = 131$ A. Then by Equation 2.101, the SCR for this machine is 0.89.

The meaning of SCR is more apparent when it is realized to be the reciprocal of the per unit x_{da}. This can be shown by referring to Figure 2.67. It has been said that the working value of the approximate synchronous reactance, x_{da}, is that corresponding to rated voltage on the OCC.

Let the short-circuit current corresponding to I_{fV} be denoted I_x. Then

$$x_{da} = \frac{V_{1B}}{I_x} = \frac{V_{LB}}{\sqrt{3}I_x} \qquad (2.102)$$

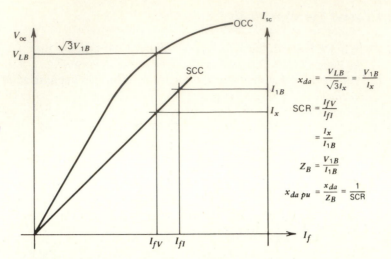

FIGURE 2.67 Relationship between approximate x_d and short-circuit ratio.

where V_{1B} is the base, or rated, phase voltage and V_{LB} is the rated line-to-line voltage of the machine. The per-unit x_{da} is given by

$$x_{da\ pu} = \frac{x_{da}}{Z_B} = \frac{x_{da}}{V_{1B}/I_{1B}} = \frac{I_{1B}}{I_x} \tag{2.103}$$

and

$$\frac{1}{x_{dapu}} = \frac{I_x}{I_{1B}}$$

Since the short-circuit characteristic is linear through the origin,

$$\frac{I_x}{I_{1B}} = \frac{I_{fV}}{I_{fI}} = \text{SCR} = \frac{1}{x_{da\ pu}} \tag{2.104}$$

SIGNIFICANCE OF SCR

A large SCR means a small x_d, and vice versa. The synchronous reactance is the sum of x_1 and X_m, and X_m is by far the larger component. Now

$$X_m = \frac{|E_{ar}|}{I_1} = \frac{mF_1}{I_1} \tag{2.105}$$

and F_1 is proportional to I_1, and by Equation 2.59, m is proportional to $\mathscr{P}_p$. Thus a large SCR means a small effective permeance of the magnetic circuit. This implies that the air gap is relatively long, radially, or that the machine is highly saturated at normal voltages. Thus more ampere turns are required on the field poles. However small x_d means that the power required to pull the machine out of step is large, by the equation for $P_{\max}$, Figure 2.50. Thus the machine is more stable.

Large machines usually require relatively large air gaps to provide sufficient mechanical clearance. Normally, a large machine would thus have a large SCR and a small machine, a smaller SCR. A large machine with a small SCR would be one with tight mechanical tolerances. It would be more efficient in the use of dc to excite the rotor. It would likely be relatively small for its rating and less costly, due to short air gap design and use of a smaller volume of core iron operated at higher flux densities. It would tend to be unstable under system transients. The small air gap means that the machine would have to be brought up to temperature carefully and started carefully. In other words, such a machine is likely to be economical and cantankerous. On the other hand, a large machine with a large SCR would tend to have a larger volume per kVA of rating. It would thus be more expensive. Its longer air gap would require more dc power to the rotor, but it would result in a more stable machine.

2.19 CALCULATING EXCITATION REQUIREMENTS FOR GIVEN OPERATING CONDITIONS

It is often important to know how much field current I_f must be supplied to the rotor winding in order to obtain the following:

1. A certain desired terminal voltage, in the case of an isolated generator.
2. A certain desired power factor, in the case of a motor or generator connected to a large system.

Knowledge of the range of field currents required by a given machine is required to determine the rating of the exciter generator. Since the excitation loss $I_f^2 R_f$ is a major loss in a synchronous machine, I_f must be known to calculate the efficiency under given load conditions. Another reason for calculating the field current of a generator is to determine how much increase in terminal voltage would occur if the machine were to lose its load, say, by the tripping of a circuit breaker.

The Saturated Synchronous Reactance Method for Finding I_f

There are many methods for calculating the field current for a given set of operating conditions, but this method has the best theoretical basis and gives excellent results. However it is necessary to know the stator leakage reactance per phase, x_1, to use the method.

Equation 2.66 defines the synchronous reactance as the sum of the leakage reactance x_1 and the magnetizing reactance X_m. The leakage reactance is practically constant. By Equations 2.65 and 2.59,

$$X_m = \left| \frac{E_{ar}}{I_1} \right| = \frac{mF_1}{I_1} = \frac{mN_{I1}|I_1|}{|I_1|} = mN_{I1} \tag{2.106}$$

[Note: $m = \omega_e N_{E1}\mathscr{P}_p = (\omega_e N_\phi k_w \mathscr{P}_p)/\sqrt{2}$, and from pp. 56–59, $N_{I1} = (3/2)(4/\pi)\,(N_\phi/p)k_w\sqrt{2}$.] Then

$$X_m = \omega_e \frac{6}{\pi p} N_\phi^2 k_w^2 \, \mathscr{P}_p \tag{2.107}$$

It is seen that the magnetizing reactance X_m depends on the degree of saturation since $\mathscr{P}_p$ is one of its component coefficients. The important thing to note here is that, by Equation 2.106, X_m **is proportional to** m.

The *unsaturated* synchronous reactance (Equation (2.100) is the sum of x_1 and the unsaturated value of X_m:

$$x_{du} = x_1 + X_{mu} \qquad (2.108)$$

Recall that x_{du} is computed from the air gap line and the short-circuit characteristic. Since X_m is proportional to m, its value under saturated conditions is given by

$$X_m = \frac{m}{m_u} X_{mu} \qquad (2.109)$$

where m_u is the value m has when the machine is unsaturated (that is, when $\mathscr{P}_p$ = effective permeance per pole of the air gap). The ratio m_u/m is defined as the *saturation factor* k_s. Equation 2.79 defines $m' \triangleq N_f m$. Then

$$k_s = \frac{m_u}{m} = \frac{N_f m_u}{N_f m} = \frac{m'_u}{m'} \qquad (2.110)$$

From Equation 2.108,

$$X_{mu} = x_{du} - x_1 \qquad (2.111)$$

Then the saturated synchronous reactance is given by

$$x_d = x_1 + \frac{X_{mu}}{k_s} = x_1 + \frac{x_{du} - x_1}{k_s} \qquad (2.112)$$

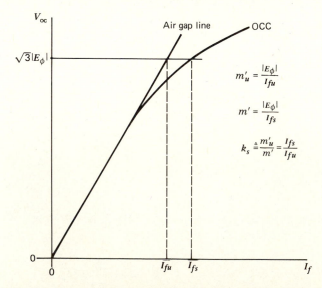

FIGURE 2.68 Relationship between m', m'_u, and the saturation factor.

Figure 2.68 shows the relationships between the saturation factor and m' and m'_u. Suppose for a given set of operating conditions E_ϕ is calculated by application of KVL to the circuit model. Since the vertical axis of the open-circuit characteristic is in line-to-line volts, the value of m' is found by dividing $|E_\phi|$ by the value of I_f corresponding to $\sqrt{3}|E_\phi|$ on the OCC. In the figure, this value of I_f is called I_{fs}:

$$m' = \frac{|E_\phi|}{I_{fs}} \tag{2.113}$$

The unsaturated m'_u is found by dividing $|E_\phi|$ by the field current I_{fu}, which corresponds to $\sqrt{3}|E_\phi|$ on the air gap line:

$$m'_u = \frac{|E_\phi|}{I_{fu}} \tag{2.114}$$

Then the saturation factor is

$$k_s = \frac{m'_u}{m'} = \frac{I_{fs}}{I_{fu}} \tag{2.115}$$

The steps in the saturated synchronous reactance method for finding I_f are:

1. Calculate E_ϕ for the given set of operating conditions by applying Kirchhoff's voltage law to the circuit model:

$$E_\phi = V_1 + jI_1x_1 \qquad \text{(generator)}$$

$$E_\phi = V_1 - jI_{1m}x_1 \qquad \text{(motor)}$$

2. Determine I_{fs} and I_{fu} corresponding to $\sqrt{3}|E_\phi|$ on the OCC and air gap line, respectively. Compute

$$k_s = \frac{I_{fs}}{I_{fu}}$$

$$m' = \frac{|E_\phi|}{I_{fs}}$$

3. Using Equation 2.100, compute x_{du} from the air gap line and the short-circuit characteristic. Then

$$x_d = x_1 + \frac{x_{du} - x_1}{k_s}$$

4. Calculate E_f:

$$E_f = V_1 + jI_1x_d \qquad \text{(generator)}$$

$$E_f = V_1 - jI_{1m}x_d \qquad \text{(motor)}$$

5. $I_f = \dfrac{|E_f|}{m'}$

Example 2.7

Calculate the excitation requirement for the alternator of Figure 2.64 when delivering rated kVA at 0.8 power factor, lagging, if x_1 is 0.10 Ω.

1. From Figure 2.54a, and Equation 2.80,

$$E_\phi = V_1 + jI_1x_1$$

Let V_1 be the reference phasor. Then

$$V_1 = \frac{2400}{\sqrt{3}} \underline{|0} = 1386 + j0 \text{ V}$$

$$I_1 = I_{1B} \underline{|-\cos^{-1} 0.8}$$

where

$$I_{1B} = \frac{kVA_B \cdot 1000}{\sqrt{3}V_{LB}} = \frac{9375 \cdot 10^3}{\sqrt{3} \cdot 2400} = 2255 \text{ A}$$

$$I_1 = 2255 \underline{|-36.87°} = 1804 - j1353 \text{ A}$$

$$E_\phi = 1386 + 0.10 \cdot 2255 \underline{|90° - 36.87°}$$

$$= 1386 + 225.5 \underline{|53.13°}$$

$$= 1386 + 135.3 + j180.4$$

$$= 1521 + j180.4 = 1532 \underline{|6.76°} \text{ V}$$

$$\sqrt{3} |E_\phi| = 2654 \text{ V}$$

2. From Figure 2.66

$$I_{fu} = 110 \text{ A}, I_{fs} = 149 \text{ A, dc}$$

$$k_s = \frac{149}{110} = 1.35$$

$$m' = \frac{|E_\phi|}{I_{fs}} = \frac{1532}{149} = 10.28 \quad \Omega$$

3. Also from Figure 2.66

$$x_{du} = 0.800$$

and

$$x_d = x_1 + \frac{x_{du} - x_1}{k_s}$$

$$= 0.100 + \frac{0.700}{1.35} = 0.169 \quad \Omega$$

4.
$$E_f = V_1 + jI_1x_d$$

$$= 1386 + j(2255 \underline{|-36.87°}) \cdot 0.619$$

$$= 1386 + 1396 \,\underline{|90° - 36.87°}$$

$$= 1386 + 838 + j1117 \text{ V}$$

$$= 2224 + j1117 = 2489 \,\underline{|26.66°} \text{ V}$$

5. $I_f = \dfrac{2489}{m'} = \dfrac{2489}{10.28} = 242$ A ∎

FOR FURTHER STUDY

THE POTIER METHOD OF MEASURING x_1

When a synchronous machine is operating as a generator with a zero power factor inductive load, the phasor/vector diagram looks like Figure 2.69. Note that all MMFs lie along the d axis and that $|E_\phi|$ is the arithmetic sum of $|V_1|$ and $|I_1|\, x_1$.

The situation relative to the OCC is shown in Figure 2.70, where point P represents the actual terminal line-to-line voltage and the actual field current required to produce that voltage at zero power factor and the particular value of armature current $|I_1|$. Noting that $I_f = F_2 / N_f$, the MMF vector diagram may be laid off along the field current axis, if all vector magnitudes are scaled by dividing by N_f. Corresponding to the resultant R, the induced voltage on a line-to-line basis ($\sqrt{3}E_\phi$) is found on the OCC. From Figure 2.69, the difference between $|E_\phi|$ and $|V_1|$ is $|I_1|\, x_1$. Therefore the difference on the voltage axis of the OCC between $\sqrt{3}\,|E_\phi|$ and $\sqrt{3}\,|V_1|$ is $\sqrt{3}\,|I_1|\, x_1$. The shaded triangle is called the *Potier triangle* after the inventor of this method. Note that the sides of this triangle are proportional to $|I_1|$.

Consider the situation at short circuit, illustrated in Figure 2.71. Here we see the Potier triangle is located on the field current axis. Note the auxiliary triangle A. The attitude of A is $\sqrt{3}\,|I_1|\, x_i$ and its hypotenuse is the air gap line. This is the clue to finding x_1 by Potier's method.

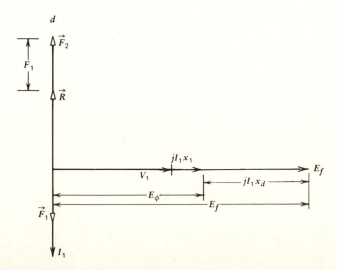

FIGURE 2.69 Phasor/vector diagram of a synchronous machine at zero power factor lagging.

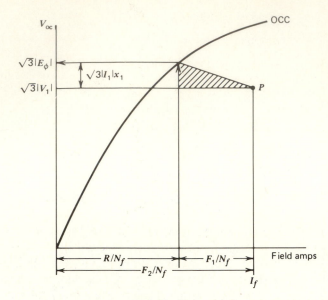

FIGURE 2.70 The Potier triangle.

Procedure

Refer to Figure 2.72.

1. With an adjustable, balanced, three-phase inductive load connected to its terminals, drive the machine at rated speed. Adjust I_f for **rated voltage** and the
   load inductance for **rated current.** Record the field current required to produce
   rated terminal voltage at rated current, zero power factor, lagging (I_{fo}).
2. Draw a horizontal line through rated volts on the OCC. Locate the point P
   corresponding to the field current of the zero power factor test.

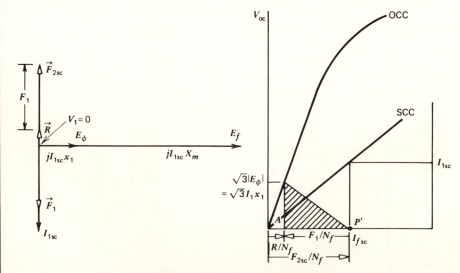

FIGURE 2.71 Conditions at short circuit.

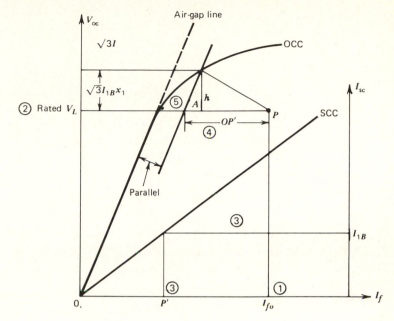

FIGURE 2.72 The Potier method of evaluating x_1.

3. Locate P' corresponding to rated current (same current used in zero power-factor test) on the SCC. The field current OP' is F_{2sc}/N_f of Figure 2.71.

4. Lay off the distance OP' to the left on P on the rated voltage line.

5. Through the point just located, draw a line parallel to the air gap line until it

6. intersects the OCC. This establishes the auxiliary triangle A of altitude h.

$$x_1 = \frac{h}{\sqrt{3}I_{1B}}$$

The reactance thus found is often called the Potier reactance rather than the leakage reactance. For cylindrical-rotor machines, it is a very good approximation for x_1. Some error arises from the fact that r_1 is neglected and that the Potier reactance does not include effects of harmonic fields.

The Potier reactance is not a very good measure of the leakage reactance of salient-pole machines. However when it is used as the value of x_1 to calculate excitation requirements by the saturated synchronous reactance method, the results are fairly good.

Approximate I_f Calculation

An often used approximate method for determining the field current employs the approximate saturated synchronous reactance x_{da} found by Equation 2.99 and further assumes that $|E_\phi|$ never varies much from the rated voltage of the machine. The process in finding I_f by this method is as follows:

1. Using $x_d = x_{da}$, apply Kirchhoff's voltage law to the circuit model to calculate E_f.

2. Taking $|E_\phi| = V_{1B}$, use the open-circuit characteristic to find the m' in Equation 2.78.

3. Then

$$I_f = \frac{|E_f|}{m'}$$

Step 2, above, requires some comment. It has been pointed out (Equations 2.92 and 2.94) that the open-circuit characteristic is a plot of $\sqrt{3}\,|E_\phi|$ as a function of R/N_f. Then if $|E_\phi|$ is assumed to be equal to rated phase voltage, the value of R required to produce that $|E_\phi|$ will be N_f times the field current corresponding to rated voltage on the open-circuit characteristic. That current has been designated I_{fV} in Figures 2.66 and 2.67. Now, from Equation 2.60,

$$m \triangleq \frac{|E_\phi|}{R} \tag{2.116}$$

and is a variable quantity. However the assumption that $|E_\phi| = V_{1B}$ gives a constant value of m:

$$m_a = \frac{V_{1B}}{N_f I_{fV}} \tag{2.117}$$

where it will be recalled that

$$V_{1B} = \frac{\text{rated line to line volts}}{\sqrt{3}} \tag{2.118}$$

In Equation 2.79, $N_f m$ was defined as m', so for this method,

$$m_a' \triangleq N_f m_a = \frac{V_{1B}}{I_{fV}} \tag{2.119}$$

and by Equation 2.78,

$$I_f = \frac{|E_f|}{m_a'} \tag{2.120}$$

▪ Example 2.8

For the synchronous machine of Figure 2.66, find:

(a) The field current required to deliver rated terminal voltage when operated as an alternator at rated kVA at a power factor of 0.8, lagging.
(b) The open-circuit voltage of the machine at this field current.
(c) The maximum kVAR the machine could deliver as a synchronous condenser operating at rated voltage, if rotor heating limits the field current to 240 A.

Solution

(a)

$$x_{da} = \frac{V_{1B}}{I_x} = \frac{2400/\sqrt{3}}{2005} = 0.691 \ \Omega$$

$$m_a' = \frac{V_{1B}}{I_{fV}} = \frac{2400/\sqrt{3}}{116} = 11.9 \ \Omega$$

From the previous example,

$$V_1 = 1386 + j0$$
$$I_1 = 2255 \underline{|-36.87°}$$
$$E_f = V_1 + jx_{da}I_1$$
$$= 1386 + j0 + j0.691 \cdot 2255\underline{|-36.87°}$$
$$= 1386 + j0 + 1558\underline{|90 - 36.87°}$$
$$= 1386 + 935 + j1247$$
$$= 2321 + j1247 = 2635\underline{|28.2°} \text{ V}$$

$$I_f = \frac{|E_f|}{m_a'} = \frac{2635}{11.9} = 221 \text{ A}$$

(b) If the machine lost its load, it would then operate on the open-circuit characteristic. With $I_f = 221$ A,

$$V_{oc} = 2960 \text{ V}, \qquad \text{line to line}$$

or

$$V_{1oc} = \frac{2960}{\sqrt{3}} = 1709 \text{ V}$$

Note that this is much less than $|E_f|$, which is 2635 V. An assumption that the model is linear would lead one to believe that the open-circuit phase voltage would be equal to E_f.

(c) Synchronous condenser:
With I_f limited to 240 A,

$$E_{f\max} = m_a' \cdot 240 = 11.9 \cdot 240 = 2856 \text{ V}$$

From Figure 2.58 for a synchronous condenser,

$$|I_1x_{da}| = |E_f| - |V_1| = 2856 - 1386 \text{ V}$$

$$I_{1max} = \frac{2856 - 1386}{x_{da}} = \frac{1470}{0.691} = 2127 \text{ A}$$

Note that this is less than rated stator current and that in this case, the synchronous condenser rating is limited by rotor heating rather than by stator heating.

Then $Q_{\max} = \sqrt{3} \, V_L I_1$, since I_1 leads V_1 by 90°:

$$Q_{\max} = \sqrt{3} \cdot 2400 \cdot 2127 = 8.84 \quad \text{MVAR} \quad \blacksquare$$

2.20 PERMANENT-MAGNET ALTERNATORS

Section 2.2 discussed the construction of both salient-pole and cylindrical rotors where the rotor MMF is produced by dc current flowing in the field windings. The revolving field may also be produced by constructing the rotor utilizing permanent-magnet materials. Such alternators are usually designed for specific applications, rather than as general-purpose machines, and are limited to a few kilowatts. For example, the head-

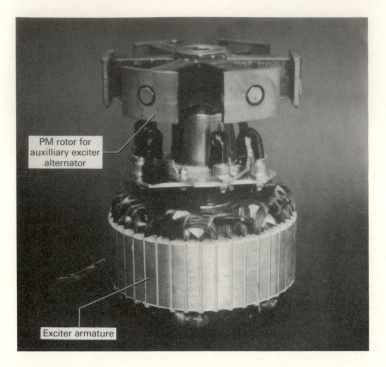

FIGURE 2.73 Field exciter portion of rotor assembly for a synchronous alternator. (Photo courtesy of Kato Engineering/Reliance Electric Company.)

lights of a riding lawnmower are often powered by such an alternator. A second stator winding may utilize the same revolving field to charge the battery (through a diode). Many aviation alternators are constructed with an integral PM auxilliary unit to maintain essential power in case of main unit failure.

Figure 2.73 illustrates another application for PM alternators. As is mentioned in Section 2.2, most modern synchronous alternators are provided with brushless exci-

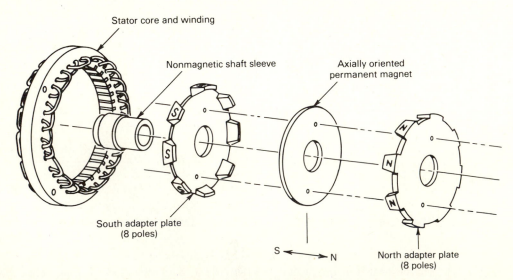

FIGURE 2.74 Sixteen-pole PM alternator for revolving field exciter. The approximate size is a 10 in. diameter cylinder 2 in. thick. (Courtesy of MagneTek, Inc., Century Electric Division.)

tation. That is, a fraction of the generated alternating current is rectified and supplied to the stationary field of a small shaft-mounted alternator, the output of which is again rectified to supply the main field current. This "bootstrap" scheme depends on the remanent magnetism of the rotor poles. The dependence can be eliminated and proper generation startup assured by adding a small PM alternator. The PM rotor assembly pictured in Figure 2.73 requires one magnet per pole. Figure 2.74 shows an exploded view of a 16-pole rotor design used to supply field excitation for alternators in the range of 150 to 2500 kW. Note the single disc magnet with axially oriented field. The rotor unit is assembled, including filling all voids with epoxy, before the magnet is magnetized.

The characteristics of a PM alternator are similar to those of wound field alternators operated with constant field excitation. This assumes that the machine is properly designed so that the demagnetizing effects of temperature and armature reaction are negligible. As with any permanent-magnet machine, assembly methods must not cause significant demagnetization. When Alnico is used, it is often necessary to magnetize the rotor assembly in place. The reader is referred to Section 5.14 for a discussion of magnet materials and their properties.

2.21 SINGLE-PHASE SYNCHRONOUS ALTERNATORS

Up to this point we have considered only synchronous machines with polyphase stator windings. Certain applications, usually restricted to less than 10 kVA, may be better served by a single-phase alternator. Examples are emergency power for a single-family residence, portable power for construction tools, and other cases where the intended load is purely single phase. As an alternative to the specialized single-phase unit, a 12-lead, three-phase winding may be reconnected to operate as single phase. The possible configurations and derating factors are shown in Figure 2.75.

The rotor of a single-phase alternator would normally be identical to a salient-pole rotor intended for a three-phase synchronous machine. The stator laminations may also be identical, though some slots will remain unfilled. If the stator winding is simply one phase of a three-phase winding, the machine will have only one third the capacity of the three-phase machine of the same frame size. If the coils of a lap winding are short-pitched by n slots (the number of coils per phase group), then each slot will contain one coil side. Then the number of turns per coil or the conductor cross section may be doubled, doubling the kVA rating. Such an alternator would thus have two thirds the capacity of the same-size three-phase machine, and one third of the stator slots would remain unfilled (see Figure 2.76).

In this section it is assumed that the rotor poles are shaped so that the rotor magnetomotive force is still given by Equation 2.2. That is, $\mathscr{F}_2(\theta_2)$ is a constant-amplitude, sinusoidally distributed field which rotates with the rotor. This field, acting alone at no load, would induce a sinusoidal voltage E_f in the stator coils. E_f, which at no load equals E_ϕ and V_1, always lags $\vec{F}_2$ by 90°.

Under load, a sinusoidal stator current flowing through a distributed single-phase winding would produce a stationary "breathing" MMF identical to one phase of a three-phase machine. The expression, as derived in Section 2.5 for phase a, is

$$\mathscr{F}_1 = F_{a1} \cos \omega_e t \cos \theta_{1e} \qquad (2.121)$$

where $F_{a1} = (4/\pi) n N_c k_w \sqrt{2} I_1$.

As is illustrated in Figure 2.77, the stator MMF described by Equation 2.121 can

		Winding Configuration	Voltage	p.u., kVA
3φ	Series Y	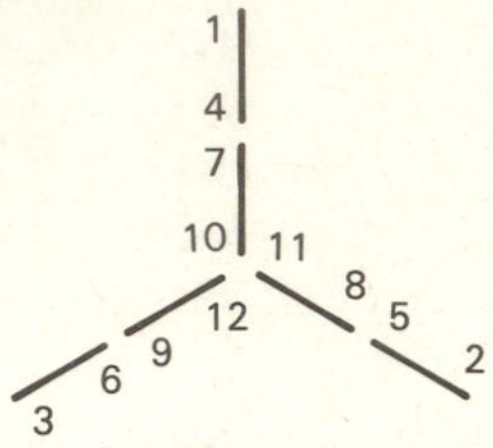	240/416	1.00
1φ	Parallel Y	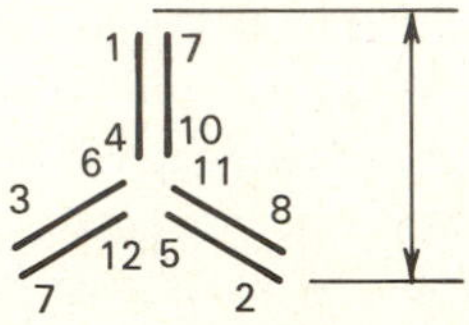	208	0.58
	Series delta	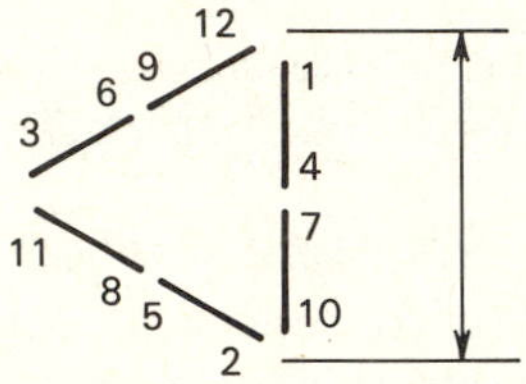	240	0.50
	Double delta	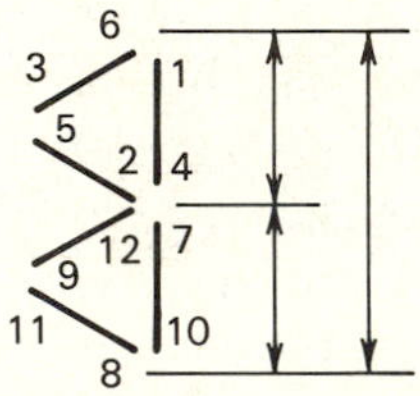	120/240	0.50
	Parallel dogleg	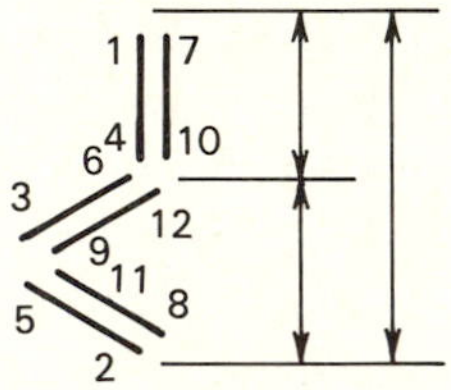	120/240	0.67

FIGURE 2.75 Connections for obtaining single-phase power from a 12-lead, three-phase alternator.

be conceptually and mathematically decomposed into two constant-amplitude MMFs, a "forward field" and a "backward field," which rotate at $\pm$ synchronous speed. These two components are

$$\mathscr{F}_{1f} = \tfrac{1}{2} F_{a1} \cos(\omega_e t - \theta_{1e}) \qquad (2.122)$$

and

$$\mathscr{F}_{1b} = \tfrac{1}{2} F_{a1} \cos(\omega_e t + \theta_{1e}) \qquad (2.123)$$

so that $\mathscr{F}_1 = \mathscr{F}_{1f} + \mathscr{F}_{1b}$. $\qquad (2.124)$

FIGURE 2.76 Single-phase alternator stator and rotor. (Components courtesy of Kato Engineering/Reliance Electric.)

It should be noted that the forward MMF rotates at ω_e in synchronism with the rotor, while the backward (or countersynchronous) MMF rotates at $-\omega_e$ with respect to the stator.

The resultant MMF is the sum of $\vec{F}_1$ and $\vec{F}_2$. However, since this is a salient-pole machine, the permeance of the magnetic circuit is not uniform. It is therefore necessary to consider the d axis and q axis projections of $\vec{R}$. If it is assumed that the reluctance of the quadrature axis flux path is sufficiently greater than that of the direct axis, so that the q axis flux may be neglected, then

$$\phi = \mathcal{P}_d R_d \qquad (2.125)$$

Thus E_ϕ, the voltage induced in the stator winding, lags $\vec{F}_2$ by 90°. If, as in Figure 2.41, it is assumed that I_1 lags E_ϕ by θ_ϕ^I, then Figure 2.78 illustrates conditions in the alternator at the instant when I_1 is a maximum. Note that taking the zero time reference

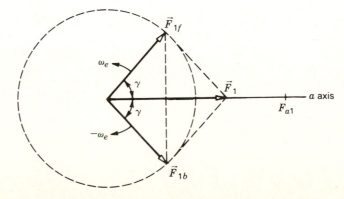

FIGURE 2.77 The stator MMF of a single-phase alternator, decomposed into the forward (synchronous) and backward (countersynchronous) fields.

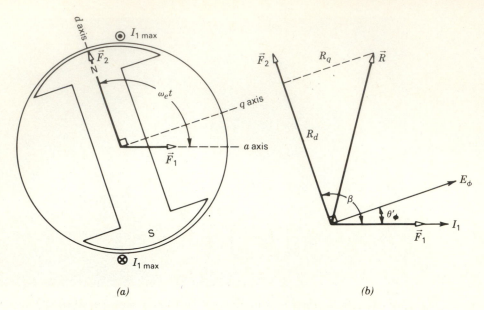

FIGURE 2.78 Conditions in a single-phase alternator at the instant when I_1 is a maximum.

when the rotor's north pole aligns with the a axis means that $\omega_e t = \beta$ (the angle between $\vec{F}_1$ and $\vec{F}_2$) in the figure.

Figure 2.79 shows conditions when the rotor of Figure 2.78 moves an additional angle γ. For this figure $\beta_f = \beta$ and $\omega_e t = \beta_f + \gamma$. Solving for γ,

$$\gamma = \omega_e t - \beta_f \tag{2.126}$$

Note that F_1 remains aligned with the a axis, but $\vec{F}_{1f}$ and $\vec{F}_{1b}$ have rotated $+\gamma$ and $-\gamma$, respectively. As the machine rotates, the angle between $\vec{F}_2$ and $\vec{F}_{1f}$ remains constant at β_f, but the angle between $\vec{F}_2$ and $\vec{F}_{1b}$ is

$$\beta_f + 2\gamma = 2\omega_e t - \beta_f$$

The direct axis component of the resultant MMF can be written as

$$R_d = F_2 + F_{1f} \cos \beta_f + F_{1b} \cos(2\omega_e t - \beta_f) \tag{2.127}$$

Substituting into Equation 2.125 and ignoring nonlinearities of the magnetic circuit,

$$\phi = \mathcal{P}_d[F_2 + F_{1f} \cos \beta_f + F_{1b} \cos(2\omega_e t - \beta_f)] \tag{2.128}$$

Since this flux is rotating at ω_e radians per second past the stator winding, the instantaneous stator winding flux linkage is

$$\lambda_1 = \sqrt{2}N_{E1}\phi \cos \omega_e t$$

$$= \sqrt{2}N_{E1}\mathcal{P}_d[(F_2 + F_{1f} \cos \beta_f) \cos \omega_e t + F_{1b} \cos(2\omega_e t - \beta_f) \cos \omega_e t]$$

$$= \sqrt{2}N_{E1}\mathcal{P}_d\{(F_2 + F_{1f}\cos \beta_f)\cos \omega_e t + (\tfrac{1}{2})F_{1b}[\cos(3\omega_e t - \beta_f) \tag{2.129}$$

$$+ \cos(\omega_e t - \beta_f)]\}$$

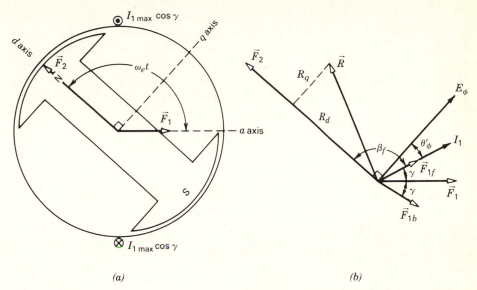

(a)   *(b)*

FIGURE 2.79 Conditions when the rotor of Figure 2.77 moves an additional angle γ.

And by Faraday's law,

$$e = \frac{-d\lambda_1}{dt} \tag{2.130}$$

Thus the voltage induced in the stator winding will contain both line frequency and third-harmonic terms. The third-harmonic term is a result of the direct-axis MMF being modulated by the backward stator MMF. Unlike slot harmonics, it cannot be eliminated by short-pitching the stator coils. The amplitude of the third-harmonic voltage is reduced by the self-inductance of the field winding, eddy currents in the rotor, and double-frequency currents that will flow in the damper bars.

2.22 "BRUSHLESS DC" MOTORS

A dc motor has beautiful control characteristics but has the disadvantages of high construction and maintenance costs. These disadvantages can be attributed largely to the commutator, including the carbon brushes which carry current to the rotor windings through their sliding contact with the commutator surface. In a dc motor the commutator acts to maintain a space displacement between the rotor and stator magnetic poles of 90 electrical degrees, as shown in Figure 1.13, Section 1.7. It does this by switching connections to the rotor coils to keep the rotor current pattern stationary in space, even as the rotor revolves. When the current is reversed in a given coil, there is a troublesome $L(di/dt)$ voltage that tends to cause sparking at the brush surface, which in turn helps wear the brushes out. One of the reasons for having a large number of coils in the rotor winding (the armature winding) is to keep the inductance per coil low.

A brushless dc machine is synchronous machine operated in such a way as to behave like a dc machine. An electronic inverter takes input power from a dc source and supplies polyphase alternating currents to the stator (armature) windings. Sensors signal the position of the rotor (field) of the synchronous machine to the electronics controlling the switching elements in the inverter. The frequency and phase angles of the stator currents are controlled in such a way that there is always a constant angular displacement β between the poles of the rotating stator field $\vec{F}_1$ and the field poles $\vec{F}_2$ of the rotor. This is the same condition that exists in a dc machine, except that in a dc machine the field poles are on the stator and the armature is on the rotor. Compare Figure 1.13 with Figure 1.15a. It will be seen that in the case of the dc machine, β is 90 electrical degrees. This is approximately true in the brushless motor also. In both machines, β is sometimes varied to reduce voltage transients occurring when the armature coils are switched.

A schematic diagram of a brushless dc machine is shown in Figure 2.80. The switching devices in the inverter, S1 through S6, are shown as SCRs (silicon-controlled rectifiers), but they may be power transistors in small brushless motors or power FETs (field effect transistors) in motors of intermediate power ratings. The "flyback" diodes connected backward across the SCR protect them from the $L(di/dt)$ voltage induced in the armature windings each time the current is switched off. The diodes carry currents due to the energy that was stored in the magnetic field of phase windings whose current is being switched off. They are usually Shockley diodes, which turn on fast enough to prevent voltage spikes that might otherwise cause ratio and TV interference. Some FETs designed for handling inductive loads have flyback diodes built in.

Figure 2.81 shows the waveforms of the currents supplied to the phase windings. The heavy lines are idealized waveforms. More realistic waveforms, showing the effects of winding inductances, are indicated by the dotted curves. The idealized currents are sufficiently accurate for developing engineering equations of the system. The flat tops of the current waveforms are due to the inductor L in Figure 2.80. The constant current *during the pulses* is characteristic of the "current source" inverter

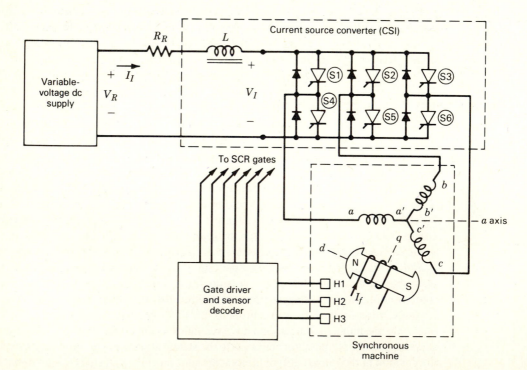

FIGURE 2.80 Schematic diagram of a "brushless dc" machine fed by a current-source inverter.

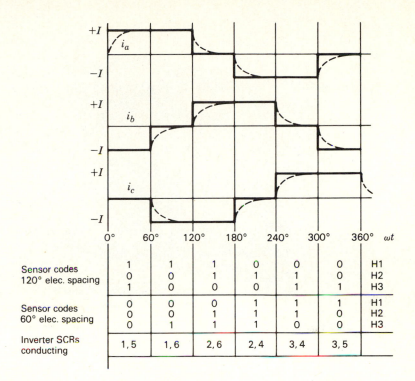

FIGURE 2.81 Phase current waveforms, sensor codes, and switching-device firing order in a brushless dc machine.

	0°	60°	120°	180°	240°	300°	360°	ωt
Sensor codes 120° elec. spacing	1	1	1	0	0	0		H1
	0	0	1	1	1	0		H2
	1	0	0	0	1	1		H3
Sensor codes 60° elec. spacing	0	0	0	1	1	1		H1
	0	0	1	1	1	0		H2
	0	1	1	1	0	0		H3
Inverter SCRs conducting	1, 5	1, 6	2, 6	2, 4	3, 4	3, 5		

(CSI). This does not imply that the amplitude of the pulses, I, never varies. The value of I depends on the rectifier output voltage V_R, the counter voltage of the synchronous motor, and the internal resistance of the rectifier. Usually, a CSI is used with large machines. In such cases the angle β may be made sufficiently less than 90° to cause the synchronous machine to operate at leading power factor over the expected load range. Under these conditions the flyback diodes may not be required in the inverter circuit.

In the small brushless dc machines a *voltage source* inverter is used. The inductor L is eliminated. The rotor is usually provided with ceramic field magnets, as in Figure 2.82. The solid-state switching devices and the synchronous-machine windings are often protected from overcurrent by pulse modulating the input voltage to the inverter. The modulation may be effected by a chopper replacing the inductor L in Figure 2.80, or the switching elements S4, S5, and S6 of the inverter itself may be switched on and off. The ratio of off to on time of the voltage pulses determines the average voltage applied to the inverter (see Figure 5.19 in Chapter 5), and this ratio is controlled by feedback from a current sensor in series with the inverter input. This method of controlling the voltage level may also be used to vary the speed of the motor. Alternatively, the torque of the motor may be controlled by comparing the current feedback signal with an adjustable reference to develop an error signal used to control the pulse width modulator. For very small machines, the inverter, modulator, and switching logic may be purchased on a single chip.

All of these pulse modulation schemes obviously result in a ripple superimposed on the waveforms of Figure 2.81. It is important that this ripple by minimized to keep machine losses low. A relatively high pulse frequency is used, typically 18 kHz. This frequency is out of the audible range for most people.

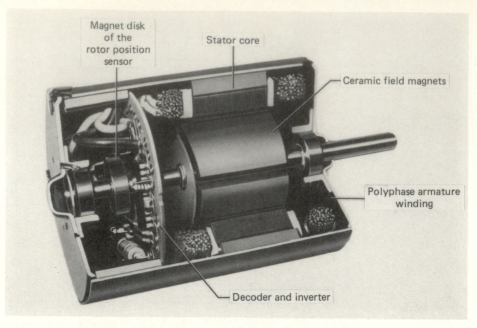

FIGURE 2.82 Photograph of a cutaway fractional-hp, brushless dc motor. (Courtesy of Fasco Industries, Inc., Motor Division.)

Figure 2.83 shows a synchronous motor with a shaft position encoder to permit it to be controlled as a brushless dc machine. The particular encoder consists of a disk having a 180° (electrical) slot which exposes three light-sensitive sensors (H1, H2, H3) in sequence. Infrared light is usually employed. When illuminated, each sensor sends a logical "1" to the decoder and a "0" when dark. When decoded, these signals turn on the SCRs of the inverter in the pattern shown in Figure 2.81 for 120° (electrical) spacing of the sensors, resulting in the current waveforms indicated. When the rotor is in the position shown, the sensor code is switching from 101 to 100, i_a is in the middle of its positive pulse, and i_b and i_c are in transition to and from zero, respectively. As a result, the stator MMF vector $\vec{F}_1$ is aligned with the axis of phase a. As the rotor turns, the disk turns, and the north and south poles of $\vec{F}_1$ step around the inside surface of the stator, pushing the corresponding rotor poles ahead of them, thus producing motor torque. Note that if the positions of H1, H2, and H3 remain fixed, the instantaneous positions of the poles of $\vec{F}_1$ will have a fixed relationship to the disk. It is thus possible to place an index mark on the disk which will always point to the north pole of the stator field.

It will be seen that the speed of the stator field poles, ω_s, is not constant in a brushless dc machine as it is when a synchronous machine is connected to a constant-frequency line. The frequency of the stator currents in a brushless dc machine depends on the shaft speed. This is what keeps the poles of the rotating stator field at a constant angular displacement from the poles of the rotor field ($\vec{F}_2$). The displacement between the two fields is related to the angle γ, measured between the axis of the stator field and the q axis of the rotor. The q axis lags the north pole (d axis) if the rotor by 90° electrical. The angle γ may be adjusted by rotating the commutating disk on the shaft relative to the rotor poles.

Fractional-horsepower brushless dc machines usually have permanent-magnet field poles. Hall effect devices are usually employed to sense rotor position when permanent magnets are used. In addition to the Hall element, these sensors often contain the

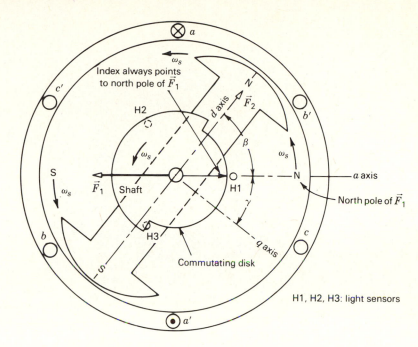

Situation when $\omega t = 60°$: sensor code switching from 101 to 100

FIGURE 2.83 Schematic representation of a synchronous motor operated as a "brushless dc" machine; γ is the angle between the disk index and the q axis.

electronics necessary to delivery a TTL (transistor-transistor-logic) logical "one" when under the influence of a north magnetic pole and a "zero" otherwise. Figures 2.84a and b illustrate two possible sets of locations for the Hall effect sensors in four-pole PM machines. In Figure 2.84a the sensors are 120° apart electrically but are spaced 60° apart mechanically. Sensor H1 is on an axis of the a-phase winding, H2 is on the following c-phase axis, and H3 is on the b-phase axis. These axes may be located with a magnetic compass if a small dc current is fed into the phase windings, one at a time. The rotor is shown in its position for $\omega_e t = 0$ in Figure 2.83a, with $\gamma = 0$. The sensor codes are the same as in the previous two-pole machine.

Figure 2.84b illustrates a similar machine with a sensor spacing of only 30° mechanical (60° electrical). The codes produced by this arrangement are different, as indicated in Figure 2.81, and require a different decoder. However, decoders have been developed that will handle both situations. Here H3 is located on the $-a$ axis, that is, 90° mechanical in the direction of rotation from the a axis. The rotor is shown at the instant $\omega_e t = 60°$, $\gamma = 0$. If a nonzero γ is necessary, the sensors may all be shifted in a direction opposite to rotation (for $+\gamma$) by an angle of γ electrical degrees.

A cutaway of a brushless dc servomotor with rear-mounted brushless tachometer is shown in Figure 2.85a. Figure 2.85b is a close-up of the shaft position sensor, consisting of a permanent-magnet disk and three Hall effect sensors spaced 60° apart. The use of a separate magnet disk to trigger the Hall effect devices avoids interference from the stator magnetic fields and makes for easier adjustment of the angle γ.

Dr. Donald Novotney and his students at the University of Wisconsin have developed models for the brushless dc machine and other systems involving inverter-driven machines. Figure 2.86 is the Novotny–Abbas circuit model of a CSI-driven brushless dc machine, and Figure 2.87 is the corresponding phasor/vector diagram. Figure 2.80

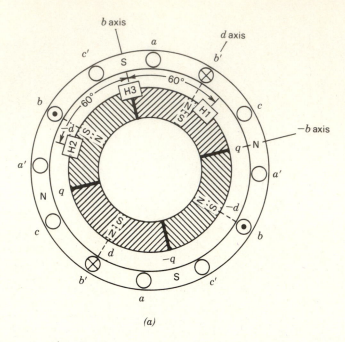

(a)

FIGURE 2.84a. Four-pole PM brushless dc motor with 60° (mechanical) spacing of Hall effect sensors set for γ = 0. Rotor shown at $\omega_e t = 0$, $\omega_s t = 0$. At this instant, the sensor code is switching from 001 to 101.

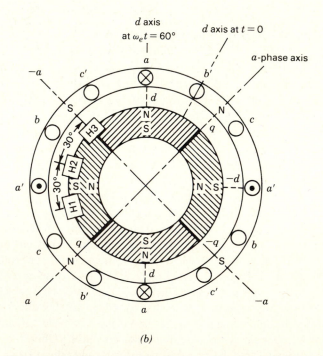

(b)

FIGURE 2.84b. 30° (mechanical) spacing of rotor position sensors in a four-pole PM brushless motor, γ = 0. Rotor shown at $\omega_e t = 60°$, $\omega_s t = 30°$. Sensor code switching from 000 to 001.

FIGURE 2.85 Brushless dc servomotor with tachometer. (*a*) Cutaway view. (*b*) Detail of rotor position sensor, showing magnet disk and Hall effect devices. (Courtesy of Pacific Scientific.)

defines V_R and R_R. It will be noted that the fictitious reactance X_c merely allows for the phase and magnitude differences between V_1, the actual terminal voltage of the synchronous machine, and E_{ge}, the effective countervoltage of the equivalent dc machine. In the model, the armature current of the equivalent dc machine, I_{ae}, is made equal to I_{1m}, the actual line current of the synchronous machine. To behave like direct-current quantities, E_{ge} must be in phase with I_{ae} so that their product equals the converted power per phase without the introduction of a power factor.

The synchronous motor's line current is shown as leading its terminal voltage in the figure. This is a desirable condition when the switching elements in the inverter are SCRs, because complex turn-off circuits are not necessary, as they would be with lagging power factor. Obviously, a leading power factor is not possible unless $\gamma > \delta$. Because they can be made to have large current ratings, SCRs are required in inverters

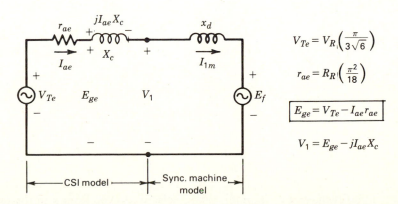

$$V_{Te} = V_R\left(\frac{\pi}{3\sqrt{6}}\right)$$

$$r_{ae} = R_R\left(\frac{\pi^2}{18}\right)$$

$$\boxed{E_{ge} = V_{Te} - I_{ae}r_{ae}}$$

$$V_1 = E_{ge} - jI_{ae}X_c$$

FIGURE 2.86 Novotny–Abbas model of one phase of a CSI-driven "brushless dc" machine.

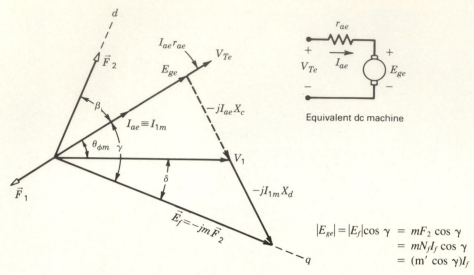

FIGURE 2.87 Phasor/vector diagram of the brushless dc machine.

driving large machines. A nonzero value of γ is thus important in brushless dc motors using SCR inverters.

The converted power per phase of the actual synchronous machine is given by

$$|I_{1m}| \, |E_f| \cos \gamma = I_{ae} \, |E_f| \cos \gamma \qquad (2.131)$$

where I_{ae} is numerically equivalent to the armature current of a dc machine having the same characteristics as the brushless dc machine being modeled. Let E_{ge} be the effective voltage induced in the armature of the equivalent dc machine. Then

$$E_{ge} I_{ae} = I_{ae} \, |E_f| \cos \gamma \qquad (2.132)$$

and the magnitude of E_{ge} is given by

$$E_{ge} = |E_f| \cos \gamma \qquad (2.133)$$

There remain the problems of relating the terminal voltage and internal resistance of the equivalent dc machine to actual quantities in the brushless dc system. The rectangular current waveforms supplied to the synchronous machine by the inverter may be subjected to Fourier analysis. When this is done, the amplitude of the fundamental-frequency component of the current is found to be

$$I_{1 \text{ max}} = \frac{2\sqrt{3}}{\pi} I_I \qquad (2.134)$$

where I_I is the dc output current of the rectifier supplying the inverter. All of the current harmonics produce stator MMF fields with speeds different from that of the rotor. The poles of these fields, as they pass the rotor poles, alternately attract and repel the rotor. These harmonic fields cause core losses in the rotor and stator iron and cause vibration and noise, but their average torque production is zero. Thus, as far as useful torque is concerned, only the rms fundamental-frequency current need be considered. For this reason, the synchronous-machine phase current is taken to be

$$I_{ae} \equiv I_{1m} = \frac{I_{1\ max}}{\sqrt{2}} = \frac{\sqrt{6}}{\pi}I_I \qquad (2.135)$$

The effective terminal voltage (V_{Te}) of the equivalent dc motor may now be found by conservation of power in the steady state, neglecting losses in the inverter switching elements:

$$V_R I_I = 3V_{Te} I_{1m} \qquad (2.136)$$

since V_{Te} is in phase with I_{1m} and their product is the power input to only one of three phases. Then

$$V_R I_I = 3V_{Te}\frac{\sqrt{6}}{\pi}I_I \qquad (2.137)$$

or

$$V_{Te} = V_R\left(\frac{\pi}{3\sqrt{6}}\right) \qquad (2.138)$$

If a short circuit is placed across the circuit at E_{ge}, making E_{ge} zero, then

$$I_{ae(sc)} = \frac{V_{Te}}{r_{ae}}$$

or

$$r_{ae} = \frac{V_{Te}}{I_{ae(sc)}} = \frac{V_R(\pi/3\sqrt{6})}{(\sqrt{6}/\pi)I_{I(sc)}} = \frac{V_R}{I_{I(sc)}} \cdot \frac{\pi^2}{18} \qquad (2.139)$$

But since $V_R/I_{I(sc)}$ is the internal resistance of the rectifier, R_R,

$$r_{ae} = \left(\frac{\pi^2}{18}\right)R_R \qquad (2.140)$$

Characteristics of the Brushless dc Motor

A "brushless dc motor" has the characteristics of a shunt or PM dc motor, depending on how the rotor (field) poles are excited. Consider first a machine having its rotor poles excited by windings carrying a direct current I_f. In this case both the counter-voltage of the dc machine, E_g, and the effective countervoltage of the brushless dc machine are saturating functions of their respective field currents, as is indicated by the magnetization curves of Figure 2.88. The countervoltages E_g of a dc machine and E_{ge} of the brushless motor are both proportional to rotor speed; however E_{ge} also depends on cos γ.

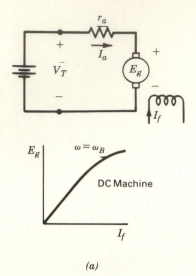

$$E_g = K_a\,\phi\omega = V_T - I_a r_a$$

$$= K_a\,\mathscr{P}_p N_f I_f \omega$$

$$= K'\,I_f\omega$$

$$\omega = \frac{V_T - I_a r_a}{K'\,I_f}\ \text{rad/s}$$

$$\tau = \frac{E_g I_a}{\omega} = K_a\,\phi\,I_a$$

$$= K'\,I_f I_a$$

Note: K' varies with saturation.

(a)

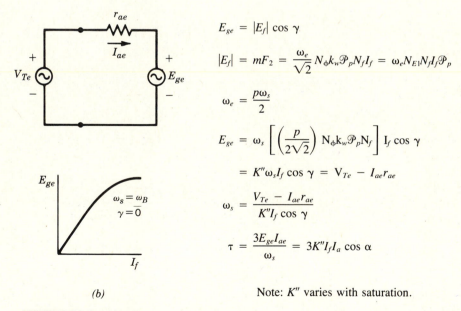

$$E_{ge} = |E_f|\cos\gamma$$

$$|E_f| = mF_2 = \frac{\omega_e}{\sqrt{2}}N_\phi k_w \mathscr{P}_p N_f I_f = \omega_e N_{E1} N_f I_f \mathscr{P}_p$$

$$\omega_e = \frac{p\omega_s}{2}$$

$$E_{ge} = \omega_s\left[\left(\frac{p}{2\sqrt{2}}\right)N_\phi k_w \mathscr{P}_p N_f\right]I_f\cos\gamma$$

$$= K''\omega_s I_f\cos\gamma = V_{Te} - I_{ae}r_{ae}$$

$$\omega_s = \frac{V_{Te} - I_{ae}r_{ae}}{K''I_f\cos\gamma}$$

$$\tau = \frac{3E_{ge}I_{ae}}{\omega_s} = 3K''I_f I_a\cos\alpha$$

(b)   Note: K'' varies with saturation.

FIGURE 2.88 Brushless dc motor compared to a dc motor. (a) dc machine. (b) Brushless dc machine.

By Kirchhoff's voltage law,

$$E_{ge} = V_{Te} - I_{ae}r_{ae} \tag{2.141}$$

and it has been shown that $E_{ge} = |E_f|\cos\gamma$.

Now E_f is simply the voltage model of the rotor MMF vector $\vec{F}_2$ and is given by

$$E_f = -jm\vec{F}_2 \tag{2.142}$$

where

$$m = \omega_e N_{E1}\mathscr{P}_p$$

and

$$|\vec{F_2}| = N_f I_f$$

Recall that $\mathscr{P}_p$ varies with saturation, thus m varies with the degree of saturation of the synchronous machine. Also, in a brushless dc motor, the electrical angular frequency ω_e is proportional to the shaft speed ω_s:

$$\omega_e = \frac{p\omega_s}{2} \tag{2.143}$$

Then

$$E_{ge} = K''\omega_s I_f \cos \gamma \tag{2.144}$$

where

$$K'' = \frac{pN_{E1}N_f\mathscr{P}_p}{2}$$

The speed equation for the brushless motor is obtained by combining Equations 2.141 and 2.144:

$$\omega_s = \frac{V_{Te} - I_{ae}r_{ae}}{K''I_f \cos \gamma} \tag{2.145}$$

where the quantity $K''I_f \cos \gamma$ may usually be found from Equation 2.144. If losses in the synchronous machine are neglected, the developed torque may be found by assuming the mechanical output power to be equal to the electrical input:

$$\tau_d = \frac{P_{\text{out}}}{\omega_s} = \frac{3V_1 I_{ae} \cos \theta_{\phi m}}{\omega_s} = \frac{3I_{ae}E_{ge}}{\omega_s} \tag{2.146}$$

or

$$\tau_d = 3K''I_{ae}I_f \cos \gamma \tag{2.147}$$

For constant γ, the speed–torque relationship may be found by choosing values of I_{ae} and calculating the corresponding values of ω_s and τ_d. **USE CAUTION** in attempting to determine torque and speed as functions of the angle γ, because I_{ae} is a function of γ, and so is E_{ge}.

In a dc motor, a reduction in field current causes an increase in speed. In a synchronous motor, a decrease in field current does not change the speed, but it does cause the power factor to become more lagging. How does a change in I_f affect a brushless dc motor? A reduction in I_f initially causes a decrease in E_f, and hence in

E_{ge}. This results in a disproportionate increase in I_{ae} because r_{ae} is quite small. The torque increases, causing an acceleration in ω_s and increasing frequency of the inverter currents. With increasing ω_e, x_d increases. If the load torque remains constant, the final value of I_{ae}, after the speed stops rising, will have to be greater than before the change in I_f to compensate for the reduction of I_f. Thus the $I_1 x_d$ drop will be larger, making for a larger δ and a more lagging current. Then for constant γ, a decrease in field current results in both a higher speed and a more lagging power factor.

PM Brushless dc Machines

In a PM machine, F_2 may be considered constant, and the rotor position sensors are likely to be placed so that $\gamma = 0$; that is, the displacement between the armature and field MMFs is maintained at 90 electrical degrees. In small machines, the phase winding resistance r_1 is not a negligible factor. The phasor diagram for $\gamma = 0$, including the $I_{ae}r_1$ drop, is shown in Figure 2.89.

Under these circumstances, the voltage drop across a fictitious *capacitive* reactance X_c exactly compensates for the drop across the synchronous reactance x_d, and V_{Te} is in phase with E_f. The speed equation may then be derived from the alternative expressions for $|E_f|$:

$$|E_f| = V_{Te} - I_{ae}(r_{ae} + r_1) = mF_2 = \left(\omega_s \frac{p}{2} N_{E1} \mathscr{P}_p \right) F_2 \qquad (2.148)$$

If the effect of the armature reaction (F_1) is neglected, then $\mathscr{P}_p F_2$ is equal to ϕ, the field flux per pole. Define

$$K_{PM} \triangleq \frac{p}{2} N_{E1} \phi = \frac{|E_f|}{\omega_s} \qquad (2.149)$$

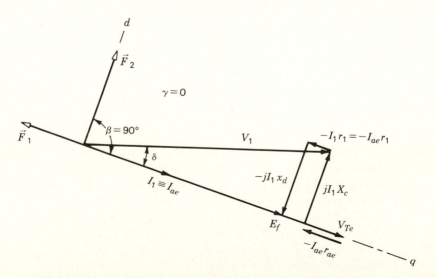

FIGURE 2.89 Phasor diagram of a brushless dc machine operated with $\gamma = 0$.

Then

$$\omega_s = \frac{V_{Te} - I_{ae}(r_{ae} + r_1)}{K_{PM}} \tag{2.150}$$

For zero γ, the positions of the rotor position sensors are such that the armature current is forced to be in phase with E_f. The developed mechanical power is thus given by

$$\text{DMP} = 3I_{ae}|E_f| = 3I_{ae}K_{PM}\omega_s \quad \text{W} \tag{2.151}$$

The torque developed by the motor is found by dividing the DMP by ω_s:

$$\tau_d = 3K_{PM}I_{ae} \quad \text{newton-meters} \tag{2.152}$$

2.23   THE CAPABILITY CURVE

The capability curve of a synchronous machine shows the limits placed on the electrical watts and VARs by the permissible temperature rise of the windings and by the mechanical system connected to the shaft, assuming operation at rated terminal voltage. Capability curves are usually provided for alternators; however they can be drawn for motors, too. In the case of an alternator, the power limit is determined by the prime mover rating, and is fairly definite. The two other limits are imposed by rotor heating (maximum permissible I_f) and by stator heating (maximum permissible I_1).

The capability curve is based on the phasor diagram of the synchronous machine. It will be recalled that the complex power in an ac system is given by

$$S = P + jQ \quad \text{VA} \tag{2.153}$$

where for three phase, the power is given by

$$P = 3V_\phi I_\phi \cos\theta_\phi = \sqrt{3}V_L I_L \cos\theta_\phi \quad \text{W} \tag{2.154}$$

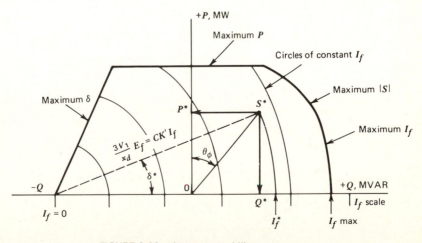

FIGURE 2.90 Alternator capability curve.

and the reactive volt-amperes are given by

$$Q = 3V_\phi I_\phi \sin \theta_\phi = \sqrt{3} V_L I_L \sin \theta_\phi \text{ VAR} \qquad (2.155)$$

By convention, Q is positive for *lagging* current.

Figure 2.90 shows a typical capability curve for an alternator, plotted on the S plane, where P is the vertical axis and Q is the horizontal axis. Operation within the boundaries of the curve is safe from the standpoints of heating and stability. Once an operation point S^* is located, corresponding to a desired power P^* and VAR output Q^*, the following information is available:

1. If S^* is inside the capability curve, the machine will not get too hot and will not be likely to fall out of synchronism.
2. A line drawn from S^* to the origin of the I_f axis is at an angle δ^* from that axis.
3. The length of the line of 2, above, is a measure of the field current required to operate at the designated values of $P = P^*$ and $Q = Q^*$ with rated terminal voltage.
4. A line from S^* to the origin of the Q axis will be at the power factor angle θ_ϕ from the vertical axis.

Figure 2.91 demonstrates how the complex output power of a synchronous machine in the generator mode can be related to its phasor diagram if each voltage phasor is multiplied by the quantity $3|V_1|/x_d$. For the purpose of the capability curve, it is assumed that $|V_1|$ is V_{1B}, the rated phase voltage. It will be recognized that

$$|I_1| x_d \left(\frac{3|V_1|}{x_d} \right) = 3|V_1||I_1| = |S| \text{ VA} \qquad (2.156)$$

which is the magnitude of the complex power. Also

$$x_d (|I_1| \cos \theta_\phi) \left(\frac{3|V_1|}{x_d} \right) = 3|V_1||I_1| \cos \theta_\phi = P \quad \text{W} \qquad (2.157)$$

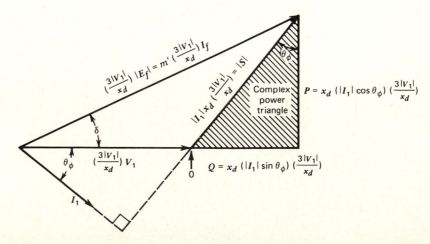

FIGURE 2.91 Synchronous machine phasor diagram modified to form a complex-power triangle.

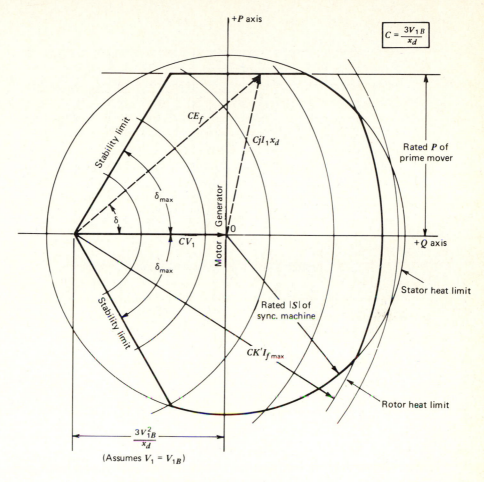

FIGURE 2.92 Construction of the capability curve.

and

$$x_d\left(|I_1|\ \sin\ \theta_\phi\right)\left(\frac{3|V_1|}{x_d}\right) = 3|V_1||I_1|\ \sin\ \theta_\phi = Q \quad \text{VAR} \tag{2.158}$$

Figure 2.92 shows how the a capability curve is constructed. The procedure is illustrated by the following example.

■ **Example 2.9**

Draw the capability diagram of the machine of Figure 2.66, assuming maximum field current to be 240 A and using the approximate values $x_d = x_{da} = 0.691\ \Omega$ and $m' = m'_a = 11.9\ \Omega$ of Example 2.8. The machine is driven by a 7500-kW turbine, and the maximum power angle for safe, stable operation is 60°. Refer to Figure 2.93 for the solution.

1. Locate a suitable center for the diagram on a sheet of graph paper and mark off the P and Q scales.
2. Draw a circle around this center, having a radius equal to the rated $|S|$ of the machine, 9,375 kVA. This is the stator heat limit curve.

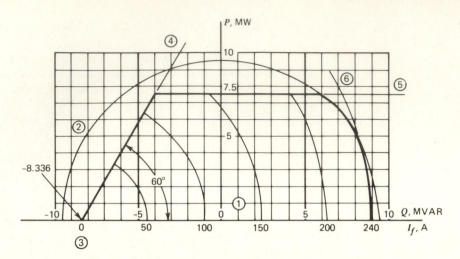

FIGURE 2.93 Capability curve for alternator of Figure 2.66.

3. Locate the center for field current circles at $Q = -3V^2_{1B}/x_d$.
   $Q = -[3 \cdot (2400/\sqrt{3})^2/0.691] = -8336$ kVA. Note that this is the origin of the modified phasor diagram of the machine and that 8336 kVA $\underline{|0}$ is CV_1, where $|V_1| = V_{1B}$ and C is $3V_{1B}/x_d$.

4. From the point just located at $Q = -CV_{1B}$, draw a line at angle $\delta_{max} = 60°$ above the Q axis. This is the stability limit (the limiting angle between CV_1 and CE_f).

5. Draw a horizontal line at P = maximum output power of the prime mover = 7500 kW. This is the power limit.

6. The rotor will get too hot if the field current exceeds 240 A dc. The corresponding $E_{f\ max}$ is $m' \cdot 240 = 11.9 \cdot 240 = 2856$ V. Then $CE_{f\ max} = (3 \cdot V_{1B}/x_d) \cdot m'I_{f\ max} = (3 \cdot 1386/0.691)2856 = 17,181$ kVA. Draw the rotor heat limit as an arc of a circle with radius 17,181 kVA and centered at $Q = -8336$ kVAR.

7. Draw circles of constant-field current, say, for 50, 100, 150, and 200 A, at radii of $CK'I_f = 71.6I_f$ kVA, and mark off the I_f scale.

2.24 SHORT-CIRCUIT CURRENT TRANSIENTS IN SYNCHRONOUS ALTERNATORS

If an initially unloaded alternator is suddenly shorted at its terminals, there are several components in each phase current:

Current Component	Current Symbol	Time Constant
1. A dc current that decays rapidly		T_a
2. A double-frequency current that also decays rapidly		T_a
3. A very high, fundamental-frequency component that dies away in a few cycles. The "subtransient".	I''	T''_d
4. A high fundamental-frequency component that decays in a few seconds. The "transient".	I'	T'_d
5. The steady-state current, given by $|E_f|/x_d$	I_{ss}	

The reasons for these components can be understood in terms of the concept of "trapped flux," a concept that is related to Lenz's law. Lenz's law states that a change in flux induces voltages in such directions that the currents produced by these voltages in closed circuits would tend to oppose the flux change. Thus when an alternator is short-circuited, currents will flow in the three phase windings, the field winding, and the damper windings, which will try to maintain the flux conditions in the machine exactly as they were at the instant of the short circuit. In other words, the flux is trapped in the machine. Of course, the energy stored in the magnetic field is quickly used up as $\int i^2 r \, dt$ in the windings, so the flux does not remain trapped long but begins to decay immediately.

With no initial stator current, $\vec{F}_1 = 0$ and $\vec{R} = \vec{F}_2$ and $\phi = \mathscr{P}_p \vec{F}_2$ along the d axis of the rotor. At the instant of short circuit, the peak air gap flux will be at some angle θ_0 from the axis of phase a. After the short, the flux is trapped in that position, as indicated in Figure 2.94. Currents in both the rotor and stator must flow so as to maintain this flux. This will require dc currents in each stator phase proportional to the cosine of the angle that the d axis had with the axis of that phase at the instant of the short. This is indicated in Figure 2.94 by a positive dc current in phase a, a negative dc current in phase c, and essentially zero current in phase b, for the particular value of θ_0 illustrated. The rotor current, however, will have to alternate, so that a MMF pulse is produced to maintain the trapped flux each time a rotor pole passes through it. This ac current in the rotor is at rated frequency, assuming the speed remains constant. One can think of this phenomenon in terms of the trapped flux inducing an ac voltage in the field winding. The exciter generator usually has an extremely small ac impedance, so the field winding is essentially shorted, as far as this transient ac current is concerned. By Lenz's law, the ac rotor currents that result tend to oppose the collapse of ϕ.

Obviously, these rotor MMF pulses tend to cause pulsations of the trapped flux at twice the rated frequency. Again, by Lenz's law, double-frequency currents must flow in the stator windings to oppose these pulsations.

Since both the dc and double-frequency components of the phase currents exist to maintain the trapped air gap flux, they die out together. This flux decays rapidly, and these two components of the phase current are negligibly small after a few cycles.

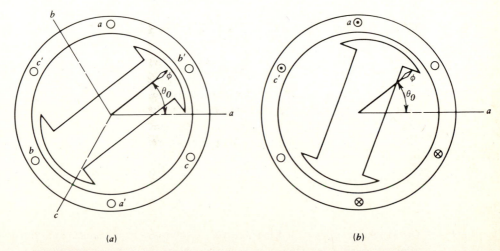

(a)   (b)

FIGURE 2.94 Air gap flux in shorted generator. (a) Instant of short circuit. (b) Fraction of a cycle after short circuit.

In the meantime, the magnetic state of the machine is moving toward the steady-state condition discussed in relation to the short-circuit test. Eventually, the current will be E_f/jx_d and the armature MMF will directly oppose that of the field, as in Figure 2.65. Now the flux inside the rotor is also trapped. As the armature MMF $\vec{F}_1$ opposes the field flux, three things happen:

1. A dc Lenz's law voltage is induced in the field winding to prevent any reduction in flux. The polarity of this voltage adds to that of the exciter, with the result that there is a sudden increase in field current.
2. Currents circulate in the damper windings in such a direction as to oppose the decrease in flux. The MMF of these currents is in the same direction along the *d* axis as that of the increased field current.
3. The armature MMF must initially be sufficient to balance the Lenz's law currents of both the field and damper windings to prevent any initial change in air gap flux. In other words, except for those pulsations necessary to maintain the trapped air gap flux, the internal trapped flux of the rotor must initially be diverted to leakage. A very high, rated-frequency armature current is required to achieve these results—about 5 to 10 times the rated current. This is called the *subtransient current*.

The time constant of the damper windings is rather short. The time constant of the subtransient current, T_d'', is about 0.03 s. However it takes longer for the incremental field current, discussed in 1, above, to decay. This time constant is not that of the field alone but of the circuit consisting of the armature and field windings coupled together. It is called the *transient time constant* T_d', and is in the order of 1 s. During this period, a rated-frequency current, somewhat less than the subtransient current, must flow in the stator to balance the increased rotor MMF. This current component is called the *transient current I'*.

If the dc and double-frequency components are neglected, the equation for the rms short-circuit current is of the form

$$I_{sc}(t) = (I'' - I')e^{-t/T_d''} + (I' - I_{ss})e^{-t/T_d'} + I_{ss} \qquad (2.159)$$

For an initially unloaded generator, the transient, subtransient, and steady-state currents are defined in terms of the open-circuit voltage E_{f0} and reactances that are constants of a given machine:

$$\text{Steady state: } I_{ss} = \frac{E_{f0}}{x_d} \qquad (2.160)$$

$$\text{Transient: } \quad I' = \frac{E_{f0}}{x_d'} \qquad (2.161)$$

$$\text{Subtransient: } I'' = \frac{E_{f0}}{x_d''} \qquad (2.162)$$

The reactance x_d'' is called the direct-axis subtransient reactance, or simply the *subtransient reactance*, while x_d' is similarly called the *transient reactance*.

■ **Example 2.10**

A 100-MVA, 16-kV, three-phase, 60-Hz synchronous alternator is delivering rated voltage, open-circuited, when a balanced, three-phase short occurs at its terminals. The machine has the following constants, the reactances expressed in per unit on a 100-MVA base:

$$x_d = 1.0 \qquad x_d' = 0.3 \qquad x_d'' = 0.2$$

$$T_d'' = 0.03 \quad \text{s} \qquad T_d' = 1.00 \quad \text{s}$$

Neglecting dc and double-frequency components of the current,

(a) Find the initial current.
(b) Find the current at the end of two cycles, and at the end of 10 s.

Solution

(a)

$$\text{Initial current} = I'' = \frac{E_{f0}}{x_d''}$$

In per-unit, since E_{f0} = rated value,

$$I_{pu}'' = \frac{1}{0.2} = 5 \text{ per unit}$$

$$I_B = \frac{100 \cdot 10^6}{\sqrt{3}\ 16{,}000} = 3610 \text{ A}$$

Then $I'' = 18{,}050$ A.

(b)

$$I' = \frac{E_{f0}}{x_d'} = \frac{1}{0.3} = 3.33 \text{ per unit}$$

$$I_{ss} = \frac{E_{f0}}{x_d} = 1.0 \text{ per unit}$$

$$I(t) = 1.67e^{-t/0.03} + 2.33e^{-t} + 1 \text{ per unit}$$

At $t = \dfrac{2}{60}$ s,

$$I(0.0333) = 167e^{-1.11} + 2.33e^{-0.0333} + 1$$
$$= 0.55 + 2.25 + 1 = 3.80 \text{ per unit}$$

or 13,720 A.

At $t = 10$ s

$$I(t) = 1.67e^{-333} + 2.33e^{-10} + 1$$
$$= 1.0001 \text{ per unit}$$

■

When a synchronous machine is carrying a load before the short circuit occurs, the equivalent circuit for the calculation of the *subtransient* current is a voltage source in series with the subtransient reactance. The source voltage is $E_i'' = V_1 + jI_1x_d''$. Then for a machine that was under load before the short, the initial short-circuit current is given by $I'' = |E_i''|/x_d''$. The result is not greatly different from that obtained under the assumption that the machine was unloaded.

PROBLEMS

2.1. Make tables of synchronous speed in radians per second and revolutions per minute for $p = 2, 4, 6, 8, 10, 12, 14,$ and 114 and (a) $f = 60$ Hz; (b) $f = 50$ Hz.

2.2. A synchronous machine has a cylindrical rotor. The effective turns per pole is $N_f = 150$. If the dc field current is 250 A and the rotor has four poles, write an expression for the rotor MMF as a function of angular distance around the rotor (a) in mechanical degrees; (b) in electrical degrees.

2.3. In a certain 14-pole, three-phase alternator, the inside diameter of the stator is 36 in. There are three slots per role per phase. The winding is double layer. The coil pitch is seven slots. (a) Find γ and ρ in electrical degrees, mechanical degrees, and inches on the inside surface of the stator. (b) Draw a set of 18 slots for 2 of the 14 poles (as in Figure 2.18) and locate the coil sides for all three phases in these slots by placing letters a, a', b, b', c, c' in the appropriate locations.

2.4. When balanced, three-phase, 60-Hz currents flow in the stator of a six-pole synchronous machine, the peak fundamental field of one phase is 400 ampere-turns. (a) Write the expression for $\mathcal{F}_a$ per Equation 2.5. (b) Write the expression for $\mathcal{F}_1$, the rotating stator MMF field. (c) What is the speed of this field in radians per second? Revolutions per minute?

2.5. For a certain operating condition in the machine of Problem 2.4, the peak rotor MMF is 600 ampere-turns and β is 120°. Find $|\vec{R}|$, δ_{SR}, and δ_{FR}.

2.6. The outside diameter of the rotor of an alternator is 29 in. and the axial length is 60 in. The machine has four poles, and the flux density at the outside surface of the rotor is given by $1.2 \cos \theta_{2e}$ T, where θ_{2e} is in electrical degrees. (a) Find the flux per pole. (b) If $|\vec{R}|$ is 22,000 ampere-turns, calculate $\mathcal{P}_p$.

2.7. If in a given machine $|\mathcal{F}_2| = 800$ ampere-turns and $|\mathcal{F}_1| = 300$ ampere-turns, $\beta = 126.9°$ and $\mathcal{P}_p = 1.505 \cdot 10^{-4}$ Wb/ampere-turn (i.e., H/turn^2), find the air gap flux per pole produced by the resultant MMF $\vec{R}$.

2.8. In the machine of Problem 2.3, the air gap flux per pole is 0.15 Wb under a certain load condition. The coils have one turn each and the speed corresponds to $f = 60$ Hz. Find (a) the pitch factor, (b) the distribution factor, (c) the winding factor, (d) the speed, (e) $|E_\phi|$, the voltage induced in each phase winding, and (f) $|E_\phi|$ on a line-to-line basis, assuming a Y connection.

2.9. A two-pole, three-phase 2300-V synchronous machine has 42 stator slots. Each slot has only two conductors in a double-layer winding, so $N_c = 1$

turn. The coil pitch is 17 slots. (a) Find the slot pitch and coil pitch in electrical degrees. (b) Find the pitch and distribution factors.

2.10. The machine of Problem 2.9 has each phase winding connected in two parallel paths. That is, the two phase groups are connected in parallel rather than in series. Since N_ϕ is defined as the number of series turns per phase, in this case it is given by $\frac{1}{2}pnN_c$. (a) Find N_ϕ for this machine. (b) Find ϕ, the flux per pole, required to generate a phase voltage of 1386 V (i.e., $2400/\sqrt{3}$) at 3600 rev/min.

2.11. Find the rms coil voltage in the machine of Problem 2.6 if the frequency is 60 Hz, the coil pitch is 150° electrical, and each coil has two turns.

2.12. Define E_f, E_{ar}, and E_ϕ in words.

2.13. In a given alternator, the leakage reactance per phase is 2.00 Ω and the resistance of each phase winding is 0.100 Ω. (a) If the load on this alternator is 500 kVA at 0.800 power factor, lagging, and the terminal voltage is 2300 V, line-to-line, find E_ϕ, the voltage induced in the winding by the air gap flux. (b) If the machine has a winding similar to that of Problem 2.3 and each coil has one turn, find the air gap flux per pole at 60 Hz.

2.14. Neglect losses in the following problem: (a) How much torque is required to drive a 300-MW generator at 3600 rev/min at rated load? Express in pound-feet and newton-meters; (b) What is the output horsepower of the turbine? (c) If the terminal voltage is 20,000 V, line-to-line and the power factor is 0.8, lagging, what is $|I_1|$?

2.15. A 2300-V, 500-kVA, three-phase alternator has a synchronous reactance of 8.00 Ω and a leakage reactance of 2.00 Ω. The stator winding resistance for each phase is 0.100 Ω. If the machine is operating at rated load and voltage, and the load power factor is 0.800, lagging: (a) Find the rms magnitude of the voltage being induced in each phase. (b) Find E_f, the phasor representing the rotor MMF in the circuit model, assuming the phase terminal voltage V_1 to be the reference phasor. (c) What is the magnetizing reactance? (d) If the ratio $|E_\phi|/R$ is 0.280 and N_f is 150 turns, find the dc rotor field current I_f. (e) What is δ?

2.16. A six-pole, 60-Hz alternator has a synchronous reactance of 4 Ω and a three-phase terminal voltage of 2300 V. The field current is adjusted so that $E_f = 2262$ V and $\delta = 25.11°$. Find (a) the output power, (b) the torque required to drive the machine, (c) jI_1x_d as a phasor, with V_1 at 0°, (d) the power factor, and (e) $|I_1|$.

2.17. A 375-MVA, 20,000-V, three-phase alternator has a per-unit synchronous reactance of 1.2. (a) What is x_d in ohms? (b) Find E_f and δ when the machine is delivering 300 MW at 20,000 V and a power factor of 0.8, lagging.

2.18. If the machine of Problem 2.15 has 10 poles and the frequency is 60 Hz, find the torque being developed.

2.19. A small industrial plant has a total electrical load of 300 kW at 0.75 power factor, lagging. There is need to add a 50-hp pump to be driven at more or less constant speed. If the pump is driven by a synchronous motor capable of operation at 0.8 power factor, leading, what will be the new total load and power factor? Assume no losses in the motor.

2.20. The field current of a synchronous motor is increased, causing the line current to decrease. Was the machine operating at leading, lagging, or unity power factor before the change?

2.21. A three-phase synchronous motor draws 96 A from the line at unity power factor. (a) Assuming constant line voltage, what would be the line current at a power factor of 0.8, leading? (b) At 0.8, lagging? (c) As the power factor becomes more leading, what would happen to the line voltage, usually?

2.22. A synchronous motor is drawing 100 A from a 208-V, three-phase line at unity power factor, with the field current adjusted to 0.900 A. The synchronous reactance x_d is 1.30 Ω. (a) Find the power angle δ. (b) Find m'. (c) Assuming no change in mechanical load, approximately what value of field current will result in a power factor of 0.8, leading?

2.23. In Problem 2.22, the field current is adjusted for unity power factor at 100 A line current. Suppose the mechanical load is reduced to half its original value. (a) Find the new line current and power factor for the same field current. (b) What field current would be required for unity power factor under the new conditions?

2.24. Given the accompanying curves (p. 151) for a 12.6-kV, 150-MW, 0.85 power-factor, four-pole, 60 Hz synchronous machine:

(a) Find the rated current.
(b) Find the approximate synchronous reactance by the method of Equation 2.99.
(c) Find the unsaturated synchronous reactance.
(d) Find the short-circuit-ratio.
(e) Find the field current under rated conditions, 0.85 power factor, lagging, by the approximate method of page 000. (Generator mode.)
(f) What is m'_a?
(g) Find the open-circuit, line-to-line terminal voltage of this machine for the field current of (e), above.
(h) Find the voltage regulation.
(i) What is the maximum power available from this machine with rated terminal voltage and the field current found in part (e)?

2.25. If the leakage reactance of the machine of Problem 2.24 is 0.200 Ω, find the field current required to produce rated terminal voltage at a load of 175 MVA, 0.800 power factor, lagging, using the saturated synchronous reactance method.

2.26. Compute and plot a V curve for the alternator of Problem 2.24 for a load of 100 MW. Use the approximate methods involving x_{da} and m'_a.

2.27. For the 1000-kVA machine whose characteristics are given in Appendix C, calculate the field current required to operate the machine at full load and rated terminal voltage (a) as a motor at 0.800 power factor, leading; (b) as a generator at 0.800 power factor, lagging. Use the given Potier reactance

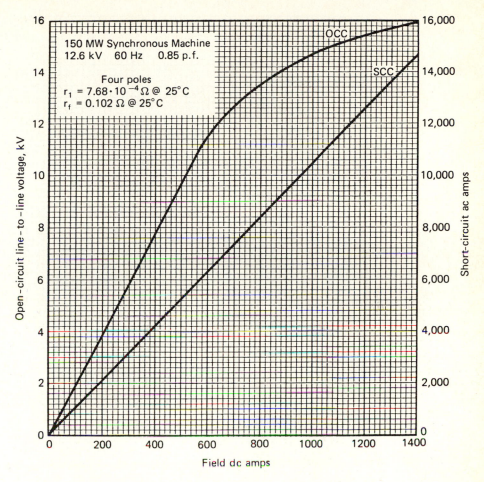

Synchronous machine curves for Problems 2.24, 2.25, 2.26, and 2.28.

as x_1 in the saturated synchronous reactance method. Compare the two values of field current and explain. (Do not use the method of Appendix C.)

2.28. The accompanying characteristic curves are for a 68.8-kVA, 240-V, three-phase, 60-Hz synchronous alternator having four poles. The leakage reactance x_1 is 0.0510 ohm.

(a) Find the rated current.
(b) Find the unsaturated synchronous reactance in ohms and in per-unit.
(c) Find the unsaturated magnetizing reactance in ohms.
(d) Find the short-circuit ratio.
(e) Find x_{da} (Equation 2.99) and m'_a.

2.29. When the machine of Problem 2.28 is operating as a generator at rated voltage and current, 0.8 power factor, lagging, (a) find the required field current, using the saturated synchronous-reactance method; (b) compare x_d under these conditions with its unsaturated value.

2.30. Do Problem 2.29 by the approximate method. Assuming the field current

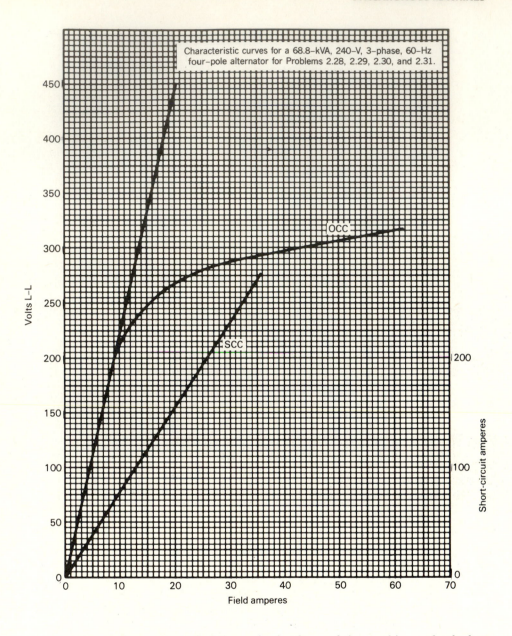

Characteristic curves for a 68.8-kVA, 240-V, 3-phase, 60-Hz four-pole alternator for Problems 2.28, 2.29, 2.30, and 2.31.

to remain constant, find the no-load voltage of the machine and calculate the voltage regulation.

2.31. Repeat Problem 2.29 for a leading power factor of 0.8.

2.32. A brushless dc motor consists of a 5-kVA, four-pole, three-phase, synchronous motor, rated at 220 V, 50 Hz, 0.8 power factor, leading, driven by a CSI rectifier-inverter having an internal resistance of 0.184 Ω. The motor has a synchronous reactance of 9.00 Ω. The maximum dc voltage output of the rectifier is 250 V and the maximum current output from the inverter is $I_I = 20.0$ A.

   (a) Find γ when the synchronous machine is operating under rated conditions. Also find V_R and I_I.

(b) For the same γ and I_f as in (a), plot speed vs. developed torque for this brushless dc machine. (*Hints:* Calculate r_{ae}, V_{Te}, and the constant $(K''I_f \cos \gamma)$. Choose values of I_{ae} and insert into the expressions for ω_s and τ_d.)

(c) Plot the speed–torque curve for the same V_{Te}, but with the field current reduced to 75% of its previous value.

(d) Repeat (b), but with half the value of V_R.

2.33. A 180-W brushless dc motor has a four-pole PM field and a Y-connected, three-phase stator winding. The placing of the rotor-position sensors is such that $\gamma = 0°$. To determine K_{PM}, the shaft is driven at 3,000 rev/min, and the peak value of the fundamental component of the open-circuit terminal voltage is measured to be 17.32 V. The inverter supplies flat-topped current pulses to the motor windings of the form shown in Figure 2.80. The effective internal resistance of the inverter is $R_R = 0.200 \ \Omega$. The synchronous reactance of the motor is $0.500 \ \Omega$ and its winding resistance per phase is $r_1 = 0.100 \ \Omega$. Find the speed and developed torque of the motor when the dc voltage input to the inverter is 24 V and the effective armature current is 6.25 A. What is the amplitude of the inverter output current pulses with this load on the motor?

2.34. Construct a capability curve for the machine of Problem 2.24 if it is driven by a prime mover capable of an output of 200,000 hp. Let δ_{max} be 50°.

2.35. A three-phase synchronous machine has a double-layer stator winding connected in Y. There are 12 slots per pole. Each coil has two turns and is pitched 10 slots. The rms phase current is 30 A. (a) Find the pitch, distribution, and winding factors. (b) What is the amplitude of the fundamental rotating stator field F_1? (c) Find N_g and N_{i1}.

2.36. For the winding of Problem 2.35, find k_{wv} for the fifth and seventh harmonics.

2.37. What are the two lowest slot harmonics for the winding of Problem 2.35?

2.38. A cylindrical rotor for a 9750-kVA, two-pole, 60-Hz alternator has five concentric coils per pole. Each coil has 31 turns. The coil pitches are, respectively, 71°, 99°, 122.14°, 145.28°, and 168.43°. (a) Calculate N_f, the effective turns per pole. (b) If the field current in these coils is 150 A dc, find the peak amplitude of the fundamental component of the rotor MMF.

2.39. A 393-MVA, 24,000-V alternator has the following reactances and time constants (reactances are given in per-unit):

$$x_d = 1.845 \qquad x_d' = 0.332 \qquad x_d'' = 0.245$$

$$T_a = 0.20 \ s \qquad T_d' = 0.83 \ s \qquad T_d'' = 0.04 \ s$$

The machine is operating at no load when a three-phase fault occurs at its terminals. Previous to the fault, the line-to-line terminal voltage was 24,000 V. (a) Write an expression for the per-unit rms fault current as a function of time, neglecting dc and double-frequency components. (b) What is the current in amperes after 0.500 s?

3

TRANSFORMERS

3.1 WHY TRANSFORMERS ARE ESSENTIAL TO POWER SYSTEMS

Transformers make large power systems possible. To transmit hundreds of megawatts of power efficiently over long distances, very high line voltages are necessary—in the range of 161 to 1000 kV. At the time of this writing, however, the highest practical design voltage for large generators is about 25 kV. How can electrical power be generated at 25 kV and transmitted at a much higher voltage? Transformers provide the solution to this problem. They can step voltages up or down with very little loss in power. Connecting a step-up transformer between the generator and a transmission line permits a practical design voltage for the generator and at the same time an efficient transmission line voltage. Step-down transformers connected between the transmission line and the various electrical loads connected to it permit the transmitted power to be used at safe voltages. The large power systems existing today could not have been developed without them.

Ideally, a transformer should not introduce any change in power factor and should have zero internal power loss. If P_1 is the input power to a three-phase transformer and P_2 its output power, the following relationship should hold for an ideal, three-phase transformer:

$$\sqrt{3}V_{L1}I_{L1} \cos \theta_\phi = P_1 = P_2 = \sqrt{3}V_{L2}I_{L2} \cos \theta_\phi \qquad (3.1)$$

In this relationship, V_{L1} is the line-to-line voltage at the input terminals of the transformer and I_{L1} is the input line current. The line voltage and current at the output terminals are V_{L2} and I_{L2}. The angle between phase current and phase voltage is θ_ϕ, assumed to be unchanged between input and output, and $\cos \theta_\phi$ is the power factor. (See Appendix B for details of three-phase power calculations.)

Solving Equation 3.1 for the current ratio,

$$\frac{I_{L1}}{I_{L2}} = \frac{V_{L2}}{V_{L1}} \qquad (3.2)$$

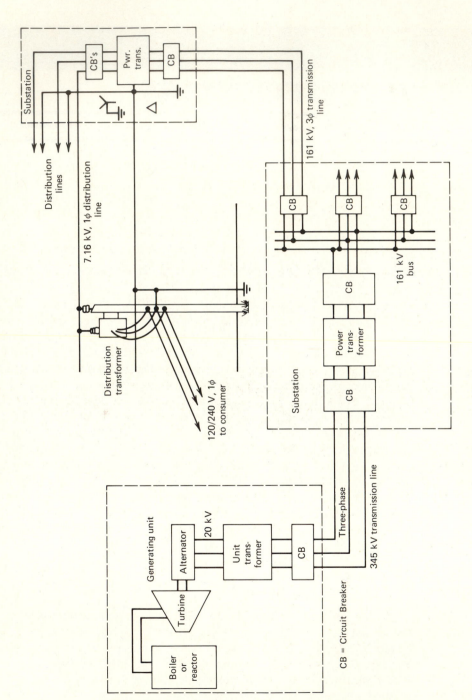

FIGURE 3.1 Transformer applications.

It is seen that, *as a transformer steps up the voltage, it reduces the current proportionately*. Since the I^2R losses in transmission lines are proportional to square of the line current, it is obvious that the high transmission line voltages obtainable with transformers increase the efficiency of a power system by reducing the transmission line currents.

3.2 TYPES OF POWER SYSTEM TRANSFORMERS

The elements of a power transmission and distribution system are shown in Figure 3.1. Alternators for large systems have ratings in the range of 30 to 1700 MVA. The output terminals of alternators are usually connected directly to a *unit transformer* of equal rating. The *unit transformer* steps the voltage of the alternator up to the desired transmission voltage. Figure 3.2 is a photograph of a typical unit transformer. At the receiving ends of the transmission lines there are substations, at each of which there are one or more *power transformers*. These reduce the voltage to the subtransmission levels, 69 to 230 kV. The subtransmission circuits fan out from the substation to other substations located at load centers. At the load center substations, other *power transformers* further reduce the voltage to distribution levels, say, 2.3 to 69 kV. Distribution circuits go to industrial loads or residential districts, where the voltage is reduced to the final utilization voltage, 120/240 V single phase, or 120/208 V or 240 V or 480 V or higher three phase. The local transformers performing this final voltage reduction

FIGURE 3.2 A 1300-MVA unit transformer, 24-kV Δ to 345-kV Y. (Courtesy of Westinghouse Electric Corporation.)

FIGURE 3.3 A pole-mounted, 25-kVA distribution transformer. (Courtesy of Westinghouse Electric Corporation.)

are called *distribution transformers*. A typical single-phase distribution transformer is shown in Figure 3.3.

Three-phase power may be transformed by using either two or three *single-phase transformers,* or by a single *three-phase transformer*. When a set of single-phase transformers is employed to transform three phase, it is called a *three-phase bank* of transformers.

Transformers are also used in measuring voltage, current, and power flow in power systems. *Potential transformers* ("pot transformers") are single-phase transformers of special design, which step down the voltage to be measured to a safe value (usually for the operation of 0–150 V meters) with negligible phase shift. *Current transformers* step down currents with negligible phase shift and have insulation adequate to isolate metering equipment and personnel from the line voltage. They are usually designed to operate 0–5 A meters. The secondaries of both potential and current transformers are usually grounded. Low phase shifts are necessary to permit accurate measurement of watts, vars, and impedance. Such measurements depend on phase relationships between voltages and currents.

3.3 ELEMENTS OF A TRANSFORMER

A transformer has three basic elements: a primary winding, a secondary winding, and a core. Figure 3.4 shows schematically the arrangement of these elements. The two windings are insulated from each other and from the core. By definition, the *primary winding* is the one *connected to the electrical source*. The *secondary winding* is the *output winding*. There may be more than one secondary winding, each connected to a different load, or interconnected to provide different output voltages. The *core* is a

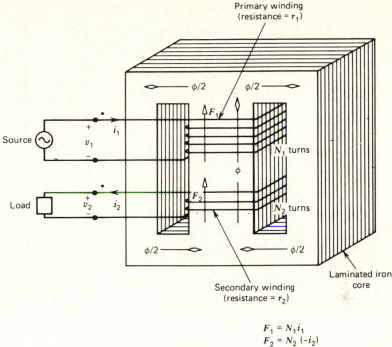

$$F_1 = N_1 i_1$$
$$F_2 = N_2 (-i_2)$$

FIGURE 3.4 Elements of a transformer.

high-permeance magnetic circuit that links all of the transformer's windings. An alternating flux is set up in the core when the primary is connected to an ac voltage source. This flux induces voltage in all windings. The voltage induced in each winding, in accord with Faraday's law, is proportional to the number of turns in that winding. The voltage induced in the primary is nearly equal to the applied voltage, and the voltage at the secondary-winding terminals also differs by only a few percent from the voltage induced into that winding. Thus the primary-to-secondary voltage ratio is essentially equal to the *ratio of the numbers of turns* in the two windings, given the symbol *a*:

$$a \triangleq \frac{N_1}{N_2} \cong \frac{v_1}{v_2} \tag{3.3}$$

where N_1 is the number of primary coil turns, N_2 is the number of secondary turns, and v_1 and v_2 are the voltages at the winding terminals. By selecting the proper turns ratio, the transformer designer can determine the ratio of input to output voltages to meet the requirements of the power system.

"Dry type" transformers operate in air. Most power transformers are immersed in a tank of oil. The oil is a better insulator than air. Also, convection currents in the oil help carry heat away from the windings and the core. The ends of the windings are brought to a terminal block, from which leads are brought to the outside of the tank through insulating bushings mounted in holes in the sides of the tank or in the lid. Three high-voltage bushings are seen clearly in Figure 3.2.

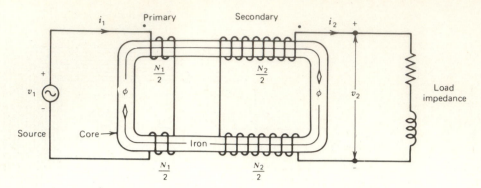

FIGURE 3.5 An elementary, two-winding, core-form transformer.

In both Figures 3.4 and 3.5, the windings are separated from each other to avoid confusion in the drawing. In actual practice, the low-voltage winding is placed next to the core and extends most of the length of the core leg. The high-voltage windings surround the low-voltage windings. This simplifies the problem of insulating the high-voltage windings from the core. A cutaway shell form transformer is shown in Figure 3.6. This transformer is designed for line-to-line connection—hence two high-voltage bushings rather than the single one shown in Figure 3.3.

The core structure must provide a high-permeance path for the magnetic flux linking the transformer windings and must also contain insulating materials inserted in such a way as to block the flow of eddy currents. Most commonly, the core is formed of a stack of thin, silicon–steel laminations (each about 0.014 in. thick for 60-Hz operation) which are electrically insulated from each other. The insulation may be provided by a varnish, but often the iron oxide coating which results from heat treatment of the laminations is sufficient. The laminations must usually be heat treated to have the specified magnetic properties.

Alternatively, the steel may be formed into a thin ribbon and the core formed by winding enough of this ribbon on a mandrel to result in the required cross-sectional area. The wound core is heat treated and the mandrel removed. During heat treatment, the insulating material cements the core into a more-or-less solid mass.

The iron in laminated cores, whether stacked or wound, has a crystalline structure. Recently, *amorphous metal* cores have been developed in which the iron atoms are randomly distributed. A molten alloy of iron, silicon, and boron is caused to spill out of a furnace in a thin, ribbonlike stream, which is chilled at the rate of about 10^6 Farenheit degrees per second. Crystals do not form during such fast chilling. The resulting glasslike material has inherently low hysteresis loss and a high resistivity, thus impeding the flow of eddy currents. Cores wound of amorphous-metal ribbon have less than one fourth the core loss of the best steel laminations on a per-pound basis. However the fraction of the cross-sectional area which is actually ferromagnetic (the *stacking factor* of Appendix A) is only about 0.80, compared to about 0.96 for laminated steel cores. This means an amorphous core must be larger than a laminated steel core for the same transformer kVA rating, and a larger core requires a larger tank and thus more oil. Amorphous-metal cores have additional disadvantages in that the material is hard and brittle. The ribbon is difficult to cut, and pieces of core have broken off in transformers subjected to drop tests. However it has been shown that the higher initial cost of transformers with

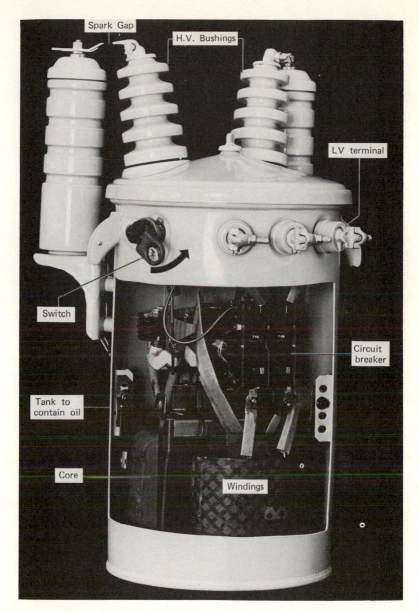

FIGURE 3.6 Cutaway view of a shell-form distribution transformer. (Courtesy of Westinghouse Electric Corporation.)

amorphous-metal cores is more than offset by the savings that accrue from reduced energy loss during the life of the transformer.

3.4 FARADAY'S LAW AND LENZ'S LAW

Transformers operate on the basis of Faraday's law:

$$e = \pm \frac{d\lambda}{dt} \tag{3.4}$$

where e is the instantaneous voltage induced by a magnetic field and λ is the number of flux linkages between the field and the electric circuit in which the voltage is being induced. The sign depends on Lenz's law and on which circuit terminal is taken to be positive.

The Dot Convention

In drawing schematic diagrams involving mutual inductance, dots are placed on terminals of the coil symbols to indicate the relative polarities of the voltages induced in the coils by changes in the mutual flux. The dots indicate those coil terminals that will have the same instantaneous polarity for voltages induced by any mutual-flux variation. Once a dot is assigned arbitrarily to a terminal of a given coil, the dotted terminals of all other coils coupled to it are determined by Lenz's law and cannot be chosen at will.

In Figure 3.4, 3.5, and 3.7, the dotted terminals of both windings are taken as the positive terminals. This is indicated by the signs associated with v_1, v_2, e_1', and e_2'. Lenz's law states that the polarities of induced voltages are such that they tend to oppose any *change* in flux. If, in these figures, the core flux ϕ is increasing ($d\lambda/dt > 0$), currents would have to flow out of the dotted terminals of both windings to produce MMFs in opposition to this flux change. Lenz's law requires that the dotted coil terminals be positive, relative to the undotted terminals, in order to drive current out of the dotted terminals when $d\lambda/dt$ is positive. Thus positive voltages are associated with positive $d\lambda/dt$. The positive sign should be used in Equation 3.4 when Faraday's law is applied to the transformers of these figures. If the *lower* terminals of the two windings had been taken as positive, the *minus* sign would be used in Equation 3.4.

A corollary to Lenz's law is that currents flowing into the dotted terminals of mutually coupled coils will produce MMFs in the same direction around the magnetic circuit. This provides a means for locating dots which is simpler than the application of Lenz's law. *In this book, a magnetomotive force will be considered positive when produced by a current flowing into a dotted coil terminal.* Thus in Figures 3.4 and 3.5, $\mathcal{F}_1 = +N_1 i_1$ but $\mathcal{F}_2 = -N_2 i_2$, because i_1 is taken to be flowing into the dotted primary terminal while i_2 is taken to be flowing out of the dotted secondary terminal.

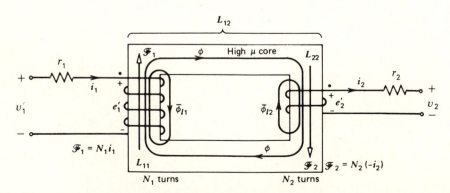

FIGURE 3.7 Electrical and magnetic quantities in a transformer.

Flux Linkages

In the case of a current-carrying *coil* of wire, λ is the sum of the flux linkages of all the turns of the coil. Reference to Figure A.1 (Appendix A) will show that not all turns are linked by the same amount of flux. Then for a coil of N turns,[1]

$$\lambda = \sum_{i=1}^{N} \phi_i \text{ Wb-turns} \qquad (3.5)$$

where ϕ is webers of flux linking the ith turn. Taking the positive sign in Equation 3.4, the voltage induced in a single turn is

$$e_i = \frac{d\phi_i}{dt} \qquad (3.6)$$

since the λ of a single turn is equal to $1 \cdot \phi_i$. By Kirchhoff's voltage law, the induced voltage in the whole coil is

$$e = \sum_{i=1}^{N} e_i = \sum_{i=1}^{N} \frac{d\phi_i}{dt} = \frac{d}{dt} \sum_{i=1}^{N} \phi_i = \frac{d\lambda}{dt} \qquad (3.7)$$

The Concept of Effective Flux $\overline{\phi}$

It is often useful to replace the concept of λ with that of an equivalent effective flux which links all of the turns of a winding. This effective flux $\overline{\phi}$ is defined as follows:

$$\overline{\phi} \triangleq \frac{\lambda}{N} \qquad (3.8)$$

Then

$$\lambda = N\overline{\phi} \qquad (3.9)$$

and Faraday's law becomes

$$e = N\frac{d\overline{\phi}}{dt} \qquad (3.10)$$

3.5 MAGNETIC FLUXES OF A TRANSFORMER

The difference between the total flux of a winding and the mutual flux between it and another winding is called the *leakage flux*. Leakage flux has a complex pattern similar

---

[1]Do not confuse the word "turn" with the word "winding." A "winding" consists of many "turns." In Figure A.1 the whole coil is the winding, consisting of 13 turns.

to that of Figure A.2 in Appendix A. However it may be represented by an equivalent effective flux $\overline{\phi}_l$. If the mutual flux between the primary and secondary windings is ϕ, then the leakage fluxes are given by

$$
\begin{aligned}
\overline{\phi}_{l1} &= \overline{\phi}_1 - \phi \\
\overline{\phi}_{l2} &= \overline{\phi}_2 - \phi
\end{aligned}
\tag{3.11}
$$

The total flux of either of the two windings may be expressed as the sum of the mutual and the leakage fluxes:

$$
\begin{aligned}
\overline{\phi}_1 &= \overline{\phi}_{l1} + \phi \\
\overline{\phi}_2 &= \overline{\phi}_{l2} + \phi
\end{aligned}
\tag{3.12}
$$

The coupled circuit of Figure 3.7 illustrates these relationships between the mutual and the leakage fluxes.

The core has a very high permeance while most of each leakage flux path is in the nonferromagnetic material surrounding the windings. These facts lead to the following important results:

1. The leakage flux is small compared to the mutual flux, contributing only about 1 to 7 percent of the flux linkages of each winding.
2. The leakage paths do not saturate, so that the leakage flux of each winding is proportional to the current in that winding.
3. The mutual flux is confined almost totally to the iron core.

Transformers operating at high voltages tend to have more leakage, because the windings have to be separated from each other farther to provide sufficient insulation.

The flux linkages of the primary and secondary may be expressed in terms of the leakage and mutual fluxes:

$$
\begin{aligned}
\lambda_1 = N_1\overline{\phi}_1 &= \lambda_{l1} + \lambda_{m1} \\
&\triangleq N_1\overline{\phi}_{l1} + N_1\phi \\
\lambda_2 = N_2\overline{\phi}_2 &= \lambda_{l2} + \lambda_{m2} \\
&\triangleq N_2\overline{\phi}_{l2} + N_2\phi
\end{aligned}
\tag{3.13}
$$

where λ_{l1} and λ_{l2} are the primary and secondary flux linkages due to leakage fluxes, and λ_{m1} and λ_{m2} are the linkages between the mutual flux ϕ and these windings.

3.6 TRANSFORMER INDUCTANCES

It must be observed that electrical engineers are always trying to develop circuit models for the devices they study. They then often discuss the models rather than the devices themselves, even using concepts such as the "impedance" of free space! The reason for this tendency is that a circuit analog helps them to organize their thinking about a given device and allows them to bring the tremendous power of circuit theory to bear on device-related problems.

Charles P. Steinmetz (1865–1923) developed the circuit model that is universally used for the analysis of iron core transformers at power frequencies. His model has many advantages over those resulting from straightforward application of linear circuit theory, primarily because the iron core exhibits saturation and hysteresis and is thus definitely nonlinear. However it is a good idea to consider transformers first from the point of view of basic linear circuit theory to better appreciate the Steinmetz model. ("Linear" implies that the inductances and resistances in the circuit model have constant values; that is, their values are independant of the voltages across them or the currents flowing through them.)

To begin the linear analysis, consider the schematic representation of a transformer in Figure 3.7, in which the flux components have been clearly defined. The two windings may be treated as voltage sources having respective internal voltages of e_1' and e_2' and internal resistances of r_1 and r_2, representing the actual ac resistances of the winding conductors. Then by Thevenin's theorem these resistances may be considered separate circuit elements, as shown in Figure 3.7, leaving a hypothetical transformer with zero-resistance windings.

Now apply Kirchhoff's voltage law to the primary winding of the transformer of Figure 3.7:

$$v_1 = i_1 r_1 + e' \quad \text{V} \tag{3.14}$$

where e_1' is the voltage induced in the N_1 turns of the primary winding by the total flux linking that winding. This equation is equivalent to the linear-circuit equation

$$v_1 = i_1 r_1 + L_{11}\frac{di_1}{dt} - L_{12}\frac{di_2}{dt} \tag{3.15}$$

where

$$e_1' = L_{11}\frac{di_1}{dt} - L_{12}\frac{di_2}{dt} \tag{3.16}$$

Applying Faraday's law to Equation 3.13 for λ_1,

$$e_1' = \frac{d\lambda_1}{dt} = \frac{d}{dt}(N_1\phi + N_1\overline{\phi}_{l1}) = e_1 + e_{l1} \quad \text{V} \tag{3.17}$$

where e_{l1} is the voltage induced by the primary leakage flux and e_1 is the voltage induced by the mutual flux ϕ:

$$e_1 = N_1\frac{d\phi}{dt} \qquad e_{l1} = N_1\frac{d\overline{\phi}_{l1}}{dt} \tag{3.18}$$

Figure 3.8 shows the circuit relationships among these voltages.

The MMFs of the primary and secondary are (observing the dot convention)

$$\mathscr{F}_1 = N_1 i_1 \qquad \mathscr{F}_2 = -N_2 i_2 \quad \text{A-turns} \tag{3.19}$$

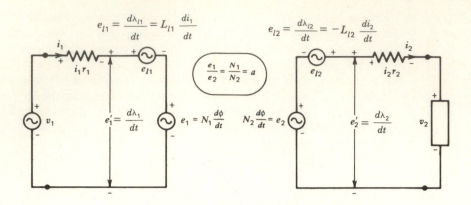

$$v_1 \cong e_1' \cong e_1 = ae_2 \cong ae_2' \cong av_2$$

$$\frac{v_1}{v_2} \cong a \equiv \frac{N_1}{N_2}$$

FIGURE 3.8 Primary and secondary transformer voltages.

The primary and secondary MMFs $\mathscr{F}_1$ and $\mathscr{F}_2$ act simultaneously on the mutual flux path (i.e., the core) to produce the mutual flux ϕ. Let $\mathscr{P}_c$ be the core permeance. Then let

$$R = \mathscr{F}_1 + \mathscr{F}_2 \quad \text{A-turns} \tag{3.20}$$

so that

$$\phi = \mathscr{P}_c R = (\mathscr{F}_1 + \mathscr{F}_2)\mathscr{P}_c$$
$$= N_1\mathscr{P}_c i_1 - N_2\mathscr{P}_c i_2 \quad \text{Wb} \tag{3.21}$$

Since most of the leakage flux paths are outside of the core, the effective flux may be considered to be *proportional* to the MMF of the winding. *This point is important in the development of the Steinmetz model.* Let the constant of proportionality be $\mathscr{P}_{l1}$:

$$\overline{\phi}_{l1} = \mathscr{F}_1 \cdot \mathscr{P}_{l1} = N_1 i_1 \mathscr{P}_{l1} \quad \text{Wb} \tag{3.22}$$

Substituting Equations 3.20 and 3.21 into Equation 3.17:

$$e_1' = \frac{d}{dt}\{N_1[N_1\mathscr{P}_c i_1 - N_2\mathscr{P}_c i_2) + N_1\mathscr{P}_{l1}i_1]\}$$
$$= \frac{d}{dt}[N_1^2(\mathscr{P}_{l1} + \mathscr{P}_c)i_1 - N_1 N_2 \mathscr{P}_c i_2] \tag{3.23}$$

Since this is a *linear* analysis, it will be assumed that all of the quantities inside the brackets of Equation 3.23 are constants, with the exception of the currents. Then this equation may be written

$$e_1' = N_1^2(\mathscr{P}_{l1} + \mathscr{P}_c)\frac{di_1}{dt} - N_1 N_2 \mathscr{P}_c \frac{di_2}{dt} \tag{3.24}$$

By comparing Equation 3.24 with Equation 3.16, it may be seen that

$$L_{11} = N_1^2 \mathcal{P}_{l1} + N_1^2 \mathcal{P}_c \text{ and } L_{12} = N_1 N_2 \mathcal{P}_c \quad \text{H} \tag{3.25}$$

There is a great deal of information packed into Equation 3.23. First, the self- and mutual inductances are related to the numbers of turns and the magnetic permeances of the transformer. Second, the primary flux linkage is related to the inductances, since (see Equation 3.17) the quantity inside the brackets of Equation 3.23 is λ_1:

$$\lambda_1 = L_{11} i_i - L_{12} i_2 \quad \text{Wb-turns} \tag{3.26}$$

Finally, note that the self-inductance of the primary may be divided into two components (See Equation 3.25.), which will be defined as L_{l1}, the primary *leakage inductance*, and L_{m1}, the primary *magnetizing inductance*:

$$L_{l1} = N_1^2 \mathcal{P}_{l1} \qquad L_{m1} = N_1^2 \mathcal{P}_c$$
$$= a L_{12} \quad \text{H} \tag{3.27}$$

In terms of these inductances, the primary flux linkage becomes

$$\lambda_1 = \lambda_{l1} + \lambda_{m1} = L_{l1} i_1 + (L_{m1} i_1 - L_{12} i_2) \quad \text{Wb-turns} \tag{3.28}$$

Now L_{l1} depends on the permeance of the leakage flux path; and as was pointed out in the discussion of Equation 3.21, this component of L_{11} may be considered linear. On the other hand, L_{m1} depends solely on the core permeance, and in iron core transformers it is potentially a *non*linear quantity. Steinmetz saw that the linear part of transformer analysis could be separated from the nonlinear part.

In the secondary circuit of Figure 3.7 the direction of the current i_2 relative to the terminal voltage v_2 is such that the KVL equation for the secondary is

$$v_2 = -i_2 r_2 - L_{22} \frac{di_2}{dt} + L_{12} \frac{di_1}{dt}$$
$$= -i_2 r_2 + e_2' \quad \text{V} \tag{3.29}$$

where e_2' is the voltage induced in the secondary by the total flux linkage of that winding. As in the primary, this voltage will be taken as the sum of two components: e_2, induced by the core flux, and e_{l2}, induced by the effective secondary leakage flux ϕ_{l2}:

$$e_2' = e_2 + e_{l2} = \frac{d}{dt} N_2 (\phi + \overline{\phi}_{l2}) = \frac{d\lambda_2}{dt}$$

$$= \frac{d}{dt} N_2 [(N_1 i_1 - N_2 i_2) \mathcal{P}_c - N_2 i_2 \mathcal{P}_{l2}]$$

$$= -(N_2^2 \mathcal{P}_{l2} + N_2^2 \mathcal{P}_c) \frac{di_2}{dt} + N_1 N_2 \mathcal{P}_c \frac{di_1}{dt} \tag{3.30}$$

$$= -(L_{l2} + L_{m2}) \frac{di_2}{dt} + L_{12} \frac{di_1}{dt}$$

$$= -L_{22} \frac{di_2}{dt} + L_{12} \frac{di_1}{dt}$$

It is seen, then, that

$$L_{22} = N_2^2 \mathscr{P}_{l2} + N_2^2 \mathscr{P}_c = L_{l1} + L_{m2} \tag{3.31}$$

Note that

$$L_{m2} = \frac{L_{12}}{a}$$

Also

$$\lambda_2 = -L_{22}i_2 + L_{12}i_1$$

$$\lambda_{l2} = -L_{l1}i_2 \tag{3.32}$$

$$\lambda_{m2} = -L_{m2}i_2 + L_{12}i_1 \quad \text{Wb-turns}$$

Figure 3.8 clarifies the voltage relationships derived in this analysis of a transformer.

Ratio of Voltages Induced by Core Flux

Another insight we have from Steinmetz is that the ratio of e_1 to e_2 is exactly equal to the turns ratio a (see Equation 3.3 and Figure 3.8). This is shown by the ratios of the expressions for these voltages found in Equations 3.18 and 3.30:

$$e_1 = N_1 \frac{d\phi}{dt} \quad e_2 = N_2 \frac{d\phi}{dt} \quad \text{V} \tag{3.33}$$

$$\frac{e_1}{e_2} = \frac{N_1}{N_2} = \stackrel{\Delta}{=} a \tag{3.34}$$

This linear relationship is true in spite of the fact that the flux inducing these voltages flows in the nonlinear element of the transformer, the core!

Coefficient of Coupling

Each winding in a transformer has a *coupling factor* associated with it, which indicates its effectiveness in affecting the mutual flux. For the primary and secondary windings, these factors (k_1 and k_2, respectively) may be expressed in terms of the leakage and mutual permeances and hence in terms of the self- and magnetizing inductances:

$$k_1 = \frac{\mathscr{P}_c}{\mathscr{P}_c + \mathscr{P}_{l1}} \quad k_2 = \frac{\mathscr{P}_c}{\mathscr{P}_c + \mathscr{P}_{l2}}$$

$$L_{m1} = k_1 L_{11} = N_1^2 \mathscr{P}_c \quad L_{m2} = k_2 L_{22} = N_2^2 \mathscr{P}_c \tag{3.35}$$

This concept allows the mutual inductance to be expressed in terms of the self-inductances of the windings:

$$L_{12} = N_1 N_2 \mathscr{P}_c = \sqrt{L_{m1}L_{m2}} = \sqrt{k_1 k_2} \sqrt{L_{11}L_{22}}$$

$$= k\sqrt{L_{11}L_{22}} \tag{3.36}$$

where k is the *coefficient of coupling* of the transformer and is related to the coupling factors by

$$k = \sqrt{k_1 k_2} \tag{3.37}$$

Transformer Reactances

Since transformers in power systems normally operate at constant frequency (60 Hz in the United States and Canada, 50 Hz almost everywhere else, $16\frac{2}{3}$ Hz on some railroads, 400 Hz on Navy ships, etc.), transformer circuit inductances are usually specified in terms of their corresponding reactances at rated frequency. Recall that inductive reactance is given by

$$X = \omega L = 2\pi f L \quad \Omega \tag{3.38}$$

where ω is the angular frequency in radians per second and X is the reactance in ohms of an inductor of L henries at a frequency of f hertz. Here is a list of inductances and their corresponding reactances and symbols:

Inductance (henries)	Reactance (ohms)
L_{11} Primary self-inductance	X_{11} Primary reactance
L_{l1} Primary leakage inductance	x_1 Primary leakage reactance
L_{m1} Primary magnetizing inductance	X_m Primary magnetizing reactance
L_{22} Secondary self-inductance	X_{22} Secondary reactance
L_{l2} Secondary leakage inductance	x_2 Secondary leakage reactance
L_{m2} Secondary magnetizing reactance	X_{m2} Secondary magnetizing reactance (not used in the Steinmetz model)
L_{12} Mutural inductance	X_{12} Mutual reactance

In addition, r_1 is the resistance of the primary winding in ohms and r_2 is the resistance of the secondary winding.

Voltage relationships in a transformer in terms of reactances and rms phasor voltages and currents are shown in Figure 3.9.

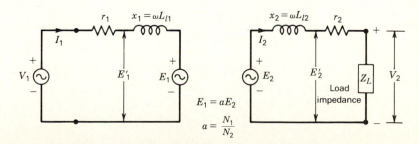

FIGURE 3.9 Transformer voltage relationships in terms of rms phasors.

■ **Example 3.1**

(A 10-kVA, 100-V to 1000-V, 60-Hz transformer has the following inductances, winding resistances and rated-frequency reactances:

$$L_{11} = 53.05 \text{ mH} \qquad X_{11} = 20.0 \ \Omega \qquad L_{22} = 5.305 \text{ H} \qquad X_{22} = 2000 \ \Omega$$

$$L_{l1} = 0.1326 \text{ mH} \qquad x_1 = 0.0500 \ \Omega \qquad L_{l2} = 13.26 \text{ mH} \qquad x_2 = 5.00 \ \Omega$$

$$r_1 = 0.0100 \ \Omega \qquad\qquad\qquad\qquad\qquad\qquad\qquad r_2 = 1.00 \ \Omega$$

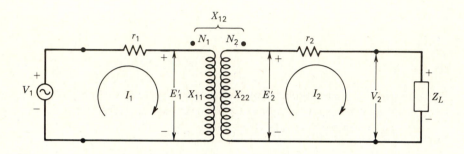

What 60-Hz voltage must be applied to the primary terminals of this transformer to result in rated voltage at the secondary terminals when the transformer is supplying rated kVA to its load at unity power factor? (The inductance and impedance values given are typical of an iron core having these kVA and voltage ratings.) Also calculate the coupling factors and the coefficient of coupling.

Solution. To find the mutual reactance X_{12},

$$X_{m1} = X_{11} - x_1 = 20.0 - 0.05 = 19.95 \ \Omega$$

$$X_{m2} = X_{22} - x_2 = 2000 - 5.0 = 1995 \ \Omega$$

$$X_{12} = \sqrt{X_{m1}X_{m2}} = \sqrt{19.95 \cdot 1995} = 199.5 \ \Omega$$

Let V_2 be the reference phasor. Then

$$V_2 = 100 \ \underline{|0°} \text{ V}$$

I_2 is rated current, in phase with V_2 (unity power factor):

$$I_2 = \frac{\text{rated kVA} \cdot 1000}{\text{rated } V_2} \underline{|0°} = \frac{10{,}000}{1{,}000} \underline{|0°} = 10 \ \underline{|0°} \text{ A}$$

The load impedance is given by $Z_L = V_2/I_2 = 100 + j0\Omega$. The simultaneous KVL equations for the primary and secondary circuits are

$$V_1 = (r_1 + j\omega L_{11})I_1 - j\omega L_{12}I_2$$

$$= (0.01 + j20)I_1 - j199.5 \cdot 10 \tag{1}$$

$$0 = (Z_L + r_2 + j\omega L_{22})I_2 - j\omega L_{12}I_1$$

$$= (101 + j2000) \cdot 10 - j199.5 \cdot I_1 \tag{2}$$

Solving (2) for I_1,

$$I_1 = \frac{1010 + j20,000}{j199.5} = 100.25 - j5.06 \quad \text{A}$$

From (1),

$$V_1 = (0.01 + j20)(100.25 - j5.06) - j199.5 \cdot 10$$

$$= 1.0025 + j2005 - j0.0506 + 101.2 - j1995$$

$$= 102.2 + j9.95 = 102.7 \,\underline{|5.56°} \quad \text{V} \qquad \text{Answer}$$

Note that this method assumes linearity and involves small differences between large numbers, so its accuracy is questionable. The coupling coefficient may be found from Equation 3.24:

$$k_1 = \frac{L_{m1}}{L_{11}} = \frac{\omega L_{m1}}{\omega L_{11}} = \frac{X_m}{X_{11}} = \frac{19.95}{20} = 0.9975$$

$$k_2 = \frac{1995}{2000} = 0.9975$$

$$k = \sqrt{k_1 k_2} = 0.9975 \qquad \text{Answer}$$

3.7 REAL-WORLD TRANSFORMERS ARE NEARLY IDEAL

It has been shown that the resistances of transformer windings can be treated as separate circuit elements by application of Thevenin's theorem. It has also been demonstrated that the self-inductances of the windings may each be separated into two parts: a linear *leakage inductance* and a *magnetizing inductance* which is nonlinear because it depends on the core permeance. Two additional concepts will be useful in understanding the Steinmetz transformer model. The first is that of the *ideal transformer*.

Power system transformers are nearly ideal. They have efficiencies of about 97 percent or better and their internal voltage drops are only about 5 percent. Understanding exactly why transformers are not quite perfect will permit prediction of their performance with great precision.

Voltage Ratio

Transformer nameplates are marked with rated voltages for the primary and secondary (or secondaries). Manufacturers of transformers for use in power systems agree to a standard which requires that the ratios of *rated* voltages must be equal to the ratios of the numbers of turns of the windings. For example, if a two-winding transformer is rated "2400V–240V," the turns ratio must be 10. Because there are internal impedance drops, this means that if the voltage at the low-voltage terminals is exactly 240 V, the voltage at the high-voltage terminals will probably be found to be a little more than 2400 V.

Figure 3.8 may be redrawn to illustrate transformer voltages in the steady state, when voltages and currents are sinusoids and the inductances are treated as reactances. The result is shown in Figure 3.9. It is evident by Kirchhoff's law that

$$E_2 = V_2 + I_2(r_2 + jx_2) \tag{3.39}$$

and

$$V_1 = E_1 + I_1(r_1 + jx_1) \tag{3.40}$$

Since $E_1 = e_{1\text{max}}/\sqrt{2}$, $E_2 = e_{2\text{max}}/\sqrt{2}$, and $e_1 = a \cdot e_2$, then

$$E_1 = a \cdot E_2 = \left(\frac{N_1}{N_2}\right) \cdot E_2 \tag{3.41}$$

The actual voltage ratio of a transformer is V_2/V_2, which is obviously different from the turns ratio. What is more, the actual voltage ratio will depend on the current. However if there were no leakage flux in the transformer, x_1 and x_2 would be zero. If the windings had no resistance, r_1 and r_2 would be zero and the transformer would have no copper loss. Then by Equations 3.39 and 3.40, V_1 would equal E_1 and V_2 would equal E_2, and the voltage ratio would be the same as the turns ratio. This would be an ideal situation. As a matter of fact, in a practical transformer the voltage ratio will differ from the turns ratio by only a few percent unless the load has a greatly lagging power factor.

Current Ratio

The purpose of power transformers is to increase transmission voltages, so that a given amount of kVA can be transmitted at lower current. It has been pointed out (Equation 3.2) that ideally the current ratio of a transformer would be the inverse of the voltage ratio.

To understand the actual current transformation process, consider the total MMF applied to the core, as stated in Equation 3.13 and restated here:

$$R = \mathcal{F}_1 + \mathcal{F}_2 \quad \text{A-turns} \tag{3.42}$$

where $\mathcal{F}_1 = N_1 i_1$ and $\mathcal{F}_2 = -N_2 i_2$. Then

$$R = N_1 i_1 - N_2 i_2 \tag{3.43}$$

Solving for the input current to the primary,

$$i_1 = \frac{R}{N_1} + \frac{N_2}{N_1} i_2 \quad \text{A} \tag{3.44}$$

The term R/N_1 has the dimension of amperes since R is an ampere-turns. This component of the primary current is called the *exciting current* and will be assigned the symbol i_{ex}. It is the current required to produce the necessary mutual flux in the core. In doing this the core losses must also be overcome. This means that the exciting current must have an inductive component, lagging e_1 by $90°$, to produce the flux; and, in addition, a power component, in phase with e_1, to supply the core losses. The exciting current is the only current to flow in the primary winding when the secondary terminals are open (i.e., when $i_2 = 0$). Equation 3.43 may be restated as

$$i_1 = i_{ex} + \frac{i_2}{a} \tag{3.45}$$

or, in terms of phasors,

$$I_1 = I_{ex} = \frac{I_2}{a} \tag{3.46}$$

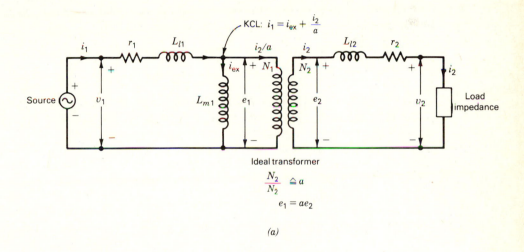

(a)

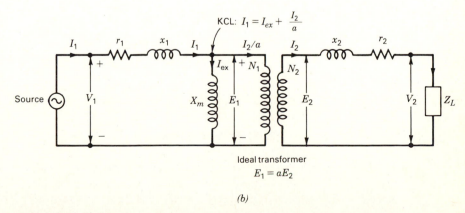

(b)

FIGURE 3.10 Circuit model of a transformer, assuming magnetic linearity. (a) Schematic diagram. (b) In terms of rms phasor quantitites.

At full load; that is, when I_2 is at its rated value, the quantity I_2/a is much larger than I_{ex}, and

$$I_1 \approx \frac{I_2}{a} \tag{3.47}$$

In an ideal transformer, I_{ex} would be zero, implying a core with infinite permeability and zero core loss. The primary current would then be exactly equal to the inverse of the turns ratio, as in Equation 3.2. In fact, for loads greater than about half the rated load, a practical transformer is very close to the ideal as far as the current ratio is concerned.

A Linear Circuit Model

It is now possible to draw a complete *linear* circuit model of a two-winding transformer as in Figure 3.10a. The corresponding sinusoidal, steady-state model in terms of rms phasor voltages and currents is shown in Figure 3.10b. This model would be useful in analyzing an audio transformer, so designed as to keep the core unsaturated, in the low- and mid-frequency ranges. At high frequencies, the winding capacitances would have to be considered.

■ Example 3.2

A certain power transformer is connected between a transmission line and a load. The secondary terminal voltage is $707 \sin 377t$ and the load current is $i_2 = 141.4 \sin(377t - 30°)$. The primary winding has 300 turns and a resistance of 2.00 Ω. The secondary winding has 30 turns and a resistance of 0.0200 Ω. The leakage inductance of the primary is 0.0300 H while that of the secondary is $3.00 \cdot 10^{-4}$ H. The exciting current of this transformer is $0.707 \sin(377t - 80°)$. Find the turns ratio a, the primary and secondary induced rms voltages E_1 and E_2, and the primary current and terminal voltages. Compare the actual voltage and current ratios with the turns ratio.

Solution. The rms phasor notation will be used to avoid involvement in trigonometric identities. Then the secondary voltage and current are

$$V_2 = \frac{707}{\sqrt{2}} \underline{|0°} = 500 \underline{|0°} = 500 + j0 \text{ V}$$

$$I_2 = \frac{141.4}{\sqrt{2}} \underline{|-30°} = 100 \underline{|-30°} = 86.6 - j50 \quad \text{A}$$

The secondary $I_2 r_2$ drop is

$$0.0200 (86.6 - j50) = 1.732 - j1.00 \quad \text{V}$$

It has been noted that positive secondary current sets up a negative magnetomotive force. The direction of the secondary leakage flux is in opposition to the core flux. Therefore by Equation 3.32, the secondary leakage flux linkage is

$$\lambda_{12} = 3.00 \cdot 10^{-4} (-i_2) = -0.0424 \sin(377t - 30°) \text{ Wb-turns}$$

The secondary leakage flux voltage is

$$e_{l2} = \frac{d\lambda_{l2}}{dt} = -0.0424 \cdot 377 \cos(377t - 30°)$$

$$= -15.98 \cos(377t - 30°) = -15.98 \sin(377t - 30° + 90°)$$

Then

$$E_{l2} = \frac{-15.98}{\sqrt{2}} \underline{|90° - 30°} = -j11.31 \underline{|-30°} = -11.31 \underline{|+60°}$$

Pause to see what has happened. Note that

$$-E_{l2} = 3.00 \cdot 10^{-4} \cdot 377 \cdot jI_2 = j\omega 3.00 \cdot 10^{-4} I_2$$

where $\omega = 377$ rad/s $(f = ?)$ and $3.00 \cdot 10^{-4}$ is the secondary leakage inductance L_{l2}. Then $\omega L_{l2} = x_2$ is the secondary *leakage reactance*. On this basis

$$x_2 = \omega L_{l2} = 377 \cdot 3.00 \cdot 10^{-4} = 0.1131 \quad \Omega$$

$$-E_{l2} = I_2 j x_2 = (100 \underline{|-30°}) \cdot 0.1131 \underline{|90°}$$

$$= 11.31 \underline{|+60°} = 5.655 + j9.79 \quad V$$

Applying Kirchhoff's voltage law to the secondary (Figure 3.9),

$$E_2 = V_2 + I_2 r_2 - E_{l2}$$

$$\equiv V_2 + I_2(r_2 + jx_2)$$

$$= 500 + j0 + (1.732 - j1.00) + (5.655 + j9.79)$$

$$= 507.4 + j8.79 = 507.46 \underline{|0.992°} \quad V$$

Now

$$a = \frac{N_1}{N_2} = \frac{300}{30} = 10$$

By Equation 3.34,

$$E_1 = aE_2 = 5074.6 \underline{|0.992°}$$

$$= 5074.6 + j87.9 \quad V$$

By Equation 3.46,

$$I_1 = \frac{I_2}{a} + I_{ex}$$

$$I_{ex} = \frac{0.707}{\sqrt{2}} \underline{|-80°} = 0.5 \underline{|-80°}$$

$$= 0.0868 - j0.492 \quad \text{A}$$

$$I_1 = \frac{86.6 - j50}{10} + 0.0868 - j0.492$$

$$= 8.75 - j5.49 = 10.33 \,\underline{|-32.1°} \quad \text{A}$$

$$\text{Actual current ratio} = \frac{I_2}{I_1} = \frac{100}{10.33} = 9.68$$

$$\text{Compare with } a = 10$$

By Kirchhoff,

$$V_1 = E_1 + I_1 r_2 + E_{l1}$$

$$E_{l1} = j\omega L_{l1} I_1$$

$$= j \cdot 377 \cdot 0.0300 \cdot 10.33 \,\underline{|-32.1°}$$

$$= 116.83 \,\underline{|90° - 32.1°}$$

$$= 62.1 + j99.0 \quad \text{V}$$

$$I_1 r_2 = (8.75 - j5.49) \cdot 2.00$$

$$= 17.5 - j11.0 \quad \text{V}$$

$$V_1 = (5074.6 + j87.9) + (17.5 - j11.0) + (62.1 + j99.0)$$

$$= 5154.2 + j175.9 = 5157.2 \,\underline{|1.95°} \quad \text{V}$$

The actual voltage ratio is

$$\frac{V_1}{V_2} = \frac{5157.2}{500} = 10.31 \qquad \blacksquare$$

3.8 THE IDEAL TRANSFORMER

Power system transformers are very nearly ideal. Their efficiencies are about 97 percent and better, and their internal voltage drops are only about 5 percent. Many engineering calculations can be carried out with satisfactory accuracy on the assumption that transformers are "ideal." An ideal transformer would have windings with zero resistance and a lossless, infinite-permeability core. The efficiency would be 100 percent. Infinite permeability would result in zero exciting current and no leakage flux. Then

$$\lambda_1 \equiv N_1 \phi \qquad \lambda_2 \equiv N_2 \phi$$

and

$$e_1 = e_1' \qquad e_2 = e_2'$$

Since the resistances of the windings would be zero there would be no voltage drops between the terminal voltages and the induced voltages:

$$v_1 = e_1 = e_1' \qquad v_2 = e_2 = e_2',$$

then

$$\frac{v_1}{v_2} = a \tag{3.48}$$

Since $i_{ex} = 0$,

$$i_1 = \frac{i_2}{a} \tag{3.49}$$

Figure 3.11 shows an ideal transformer connected between a generator and a load. Let $v_2 = \sqrt{2}|V_2| \sin \omega t$, where $|V_2|$ is the rms secondary terminal voltage. Then $i_2 = \sqrt{2}(|V_2|/|Z|) \sin(\omega t - \theta_\phi)$, where θ_ϕ is the impedance angle of the load. In phasor terms,

$$I_2 = \frac{V_2}{Z} = |I_2| \underline{/-\theta_\phi} \tag{3.50}$$

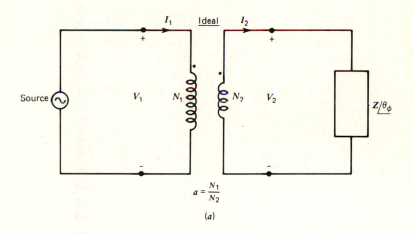

$$a = \frac{N_1}{N_2}$$

(a)

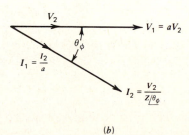

(b)

FIGURE 3.11 Load connected to source by an ideal transformer. (a) Schematic diagram. (b) Phasor diagram.

where

$$|I_2| = \left| \frac{V_2}{Z} \right| \quad \text{and} \quad Z = |Z| \underline{|\theta_\phi}$$

From Equation 3.48,

$$v_1 = \sqrt{2} \, a \, V_2 \sin \omega t$$

or in rms,

$$\boxed{V_1 = aV_2} \tag{3.51}$$

From Equation 3.49,

$$i_1 = \sqrt{2} \frac{|I_2|}{a} \sin(\omega t - \theta)$$

$$= \sqrt{2}|I_1| \sin(\omega t - \theta)$$

In phasor terms,

$$\boxed{I_1 = \frac{I_2}{a}} \tag{3.52}$$

The impedance seen by the generator is given by

$$Z_{\text{in}} = \frac{V_1}{I_1} = \frac{aV_2}{I_2/a} = a^2 \frac{V_2}{I_2}$$

or, from Equation 3.50,

$$\boxed{Z_{\text{in}} = a^2 Z} \tag{3.53}$$

Equations 3.51, 3.52, and 3.53 are the basic relations describing the circuit behavior of an ideal transformer.

■ **Example 3.3**

A impedance of $2 \underline{|36.9°}$ is connected to the secondary terminals of a 2400-240-V transformer. The primary is connected to a 2200-V line. Find the secondary current, the primary current, the impedance as seen by the line, the output power and volt-amperes and the input power and volt-amperes. Assume the transformer to be ideal.

Solution

> **1.** It is standard practice to make the ratio of *rated* terminal voltages equal to the actual turns ratio. Then

$$a = \frac{2400}{240} = 10$$

2. By Equation 3.51,

$$|V_2| = \frac{|V_1|}{a} = \frac{2200}{10} = 220 \text{ V}$$

3. Let $V_2 = 220 \underline{|0°}$ V. Then

$$I_2 = \frac{220 \underline{|0°}}{2 \underline{|36.9°}} = 100 \underline{|-36.9°} \text{ A}$$

4. By Equation 3.57,

$$I_1 = \frac{I_2}{a} = 11.0 \underline{|-36.9°} \text{ A (Note that } a \text{ is a scalar.)}$$

5.

$$Z_{in} = \frac{V_1}{I_1} = \frac{aV_2}{I_1} = \frac{2200 \underline{|0}}{11 \underline{|-36.9°}} = 200 \underline{|36.9°} \text{ }\Omega$$

or by Equation 3.53,

$$Z_{in} = a^2Z = 10^2 \cdot 2 \underline{|36.9°}$$

$$Z_{in} = 200 \underline{|36.9°} \text{ }\Omega$$

6. $|S_2| = |V_2| |I_2| = 220 \text{ V} \cdot 110 \text{ A} = 24.2 \quad \text{kVA}$

$P_2 = |S_2| \cos \theta_2 = 24.2 \cdot 0.8 = 19.36 \quad \text{kW}$

$|S_1| = |V_2| |I_1| = 2200 \cdot 11 = 24.2 \quad \text{kVA}$

$P_1 = |S_1| \cos \theta_1 = 24.2 \cos 36.9° = 19.36 \quad \text{kW}$

■ **Example 3.4**

A transformer is to be used to match a 4-Ω loudspeaker in a paging system to a 500 Ω audio line. What should be the turns ratio of the transformer, and what will be the voltage levels at the primary and secondary terminals when 10 W of audio power is delivered to the speaker? Assume the speaker to act as a resistive load (a poor assumption) and that the transformer is ideal (also a rash assumption for an audio transformer).

Solution

1. $Z_{in} = a^2Z$
   $500 = a^2 4$
   $a = \sqrt{500/4} = 11.18$
2. $P_2 = 10 \text{ W} = V_2^2/4 \text{ }\Omega$
   $V_2 = \sqrt{40} = 6.32 \text{ V}$
   $V_1 = aV_2 = 70.7 \text{ V}$

■

3.9 CORE EXCITATION

The final concept involved in the Steinmetz transformer model is a scheme for handling the nonlinearity of the core and the core losses. In Example 3.2, almost everything that was done was legitimate, and it was done on a *linear* basis. The primary and secondary winding resistances are linear elements, as are the two leakage inductances— very, very nearly. And the ratio of E_1/E_2 is exactly equal to the turns ratio a. The only illegitimate thing was the assumption that the exciting current i_{ex} is sinusoidal. However an interesting insight into the role of the primary magnetizing inductance L_{m1} will result from considering core permeance to be constant, thus assuming that the magnetizing inductances are linear.

From Equation 3.18,

$$e_1 = N_1 \frac{d\phi}{dt} \tag{3.54}$$

Now from Equations 3.21, 3.44, and 3.45, $\phi = \mathcal{P}_c R$ and $R = N_1 e_{ex}$. Then

$$e_1 = N_1 \frac{d\mathcal{P}_c R}{dt} = N_1 \frac{d}{dt}(\mathcal{P}_c N_1 i_{ex}) = N_1^2 \mathcal{P}_c \frac{di_{ex}}{dt} \tag{3.55}$$

and from Equation 3.27, the quantity $N_1^2 \mathcal{P}_c$ is the primary magnetizing inductance L_{m1}, so Equation 3.55 may be written

$$e_1 = L_{m1} \frac{di_{ex}}{dt} \tag{3.56}$$

The implication of this equation is that the exciting current, and hence the core flux results from the application if the voltage e_1 to the primary magnetic L_{m1}.

The exciting current i_{ex} is easy to measure. By Equation 3.44, th is equal to the primary current when the secondary is open-circuited. With rated sinusoidal voltage applied to one winding of a power system transformer, the exciting current has the horrible waveform shown in Figure 3.12. Does this mean that the core

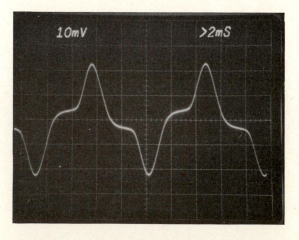

FIGURE 3.12 Osciallogram of a transformer's exciting current.

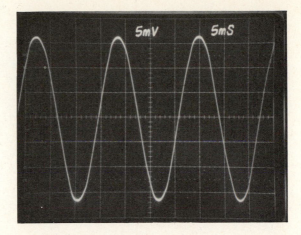

FIGURE 3.13 Flux waveform for the transformer of Figure 3.12.

flux is nonsinusoidal? Not at all! When core flux was measured for the same laboratory conditions as those for Figure 3.12, the flux waveform was that shown in Figure 3.13—not a bad sine wave. How can this be?

Well, Faraday's law is a most fundamental law. It not only tells how much voltage is induced by a changing flux but also relates the magnitude of the flux to the applied voltage. It has been pointed out that the induced primary voltage e_1 is nearly equal to v_1, the source voltage applied to the terminals. Then if v_1 is sinusoidal, e_1 is bound to be nearly sinusoidal also. In this case the flux is obtained as follows:

$$\text{Faraday's law: } e_1 = N_1 \frac{d\phi}{dt}$$

$$\phi = \frac{1}{N_1} \int e_1 dt$$

Let

$$e_1 = \sqrt{2} E_1 \sin(\omega t + \alpha) \tag{3.57}$$

Then

$$\phi = -\frac{\sqrt{2} E_1}{N_1 \omega} \cos(\omega t + \alpha) + \phi_c \tag{3.58}$$

The constant of integration, ϕ_c, is a transient flux that quickly decays as the result of excess eddy current and primary $i_1^2 r_1$ loss, factors that are too complex to be considered at this point. After a few cycles, the flux becomes

$$\phi = -\frac{\sqrt{2}}{\omega} \frac{E_1}{N_1} \cos(\omega t + \alpha) \tag{3.59}$$

and

$$\phi_{max} = \frac{\sqrt{2}}{\omega} \frac{E_1}{N_1} \cong \frac{1}{\sqrt{2}\pi f} \frac{V_1}{N_1} \tag{3.60}$$

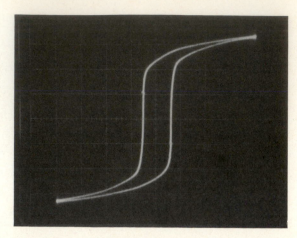

FIGURE 3.14 Plot of flux as a function of exciting current for the transformer of Figure 3.12.

In other words, *the peak flux in the core under steady-state conditions is proportional to the volts per turn of the winding and is inversely proportional to the frequency.* Note also that if the voltage is sinusoidal, so is the flux, but that the flux lags the voltage by 90°.

The startling thing about the above analysis is that no mention was made of the primary current or MMF. As long as the $i_1 r_1$ and $i_1 x_1$ drops are small compared to v_1, the flux depends on the applied *voltage,*[2] not the primary current! The exciting current is the *dependent* variable. When the flux waveform of Figure 3.12 is applied to the vertical axis of an oscilloscope and the exciting-current (i_{ex}) waveform of Figure 3.11 is applied to the horizontal axis, the typical hysteresis loop of Figure 3.14 appears on the screen. Remember that the MMF acting on the core is $N_1 i_{ex}$. So the oscillogram of Figure 3.13 is a scaled plot of ϕ versus R, the sum of $\mathcal{F}_1 + \mathcal{F}_2$. It is the ac hysteresis loop for the ferromagnetic core. An ac hysteresis loop includes the effects of eddy currents in the core, as well as hysteresis. This curve shows why the exciting current *must* be nonsinusoidal when the core flux is sinusoidal. The exciting current must vary in such a way as to overcome the effects of saturation, hysteresis, and eddy currents. Obviously, these phenomena cannot be modeled by the simple, linear inductor L_{m1} as in Figure 3.10a.

The Steinmetz model approaches the problem of representing core excitation by first dividing it into two parts: magnetization and core losses. The exciting current is, in concept, separated into two components: a *magnetizing current* and a *core loss current*. The process is shown in Figure 3.15. When the sinusoidal flux waveform is plotted against the hysteresis loop, a curve of the exciting current i_{ex} is obtained. Then a magnetization curve is plotted within the hysteresis loop on the same grid to represent the characteristic of the core without losses. This is shown as a broken-line curve in Figure 3.15. It was pointed out in Chapter 1 that these losses are proportional to the area of the hysteresis loop. The magnetization curve, having zero area, is thus the hypothetical characteristic of a core which saturates but has no losses. The sinusoidal core flux wave, when projected on this curve produces the dashed current waveform labeled i_ϕ. This is the magnetizing current. It is nonsinusoidal and contains a large third-harmonic component. Note that the peaks and zero crossings of i_ϕ lag the induced

---

[2]The same sort of thing was implied in the analysis of synchronous machines. The flux depends on E_ϕ, which is not much different from V_1 if $I_1 r_1$ and $I_1 x_1$ are small.

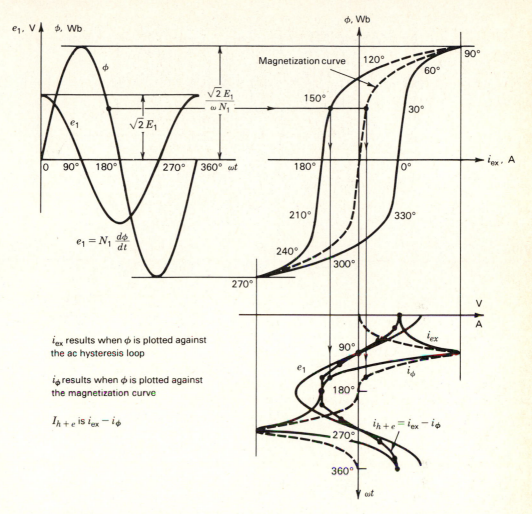

FIGURE 3.15 Derivation of core-loss and magnetizing components of the exciting current under sinusoidal flux conditions.

voltage e_1 by 90°. When the magnetizing current is subtracted, point by point, from the exciting current, an odd, flat-topped, nonsinusoidal current waveform results, labeled i_{h+e} in the figure. The zero crossings of this current are in phase with e_1. It is the current due to the hysteresis and eddy current losses in the core, and if the hysteresis loop had zero area, this current would be zero. Hence, according to this concept, the exciting current is visualized as the sum of two components: a magnetizing current that lags the induced voltage e_1 by 90°, and a core loss current i_{h+e} in phase with e_1.

As the second step, the Steinmetz model replaces these nonsinusoidal currents with sinusoidal currents having equivalent rms values, such that

$$I_{ex} = I_\phi + I_{h+e} \quad \text{A} \tag{3.61}$$

where each current is an rms phasor quantity. These currents may be measured by the voltmeter-ammeter-wattmeter method when the secondary current I_2 is zero. (See Sections 3.18 and 3.19.) As is shown in Figure 3.16a, the transformer core loss is

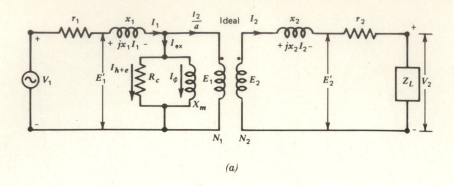

(a)

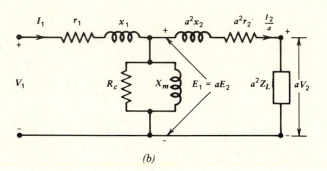

(b)

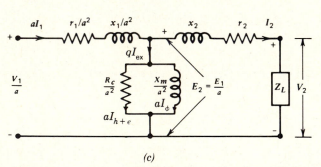

(c)

FIGURE 3.16 Development of the Steinmetz transformer model. (a) Model including an ideal transformer. (b) "Exact" model referred to the primary. (c) "Exact" model referred to the secondary.

represented by the current I_{h+e} flowing through a resistor R_c connected across E_1. The resistance of R_c is given by

$$R_c = \frac{P_{h+e}}{I_{h+e}^2} \qquad (3.62)$$

The magnetizing current, lagging E_1 by 90°, is represented by an inductive reactance X_m across E_1, having a value

$$X_m = \frac{\text{core VARS}}{I_\phi^2} \approx \omega L_{m1} \tag{3.63}$$

If the core were linear and lossless, X_m would be given correctly by ωL_{m1}.

FOR FURTHER STUDY

Current Inrush

The constant of integration ϕ_c in Equation 3.58 results in a sudden inrush of primary current when a transformer is first connected to the line, unless the connection is made at the instant of time in the line voltage cycle at which the voltage is maximum. Assume that, when the transformer was last disconnected from the line, a small residual flux ϕ_R remained in the core. (The residual flux density for transformer iron is always quite small.) The constant of integration ϕ_c may be found from this initial condition at $t = 0$:

$$\phi_R = -\frac{\sqrt{2}E_1}{N_1\omega} \cos \alpha + \phi_c$$

$$= -\phi_{\text{max}} \cos \alpha + \phi_c$$

Then

$$\phi_c = \phi_R + \phi_{\text{max}} \cos \alpha$$

From Equation 3.57, note that α is the angle of the voltage sinusoid at $t = 0$. For example, if $\alpha = 0$, then by Equation 3.57, $e_1 = 0$ at $t = 0$, and de_1/dt is positive. This is the worst case for positive ϕ_R, because ϕ_c is largest:

$$\phi_c = \phi_R + \phi_{\text{max}} \cdot 1$$

Under these circumstances, the core flux, by Equation 3.58, becomes

$$\phi = -\phi_{\text{max}} \cos(\omega t + 0) + \phi_R + \phi_{\text{max}}$$

The maximum value attained by the flux is then $2\phi_{\text{max}} + \phi_R$, or over twice the normal flux! The core is driven far into saturation, with the result that the exciting current i_{ex} has a very high peak value. In fact, the rms *exciting* current may greatly exceed the rated primary current of the transformer. However, this "inrush current" transient lasts only a few cycles.

There are at least two reasons for the rapid decay of the inrush current. First, the high current results in large voltage drops in r_1 and x_1, the primary-winding resistance and leakage reactance, so that e_1 is reduced below its normal value during the transient. Second, ϕ_c represents a dc flux component—a magnetization of the core. The core is gradually demagnetized as the core is cycled through successive hysteresis loops, and eventually the magnetizing current returns to its normal value, a few percent of rated primary current.

To obtain no transient inrush current, ϕ_c should be zero:

$$\phi_c = \phi_R + \phi_{\text{max}} \cos \alpha = 0$$

or

$$\cos \alpha = - \frac{\phi_R}{\phi_{\max}}$$

Since ϕ_R is usually quite small, $\cos \alpha \cong 0$, and $\alpha \cong n\pi/2$. In other words, if the transformer is connected to the line near a positive or negative voltage maximum, inrush will be minimized.

It is usually impractical to attempt to connect a transformer at a predetermined time in the voltage cycle. The value of α is thus rather random, and a transformer may or may not experience a large inrush current when first connected. A heavy inrush is accompanied by a burst of 120 Hz hum, arising largely from magnetostriction of the core.

3.10 THE STEINMETZ "EXACT" CIRCUIT MODEL OF A TRANSFORMER

Now for the grand anticlimax! All that remains to be done to obtain the Steinmetz transformer model from the one thus far derived (Fig. 3.16a) is to use the ideal transformer relationships to reflect impedances from one side to the other and then get rid of the ideal transformer. In Figure 3.16a, the impedance connected to the secondary of the ideal transformer is

$$jx_2 + r_2 + Z_L$$

By Equation 3.53, this impedance as seen looking into the primary terminals of the ideal transformer appears to be

$$ja^2x_2 + a^2 r_2 + a^2Z_L$$

as shown in Figure 3.16b. This version of the Steinmetz model is called the *exact equivalent circuit referred to the primary*.

Alternatively, the impedance elements on the primary side of the ideal transformer in Figure 3.16a may be referred to the secondary as in Figure 3.16c. The result is called the *exact equivalent circuit referred to the secondary*. It is instructive to check the validity of the KVL and KCL equations implied by this version of the Steinmetz model. From Equation 3.46,

$$I_1 = I_{ex} + \frac{I_2}{a}$$

Dividing through by a and rearranging,

$$I_2 = aI_1 - I_{ex}$$

This is a correct KCL equation for the top center node. This expression may be rewritten

$$I_2 = aI_1 - aI_{h+e} - aI_\phi \qquad (3.64)$$

Similarly, the primary KVL equation is, from Equation 3.40,

$$E_1 = V_1 - I_1 (r_1 + jx_1)$$

which, when divided through by the turns ratio a, gives

$$\frac{E_1}{a} = \frac{V_1}{a} - \frac{I_1}{a}(r_1 + jx_1)$$

which is equivalent to

$$E_2 = \frac{V_1}{a} - aI_1\left(\frac{r_1}{a^2} + \frac{jx_1}{a^2}\right) \tag{3.65}$$

Observe that this is a correct KVL equation for the left branch of the exact equivalent circuit referred to the secondary. The resistive element of the shunt branch is given by

$$\frac{E_1/a}{aI_{h+e}} = \frac{E_1}{a^2 I_{h+e}} = \frac{R_c}{a^2}$$

and so on. The student may wish to check the rest of both versions of the model. Figure 3.17 is a complete phasor diagram of a two-winding transformer, based on the Steinmetz concept.

It is well to pause at this point to consider what has been done in the development of a circuit model of an iron core transformer. Step by step, the characteristics that make a real transformer less than ideal have been reassigned roles as external circuit elements. Winding resistances are now replaced by r_1 and r_2. The leakage flux of each winding has been replaced by an external reactance x_1 or x_2. The core losses have been replaced by an external shunt resistor R_c, and the MMF required to produce the flux determined by Equation 3.59 has been represented by X_m. What is left is a transformer with no copper loss and no core loss; that is, its efficiency is 100 percent. Also, the remaining transformer has no leakage flux and requires zero current to magnetize its core, implying that the permeance of its core is infinite. In short, the remaining transformer is ideal. The real transformer has been replaced by an ideal transformer and an arrangement of external circuit elements.

The student will object that this model rides roughshod over the inherent nonlinearity of the transformer. However this is not the case. The most frequent engineering problems involving transformers involve the magnitudes of the internal voltage drops

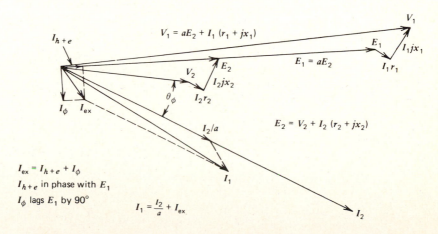

FIGURE 3.17 Rms phasor diagram for the "exact" circuit model of a transformer.

and the transformer efficiency. The winding-resistance and leakage-reactance voltage drops are accurately represented by the linear elements r_1, r_2, x_1, and x_2. The winding copper losses $I_1^2 r_1$ and $I_2^2 r_2$ are also well represented. By making I_{h+e} represent the rms value of the nonsinusoidal i_{h+e}, the core loss is also accurately modeled.

In analyzing the nonlinearity of the core, one must know the flux waveform and then go to the characteristic curves of the core material. The flux waveform is quite accurately predicted by the model as a result of the fact that the winding resistances and leakage reactances are very small. By Kirchhoff's voltage law,

$$E_1 = V_1 - I_1(r_1 + jx_1)$$

$$= V_1 - (I_{h+e} + I_\phi + \frac{N_2}{N_1} I_2)(r_1 + jx_1) \tag{3.66}$$

Although I_{j+e} and I_ϕ are considered sinusoidal when in fact they are not, the effect on E_1 is negligible in all but very special transformers. As a practical matter, e_1 has the same waveform as v_1, with a very slight reduction in magnitude and shift in phase. This permits calculation of e_1 from the model on a linear basis. The core flux may then be found by Equation 3.59.

Example 3.5

A 16.67-kVA, 7200–120-V, 60-Hz, two-winding transformer has the following circuit model impedances with the 7200-V winding as the primary:

$$R_c = 311,000 \ \Omega \qquad X_m = 58,000 \ \Omega$$

$$R_1 = 18.7 \ \Omega \qquad x_1 = 77.8 \ \Omega \qquad r_2 = 0.00519 \ \Omega \qquad x_2 = 0.0216 \ \Omega$$

The voltage applied to the primary terminals is adjusted so that the secondary terminal voltage is 120.0 V when the load on the secondary is 16.67 kVA at unity power factor. Find the primary current and the required primary terminal voltage. Find the core and copper losses and the transformer's efficiency under these conditions.

Solution. Let V_2 be the reference phasor. Then $V_2 = 120.0 \ \underline{/0°}$. At unity power factor,

$$I_2 = \frac{16.67 \cdot 10^3}{120.0} \ \underline{/0°} = 138.9 \ \underline{/0°} \ \text{A}$$

Secondary induced voltage $E_2 = V_2 + I_2(r_2 + jx_2)$

$$= 120.0 + 138.9 \ \underline{/0°}(0.00519 + j0.0216)$$

$$= 120.7 + j3.00 = 120.74 \ \underline{/1.424°} \ \text{V}$$

Primary induced voltage $E_1 = aE_2$, where $a = \frac{7200}{120} = 60$

$$E_1 = 7244 \ \underline{/1.424°} \ \text{V}$$

Core loss current $I_{h+e} = \frac{E_1}{R_c} = \frac{7244 \ \underline{/1.424°}}{311,000} = 0.023 \ \underline{/1.424°}$

$$= 0.023 + j0.0006 \ \text{A}$$

$$\text{Core loss } P_{h+e} = |I_{h+e}|^2 R_c = 169 \quad \text{W} \qquad \text{Answer}$$

$$\text{Magnetizing current } I_\phi = \frac{E_1}{jX_m} = \frac{7244 \,\underline{|1.424°}}{54{,}800 \,\underline{|90°}} = 0.132 \,\underline{|-88.58°}$$

$$= 0.0003 - j0.132 \text{ A}$$

$$\text{Exciting current } I_{ex} = I_{h+e} + I_\phi \approx 0.023 - j0.131 \quad \text{A}$$

$$\text{Primary current } I_1 = I_{ex} + \frac{I_2}{a} = (0.023 - j0.131) + \frac{138.9}{60} \,\underline{|0°}$$

$$= 2.338 - j0.131 = 2.342 \,\underline{|-3.21°} \text{ A} \qquad \text{Answer}$$

(If an ideal transformer had been assumed, the primary current would have been 2.315 $\underline{|0°}$, a magnitude of 1.2 percent and an angle error of 3.21°).

Required primary terminal voltage

$$V_1 = E_1 + I_1 (r_1 + jx_1)$$

$$V_1 = 7244 \,\underline{|1.424°} + 2.342 \,\underline{|-3.21°}(18.7 + j77.8)$$

$$= 7242 + j180.0 + 2.342 \,\underline{|-3.21°}(80.0 \,\underline{|76.48°})$$

$$= 7303 \,\underline{|2.82°} \quad \text{V} \qquad \text{Answer}$$

(If an ideal transformer had been assumed, the required primary voltage would have been 7200 $\underline{|0°}$, a magnitude error of -1.3 percent and an angle error of $-2.82°$.)

$$\text{Core loss } P_{h+e} = I_{h+e}^2 R_c = (0.023)^2 \cdot 311{,}000 = 169 \quad \text{W} \qquad \text{Answer}$$

$$\text{Copper loss } P_{Cu} = I_1^2 r_1 + I_2^2 r_2 = 102.6 + 100.1 = 202.7 \quad \text{W} \qquad \text{Answer}$$

$$\text{Efficiency } \eta = \frac{\text{output watts}}{\text{output + losses}} = \frac{16{,}670}{16{,}670 + 169 + 203} = 97.8\% \text{ Answer} \blacksquare$$

3.11 THE STEINMETZ APPROXIMATE TRANSFORMER MODEL

The so-called "exact" circuit model of the iron core transformer just derived is more accurate than necessary for most engineering calculations. In addition, it suffers from the fact that the leakage reactances x_1 and x_2 cannot be measured independently. (An experienced designer can estimate them from the geometry of the core, windings, and tank of the transformer, sometimes using complex field-mapping computer programs to do so.) For the vast majority of problems involved in the application of transformers, the simpler "approximate" Steinmetz model is employed. The two versions of this simplified model are shown in Figure 3.18. The rationale for these approximate equivalent circuits is that the voltage drop in the impedance $r_1 + jx_1$ is small, even at full load. Also I_{ex} is so small that the drop it produces in $r_1 + jx_1$ is negligible in a practical sense. Therefore it matters little whether the shunt branch of $R_c \parallel X_m$ is connected before the primary series impedance or after it. The core loss and magnetizing currents are not greatly affected. Connecting it right at the input terminals has the great ad-

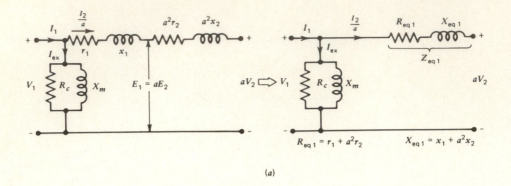

(a)

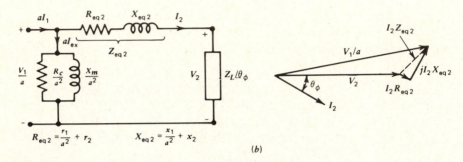

(b)

FIGURE 3.18 Approximate circuit models of an iron core transformer. (*a*) Approximate transformer equivalent circuit referred to the primary. (*b*) Approximate transformer equivalent circuit referred to the secondary.

vantage of permitting the two series impedances to be combined into one complex impedance.

The value of this *equivalent impedance* of a particular transformer depends, of course, on whether the model used is referred to the primary or secondary. If referred to the primary,

$$Z_{eq\ 1} = (r_1 + a^2 r_2) + j(x_1 + a^2 x_2)$$

$$\equiv R_{eq\ 1} + jX_{eq\ 1} \qquad\qquad (3.67)$$

where

$$R_{eq\ 1} \overset{\triangle}{=} r_1 + a^2 r_2$$

$$X_{eq\ 1} \overset{\triangle}{=} x_1 + a^2 x_2$$

If referred to the secondary,

$$Z_{eq\ 2} = \left(\frac{r_1}{a^2} + r_2\right) + j\left(\frac{x_1}{a^2} + x_2\right)$$

$$\equiv R_{eq\ 2} + jX_{eq\ 2} \qquad\qquad (3.68)$$

where

$$R_{eq\ 2} \overset{\triangle}{=} \frac{r_1}{a^2} + r_2$$

$$X_{eq\ 2} \overset{\triangle}{=} \frac{x_1}{a^2} + x_2$$

The student should satisfy him or herself that

$$\frac{Z_{eq\ 1}}{Z_{eq\ 2}} = \frac{R_{eq\ 1}}{R_{eq\ 2}} = \frac{X_{eq\ 1}}{X_{eq\ 2}} = a^2 \qquad (3.69)$$

Example 3.6

Find the primary terminal voltage in Example 3.1 by two additional methods: (a) Use the "equivalent tee" of basic circuit theory. (b) Use the Steinmetz approximate equivalent circuit referred to the primary. Compare the results obtained and discuss.

Solution

(a) The following "tee" circuit is equivalent to a *linear* transformer:

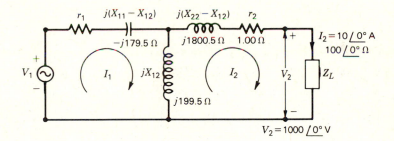

This equivalent circuit differs from the Steinmetz model in that the primary and secondary quantities are all actual values, whereas in the Steinmetz model one set of terminal quantities is "referred" to the other winding. In Example 3.1, the secondary current I_2 was found to be 10 $\underline{/0°}$ A. Let V_{12} be the voltage across X_{12}. Applying Kirchhoff's voltage law to the right-hand mesh,

$$V_{12} = V_2 + I_2[r_2 + j(X_{22} - X_{12})] = 1000\ \underline{/0°} + 10\ \underline{/0°}(1 + j1800.5)$$

$$= 1010 + j18,005$$

Kirchhoff's current law at the top, center node, yields

$$I_1 = I_2 + \frac{V_{12}}{jX_{12}}$$

$$I_1 = 10 + \frac{1010 + j18{,}005}{j199.5}$$

$$= 100.25 - j5.06$$

By Kirchhoff's voltage law,

$$V_1 = V_{12} + I_1[r_1 + j(X_{11} - X_{12})] = 1010 + (100.25 - j5.06)(0.01 - j179.5)$$

(The rectangular form of I_1 is used to maintain accuracy which might be lost in converting to polar form.)

$$V_1 = 1010 + j18{,}005 + 1.00 - j17{,}995 - j0.0506 - 908.3$$
$$= 103.2 \ \underline{|5.56°} \qquad \text{Answer}$$

The above process has been fairly ridiculous, involving retention of insignificant figures and small differences between large numbers. The accuracy of the result is certainly open to question. It is surprising to obtain 1000 V at the output terminals of a passive circuit with only 103.2 V in. This result is achieved by partial resonance between the capacitor $j(X_{11} - X_{12})$ and the inductive reactance of the parallel circuit consisting of jX_{12} and the right-hand branch. It would be possible to obtain this result in the laboratory with high-Q inductors designed so as not to saturate at the voltages involved. It would be a very expensive way to build a transformer.

(b) The following is the Steinmetz approximate circuit model for the same transformer, referred to the primary:

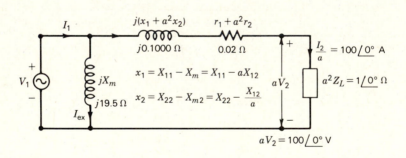

The core loss resistor R_c was not specified and is omitted. In any case, it is not involved in Steinmetz's approximate solution to the problem.

From Example 3.1, the turns ratio $a = 100/1000 = 0.1$. Then

$$Z_{eq\ 1} = r_1 + a^2 r_2 + j(x_1 + a^2 x_2)$$

$$= 0.0100 + 0.01 \cdot 1.00 + j(0.500 + 0.01 \cdot 5.00)$$

$$= 0.0200 + j0.1000 \ \Omega$$

Then by Kirchhoff's voltage law,

$$V_1 = aV_2 + \left(\frac{I_2}{a}\right) Z_{eq\ 1}$$

$$= 0.1 \cdot 1000 \,\underline{|0°} + \frac{10 \,\underline{|0°}}{0.1} (0.0200 + j0.1000)$$

$$= 100 + 2.00 + j10.00 = 102.5 \,\underline{|5.60°} \text{ V} \qquad \text{Answer}$$

This answer is the most easily obtained of the three and is the most accurate, even if this had been a linear transformer. Since the core is actually nonlinear, X_{12} is nonlinear, and as a result every reactive element in the "equivalent tee" is nonlinear. The genius of the Steinmetz model is that the reactive part of $Z_{eq\,1}$ or $Z_{eq\,2}$ is due to leakage flux and is thus linear, making a linear circuit solution legal. ∎

3.12 RELATIVE MAGNITUDES OF THE CIRCUIT MODEL CONSTANTS: THE PER-UNIT SYSTEM

As in the case of the synchronous machine, some feeling for the effects of winding resistance, leakage reactance, and so on, on the performance of a transformer is obtained by comparison with rated or "base" quantities. Each two-winding transformer has three essential ratings:

Rated primary voltage $= V_{1B}$
Rated secondary voltage $= V_{2B}$
Rated volt-amperes $= S_B$

These result in two sets of base quantities, one for the primary and one for the secondary:

PRIMARY		SECONDARY	
Primary rated volt-amperes	$\equiv S_B \equiv$	Secondary rated volt-amperes	
$V_{1B} \equiv$ rated V_1		$V_{2B} \equiv$ rated V_2	
$I_{1B} \overset{\triangle}{=} \dfrac{S_B}{V_{1B}}$		$I_{2B} \overset{\triangle}{=} \dfrac{S_B}{V_{2B}}$	(3.70)
$Z_{1B} \overset{\triangle}{=} \dfrac{V_{1B}}{I_{1B}}$		$Z_{2B} \overset{\triangle}{=} \dfrac{V_{2B}}{I_{2B}}$	(3.71)

Relationships Between Base Quantities

By convention,

$$\frac{V_{1B}}{V_{2B}} = a \qquad\qquad (3.72)$$

By Equations 3.70,

$$\frac{I_{1B}}{I_{2B}} = \frac{1}{a} \qquad\qquad (3.73)$$

Substituting Equations 3.72 and 3.73 into Equations 3.71,

$$\frac{Z_{1B}}{Z_{2B}} = a^2 \tag{3.74}$$

Transformer Losses

It was pointed out in Chapter 1 that maximum efficiency occurs when those losses that vary as the square of the current are equal to the constant losses. *The sum of hysteresis and eddy current losses is called the "core loss" of the transformer.* The core loss depends on the peak value of the sinusoidal core flux. As long as the voltages at the transformer primary and secondary remain within a few percent of rated value, the core loss is essentially constant. *Thus the core loss is considered to be the constant loss of the transformer.*

The sum of the primary and secondary I²R losses is called the "copper loss":

$$P_{Cu} = I_1^2 r_1 + I_2^2 r_2 \tag{3.75}$$

If it is assumed that $I_1 = I_2/a$, then

$$P_{Cu} = \left(\frac{I_2}{a}\right)^2 r_1 + I_2^2 r_2$$

$$= I_2^2 \left(\frac{r_1}{a^2} + r_2\right) = I_2^2 R_{eq\ 2} \tag{3.76}$$

$$= \left(\frac{I_2}{a}\right)^2 (r_1 + a^2 r_2) = \left(\frac{I_2}{a}\right)^2 R_{eq\ 1}$$

$$\cong I_1^2 R_{eq\ 1} \tag{3.77}$$

The *full-load* efficiency of a 60-Hz transformer *with a unity-power-factor load* is about 98 percent. In other words, the full-load losses are about 2 percent of rated kVA. Furthermore, *if* the transformer is designed for maximum efficiency at rated current and voltage (i.e., rated kVA), the *full-load* copper loss will be equal to the core loss. Thus one would expect the core loss at rated voltage to be about 1 percent of rated kVA, and the copper loss at rated current to be about 1 percent of rated kVA. Let the core loss be P_{h+e}. Then in per-unit (pu) values,

$$P_{h+e\ pu} = \frac{P_{h+e}}{S_B} \tag{3.78}$$

and a normal value for $P_{i+e\ pu}$ is 0.01. In the approximate circuit model,

$$P_{h+e} = \frac{V_1^2}{R_c}$$

and

$$P_{h+e\ pu} = \frac{1}{S_B}\frac{V_1^2}{R_c} = \frac{1}{V_{1B}I_{1B}}\frac{V_1^2}{R_c}$$

Now if it is assumed that the transformer will be operated at or near rated voltage, $V_1 \cong V_{1B}$, and

$$P_{h+e\ \text{pu}} = \frac{V_{1B}}{I_{1B}} = \frac{1}{R_c} = \frac{Z_{1B}}{R_c} \tag{3.79}$$

But

$$R_{c\ \text{pu}} = \frac{R_c}{Z_{1B}}$$

so

$$P_{h+e\ \text{pu}} = \frac{1}{R_{c\text{pu}}} \tag{3.80}$$

which implies that a normal value of $R_{c\ \text{pu}}$ would be 100.

Then for the copper loss,

$$P_{\text{Cu pu}} = \frac{I_2^2 R_{\text{eq 2}}}{S_B} = \frac{I_2^2 R_{\text{eq 2}}}{V_{2B}\ I_{2B}} \tag{3.81}$$

Equation 3.81 may be written in several interesting and convenient ways. First at *full load* only, $I_2 = I_{2B}$, and

$$\text{Full-load } P_{\text{Cu pu}} = \frac{I_{2B}^2}{V_{2B}I_{2B}} R_{\text{eq 2}} = \frac{R_{\text{eq 2}}}{Z_{2B}} = R_{\text{eq 2 pu}}$$

$$= \frac{a^2 R_{\text{eq 2}}}{a^2 Z_{2B}} = \frac{R_{\text{eq 1}}}{Z_{1B}} = R_{\text{eq 1 pu}} \tag{3.82}$$

Then, in per-unit values,

$$R_{\text{eq 1 pu}} = R_{\text{eq 2 pu}} = \text{full-load } P_{\text{Cu pu}} \tag{3.83}$$

In other words:

1. A normal value for $R_{\text{eq pu}}$ is 0.01.
2. In per unit, there is no distinction between $R_{\text{eq 1 pu}}$ and $R_{\text{eq 2 pu}}$.
3. The equivalent resistance in per unit is equal to the full-load copper loss in per unit.

Transformer Currents

In the usual power transformer the magnetizing current I_ϕ is about 5 percent of full-load primary current; that is, about 0.05 per unit, but in some transformers it may run as high as 10 percent. The core loss current I_{h+e} is normally about 0.01 per unit $=$ $1/R_{C\ \text{pu}}$. Since I_{h+e} is in phase with E_1 and I_ϕ lags E_1 by 90°, Figure 3.17 shows that

$$|I_{\text{ex}}| = \sqrt{|I_\phi^2| + |I_{h+e}^2|} \tag{3.84}$$

or in per-unit values,

$$I_{\text{ex pu}} \cong \sqrt{0.05^2 + 0.01^2} = 0.0510$$

which is to say that the core loss current contributes little to the magnitude of the exciting current. The power factor of the exciting current is quite low: roughly 0.01 to 0.05, indicating that i_{ex} lags E_1 or, in the approximate circuit, lags V_1 by 75 to 80°.

Since there is little difference between the magnitudes of I_{ex} and I_ϕ, it can also be said that a normal per-unit value of I_ϕ is about 0.05. Now in the approximate model,

$$|I_\phi| = \frac{|V_1|}{X_m} \qquad \text{and} \qquad X_m = \frac{|V_1|}{|I_\phi|} \qquad\qquad (3.85)$$

so a normal value for X_m would be

$$X_m = \frac{V_{1B}}{|I_\phi|}$$

Then

$$X_{m \text{ pu}} = \frac{X_m}{Z_{1B}} = \frac{V_{1B}}{|I_\phi|} \frac{I_{1B}}{V_{1B}} = \frac{I_{1B}}{|I_\phi|}$$

and

$$X_{m \text{ pu}} = \frac{1}{|I_\phi|_{\text{pu}}} \qquad\qquad (3.86)$$

A normal per-unit value of X_m would thus be about 20.

Consider a rather normal situation in which I_2 is half its rated value and lags V_1 by 30°, and further that $I_{\text{ex pu}}$ is $0.05 \,\underline{/-75°}$ relative to V_1, Then since

$$I_1 = I_{\text{ex}} + \frac{I_2}{a} \qquad\qquad (3.87)$$

$$I_{1 \text{ pu}} = \frac{I_1}{I_{1B}} = I_{\text{ex pu}} + \frac{I_2}{aI_{1B}} \qquad\qquad (3.88)$$

By application of Equation 3.73, $aI_{1B} = I_{2B}$ and $I_{1 \text{ pu}} = I_{\text{ex pu}} + I_{2 \text{ pu}}$, so for the given situation

$$I_{1 \text{ pu}} = 0.05 \,\underline{/-75°} + 0.5 \,\underline{/-30°}$$

$$= 0.537 \,\underline{/-33.8}$$

This indicates that, at half load, the error in assuming $I_2/I_1 = a$ is in error by about 7 percent in this case.

Transformer Impedances

Equation 3.74 shows that

$$Z_{1B} = a^2 Z_{2B}$$

Equation 3.69 states that

$$Z_{eq\ 1} = a^2 Z_{eq\ 2}$$

Thus the per-unit equivalent impedance is the same, whether referred to the primary or the secondary:

$$Z_{eq\ 1\ pu} = \frac{Z_{eq\ 1}}{Z_{1B}} = \frac{Z_{eq\ 1}}{a^2 Z_{2B}} = \frac{Z_{eq\ 2}}{Z_{2B}} = Z_{eq\ 2\ pu} \tag{3.89}$$

As a result, *the equivalent series impedance may be expressed in per unit without reference to either winding*. Its value is called the *per-unit impedance* or is expressed as the *percent impedance* of the transformer.

Normal values of the following impedances have already been discussed:

$$R_{eq\ pu} \cong 0.01$$

$$R_{o\ pu} \cong 100$$

$$X_{m\ pu} \cong 20$$

These are merely ballpark values! A specific transformer may have values differing considerably from these and still be a good transformer. This is particularly true of leakage reactance. Per-unit leakage reactance, often called simply *per-unit reactance*, is given by

$$X_{eq\ pu} = \frac{X_{eq\ 1}}{Z_{1B}} = \frac{a^2 X_{eq\ 2}}{a^2 Z_{2B}} = \frac{X_{eq\ 2}}{Z_{2B}} \tag{3.90}$$

Per-unit reactance ranges from 0.015 to 0.15 in transformers used in power systems. The higher values go with high-voltage ratings. For a distribution transformer the per-unit reactance would be nearer to 0.015, while a unit transformer feeding directly into a 345-kV line might have a per-unit reactance approaching 0.15.

Another way of looking at per-unit R_{eq} and X_{eq} is in terms of full-load voltage drop. At full load, the internal voltage drop referred to the secondary is $I_{2B} Z_{eq\ 2}$. Assume the secondary voltage to be at rated value. Then the internal drop expressed as a fraction of secondary terminal voltage is

$$\frac{I_{2B} Z_{eq\ 2}}{V_{2B}} = Z_{eq\ pu} \tag{3.91}$$

Then at full load, the internal voltage drop is the per-unit impedance times rated secondary volts. *The transformer nameplate usually gives the equivalent impedance in percent (i.e., $Z_{eq\ pu} \times 100$)*. Sometimes both %R_{eq} and %X_{eq} are given, and other times only $\% \mid Z_{eq} \mid$. When only *percent impedance* is given it is often assumed that $R_{eq} \ll X_{eq}$, so that $\% \mid Z_{eq} \mid \cong \% X_{eq}$.

3.13 HOW TO FIND THE PRIMARY VOLTAGE NECESSARY TO PRODUCE A DESIRED SECONDARY VOLTAGE

It has been pointed out that, by covention, the ratio of rated primary and secondary voltages listed on the nameplate of a transformer shall be the same as the turns ratio. That is,

$$\frac{V_{1B}}{V_{2B}} = a \tag{3.92}$$

Because of the internal impedance, this means that, if the voltage at the terminals of one winding is at its rated value, the voltage at the terminals of the other winding is *not* at rated value. However, since the internal impedances are so low, the ratio of the winding terminal voltages under load is within a few percent of the turns ratio.

Applying Kirchhoff's voltage law to the circuit of Figure 3.18,

$$\frac{V_1}{a} = V_2 + I_2 Z_{eq\ 2} \tag{3.93}$$

The simplest procedure usually is to let V_2 be the reference phasor:

$$V_2 = |V_2| \underline{|0°}$$

Then

$$|I_2| = \frac{\text{Load kVA} \cdot 1000}{|V_2|} \tag{3.94}$$

and

$$I_2 = |I_2| \underline{|\pm\theta_\phi} \tag{3.95}$$

where $+\theta_\phi$ corresponds to leading power factor or capacitive load and $-\theta_\phi$ to lagging power factor or inductive load. If the *power factor* is given,

$$\theta_\phi = \pm \cos^{-1}(\text{power factor}) \tag{3.96}$$

Then

$$|V_1| = a|V_2 + I_2 Z_{eq\ 2}| \tag{3.97}$$

■ Example 3.7

A 10-kVA, 2400-240-V, single-phase transformer has the following resistances and leakage reactances. Find the primary voltage required to produce 240 V at the secondary terminals at full load, when the load power factor is

(a) 0.8 power factor, lagging
(b) 0.8 power factor, leading

$$r_1 = 3.00 \ \Omega \qquad r_2 = 0.0300 \ \Omega$$
$$x_1 = 15.00 \ \Omega \qquad x_2 = 0.150 \ \Omega$$

Solution

$$Z_{eq \ 2} = \frac{r_1}{a^2} + r_2 + j \left(\frac{x_1}{a^2} + x_2 \right)$$

$$a = \frac{V_{1B}}{V_{2B}} = \frac{2400}{240} = 10$$

$$Z_{eq \ 2} = \left(\frac{3.00}{100} + 0.0300 \right) + j \left(\frac{15.00}{100} + 0.150 \right)$$

$$= 0.0600 + j0.300$$

$$= 0.3059 \ \underline{|78.69°}$$

(a) 0.8 power factor, lagging:

1.

$$I_{2B} = \frac{S_B}{V_{2B}} = \frac{10,000}{240} = 41.7 \ \text{A}$$

With V_2 as a reference, the full-load secondary current is

$$I_2 = 41.7 \ \underline{|-\cos^{-1}0.8}$$

$$= 41.7 \ \underline{|-36.87°} \ A$$

2.

$$\frac{V_1}{a} = V_2 + I_2 Z_{eq \ 2} = 240 \ \underline{|0°} + 41.7 \cdot 0.3059 \ \underline{|78.69° - 36.87°}$$

$$= 240 + j0 + (9.506 + j8.506)$$

$$= 249.506 + j8.506$$

$$= 249.65 \ \underline{|1.952°}$$

3.

$$|V_1| = a|V_2 + I_2 Z_{eq \ 2}| = 2496.5 \ V \quad \text{Answer}$$

(b) 0.8 power factor, leading:

1.

$$I_2 = 41.7 \ \underline{|+36.87°}$$

2.

$$\frac{V_1}{a} = 240 \ \underline{|0°} + 41.7 \cdot 0.3059 \ \underline{|78.69° + 36.87°}$$

$$= 240 + j0 + 12.76 \ \underline{|115.56°}$$

$$= 240 + j0 + (-5.505 + j11.51)$$
$$= 234.50 + j11.51 = 234.78 \underline{|2.81°}$$

3.

$$|V_1| = a|V_2 + I_2 Z_{eq\ 2}| = 2347.8\ V \quad \text{Answer} \quad \blacksquare$$

Note that for lagging power factor loads, V_1 must be greater than rated value to produce rated secondary voltage. If the load is *sufficiently* leading, as in this case, V_1 may be *less* than rated value for rated V_2.

3.14 TRANSFORMER VOLTAGE REGULATION

Voltage "regulation" is a measure of how much changes in load will cause the lights to flicker on the secondary side of the transformer, due to variations in the transformer's internal voltage drop. Regulation is defined by

$$\text{Regulation} \stackrel{\triangle}{=} \left.\frac{|V_{2\ nl}| - |V_{2\ lo}|}{|V_{2\ lo}|}\right|_{|V_1| = \text{const}} \quad \text{per unit} \quad (3.98)$$

The percent regulation is of course 100 times the per-unit regulation. In the above expression, $V_{2\ lo}$ is the secondary voltage under load, and $V_{2\ nl}$ is the secondary no-load voltage ($I_2 = 0$), with primary voltage held constant at the value it had under load. Note that the *magnitudes* of the voltages, not their phasor values, determine the regulation.

The voltage regulation may be calculated for any load and for any value of $V_{2\ lo}$. However, the regulation of most interest is that corresponding to rated secondary volts at full load:

$$\text{Full-load regulation} = \left.\frac{|V_{2\ nl}| - |V_{2\ fl}|}{|V_{2\ fl}|}\right|_{|V_1| = \text{const}}$$

where the full-load secondary voltage $V_{2\ fl}$ is the rated voltage V_{2B}. When the regulation of a transformer is mentioned without qualification, full-load regulation is implied. Then

$$\text{Full-load regulation} = \left.\frac{|V_{2\ nl}| - V_{2B}}{V_{2B}}\right|_{|V_1| = \text{const}} \quad (3.99)$$

Reference to Figure 3.18 will show that when there is no load on the transformer ($Z_L = \infty$), $I_2 = 0$ and there is no IZ drop in $Z_{eq\ 2}$. Then

$$V_{2\ nl} = \frac{V_1}{a} \quad (3.100)$$

The regulation expression becomes

$$\text{Regulation} = \frac{\left|\dfrac{V_1}{a}\right| - V_{2B}}{V_{2B}} \quad (3.101)$$

It is assumed that V_1 is adjusted so that rated voltage is obtained at the secondary terminals under given load conditions. Kirchhoff's law relates V_1/a to V_2 (Equation 3.93):

$$\frac{V_1}{a} = V_2 + I_2 Z_{eq\ 2} \qquad (3.102)$$

The phasor relationship of $I_2 Z_{eq\ 2}$ relative to V_2 depends on the power factor of the load. Therefore *regulation depends on the load power factor*.

To calculate full-load regulation.

1. Write $I_{2\ fl}$ as a phasor. With V_2 as a reference,

$$I_{2\ fl} = I_{2B} \,\underline{|\pm\ \theta_\phi} \qquad (3.103)$$

2. Calculate $V_{2\ nl} = V_1/a$:

$$\frac{V_1}{a} = V_{2B} \,\underline{|0} + Z_{eq\ 2}(I_{2B} \,\underline{|\pm\theta_\phi}) \qquad (3.104)$$

3.

$$\text{Regulation} = \frac{\left|\dfrac{V_1}{a}\right| - V_{2B}}{V_{2B}} \qquad (3.105)$$

*CAUTION: The numerator of the regulation expression is a scalar, the difference of two magnitudes. It is **not** equal to the $I_2 Z_{eq\ 2}$ drop, nor is it $|I_2 Z_{eq\ 2}|$.*

■ Example 3.8

Find the full-load voltage regulation of the transformer of Example 3.7 at (a) 0.8 power factor, lagging, and (b) 0.8 power factor, leading.

Solution

(a) From Example 3.7, for $240 = V_{2B}$ V at the terminals of the secondary at full load, 0.8 power factor, lagging:

$$\left|\frac{V_1}{a}\right| = 249.65 \text{ V}$$

Then

$$\text{Regulation} = \frac{\left|\dfrac{V_1}{a}\right| - V_{2B}}{V_{2B}}$$

$$= \frac{249.65 - 240}{240} = 0.0402, \text{ or } 4.02\%$$

(b) For 0.8 power factor, leading:

$$\left| \frac{V_1}{a} \right| = 234.78 \ V$$

$$\text{Regulation} = \frac{234.78 - 240}{240} = -0.0218, \text{ or } -2.18\%$$ ■

It is interesting to note that a negative regulation is possible. The meaning of a negative regulation is that the secondary voltage *increases* when the transformer is loaded! This is the result of a partial resonance between the capacitance of the load and the leakage inductance of the transformer. Figure 3.19 shows the effect of power factor on the relative magnitudes of V_2 and V_1/a.

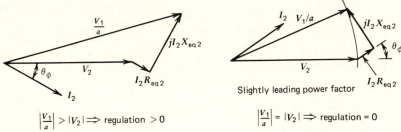

$$\frac{V_1}{a} = V_2 + I_2 R_{eq2} + I_2 \cdot jX_{eq2}$$

$$I_2 = \frac{V_2}{Z_L} = \left| \frac{V_2}{Z_L} \right| \angle{-\theta_\phi} = |I_2| \angle{-\theta_\phi}$$

$$\text{Regulation} = \left. \frac{|V_1/a| - |V_2|}{|V_2|} \right|_{|V_2| = V_{2B}}$$

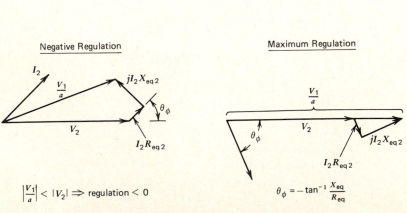

Lagging Power Factor

$$\left| \frac{V_1}{a} \right| > |V_2| \Rightarrow \text{regulation} > 0$$

Zero Regulation

Slightly leading power factor

$$\left| \frac{V_1}{a} \right| = |V_2| \Rightarrow \text{regulation} = 0$$

Negative Regulation

$$\left| \frac{V_1}{a} \right| < |V_2| \Rightarrow \text{regulation} < 0$$

Maximum Regulation

$$\theta_\phi = -\tan^{-1}\frac{X_{eq}}{R_{eq}}$$

FIGURE 3.19 Examples of transformer voltage regulation.

Regulation by Per-Unit Quantities

Transformer impedances are most often given in percent of base impedance. Full-load regulation is most easily calculated by converting these quantities into per-unit and using per-unit values throughout the calculations. The Kirchhoff's voltage law equation for the approximate equivalent circuit referred to the secondary is given in Equation 3.93. When both sides of this equation are divided by V_{2B}, the secondary base voltage, the following expression is obtained:

$$\frac{V_1}{aV_{2B}} = \frac{V_2}{V_{2B}} + \frac{I_2 Z_{eq\ 2}}{V_{2B}} \tag{3.106}$$

But since $aV_{2B} = V_{1B}$ and by Equation 3.71 $V_{2B} = I_{2B}Z_{2B}$, Equation 3.106 may be written

$$V_{1\ pu} = V_{2\ pu} + I_{2\ pu}\ Z_{eq\ pu} \tag{3.107}$$

For *full-load* regulation, $V_2 = V_{2B}$ and $|V_{2\ pu}| = 1$. Also $|I_2| = |I_{2B}|$ and $|I_{2\ pu}| = 1$. Then for this special case, Equation 3.104 becomes

$$V_{1\ pu} = 1\ \underline{/0°} + (1\ \underline{/\pm\theta_\phi})Z_{eq\ pu} \tag{3.108}$$

and the calculation is much simplified.

The expression for full-load regulation is also quite simple when converted to per-unit. If the numerator and denominator of Equation 3.101 are divided by the base secondary voltage V_{2B}, a very convenient expression is the result:

$$\text{Regulation} = \frac{|V_1/aV_{2B}| - (V_{2B}/V_{2B})}{V_{2B}/V_{2B}} = \left|\frac{V_1}{V_{1B}}\right| - 1$$
$$= |V_{1\ pu}| - 1 \tag{3.109}$$

■ Example 3.9

A transformer has 2 percent resistance and 5 percent reactance. Find its voltage regulation at full load, 0.8 power factor, lagging.

Solution

1. Find $I_{2\ pu}$:
   At full load, $|I_2| = I_{2B}$ and $|I_{2\ pu}| = 1$. Then

   $$I_{2\ pu} = 1\ \underline{/-\cos^{-1} 0.8} = 1\ \underline{/-36.87°}$$

2.
   $$Z_{eq\ pu} = 0.02 + j0.05 = 0.05385\ \underline{/68.20°}$$

3.
   $$V_{1\ pu} = 1 + I_{2\ pu}Z_{eq\ pu} \quad \text{(Equation 3.107)}$$
   $$= 1 + 0.05385\ \underline{/68.20° - 36.87°}$$
   $$= 1.0460 + j0.0280$$
   $$= 1.0464\ \underline{/1.53°}$$

4. By equation 3.109,

$$\text{Regulation} = |V_{1\ pu}| - 1$$
$$= 1.0464 - 1 = 0.0464, \text{ or } 4.64\%$$ ∎

3.15 TRANSFORMING THREE-PHASE

Three-phase power may be transformed by a bank of single-phase transformers or by a single three-phase transformer. A three-phase transformer is essentially three transformers wound on a common core. The geometry of the core is such that the fluxes of the phases share common paths. As a result, the volume of iron is less than that of three single-phase transformers of the same total rating.

Three-phase transformers do not provide the flexibility of a set of single-phase transformers. For example, one single-phase transformer in a bank may have a higher kVA rating than the others, to serve an unbalanced load. In case of failure of the transformer serving one phase, only that one transformer need be replaced; whereas it is most likely that a three-phase transformer damaged in one phase will have to be completely removed from service, at least temporarily.

However three-phase transformers, in addition to being lighter than the equivalent bank of single-phase units, take up less space, are less expensive, and involve much less external wiring. Their efficiency is slightly better. Improved construction and better means of protection from overvoltages and short circuits have made transformer failure a very rare occurrence. As a result, three-phase banks of single-phase transformers are seldom used in new installations except in distribution circuits where a combination of single and three-phase loads is to be served.

In three-phase transformation, the primary and secondary windings may be connected independently in either delta (Δ) or wye (Y). The possible combinations are Δ–Δ, Δ–Y, Y–Y, Y–Δ. Figure 3.20 shows diagrams of these four connections and how they relate to the approximate equivalent circuit of one phase.

The Y–Y connection is to be avoided unless a very solid (low impedance) neutral connection is made between the primary and the power source. If a neutral is not provided, the phase voltages tend to become severely unbalanced when the load is unbalanced. There are also troubles with third harmonics. These problems do not exist when one of the sets of windings is in Δ. When it is necessary to have Y–Y connection with a weak primary neutral, or none, each phase transformer is provided with a third winding in addition to the primary and secondary, called a *tertiary*. The tertiaries are connected in Δ. This is a relatively expensive arrangement. The connection is called "wye-delta-wye." The tertiary-winding terminals are often brought out to supply auxiliary power (e.g., lights, fans, and pumps) for the substation.

When a Y–Δ or Δ–Y connection is used, the Y is preferably on the high-voltage side, and the neutral is grounded. The transformer insulation may thus be designed for $1/\sqrt{3}$ times the line voltage rather than for the full line voltage. Sometimes it is necessary to have the Y on the low-voltage side if a neutral is required for the low-voltage circuit. A transformer chosen to provide 120/208-V, three-phase service would have its low voltage windings in Y, and most probably have its high-voltage primary windings in Δ.

Y–Δ and Δ–Y connections result in a 30° phase displacement between the primary and secondary line voltages. It is standard practice in the United States to connect

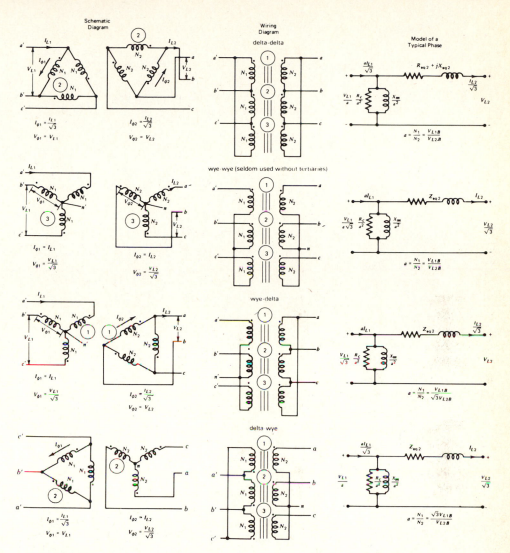

FIGURE 3.20 Basic transformer connections for three-phase. Voltages and currents shown are magnitudes, not phasors.

these transformers in such a way that the lower voltages *lag* the higher voltages by 30°. This is an important consideration when three-phase transformers are to be operated in parallel. The Y–Δ and Δ–Y transformer connections of Figure 3.20 are such that the secondary voltages lag the corresponding primary voltages by 30°.

In specifying the ratings for a three-phase transformer, the transformer is assumed to be ideal:

Example 3.10

What should be the ratings and turns ratio of a three-phase transformer to transform 10,000 kVA from 230 kV to 4160 V if the transformer is to be connected (a) Y–Δ, (b) Δ–Y, (c) Δ–Δ?

Solution

$$\text{Rated primary line current} \equiv I_{L1B} = \frac{S_B}{\sqrt{3}V_{L1B}} = \frac{10,000,000}{\sqrt{3} \cdot 230,000} = 25.1 \text{ A}$$

$$\text{Rated secondary line current} \equiv I_{L2B} = \frac{10,000,000}{\sqrt{3} \cdot 4160} = 1388 \text{ A}$$

(a) Y–Δ:

Rated kVA = $S_B/1000$ = 10,000 kVA
Rated $I_1 \equiv I_{1B} = I_{L1B}$ = 25.1 A
Rated $I_2 \equiv I_{2B} = I_{L2B}/\sqrt{3}$ = 801 A
Rated V_{L1} = 230 kV; rated V_{L2} = 4160 V
Rated $V_1 \equiv V_{1B} = 230/\sqrt{3}$ = 132.8 kV
Rated $V_2 \equiv V_{2B}$ = 4160 V
$\qquad$ Turns ratio = V_{1B}/V_{2B} = 132.8 $\cdot$ 10³/4160 = 31.9
$\qquad$ kVA per phase = 10,000/3 = 3333 kVA

(b) Δ–Y:

Rated kVA = 10,000
kVA per phase = 3333
$V_{1B} = V_{L1B}$ = 230 kV
$V_{2B} = V_{L2B}/\sqrt{3}$ = 4160/$\sqrt{3}$ = 2400 V
$I_{1B} = I_{L1B}/\sqrt{3}$ = 14.5 A
$I_{2B} = I_{L2B}$ = 1388 A
$a = V_{1B}/V_{2B}$ = 95.8

(c) Δ–Δ:

Rated kVA = 10,000
kVA per phase = 3333
V_{1B} = 230 kV
V_{2B} = 4160 V
I_{1B} = 14.5 A
I_{2B} = 801 A
a = 55.3

■ Example 3.11

A 7200 to 208-V, 50-kVA, three-phase distribution transformer is connected Δ–Y. The transformer has 1.2 percent resistance and 5 percent reactance. Find the voltage regulation at full load, 0.8 power factor, lagging.

Solution I

$$\Delta: V_{1B} = V_{L1B} = 7200 \text{ V} \qquad I_{L1B} = \frac{50,000}{\sqrt{3} \ 7200} = 4.009 \text{ A}$$

$$I_{1B} = \frac{I_{L1B}}{\sqrt{3}} = 2.315 \text{ A}$$

$$Y: V_{2B} = \frac{V_{L2B}}{\sqrt{3}} = \frac{208}{\sqrt{3}} = 120V \qquad I_{L2B} = \frac{50,000}{\sqrt{3}\ 208} = 138.8 \text{ A}$$

$$I_{2B} = I_{L2B} = 138.8 \text{ A}$$

$$a = \frac{V_{1B}}{V_{2B}} = \frac{7200}{120} = 60 \qquad Z_{2B} = \frac{V_{2B}}{I_{2B}} = 0.8646 \ \Omega$$

$$Z_{\text{eq pu}} = R_{\text{eq pu}} + jX_{\text{eq pu}} = 0.012 + j0.05 = 0.05142\underline{|76.50°}$$

$$Z_{\text{eq 2}} = Z_{2B} \cdot Z_{\text{eq pu}} = 0.8646 \cdot (0.012 + j0.05)$$

$$= 0.8646 \cdot 0.05142\underline{|76.50°}$$

$$= 0.010375 + j0.04323 = 0.04446\underline{|76.50°}$$

At full load, 0.8 power factor, lagging,

$$I_2 = 138.8\underline{|-36.87}$$

Applying Kirchhoff's voltage law to the Δ–Y equivalent circuit of Figure 3.20,

$$\frac{V_1}{a} = V_2 + I_2 Z_{\text{eq 2}}$$

$$= 120 + 138.8 \cdot 0.04446\underline{|75.50°} - 36.87°$$

$$= 120 + 6.171\underline{|39.63°} = 120 + 4.752 + j3.936$$

$$= 124.81\underline{|1.807°} \text{ V}$$

Then the no-lead secondary phase voltage is 124.81 V. (The primary line voltage is $a \cdot |V_1/a| = 60 \cdot 124.81 = 7489$ V.) The no-load secondary line voltage is $\sqrt{3} \cdot 124.81 = 216.18$ V.

$$\text{Regulation} = \frac{124.81 - 120}{120} = 0.040, \text{ or}$$

$$= \frac{\sqrt{3}(124.81 - 120)}{\sqrt{3} \cdot 120}$$

$$= \frac{216.18 - 208}{208}$$

Note that since the line voltage magnitude is a constant times the phase voltage magnitude, the regulation may be computed in terms of either line voltages or phase voltages.

Solution II (per-unit method) From Equation 3.107,

$$V_{1 \text{ pu}} = V_{2 \text{ pu}} + I_{2 \text{ pu}} \cdot Z_{\text{eq pu}}, \qquad \text{where } I_{2 \text{ pu}} = \frac{I_2}{I_{2B}}$$

$$I_{2 \text{ pu}} = \frac{138.8\ \underline{|-36.87°}}{138.8} = 1\ \underline{|-36.87°}$$

$$V_{2 \text{ pu}} = 1\ \underline{|0°}, \text{ since rated secondary voltage is assumed}$$

$$V_{1\,pu} = 1\,\underline{|0°} + (1\,\underline{|-36.87°})(0.05142\,\underline{|76.50°})$$

$$= 1 + 0.05142\,\underline{|39.63°}$$

$$= 1.0396 + j0.0328 = 1.0410\,\underline{|1.807°}$$

$$\text{Regulation} = \frac{|\,V_1/aV_{2B}\,| - (V_{2B}/V_{2B})}{V_{2B}/V_{2B}} = |\,V_{1\,pu}\,| - 1 = 0.0401 \quad\blacksquare$$

FOR FURTHER STUDY

TERTIARY WINDINGS IN Y-Y TRANSFORMERS

A Y–Y connection is used infrequently because the phase voltages become severely unbalanced when the loads are unbalanced. Furthermore, inspection of Figure 3.12 shows that the exciting current of any transformer is very nonsinusoidal. It in fact contains a very large third harmonic, which is necessary to overcome saturation in order to produce a sinusoidal flux. In the discussion of triplen harmonics in Chapter 2, it was pointed out that the three third-harmonic voltages in a balanced three-phase system are equal in magnitude and in phase with each other. By a similar argument, the same can be said for the third-harmonic phase currents. Applying Kirchhoff's current law at the neutral of a Y connection, the three line currents must sum in zero. This is no problem for the fundamental currents, which are 120° out of phase with each other. But in the case of the third-harmonic currents, three equal currents must add to zero, so they must *all* be zero. In other words, no triplen-harmonic currents can flow in a Y, if there is no connection from the neutral back to the three-phase source. As a result, the flux does not vary sinusoidally with time; in fact, it can be shown that the flux *has* a large third harmonic (about 30 percent) if there is *no* third harmonic in the magnetizing current. Since the induced voltage is proportional to both the flux amplitude and its frequency, the third harmonic component of the induced voltage is nearly as large as the fundamental. When added to the fundamental, the peak voltage is nearly twice the normal value!

There are two solutions to both the unbalance problem and the third-harmonic problem of the Y–Y connection. *(These problems do not arise when one or both sets of windings are connected in Δ.)* The most obvious is to provide a solid (low impedance) neutral connection to allow third harmonic currents to flow. This means that the neutral conductor will carry a strong 180-Hz current in a 60-Hz system, causing much trouble in telephone lines running parallel to the power line. The second solution is to install a third set of windings, called *tertiaries,* in each phase transformer, and connect these in Δ. As in the case of a Δ-connected alternator (pp. 72–73), third-harmonic currents will flow in the tertiaries to supply the needed component of the exciting current, thus restoring the flux to near-sinusoidal waveform. There will still be a third-harmonic flux component—just enough to provide sufficient third-harmonic tertiary voltage to provide the circulating current—a negative feedback phenomenon.

The Δ-connected tertiaries also couple the three phases together. Any fundamental-frequency voltage unbalance between the phases, due to secondary line current inbalance, causes a circulating fundamental-frequency current in the tertiaries. This tends to equalize the phase voltages and distributes the secondary load unbalance more evenly among the primary phases.

The tertiaries are usually designed for one third the VA rating of the transformer.

In some three-phase transformers the tertiary leads are not even brought outside the tank. In many cases the tertiaries are used to supply power for lights and motors at the substation where the transformer is installed.

In three-legged, core-type, three-phase transformers, the sum of the fluxes of the three phases must odd to zero at points at either end of the center leg. Since three-phase third harmonics cannot add to zero, there cannot be third harmonics in the mutual flux. Consequently, a core-type transformer may be operated Y–Y without tertiaries.

FOR FURTHER STUDY

THREE-PHASE TRANSFORMATION OF BALANCED IMPEDANCES—THE EQUIVALENT Y–Y

To minimize voltage unbalance and the effects of triplen harmonics in exciting current, one set of windings in a three-phase transformer or transformer bank is usually connected in Δ. However, impedance calculations in power systems are simpler if the transformers in the system are assumed to be connected Y–Y. To take advantage of this simplicity, system circuit analysts often replace all three-phase circuit models in the system with equivalent Y–Y models.

Consider the Y–Y transformer in Figure 3.21, feeding a balanced load of Z_L ohms per line-to-neutral phase through a transmission line having a series impedance per line of Z_T ohms.

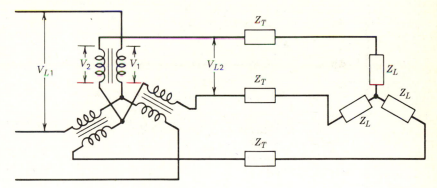

FIGURE 3.21 "Equivalent Y–Y"

The approximate equivalent circuit of the above, referred to the primary, is as shown in Figure 3.22:

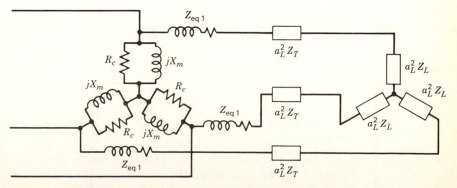

FIGURE 3.22 Circuit of Figure 3.21 referred to the transformer primary.

The turns ratio involved in referring all impedances is given by

$$a_{YY} = \frac{V_{1B}}{V_{2B}} = \frac{V_{L1B}/\sqrt{3}}{V_{L2B}/\sqrt{3}} = \frac{V_{L1B}}{V_{L2B}} \triangleq a_L$$

Note that the turns ratio to be used in referring impedances from the secondary to the primary is given by either the ratio of the rated phase voltages or rated line voltages. The ratio of the rated line voltages will be given the symbol a_L.

When the exciting branches are neglected (Their neglect is possible because of their relative high impedance.), an extremely simple equivalent circuit results:

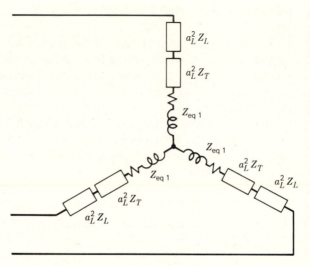

FIGURE 3.23 Exciting branches neglected.

This simplicity is the reason for using equivalent Y–Y transformers in power system analysis.

Now consider a Δ–Y transformer as shown in Figure 3.24. (The student may wish to go through a similar analysis for the Y–Δ connection.)

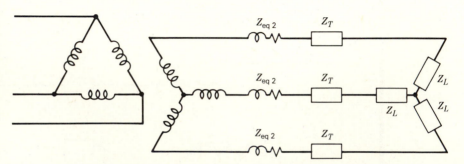

FIGURE 3.24 Delta-Y transformer feeding a balanced load through a transmission line.

In this figure the transformers are ideal, and the exciting branches are again neglected. Since the secondaries are in Y, $Z_{eq\,2}$ is the same as in the Y–Y case. In referring the secondary impedances to the primary, the turns ratio would be

$$a_{\Delta Y} = \frac{V_{L1B}}{V_{L2B}/\sqrt{3}} = \sqrt{3}\, a_L$$

When the secondary impedances are referred to the Δ primaries, the primaries appear to the incoming line as shown in Figure 3.25:

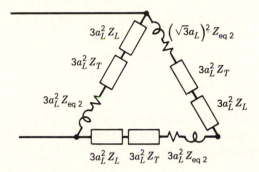

FIGURE 3.25 Secondary impedances reflected into the primary delta.

Each phase of the primary Δ appears to the circuit as the reflected secondary impedance, given by

$$Z_\Delta = a_{\Delta Y}^2 (Z_{eq\ 2} + Z_T + Z_L)$$

$$= 3a_L^2 (Z_{eq\ 2} + Z_T + Z_L)$$

Now when the Δ side of the transformer is replaced by its equivalent Y to form the equivalent Y–Y transformer, the incoming line sees three impedances in Y, each impedance equal to one third of the branch impedance of the Δ it replaces. That is, each phase of the Y has an impedance of

$$Z_Y = a_L^2 (Z_{eq\ 2} + Z_T + Z_L) = Z_{eq\ 1} + a_L^2 (Z_T + Z_L)$$

which is the same as if a Y–Y transformer had been used, as is evident by looking at Figure 3.23. This interesting result shows that, regardless of whether the transformer is connected Y–Y or Δ–Y, a Y–Y transformer may be assumed, with impedances transferred by the square of the ratio of *rated line voltages*. The same thing is true for the Y–Δ connection. Moreover, the assumption that the transformer is connected Y–Y allows the transformer internal impedance $Z_{eq\ 1}$ simply to be

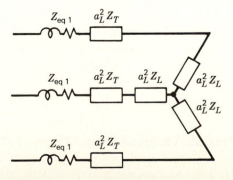

FIGURE 3.26 Y-equivalent of Figure 3.25.

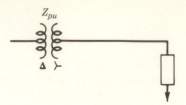

FIGURE 3.27 One-line diagram of a Δ-Y transformer feeding a load through a transmission line.

placed in *series* with the referred line and load impedances, as illustrated in Figure 3.26. Power system analyses are usually carried out in per-unit, which further simplifies the problem.

(WARNING: The equivalent Y–Y does not exhibit the 30° phase shift which actually exists between the primary and secondary circuits when one side is in Δ and the other in Y.)

THE EQUIVALENT Y–Y IN PER-UNIT SYSTEM ANALYSIS

The replacement of all transformers in a power transmission by their equivalent Y–Y$_s$ is particularly convenient when analysis is to be conducted on a per-unit basis. A base kVA is chosen for the entire system. There will be several line voltage levels in the system. Each voltage level will have its own base line current and base impedance. Base impedances are specified on a line-to-neutral basis, but base voltages are given as line-to-line voltages. At a given voltage level,

$$I_{LB} = \frac{(\text{base kVA}) \cdot 1000/3}{V_{LB}/\sqrt{3}} = \frac{S_B}{\sqrt{3}V_{LB}}$$

Then

$$Z_B = \frac{V_{LB}/\sqrt{3}}{I_{LB}} = \frac{V_{LB}^2}{S_B}$$

Note that the base impedance changes as the square of the voltage level. Then for impedances on the primary side of a three-phase transformer,

$$Z_{B1} = a_L^2\, Z_{B2}$$

Since it has been shown that impedances are referred through a transformer by the square of a_L also, line and load impedances expressed in per-unit have the same values before and after they are referred. It has been shown that the per-unit impedance of a transformer is the same, whether referred to the primary or secondary. As a result, a transformer may be represented by a simple series, per-unit impedance inserted in each line, as indicated in Figure 3.27.

3.16 THREE-PHASE TRANSFORMATION WITH TWO TRANSFORMERS

There are ways to transform three-phase with only two transformers. These schemes waste a part of the power-handling capability of the transformers involved. However

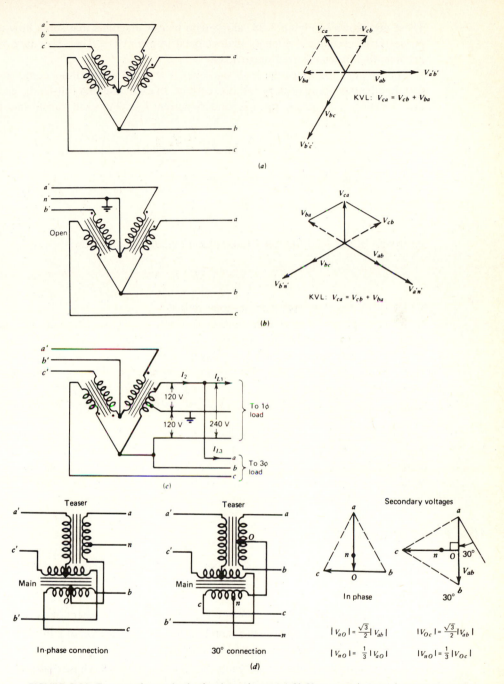

FIGURE 3.28 Two-transformer banks for three-phase. (*a*) V–V or open Δ–open Δ connection. (*b*) open Y–open Δ connection. (*c*) Modified V–V to serve a combination of three-phase and single-phase loads. (*d*) T connection.

they often may be justified on the basis of practical or economic considerations. The three most common connections are

1. V–V, or open Δ
2. Open Y–open Δ
3. T

These are shown in Figure 3.28, along with phasor diagrams that show how a three-phase set of secondary voltages is developed in each case. The secondary voltages become slightly unbalanced under load.

The loss of capacity of the V–V and open Y–open Δ connections comes from the fact that the transformers carry full line current. In each case, the full-load, secondary-line current I_{L2B} is equal to rated secondary current I_{2B}. Thus each transformer is rated at

$$S_B = V_{L2B}I_{2B} = V_{L2B}I_{L2b} \quad \text{VA}$$

for a total rating for the two transformers of

$$2S_B = 2V_{L2B}I_{L2B} \text{ VA} \tag{3.110}$$

However, at full load, the total three-phase output volt-amperes is given by

$$S = \sqrt{3}V_{LB}I_{LB} \text{ VA} \tag{3.111}$$

which is 13.4 percent less than the total rating.

In the in-phase T connection, both the main and teaser transformers may have the same ratings if no secondary neutral is required. The purchase of a special teaser transformer may thus be avoided. Again, the total rating is $2V_{LB}I_{LB}$, and the full-load output is only $\sqrt{3}V_{LB}I_{LB}$. However if a special teaser is provided, its secondary voltage rating will be $\sqrt{3}V_{L2B}/2$. The total rating of the two transformers will then be $I_{L2B}V_{L2B}(1 + \sqrt{3}/2) = 1.866 V_{L2B}I_{L2B}$, which is still slightly greater than the full-load output volt-amperes.

The reason for these discrepancies is that, in these connections, the transformers operate at power factors different from that of the load. This is a result of the fact that the load power factor angle appears between the line current and the *line-to-neutral voltage*. In each connection, the transformer windings carry line current, but in no case is the voltage across a winding the line-to-neutral voltage.

The V–V Connection

This connection is shown in Figure 3.28a. It is usually used under one of the following circumstances:

1. As a temporary measure when one phase of a Δ–Δ connection is damaged.
2. In areas where load growth is anticipated, and the addition of a third transformer to complete the Δ is expected in the future.
3. To supply loads that are a combination of a large single-phase load and a smaller three-phase load (see Figure 3.28c).
4. In instances in which this connection may be more economical in the use of materials. Examples are certain three-phase autotransformers (e.g., a starting compensator for an induction motor).

■ **Example 3.12**

A small industry requires 100 kVA of single-phase power at 0.90 power factor, lagging, at 220 V. In addition, its building has an air-conditioning compressor drive by a 20-hp motor requiring 20 kVA, 220 V, three-phase at 0.8 power factor, lagging. This combined load is to be supplied by two transformers connected in V–V. What current will flow in the secondary of each transformer?

Solution. Refer to Figure 3.28c. The motor current will flow in lines a, b, and c, and will be given by

$$I_{L3} = \frac{20,000}{\sqrt{3} \cdot 220} = 52.5 \quad \text{A}$$

$$= |I_2| \text{ of smaller transformer}$$

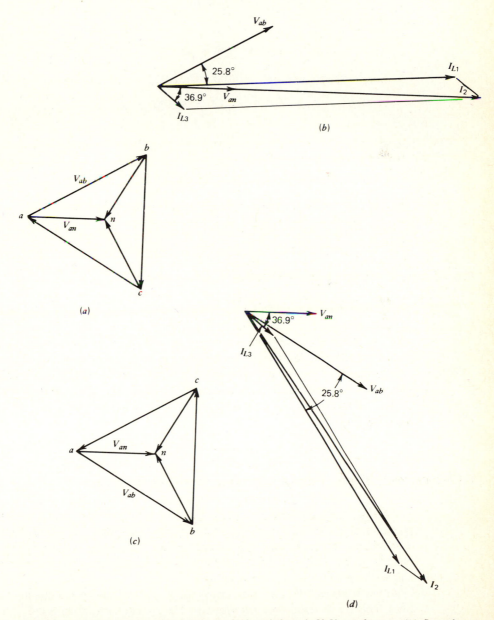

FIGURE 3.29 Phasor diagrams for Example 3.12, unbalanced, V–V transformers. (a) Secondary voltages for transformers of Figure 3.21c, a–b–c sequence. (b) Solution for secondary currents, a–b–c sequence. (c) Secondary voltages for transformers of Figure 3.21c, a–c–b sequence. (d) Solution for a–c–b sequence.

The larger transformer will carry a secondary current given by the phasor sum of the single-phase and motor line currents. This sum will depend on the angle between the two currents, and hence upon the phase sequence. There are thus two possible answers. Consider the phase sequence to be abc. The voltage relationships are as in Figure 3.29a. The motor current in line a, I_{L3} lags the line-to-neutral voltage V_{an} by the power factor angle $\cos^{-1} 0.800 = 36.9°$. For abc sequence, the line-to-line voltage across the single-phase load, V_{ab} leads V_{an} by 30°; and for acb sequence, it *lags* V_{an} by 30° (see Figure 3.29c). In either case, the single-phase load current I_{L1} lags V_{ab} by $\cos^{-1} 0.900 = 25.8°$. The secondary current of the larger transformer is given by

abc Sequence

Motor current $I_{L3} = 52.5 \underline{|-36.9°} = 42.0 - j31.5$ A

Single-phase current $I_{LI} = (100,000/220) \underline{|30° - 25.8°} = 453.3 + j33.3$A

$I_2 = I_{L3} + I_{L1} = 495.0 + j6.5 = 495 \underline{|0.75°}$ A

acb Sequence

$I_{L1} = 454.5 \underline{|-30° - 25.8°} = 255.5 - j375.9$ A

$I_2 = I_{L3} + I_{L1} = 297.5 - j407.4 = 504 \underline{|-53.9°}$ A   ■

The Open Y–Open Δ Transformer Connection

This connection is shown in Figure 3.28b. It differs from the V–V connection in that the primaries are connected from line to neutral (ground), thus permitting the use of ordinary distribution transformers for this connection. This is its greatest advantage. Only one of the two transformers may have its secondary center tap grounded, however.

This mode of three-phase transformation has some disadvantages in addition to those of the V–V connection. For one thing, the primary neutral must carry a substantial current. Also the primary three-phase load is unbalanced, even when the secondary load is balanced.

T-Connected, Three-Phase Distribution Transformers

The case or tank of a three-phase distribution transformer may contain a regular three-phase transformer with the primary and secondary windings of all three phases sharing a common core. Some three-phase distribution transformer consist of three single-phase core and coil assemblies mounted in a common tank, while others have two single-phase transformers connected in T.

In the T connection, one transformer has its primary connected directly across two lines. This is called the *main transformer*. The second transformer is called the *teaser*. Two connections are shown in Figure 3.28d. As the name implies, the "in phase" connection results in secondary voltages that are in phase with the primary voltages. The line voltage V_{ab} at the secondary terminals, for example, is in phase with $V_{a'b'}$

at the primary. In this respect the in-phase connection acts like a Y–Y or Δ–Δ connection. The "thirty degree" connection produces secondary voltages which lag the corresponding primary voltages by 30°, as do Y–Δ and Δ–Y connections.

It has been pointed out that, with a specially designed teaser, the total rating of the two transformers in a T is 1.866 $V_{L2B}I_{I2B}$, while the rated output is a little less: 1.732 $V_{L2B}I_{L2B}$. This would tend to indicate that the T connection is slightly inefficient in the use of materials. However two single-phase transformers are easier to build than a three-phase transformer, which requires a relatively complex core and coil arrangement. It is obvious that less labor is involved in making two single-phase transformers than in making three. So lower construction costs compensate for the rating disadvantage of the T connection. As a result, the T connection is an attractive alternative in the design of three-phase distribution transformers. Its greatest disadvantage is greater voltage unbalance under unbalanced load.

The Scott Connection

The Scott connection is a modified T connection that allows a three-phase circuit to be coupled to a two-phase circuit. Special single-phase transformers are used for this connection. The circuit is shown in Figure 3.30. The main transformer must be center-tapped on the three-phase side. Its turns ratio is equal to the ratio of the nominal line voltages. The turns ratio of the teaser transformer is $\sqrt{3}/2$ times the ratio of three-phase to two-phase line voltages.

An interesting application of the Scott connection is found in the Japanese railway system. The locomotives are designed to operate on single-phase, so that only one overhead wire is required. The power grid is, of course, three-phase. Scott transformer banks are used to supply two single-phase voltages from the three-phase system, one phase for northbound trains and one for southbound trains. Trains are scheduled so as to keep the load on the three-phase system as nearly balanced as possible.

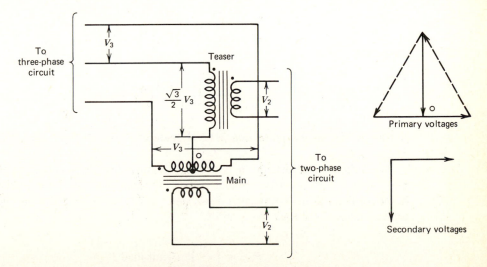

FIGURE 3.30 The Scott connection.

3.17 COMPUTING TRANSFORMER EFFICIENCY

Section 3.12 points out that the two major losses in a transformer are the copper loss and core loss. Then the input power is given by

$$P_{in} = P_{out} + P_{h+e} + P_{Cu} \qquad (3.112)$$

where

$$
\begin{aligned}
P_{out} &= |V_2| \, |I_2| \cos \theta_\phi \qquad \text{(single phase)} \\
&= |S_{out}| \cos \theta_\phi
\end{aligned}
\qquad (3.113)
$$

or

$$
\begin{aligned}
P_{out} &= \sqrt{3} \, |V_{L2}| \, |I_{L2}| \cos \theta_\phi \qquad \text{(three phase)} \\
&= 3 \, |V_2| \, |I_2| \cos \theta_\phi \\
&= |S_{out}| \cos \theta_\phi
\end{aligned}
$$

The efficiency is then

$$\eta = \frac{P_{out}}{P_{in}} = \frac{P_{out}}{P_{out} + P_{h+e} + P_{Cu}} \qquad (3.114)$$

The efficiency calculation is simplified by doing it in per-unit. Dividing numerator and denominator of Equation 3.114 by the volt-ampere rating of the transformer, S_B,

$$\eta = \frac{P_{out}/S_B}{(P_{out}/S_B) + P_{h+e \; pu} + P_{Cu \; pu}} \qquad (3.115)$$

where

$$P_{h+e \; pu} = \frac{P_{h+e}}{S_B} \qquad \text{and} \qquad P_{Cu \; pu} = \frac{P_{Cu}}{S_B} \qquad (3.116)$$

Then

$$\eta = \frac{(|S_{out}|/S_B) \cos \theta_\phi}{(|S_{out}|/S_B) \cos \theta_\phi + P_{h+e \; pu} + P_{Cu \; pu}} \qquad (3.117)$$

It was further shown in Section 3.12 (Equations 3.82 and 3.83) that, *at full load*, $P_{Cu \; pu} = R_{eq \; pu}$.

In efficiency calculations, it is usually assumed that the secondary voltage is at rated value. With constant terminal voltage, the magnitude of the secondary current is proportional to the volt-ampere load, and the copper loss is therefore proportional to the square of the load. As a result,

$$P_{Cu} = \left(\frac{|S|}{S_B}\right)^2 \cdot \text{(full-load } P_{Cu}) \qquad (3.118)$$

or in per-unit,

$$P_{\text{Cu pu}} = \left(\frac{|S|}{S_B}\right)^2 R_{\text{eq pu}}$$

Then the efficiency expression becomes

$$\eta = \frac{(|S|/S_B)\cos\theta_\phi}{(|S|/S_B)\cos\theta_\phi + P_{h+e\ \text{pu}} + (|S|/S_B)^2 R_{\text{eq pu}}} \tag{3.119}$$

and the efficiency at full load is

$$\eta_{fl} = \frac{\cos\theta_\phi}{\cos\theta_\phi + P_{h+e\ \text{pu}} + R_{\text{eq pu}}} \tag{3.120}$$

■ Example 3.13

A 2400-208-V, 300-kVA, three-phase transformer has a core loss at rated voltage of 2.70 kW. If its equivalent resistance is 1.40 percent, find the efficiency of this transformer for a load power factor of 0.9 (a) at full load, (b) at half load.

Solution

(a) Per-unit core loss is given by

$$P_{h+e\ \text{pu}} = \frac{P_{h+e}}{S_B} = \frac{2.70}{300} = 0.00900$$

Per-unit copper loss at full load = $R_{\text{eq pu}}$ = 0.0140. By Equation 3.120, for 0.900 power factor,

$$\eta_{fl} = \frac{0.900}{0.900 + 0.00900 + 0.0140} = \frac{0.900}{0.923}$$

$$= 0.975, \text{ or } 97.5\%$$

(b) At half load, $|S|/S_B = 0.5$. Then by Equation 3.119,

$$\eta_{fl} = \frac{0.5 \cdot 0.900}{0.5 \cdot 0.900 + 0.00900 + (0.5)^2 \cdot 0.0140}$$

$$= \frac{0.450}{0.450 + 0.00900 + 0.00350}$$

$$= \frac{0.450}{0.4625} = 0.973, \text{ or } 97.3\%$$

For this transformer, the efficiency is less at half load than at full load, in spite of the great reduction in copper loss at half load. Maximum efficiency would occur when $P_{\text{Cu}} = P_{h+e}$, or when

$$\left(\frac{|S|}{S_B}\right)^2 R_{\text{eq pu}} = P_{h+e \text{ pu}}$$

Then for maximum efficiency,

$$\frac{|S|}{S_B} = \sqrt{\frac{P_{h+e \text{ pu}}}{R_{\text{eq pu}}}} = \sqrt{\frac{0.00900}{0.0140}} = 0.8018$$

In other words, this transformer has its maximum efficiency at about 80 percent of full load. ∎

3.18 MEASURING TRANSFORMER QUANTITIES

The impedances of the approximate equivalent-circuit model of a transformer may be determined by two tests: an open-circuit test and a short-circuit test.

The Open-Circuit Test

The open-circuit test is conducted by applying rated voltage at rated frequency to one of the windings, with the other windings open-circuited. The input power and current are measured. For reasons of safety and convenience, the measurements are made on the low-voltage side of the transformer. Let a new turns ratio be defined as

$$a_v = \frac{N_{\text{HV}}}{N_{\text{LV}}}$$

the ratio of the turns of the high-voltage winding to those of the low-voltage winding. The equivalent circuit for the open-circuit test is as shown in Figure 3.31. Since the secondary is open, the primary current is equal to the exciting current I_{ex} and is quite small, about 0.05 per unit. The voltage drops in the primary leakage reactance and winding resistance may be neglected, and so may the primary $I_1^2 r_1$ loss. The input power is practically equal to the core loss at rated voltage and frequency:

$$P_{\text{oc}} = P_{h+e} = \frac{V_{\text{oc}}^2}{R_c} = V_{\text{oc}} I_{h+e} \tag{3.121}$$

The power factor on open circuit is

$$\cos \theta_{\text{oc}} = \frac{P_{\text{oc}}}{V_{\text{oc}} I_{\text{oc}}} \tag{3.122}$$

Since

$$I_{\text{ex LV}} = I_{\text{oc}}$$

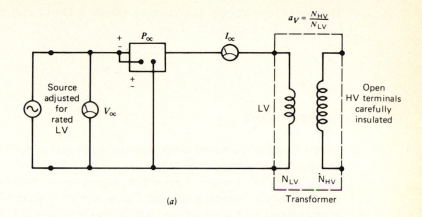

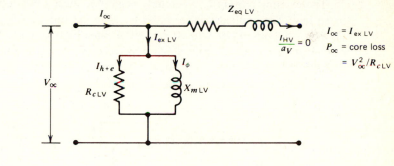

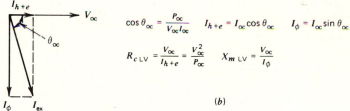

FIGURE 3.31 The open-circuit test. (*a*) Test circuit. (*b*) Equivalent circuit.

θ_{oc} is the angle by which $I_{ex\,LV}$ lags V_{oc}. The core loss current I_{h+e} is in phase with V_{oc}, while I_ϕ lags V_{oc} by 90°. Then

$$I_{h+e} = I_{oc} \cos \theta_{oc} \tag{3.123}$$

$$I_\phi = I_{oc} \sin \theta_{oc} \tag{3.124}$$

and

$$I_{oc} = \sqrt{I_{h+e}^2 + I_\phi^2} \tag{3.125}$$

The core loss current I_{h+e} may be found from Equation 3.121 or 3.123. The circuit element $R_{c\,LV}$ may then be calculated by Equation 3.121 or by

$$R_{c\,LV} = \frac{V_{oc}}{I_{h+e}} \tag{3.126}$$

The magnetizing current is given by Equation 3.124 or may be found from I_{oc} and I_{h+e}, using Equation 3.125. Then

$$X_{m\ LV} = \frac{V_{oc}}{I_\phi} \qquad (3.127)$$

The Short-Circuit Test

The short-circuit test is used to determine the equivalent series resistance and reactance. One winding is shorted at its terminals, and the other winding is connected through proper meters to a variable, low-voltage, high-current source of rated frequency. The source voltage is increased until the current into the transformer reaches rated value, at which point the meters are read and the readings recorded. To avoid unnecessarily high currents, the short-circuit measurements are made on the high-voltage side of the transformer. The circuit and effective equivalent circuits are shown in Figure 3.32.

The exciting current I_{ex} is entirely negligible during this test, because the voltage is so low. The applied voltage V_{sc} is equal to the full-load $I_{HV}Z_{eq\ HV}$ drop, which is in the neighborhood of 5 to 10 percent of rated high voltage. Even at rated voltage, the exciting current is only about 5 percent of rated current. At 10 percent rated voltage, I_{ex} would only be about 0.5 percent of the short-circuit test current, if the test is done at rated current, and may be neglected.

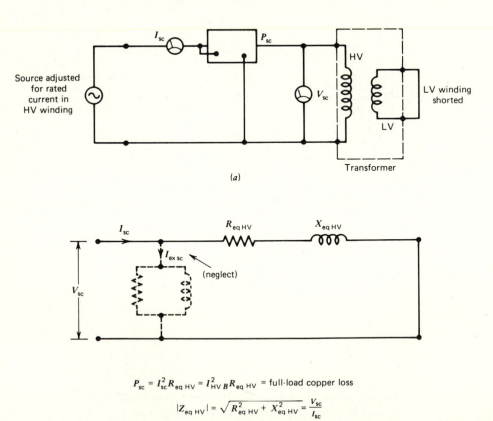

$$P_{sc} = I_{sc}^2 R_{eq\ HV} = I_{HV\ B}^2 R_{eq\ HV} = \text{full-load copper loss}$$

$$|Z_{eq\ HV}| = \sqrt{R_{eq\ HV}^2 + X_{eq\ HV}^2} = \frac{V_{sc}}{I_{sc}}$$

(b)

FIGURE 3.32 The short-circuit test. (a) Test circuit. (b) Equivalent circuit referred to HV winding.

Neglecting I_{ex}, the input power during this test is consumed in the equivalent resistance referred to the primary, $R_{eq\ HV}$. Then

$$I_{sc}^2 R_{eq\ HV} = P_{sc} \tag{3.128}$$

and

$$R_{eq\ HV} = \frac{P_{sc}}{I_{sc}^2}$$

Since rated current is used, the copper loss during the short-circuit test is equal to the full-load copper loss:

$$P_{sc} = \text{full-load } P_{Cu} \tag{3.129}$$

It is seen from Figure 3.32 that

$$|Z_{eq\ HV}| = \frac{V_{sc}}{I_{sc}} \tag{3.130}$$

$$= \sqrt{R_{eq\ HV}^2 + X_{eq\ HV}^2}$$

Then having found $R_{eq\ HV}$ by Equation 3.128,

$$X_{eq\ HV} = \sqrt{|Z_{eq\ HV}|^2 - R_{eq\ HV}^2} \tag{3.131}$$

or

$$X_{eq\ HV} = |Z_{eq\ HV}| \sin \theta_{sc} \tag{3.132}$$

where

$$\theta_{sc} = \cos^{-1} \frac{P_{sc}}{V_{sc} I_{sc}}$$

3.19 MEASUREMENTS ON THREE-PHASE TRANSFORMERS

In making open- and short-circuit tests on three-phase transformers, the quantities measured are total three-phase power, line volts, and line amperes. The impedances must be calculated on a phase basis to make any sense. Power would normally be measured by the two-wattmeter method (see Appendix B). Since the power factor is quite low for both tests, one of the wattmeter readings will be negative. Total three-phase power is the *algebraic* sum of the two readings.

Open-Circuit Test

Measurements are made on the low-voltage side and are converted to phase quantities. It is necessary to know whether the low-voltage windings are in Δ or Y. The test is made at rated low voltage and rated frequency.

Using Open-Circuit Test Data

LV in Δ	LV in Y

$$P_{\phi\ oc} = \frac{P_{oc}}{3} \qquad\qquad P_{\phi\ oc} = \frac{P_{oc}}{3}$$

$$V_{\phi\ oc} = V_{oc} \qquad\qquad V_{\phi\ oc} = \frac{V_{oc}}{\sqrt{3}}$$

$$I_{\phi\ oc} = \frac{I_{oc}}{\sqrt{3}} \qquad\qquad I_{\phi\ oc} = I_{oc}$$

$$\text{Total } P_{h+e} = P_{oc} \qquad\qquad \text{Total } P_{h+e} = P_{oc}$$

$$R_{cLV} = \frac{V_{\phi oc}^2}{P_{oc}/3} = \frac{3V_{oc}^2}{P_{oc}} \qquad\qquad R_{cLV} = \frac{V_{\phi\ oc}^2}{P_{oc}/3} = \frac{V_{oc}^2/3}{P_{oc}/3} = \frac{V_{oc}^2}{P_{oc}}$$

$$\cos\theta_{oc} = \frac{P_{oc}}{\sqrt{3}\ V_{oc}I_{oc}} \qquad\qquad \cos\theta_{oc} = \frac{P_{oc}}{\sqrt{3}V_{oc}I_{oc}}$$

$$I_{\phi} = \frac{I_{oc}}{\sqrt{3}}\sin\theta_{oc} \qquad\qquad I_{\phi} = I_{oc}\sin\theta_{oc}$$

$$X_{m\ LV} = \frac{V_{\phi\ oc}}{I_{\phi}} = \frac{\sqrt{3}V_{oc}}{I_{oc}\sin\theta_{oc}} \qquad\qquad X_{m\ LV} = \frac{V_{\phi\ oc}}{I_{\phi}} = \frac{V_{oc}}{\sqrt{3}I_{oc}\sin\theta_{oc}}$$

Note that the Δ impedances are three times those of the Y.

Short-Circuit Test

The three low-voltage terminals are shorted together. The voltage applied to the high-voltage terminals is adjusted so that rated current flows. The frequency of the voltage source is the rated frequency of the transformer. The connection (Δ or Y) of the high-voltage windings should be known.

Δ	Y

Full-load copper loss $= P_{sc}$ $\qquad\qquad$ Full-load copper loss $= P_{sc}$

$$P_{\phi\ sc} = \frac{P_{sc}}{3} \qquad\qquad P_{\phi\ sc} = \frac{P_{sc}}{3}$$

$$I_{\phi\ sc} = \frac{I_{sc}}{\sqrt{3}} \qquad\qquad I_{\phi\ sc} = I_{sc}$$

$$V_{\phi\ sc} = V_{sc} \qquad\qquad V_{\phi\ sc} = \frac{V_{sc}}{\sqrt{3}}$$

$$|Z_{eq\ HV}| = \frac{V_{\phi\ sc}}{I_{\phi\ sc}} = \frac{\sqrt{3}V_{sc}}{I_{sc}} \qquad\qquad |Z_{eq\ HV}| = \frac{V_{\phi\ sc}}{I_{\phi\ sc}} = \frac{V_{sc}}{\sqrt{3}I_{sc}}$$

$$R_{eq\ HV} = \frac{P_{\phi\ sc}}{I_{\phi\ sc}^2} = \frac{P_{sc}/3}{(I_{sc}/\sqrt{3})^2} = \frac{P_{sc}}{(I_{sc})^2} \qquad\qquad R_{eq\ HV} = \frac{P_{\phi\ sc}}{I_{\phi\ sc}^2} = \frac{P_{sc}}{3I_{sc}^2}$$

$$X_{eq\ HV} = \sqrt{|Z_{eq\ HV}|^2 - R_{eq\ HV}^2} \qquad\qquad X_{eqHV} = \sqrt{|Z_{eq\ HV}|^2 - R_{eq\ .HV}^2}$$

■ Example 3.14

Open-circuit and short-circuit tests are performed on a 50-kVA, 7200-208-V 60-Hz, Δ–Y-connected, three-phase transformer with the following results:

$$P_{oc} = 500 \text{ W} \qquad P_{sc} = 600 \text{ W}$$

$$I_{oc} = 8.00 \text{ A} \qquad I_{sc} = 4.01 \text{ A}$$

$$V_{oc} = 208 \text{ V} \qquad V_{sc} = 370 \text{ V}$$

Find the impedances of the approximate equivalent circuit referred to the high-voltage winding, the equivalent impedance referred to the low-voltage winding, the percent X and R, the voltage regulation at full load, 0.8 power factor, lagging, and the full-load efficiency at 0.8 power factor.

Solution. From the open-circuit test,

$$\text{Core loss} = P_{oc} = 500 \text{ W}$$

The low-voltage is side is Y-connected. Then

$$V_{\phi \; oc} = \frac{208}{\sqrt{3}} = 120 \text{ V} \qquad P_{\phi \; oc} = \frac{500}{3} = 166.7 \text{ W}$$

$$I_{\phi \; oc} = I_{oc} = 8.00 \text{ A}$$

$$R_{c \; LV} = \frac{V_{\phi}^2}{P_{\phi}} = \frac{120^2}{166.7} = 86.4 \; \Omega \qquad \text{or} \qquad \frac{V_{oc}^2}{P_{oc}} = \frac{208^2}{500} = 86.5 \; \Omega$$

$$\sin \theta_{oc} = \sin \cos^{-1} \frac{500}{\sqrt{3} \cdot 8.00 \cdot 208} = 0.985$$

$$I_{\phi} = I_{\phi \; oc} \sin \theta_{oc} = 8.00 \cdot 0.985 = 7.88 \text{ A}$$

$$X_{m \; LV} = \frac{V_{\phi \; oc}}{I_{\phi}} = \frac{120}{7.88} = 15.23 \; \Omega$$

From the short-circuit test,

$$\text{Full-load copper loss} = P_{sc} = 600 \text{ W}$$

The high-voltage side is Δ-connected:

$$R_{eq \; HV} = \frac{P_{\phi \; sc}}{I_{\phi \; sc}^2} = \frac{600/3}{(4.01/\sqrt{3})^2} = \frac{600}{(4.01)^2} = 37.3 \; \Omega$$

$$|Z_{eq \; HV}| = \frac{V_{\phi \; sc}}{I_{\phi \; sc}} = \frac{370}{4.01/\sqrt{3}} = 159.8 \; \Omega$$

$$X_{eq \; HV} = \sqrt{159.8^2 - 37.3^2} = 155.4 \; \Omega$$

Referring R_c and X_M to the high-voltage side,

$$a_V = \frac{N_{HV}}{N_{LV}} = \frac{V_{HVB}}{V_{LVB}} = \frac{7200}{208/\sqrt{3}} = 60$$

$$R_{c\ HV} = R_{c\ HV} \cdot a_V^2 = 86.4 \cdot 60^2 = 311{,}000\ \Omega$$

$$X_{m\ HV} = X_{m\ LV} \cdot a_V^2 = 54{,}800\ \Omega$$

The equivalent circuit referred to the high side is then

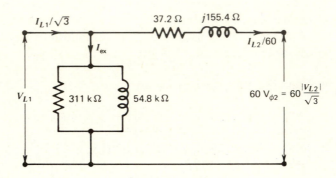

If the equivalent impedance referred to the low-voltage side is needed,

$$Z_{eq\ LV} = \frac{1}{60^2}(37.3 + j155.4) = 0.01036 + j0.0432\quad \Omega$$

In per-unit,

$$R_{eq\ pu} = P_{sc}/S_B = 600/50{,}000 = 0.01200$$

or

$$Z_{LV\ B} = \frac{208^2}{50{,}000} = 0.8653$$

$$R_{eq\ pu} = \frac{0.01036}{0.86528} = 0.01197$$

$$X_{eq\ pu} = \frac{0.0432}{0.86528} = 0.0499$$

$$Z_{eq\ pu} = 0.0120 + j0.0499 = 0.0513\ \underline{/76.48°}$$

Full load regulation in per-unit at 0.8 power factor lagging:

$$V_{1\ pu} = 1 + (1\ \underline{/-36.9})(0.0513\ \underline{/76.5°})$$

$$= 1 + 0.0513\ \underline{/39.6°}$$

$$= 1 + 0.0395 + j0.0327 = 1.0400 \;\underline{|1.8°}$$

$$\text{Regulation} = \frac{|V_{1\;pu}| - 1}{1} = 0.0400, \text{ or } 4.00\%$$

Full-load efficiency in per unit is

$$\eta = \frac{\cos\theta}{\cos\theta + R_{eq\;pu} + P_{h+e\;pu}}$$

$$= \frac{0.8}{0.8 + 0.012 + 500/50{,}000}$$

$$= \frac{0.8}{0.822} = 0.973, \text{ or } 97.3\%$$

or

$$\eta = \frac{50{,}000 \cdot 0.8}{40{,}000 + P_{h+e} + P_{Cu\;fl}}$$

$$= \frac{40{,}000}{40{,}000 + P_{oc} + P_{sc}}$$

$$= \frac{40{,}000}{40{,}000 + 500 + 600} = 97.3\%$$

3.20 AUTOTRANSFORMERS

An autotransformer is one in which one winding serves both primary and secondary functions. One of the most common forms is the variable autotransformer used to provide an adjustable ac voltage, as illustrated in Figure 3.33. Autotransformers are increasingly used to interconnect two high-voltage transmission lines operating at different voltages.

Figure 3.34 defines terms applied to the windings of an autotransformer. The "common" winding is at least a part of both the primary and secondary windings simultaneously. In a variable autotransformer the tap is movable.

The disadvantage of an autotransformer is that there is no electrical isolation between the primary and secondary circuits. However an autotransformer has several advantages in applications in which electrical isolation is not required.

The advantages of autotransformers are made evident by an example. A student living in a trailer court finds that the picture on his or her TV screen is shrinking as a result of low line voltage. The line voltage is found to be about 105 V. If the voltage were, say, 115 V, the picture would be satisfactory. The TV set draws about 5 A from the line.

The student finds an old 12-V filament transformer rated at 5 A on the secondary ($S_B = 12 \cdot 5 = 60$ VA). The primary rated voltage is 120 V. The student reconnects the filament transformer as a step-up autotransformer, with the 120-V winding as the common winding and the 12-V winding as the series winding (see Figure 3.34). The transformation ratio a is now

$$a = \frac{120}{120 + 12} = \frac{120}{132} = 0.909$$

With 105 V input, the output voltage

$$V_2 \cong \frac{V_1}{a} = 105 \cdot \frac{132}{120} = 115.5 \text{ V}$$

enough to operate the TV properly.

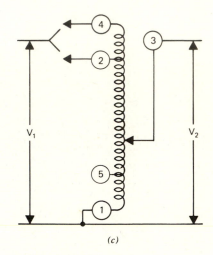

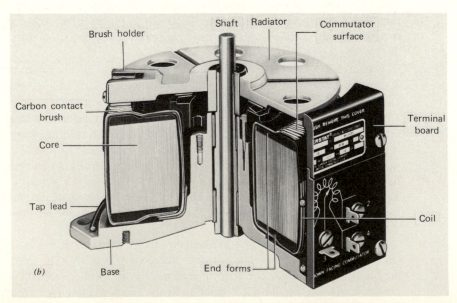

FIGURE 3.33 A variable autotransformer. (Courtesy of Superior Electric Company.) (*a*) External view, (*b*) Cutaway, (*c*) Diagram.

Is the little filament transformer overloaded? The output of the autotransformer is

$$S_{out} = 5 \text{ A} \cdot 115. \text{ V} = 578 \text{ VA!}$$

However, the 12-V common winding is carrying its rated 5 A and is not overloaded. What about the 120-V common winding?

$$I_c \cong I_2 \left(\frac{N_{se}}{N_c} \right) = 5 \left(\frac{12}{120} \right) = 0.5 \text{ A}$$

The rated current of this winding is

$$I_{cB} = \frac{60 \text{ VA}}{120 \text{ V}} = 0.5 \text{ A}$$

so both windings are operating at rated current. The volt-amperes actually being transformed are still only about 60. (Actually 105 V · 0.5 A = 52.5 VA). The remaining volt-amperes, 578 − 52.5 = 525 VA, are said to be "conducted." Note that the current drawn from the line is

$$I_1 = I_c + I_2 \cong 0.5 + 5.0 = 5.5 \text{ A}$$

assuming that I_c and I_2 are at the same power factor angle. The input volt-amperes are 105 V · 5.5 A = 578 VA, as expected. The *conducted* volt-amperes are given by

$$S_{cond} = V_1(I_1 - I_c) \cong V_1 I_2$$

$$= 105 \cdot 5 = 525 \text{ VA}$$

as before.

Thus for the cost of a 60-VA transformer, the student transforms 578 VA. In addition, the efficiency is much better than it would be for a conventional two-winding transformer. The 60-VA filament transformer has no greater losses than it would have in normal operation, but a conventional 578-VA transformer would have much larger losses.

Rating Advantage of Autotransformers

Let the rating of a conventional transformer be S_B. When reconnected as an autotransformer, let

$$V_{se} = \text{rated series winding voltage}$$

$$I_{se} = \text{rated series winding current}$$

then

$$S_B = V_{se} I_{se} \tag{3.133}$$

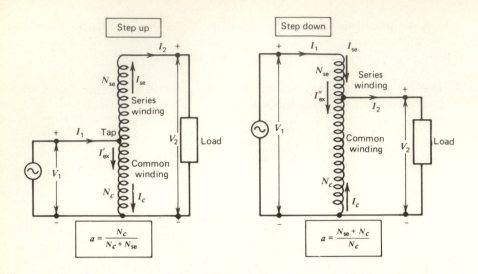

Relationships

Magnetic: $I'_{ex} \triangleq \dfrac{R}{N_c}$

$F_c + F_{se} = R$

$N_c I_c - N_{se} I_2 = R$

$I'_{ex} = I_c - \dfrac{N_{se}}{N_c} I_2$

$$\boxed{I_c = \dfrac{N_{se}}{N_c} I_2 + I'_{ex} = I_2\left(\dfrac{1-a}{a}\right) + I'_{ex}}$$

Electrical: $I_1 = I_c + I_2$

$\qquad = I_2\left(\dfrac{N_{se}}{N_c} + 1\right) + I'_{ex}$

$\qquad = \dfrac{I_2}{a} + I'_{ex}$

$\boxed{I_1 \cong \dfrac{I_2}{a}}$

$v_2 \cong (N_c + N_{se})\dfrac{d\phi}{dt}$

$v_1 \cong N_c \dfrac{d\phi}{dt}$

$\dfrac{V_1}{V_2} = \dfrac{v_1}{v_2} \cong \dfrac{N_c}{N_c + N_{se}} = a$

$\boxed{V_1 \cong aV_2}$

Magnetic: $I''_{ex} \triangleq \dfrac{R}{N_{se} + N_c}$

$F_{se} + F_c = R$

$N_{se} I_1 - N_c I_c = R$

$I''_{ex} = \dfrac{N_{se} I_1 - N_c I_c}{N_{se} + N_c}$

Electrical: $I_1 = \dfrac{I_2}{a} + I''_{ex}$

$\boxed{I_1 \cong \dfrac{I_2}{a}}$

$I_2 = I_1 + I_c = aI_1 - aI''_{ex}$

$I_2 \cong aI_1$

$I_c = I_2 - I_1 = I_2\left(\dfrac{a-1}{a}\right) - I''_{ex}$

$\boxed{I_c \cong I_2\left(\dfrac{a-1}{a}\right)}$

$v_2 \cong N_c \dfrac{d\phi}{dt}$

$v_1 \cong (N_{se} + N_c)\dfrac{d\phi}{dt}$

$\dfrac{V_1}{V_2} = \dfrac{v_1}{v_2} \cong \dfrac{N_{se} + N_c}{N_c} = a$

$\boxed{V_1 \cong aV_2}$

FIGURE 3.34 Definitions and relationships for step-up and step-down autotransformers.

Let the common-winding voltage rating be V_c. Define the voltage ratio of the auto-transformers as the ratio of rated high voltage to rated low voltage:

$$a_V = \frac{V_{se} + V_c}{V_c} \tag{3.134}$$

The volt-ampere rating of the autotransformer is

$$S_{auto} = (V_{se} + V_c)I_{se}$$

Then

$$\frac{S_{auto}}{S_B} = \frac{V_{se} + V_c}{V_{se}} = \frac{a_V}{a_V - 1} \tag{3.135}$$

Note that as the voltage ratio increases, the advantage of the autotransformer decreases. In the case of the 120- to 12-V filament transformer, $a_v = (132/120) = 1.1$, and a very great advantage was obtained.

3.21 INSTRUMENT TRANSFORMERS

Most power circuits involve voltages and currents too high to permit their measurement by the use of ordinary instruments. Instrument transformers, connected between such circuits and the instruments, permit measurements to be made safely with instruments of relatively low voltage and current ratings. The secondary circuits are usually grounded. Instrument transformers are also employed to operate relays used for power system protection, such as overcurrent relays, undervoltage relays, impedance relays, and so on. The instrument load on the secondary of a transformer is called its *burden,* and is expressed in volt-amperes at a certain power factor.

Voltage Transformers

Instrument transformers designed to step down voltages are called *voltage transformers, potential transformers,* or *pot transformers.* The secondaries are usually designed for 0–150 V meters. Voltage transformers introduce errors of two kinds into the measurements being made: (1) ratio errors and (2) phase angle errors. These errors are due primarily to the equivalent series impedance of the transformer, and they are usually quite small. For example, a voltage transformer designed to deliver 150 V to a meter load with 300 kV at its primary terminals would have a nominal ratio of 2000/1. The actual turns ratio would differ slightly from this, to compensate for the IZ_{eq} drop at some specified burden. At any other burden, the voltage ratio would be off, slightly. The maximum ratio deviation over a specified range of burdens determines the *accuracy class* of the transformer. If the maximum ratio error is ± 0.3 percent over the standard burden range, the transformer is said to be in the 0.3 accuracy class.

Phase error is a problem when watts, VARs, and impedance are to be measured. For voltage transformers, the standard symbol for phase error is γ. It is usually expressed in minutes rather than degrees. For *very* light burdens, the secondary voltage may lead the voltage being measured, but in most practical situations, γ is lagging and is considered to be negative when it lags. The phase error rarely exceeds 30′ and for a transformer of 0.3 accuracy class keeps within 7′ for most burdens. Voltage transformers are usually supplied with charts that show the ratio and phase errors as functions of burden magnitude and power factor.

Current Transformers

Current transformers step current down by the inverse of the turns ratio. Thus the secondary has more turns than the primary. A typical secondary current rating is 5 A. The transformer ratio is usually stated to include the secondary current rating; such as 1000 : 5, meaning that 1000 A in the primary results is 5 A through meters connected in series to the secondary terminals. When large currents are to be measured, the primary is most often a single turn; that is, the primary is formed by running the line through the window of the core about which the secondary is wound.

The basic cause of ratio and phase errors in current transformers is the exciting current. Equation 3.46 shows the primary current of an iron core transformer to be the sum of the exciting current and the secondary current divided by the turns ratio. The exciting current lags the primary voltage and, for most current transformer applications, so does I_2/a. That means that the primary current is greater than I_2/a, and this is the source of the ratio error. Since the exciting current is usually more lagging than the reflected secondary current, the total primary current usually lags the meter current by a small angle. The angle by which the secondary current leads the primary current is designated β, and is usually positive.

Current transformers can be dangerous. When current is passing through the primary, the secondary must always be shorted either through the current coils of meters or through the shorting switch usually provided as part of the transformer. The primary of a current transformer has few turns, and its impedance is small, even when the secondary is open, with the result that the primary current changes little if the secondary circuit is opened accidentally. However there is no longer a secondary MMF in opposition to that of the primary, and the entire primary current becomes exciting current, driving the core far into saturation in each direction on alternate half cycles. The flux waveform approaches that of a square wave having an amplitude equal to the saturation flux of the core and a sudden reversal of flux each time the core is driven from saturation in one direction to saturation in the other direction. The large $d\phi/dt$ during these flux reversals causes very high voltage spikes to be induced in the secondary winding—so high as to be dangerous to life and to the transformer insulation. Therefore extreme caution must be exercised in dealing with current-transformer secondary circuits.

PROBLEMS

3.1. A certain transformer has 50 turns in its primary winding. The leakage inductance of this winding is $8 \cdot 10^{-4}$ H. At a given instant in time, the mutual flux between the primary and secondary is 0.01 Wb and the primary current is 20 A. Find λ_1, the total primary flux linkage, at this instant.

3.2. A transformer has two windings, one of 180 turns and the other of 360 turns. With the 360-turn winding open-circuited, a sinusoidal, 60-Hz voltage of 2300 V rms is applied to the 180-turn winding. Neglect winding resistance and leakage flux (i.e., assume $v_1 = e_1' = e_1$) to (a) find the peak flux linkage of the 180-turn winding; (b) find the peak flux density in the core, if the crossectional area of the core iron is 0.04 m²; (c) and (d) repeat parts (a) and (b) with the same voltage applied to the 360-turn winding, the 180-turn winding being open.

3.3. A transformer has 180 primary turns and 360 secondary turns. The secondary is delivering 24.00 + j0 A to a load connected to its terminals. The current delivered to the primary by the supply line is 48.86 − j2.35 A. (a) Find the rms value of R, the MMF applied to the core. (b) Find the exciting current as an rms phasor. (c) If the core permeance is 7.85 · 10⁻⁵ H/turns², find the peak core flux. (d) Find the rms values of the voltages induced in the two windings by the core flux if the frequency is 60 Hz.

3.4. For the transformer and operating conditions of Problem 3.3, the permeance of the primary effective leakage flux is 2.00 · 10⁻⁷ H/turns². (a) Find the peak effective leakage flux of the primary winding. (What MMF should be used?) (b) Find the primary leakage inductance and reactance.

3.5. A two-winding transformer has 180 primary turns and a turns ratio N_1/N_2 of 0.5. The permeance of the core is 8 · 10⁻⁵ H/turns² and the primary leakage permeance is 2 · 10⁻⁷ H/turns². Find (a) the primary leakage inductance, (b) the primary magnetizing inductance, (c) the mutual inductance, (d) the coupling factor of the primary, and (e) the secondary coupling factor if the coefficient of coupling is 0.998000. (Can you find the secondary leakage inductance)?

3.6. The secondary terminal voltage of a given transformer is 707 sin 2 π 60t V. The primary has 50 turns and the secondary has 100 turns. The secondary current is 14.14 sin(377t − 36.9°) A and the exciting current is 1.414 sin(377t − 80°) A. The leakage inductance of the secondary is 2.653 mH, while that of primary is 0.6631 mH. Primary and secondary winding resistances are 0.100 and 0.400 Ω, respectively. Find the primary terminal voltage and current as rms phasor quantities. Calculate the input power and efficiency. Compare the actual terminal-voltage ratio with the turns ratio.

3.7. Three ideal transformers are connected as follows to transform three-phase power:

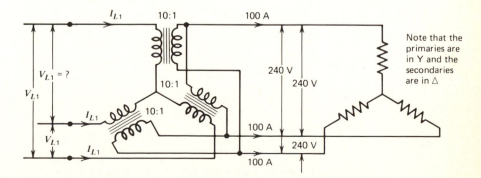

The load on the secondaries is balanced. The current in each line is 100 A (the three line currents are, of course, 120° out of phase with each other). Find (a) V_{L1}, (b) I_{L1}, (c) kVA output, and (d) kVA transformed by each transformer.

3.8.   A 23-kVA, 2300–230-V, single-phase transformer has a 2300-V winding resistance of 2.3 Ω and leakage reactance of 6.9 Ω. The leakage reactance of the low-voltage winding is 0.069 Ω and its resistance is 0.023 Ω. Find (a) the equivalent impedance referred to the high-voltage winding, (b) the equivalent impedance referred to the low-voltage winding, (c) the high-voltage and low-voltage base impedances, and (d) the per-unit resistance and reactance.

3.9.   Find the voltage regulation of the transformer of Problem 3.8 for full load at 0.8 power factor, leading, using actual quantities rather than per-unit quantities.

3.10. A single-phase, 7200–240 V, 15-kVA, 60-Hz transformer has a $Z_{eq\,2}$ of $0.0600 + j0.250$ Ω (referred to the 240-V winding). $R_{c\,LV} = 800$ Ω and $X_{m\,LV} = 160$ Ω. (a) When the transformer is delivering rated current at 240 V to a 0.8 power factor load, find the primary current and terminal voltage. (b) What error would result in calculating the primary voltage and current if the transformer were taken to be ideal?

3.11. A 15-kVA, 4160–240-V, single-phase transformer has a core loss at rated voltage of 120 W. Its per-unit resistance is 0.0100 and its per-unit reactance is 0.0600.

(a) What is the equivalent impedance of this transformer, referred to the 4160-V winding?

(b) What voltage must be applied to the 4160-V primary winding to have 240 V at the low-voltage terminals when the load on the low-voltage side is 12 kVA at 0.8 power factor, lagging?

(c) What is the efficiency under the conditions of (b) above?

(d) Using per-unit quantities, find the full-load regulation of this transformer at 0.8 power factor, leading, and at 0.8 power factor, lagging.

(e) What is the core loss of this transformer in per-unit?

(f) At what load kVA does this transformer have its maximum efficiency?

3.12. A 5-kVA single-phase dry-type transformer is rated 440–220 V. Open- and short-circuit test data are as follows:

Open-Circuit Test	Short-Circuit Test
$V_{oc} = 220$ V	$V_{sc} = 22.8$ V
$I_{oc} = 1.10$ A	$I_{sc} = 11.4$ A
$P_{oc} = 48.4$ W	$P_{sc} = 52.0$ W

(a) What is the efficiency of this transformer at full load, 0.8 power factor? At half load with the same power factor?

(b) At what load does this transformer have its maximum efficiency?

(c) Find the per-unit resistance and reactance of this transformer.

3.13. Find the voltage regulation of the transformer of Problem 3.12 at full load, 0.8 power factor, lagging, and at unity power factor.

3.14. Draw the approximate equivalent circuits referred to both high- and low-voltage windings for the transformer of Problem 3.12. Label all elements of each circuit and show their values in ohms.

3.15. Find the voltage that must be applied to the 2300-V terminals of the transformer of Problem 3.8 if the secondary voltage is 230 V and the load on the transformer is 23 kVA at a power factor of (a) 0.8, lagging, (b) 0.8, leading. Use the approximate equivalent circuit.

3.16. If the short-circuit test on a transformer is done at rated current and the open-circuit test is done at rated voltage, write a formula to calculate the fraction of rated load at which maximum efficiency will occur, in terms of P_{sc} and P_{oc}. Explain why your formula is valid.

3.17. The following are test results for a 110–7.2-kV, Δ–Y, 10-MVA, three-phase transformer. The currents given are line currents and the voltages are line-to-line voltages.

Open-Circuit Test	Short-Circuit Test
$V_{oc} = 7.20$ kV	$V_{sc} = 13.2$ kV
$I_{oc} = 16.0$ A	$I_{sc} = 52.5$ A
$P_{oc} = 72.0$ kW	$P_{sc} = 80.0$ kW

(a) Find the per-unit resistance and reactance of this transformer.

(b) Find $Z_{eq\,2}$, in ohms, referred to one phase of the Y-connected secondary.

(c) What is the full-load efficiency of this transformer at unity power factor?

(d) What is the maximum unity-power-factor efficiency of the transformer, and at what MVA load does it occur?

3.18. A 7260–240-V, single-phase transformer is rated at 10 kVA, and has an impedance referred to the primary of $Z_{eq\,1} = 100 + j400\ \Omega$. (a) What is $Z_{eq\,2}$? (b) At full load, unity power factor, what must be the voltage at the 7260-V terminals if the secondary voltage is to be at least 220 V?

3.19. A 10-kVA, single-phase, 2400–120-V, 60-Hz transformer has an equivalent impedance referred to the 120-V winding of $0.01000 + j0.0900\ \Omega$. This transformer has a core loss at rated voltage of 100 W. When rated voltage is applied to the 2400-V winding, the exciting current in that winding is found to be 0.200 A. (a) Find the full-load regulation of this transformer at a power factor of 0.8, lagging. (b) Find the full-load regulation at a power factor of 0.8, leading. (c) Find the efficiency of this transformer at full load, 0.8 power factor.

3.20. Find the maximum full-load regulation of the transformer of Problem 3.19 and the power factor at which it occurs.

3.21. If standard-open circuit and short-circuit tests were performed on the transformer of Problem 3.19, what would be the measured values of V_{oc}, I_{oc}, P_{oc}, V_{sc}, I_{sc}, and P_{sc}?

3.22. For the transformer of Problem 3.19, find (a) the per-unit impedance, (b) the per-unit core loss, and (c) the per-unit full-load copper loss.

3.23. A three-phase transformer is rated at 150 MVA, 345–275 kV. Determine the voltage and current ratings of the high-voltage and low-voltage windings of this transformer if it is connected (a) Δ–Δ, (b) Y–Δ, (c) Δ–Y. (d) If this transformer has a per-unit resistance of 0.015 and a per-unit core loss of 0.008, find the efficiency of the transformer at full load, unity power factor.

3.24. Each phase of a three-phase transformer has one winding rated at 158.77 kV, 63.9 A, and the other at 40.42 kV, 251.0 A. The per-unit resistance of this transformer is 0.015 and per-unit core loss is 0.008. (a) What are the MVA rating, core loss, and full-load copper loss of the transformer? (b) Each phase is reconnected as a 199.19–158.77-kV autotransformer, and the three phases are connected in Y, so that the line-to-line voltage ratio is 345–275 kV. A solid primary neutral connection is provided. When each winding is operated at rated current, what is the MVA rating of the auto-transformer? (Compare with the transformer of Problem 3.23) (c) Since the individual winding voltages and currents for the autotransformer are the same as they were in (a), the core and copper losses are the same as in (a). Then what is the full-load efficiency of this autotransformer at unity power factor? Compare with the transformer of Problem 3.23.

3.25. A capacitor rated at 30 kVAR, 2300 V, 60 Hz is connected to the 2300-V winding of a 230–2300-V, 30-kVA transformer. If the transformer may be considered ideal, what are the ratings of the equivalent capacitor as it appears at the 230-V terminals of the transformer? Calculate the capacitance of the capacitor itself and the capacitance appearing at the 230-V winding ter-minals.

3.26. A variable autotransformer is wound with wire capable of carrying 10 A. Rated voltage of the entire winding is 100 V, and the output voltage may be varied by means of a sliding contact from zero to 100 V. Calculate and plot as functions of output voltage (a) maximum allowable output current without exceeding 10 A in either the series or common winding, (b) the corresponding available output volt-amperes.

3.27. Open- and short-circuit tests are performed on a 500-kVA, 2300–460-V, 60-Hz, three-phase transformer with the following results:

Open-Circuit Test	Short-Circuit Test
V_{oc} = 460 V	V_{sc} = 161 V
I_{oc} = 40.0 A	I_{sc} = 125 A
P_{oc} = 3000 W	P_{sc} = 2500 W

(a) Find the per-unit resistance $R_{eq\ pu}$, the per-unit reactance $X_{eq\ pu}$, and the per-unit core loss.

(b) Find the equivalent impedance in ohms referred to one secondary wind-ing, if the transformer is connected Y–Δ.

(c) At what fraction of full load does the transformer have its maximum efficiency?

(d) What is the full-load voltage regulation of this transformer at unity power factor?

3.28. A 50-MVA, 161–69-kV, three-phase transformer has an equivalent resistance of 0.50% and an equivalent reactance of 10.0%. (a) When the transformer is delivering 50 MVA at 0.8 power factor, lagging, at 69 kV, what must be the voltage at its 161-kV terminals? (b) What is the full-load voltage regulation of this transformer at this power factor? (c) What is the full-load copper loss in watts?

3.29. A broiler production farm has a single-phase electrical heating load of 500 kW at unity power factor, and fans driven by three-phase motors requiring 75 kW at 0.8 power factor. The equipment operates at 460 V. This load is supplied by two transformers connected in V–V. Find the volt-amperes supplied by each transformer for both phase sequences.

3.30. A wattmeter, a voltmeter, and an ammeter are connected to a single-phase line through potential and current transformers. The load on the line is known to be lagging. The potential transformer has a ratio of 100 and the current transformer a ratio of 20 : 5. Assume the ratio errors are negligible. The potential transformer has a γ of $-15'$, and the current transformer has a β of $+20'$. If the voltmeter reads 100 V, the ammeter reads 4.50 A, and the wattmeter reads 360 W, what is the actual power flowing in the line? Compare with the power obtained if γ and β are neglected.

4

INDUCTION, OR ASYNCHRONOUS MACHINES

4.1 HISTORY AND APPLICATION OF INDUCTION MACHINES

The invention of induction machines in the 1880s completed the ac system of electrical power production, transmission, and utilization, which at the time was in competition with the dc system for general acceptance. The whole concept of polyphase ac, including the induction motor, was the idea of the great Yugoslavian engineer Nikola Tesla. (Tesla became a U.S. citizen in 1889.) The system was patented in 1888. The first large-scale application of the Tesla polyphase ac system was the Niagara Falls hydroplant, completed in 1895.

Today, most industrial motors of one horsepower or larger are three-phase induction machines. Induction machines require no electrical connection to the rotor windings. Instead, the rotor windings are short-circuited. Magnetic flux flowing across the air gap links these closed rotor circuits. As the rotor moves relative to the air gap flux, Faraday's-law voltages are induced in the shorted rotor windings, causing currents to flow in them. The fact that the rotor current arises from *induction*, rather than *conduction*, is the basis for the name of this class of machines. They are also called *asynchronous* (i.e., "not synchronous") machines because their operating speed is slightly less than synchronous speed in the motor mode and slightly greater than synchronous speed in the generator mode.

Induction machines are usually operated in the motor mode, so they are usually called *induction motors*. As generators, their peculiar characteristics make them suitable for only a few special applications. As motors, they have many advantages. They are rugged, relatively inexpensive, and require very little maintenance. They range in size from a few watts to about 10,000 hp. The speed of an induction motor is nearly, but not quite, constant, dropping only a few percent in going from no load to full load. The chief advantages of induction motors are:

The speed is not easily controlled.

The starting current may be five to eight times full-load current.

The power factor is low and lagging when the machine is lightly loaded.

For most applications, their advantages far outweigh their disadvantages.

4.2 CONSTRUCTION OF INDUCTION MACHINES

The *stator* core and stator windings of a three-phase induction machine are exactly like those of a synchronous machine. The only difference in construction between the two machines is in the rotor. In fact, an induction machine rotor of proper dimensions may be inserted in the bore of a synchronous machine stator, and the resulting asynchronous machine will operate perfectly well. Figure 4.1 is a cutaway view of a three-phase induction motor, and Figure 4.2 shows a vertical motor, typically used to drive pumps.

Induction machine rotors are of two types, *wound* and *squirrel cage*. In either case, the rotor windings are contained in slots in a laminated iron core which is mounted on the shaft. Examples of this construction are shown in Figure 4.3. In small machines, the rotor lamination stack is pressed directly on the shaft. In larger machines, the core is mechanically connected to the shaft through a set of spokes called a "spider."

The winding of a *wound rotor* is a polyphase winding consisting of coils placed in slots in the rotor core. It is also quite similar to the stator winding of synchronous machine. It is almost always *three* phase, and connected in Y. The three terminal

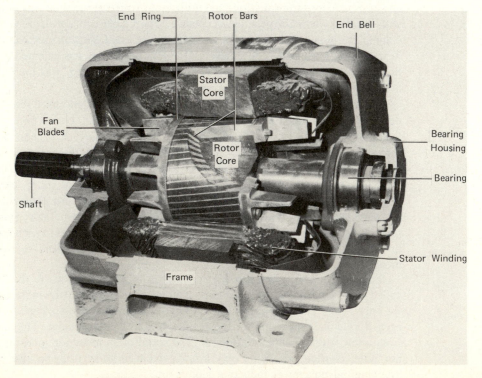

FIGURE 4.1 Cutaway three-phase squirrel cage induction motor.

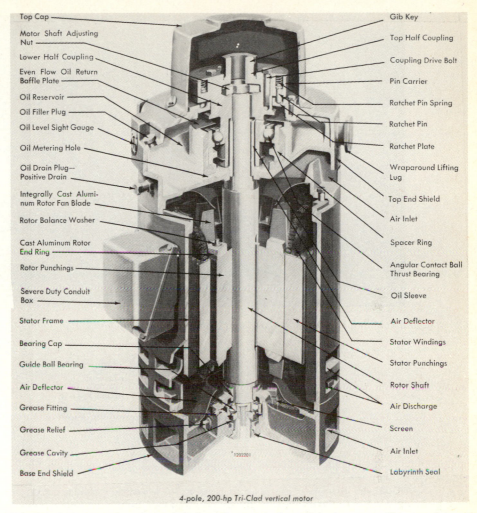

Top Cap
Motor Shaft Adjusting Nut
Lower Half Coupling
Even Flow Oil Return Baffle Plate
Oil Reservoir
Oil Filler Plug
Oil Level Sight Gauge
Oil Metering Hole
Oil Drain Plug-- Positive Drain
Integrally Cast Aluminum Rotor Fan Blade
Rotor Balance Washer
Cast Aluminum Rotor End Ring
Rotor Punchings
Severe Duty Conduit Box
Stator Frame
Bearing Cap
Guide Ball Bearing
Air Deflector
Grease Fitting
Grease Relief
Grease Cavity
Base End Shield

Gib Key
Top Half Coupling
Coupling Drive Bolt
Pin Carrier
Ratchet Pin Spring
Ratchet Pin
Ratchet Plate
Wraparound Lifting Lug
Top End Shield
Air Inlet
Spacer Ring
Angular Contact Ball Thrust Bearing
Oil Sleeve
Air Deflector
Stator Windings
Stator Punchings
Rotor Shaft
Air Discharge
Screen
Air Inlet
Labyrinth Seal

4-pole, 200-hp Tri-Clad vertical motor

FIGURE 4.2 Cutaway view of a vertical, three-phase, squirrel cage induction motor. (Courtesy of the General Electric Company.)

leads are brought to slip rings mounted on the shaft. Carbon brushes riding on these slip rings are shorted together for normal operation. Wound rotors are usually used only in large machines. External resistances are inserted into the rotor circuit, via the brushes, to improve starting characteristics. As the motor accelerates, the external resistances are gradually reduced to zero.

Squirrel cage rotor windings consist of solid bars of conducting material in the rotor slots. These *rotor bars* are shorted together at the two ends of the rotor by *end rings*. In large machines, the rotor bars may be of copper alloy, driven into the slots and brazed to the end rings. Rotors of up to about 20 in. in diameter usually have die-cast aluminum bars. The core laminations for such rotors are stacked in a mold, which is then filled with molten aluminum. In this single economical process, the rotor bars, end rings, and cooling-fan blades are cast at the same time. (See Figures 4.1 and 4.2.)

To show what the system of rotor conductors looks like, a rotor has been immersed in an acid that does not attack aluminum in order to eat away the iron core; the result is shown in Figure 4.4.

FIGURE 4.3 Examples of induction machine rotors. (*a*) Wound rotor, showing slip rings (Courtesy of General Electric Canada, Inc.). (*b*) Squirrel cage rotor for a 10-hp motor.

4.3 HOW THE INDUCTION MOTOR WORKS

When three-phase voltages are applied to the stator winding terminals, balanced, three-phase currents flow in the phase windings. As a result, a rotating MMF field is produced in the air gap of the machine. This field is exactly like that of the stator of a synchronous machine. Its speed of rotation is given by Equation 2.1, which will be repeated here:

$$\omega_s = \frac{4\pi f_1}{p} \quad \text{rad/s}$$

or   (4.1)

$$n_s = \frac{120 f_1}{p} \quad \text{rev/min}$$

where ω_s or n_s is called the *synchronous speed*, f_1 is the frequency of the stator voltages and currents, and p is the number of poles of the winding. The strength of

FIGURE 4.4 The bars and end rings of a squirrel cage rotor.

this field is proportional to the rms stator current, and may be expressed (as in Equation 2.17) by

$$F_1 = N_{I1}I_1 \quad \text{A/pole} \tag{4.2}$$

where N_{I1} is the effective stator turns per pole[1] and I_1 is the rms stator phase current.

Suppose for the moment that the rotor is being driven by an adjustable-speed motor so that it is rotating at the same speed as the stator field. The relative velocity between the stator field and rotor conductors is zero. There is a mutual flux between the rotor and stator, due to the stator MMF $\vec{F}_1$. This flux links the rotor; but the flux linking any one conductor is constant, hence $d\lambda/dt = 0$, and no voltages are induced in the rotor circuits. There are no rotor currents, and no rotor MMF field. The situation is illustrated in Figure 4.5.

If the drive motor is now turned off, the friction and windage of the two machines will cause the rotor of the induction machine to slow down. The rotor conductors

---

[1]See pp. 56–59 of Chapter 2 for the derivation of the expression: $N_{I1} = (3/2)(4/\pi)(N_{\phi1}/p)k_{w1}\sqrt{2}$, where $N_{\phi1}$ and k_{w1} are the series turns per phase and winding factor of the stator winding.

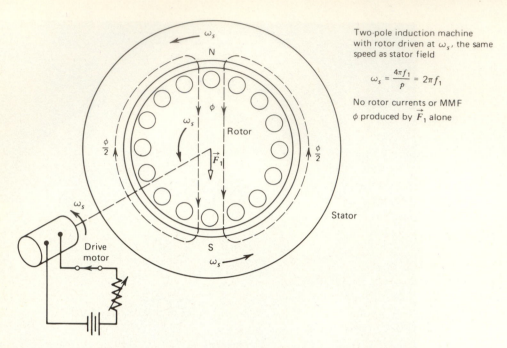

Two-pole induction machine with rotor driven at ω_s, the same speed as stator field

$$\omega_s = \frac{4\pi f_1}{p} = 2\pi f_1$$

No rotor currents or MMF
ϕ produced by $\vec{F}_1$ alone

FIGURE 4.5 Induction machine, externally driven at synchronous speed.

begin to slip backward through the stator flux field. The conductors now experience changing flux linkages, and voltages are produced in them according to Faraday's law. As a result, current flows in the closed rotor circuits. These rotor currents produce a MMF field having magnetic poles on the rotor surface.

The new situation, with the drive monitor turned off, is shown in Figure 4.6. Magnetic forces between the magnetic poles on the rotor surface and the flux field produce torque in the direction of field rotation. The rotor is wound to produce the

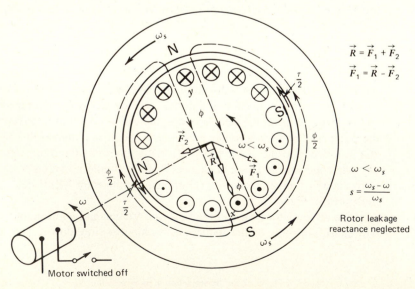

$$\vec{R} = \vec{F}_1 + \vec{F}_2$$
$$\vec{F}_1 = \vec{R} - \vec{F}_2$$

$$\omega < \omega_s$$

$$s = \frac{\omega_s - \omega}{\omega_s}$$

Rotor leakage reactance neglected

FIGURE 4.6 Lightly loaded induction motor.

same number of poles as the stator winding. As the machine slows down, the rotor conductors slip faster through the flux. Greater voltages are induced in the rotor circuits, heavier currents flow in the rotor circuits, stronger magnetic poles are produced on the rotor surface, and more torque is developed. When this torque balances friction and windage load on the machine, the rotor speed stabilizes at a value somewhat less than the field speed.

When the motor is lightly loaded as in Figure 4.6, the speed is only slightly less than synchronous speed (say, $\omega = 0.999\omega_s$). The conductors are slipping very slowly backward through the flux field, and the rotor voltages and frequency are both very low. Because the frequency is so low, rotor reactance is negligible, and the currents in the rotor conductors are practically in phase with the induced voltages. Consequently, the rotor MMF vector $\vec{F}_2$ is at right angles to the flux vector ϕ. With current flowing in the rotor, the flux is no longer due to the stator MMF $\vec{F}_1$ but rather to $\vec{R}$, which is the *resultant* of $\vec{F}_1$ and $\vec{F}_2$.

Note in Figure 4.6 that the position of the pattern of rotor currents is determined by the positions of the flux poles of the air gap field. The reason for this is that the rotor voltages are being induced by air gap fluxes. Since the air gap field is rotating at synchronous speed, the rotor current pattern is also revolving at synchronous speed. The rotor itself is rotating at *less* than synchronous speed, however. The *pattern* of rotor conductor currents is thus moving forward, relative to the rotor, at a speed of $\omega_s - \omega$ rad/s. The position of the rotor MMF vector $\vec{F}_2$ is determined by the rotor current pattern and is directed by the right-hand rule along the axis of that pattern. The following statements thus may be made about the rotation of the rotor current pattern and its MMF vector $\vec{F}_2$:

1. They rotate at synchronous speed with respect to the stator.
2. They are stationary with respect to the air gap field vectors $\vec{\phi}$ and $\vec{R}$.
3. They rotate forward at $\omega_s - \omega$ rad/s, relative to the rotor.

4.4   THE CONCEPT OF SLIP

The magnitudes and frequency of the rotor voltages depend on the speed of the *relative* motion between the rotor and the flux crossing the air gap, ϕ. The field speed is given by Equation 4.1. Let the angular speed of the rotor, in radians per second (rad/s) be assigned the symbol ω, and let n be its speed in revolutions per minute (rev/min). The *slip speed* expresses the speed of the rotor *relative to the field*, and is given by

$$\text{Slip speed} = \omega - \text{ rad/s}$$

or   (4.3)

$$= n_s - n \quad \text{rev/min}$$

The *per-unit slip*, usually called simply *the slip*, is a very useful quantity in studying induction machines. It is given the symbol s and is defined as follows:

$$s \stackrel{\Delta}{=} \frac{\omega_s - \omega}{\omega_s} = \frac{n_s - n}{n_s}$$ (4.4)

4.5 THE FREQUENCY OF ROTOR VOLTAGES AND CURRENTS

Consider a typical pair of rotor bars, labeled x and y in Figure 4.6. As the rotor slips backward through the flux field, the flux linking these bars will vary cyclically. In this figure, these conductors are shown at the instant the rate of change of flux linkages is a maximum, and hence the voltage induced in the rotor circuit composed of these two bars and the end rings is at its peak. When y has slipped to the position now occupied by x, the polarity will have reversed. Conductor y will have moved *relative to the field* a distance of one pole pitch, and one half cycle of rotor voltage will have been generated. Thus one cycle of rotor voltage is generated as a given rotor conductor slips past two poles of the air gap flux field. In other words, *one cycle of rotor voltage corresponds to 360 electrical degrees of slip.* Then the frequency of the rotor voltages and currents is given by

$$f_2 = \text{pole pairs slipped per second} = \frac{p}{2} \cdot (\text{slip speed in rev/s})$$

Equation 4.3 is divided by 2π to obtain revolutions per second. Then

$$f_2 = \frac{p}{2} \frac{\omega_s - \omega}{2\pi} \tag{4.5}$$

By Equation 4.4, the slip speed

$$\omega_s - \omega = s\omega_s \tag{4.6}$$

and by Equation 4.1, $\omega_s = 4\pi f_1/p$. Substituting these into Equation 4.5,

$$f_2 = sf_1 \tag{4.7}$$

4.6 THE INDUCTION MOTOR UNDER LOAD

When a heavy mechanical load is placed on the induction motor, the rotor slows down still more; that is, the slip increases somewhat. For most motors, the full-load slip will be about 0.03. The frequency of the rotor voltages and currents will still be quite small (say, about 2Hz), but the effect of leakage reactance can be neglected no longer. There will be a time phase lag between the rotor currents and voltages. If L_{l2} is the effective leakage inductance of a typical rotor circuit composed of a pair of rotor bars and their end ring connection, the reactance of this circuit will be

$$\begin{aligned} X_R &= 2\pi f_2 L_{l2} \\ &= 2\pi s f_1 L_{l2} \end{aligned} \tag{4.8}$$

The phase angle of lag between maximum voltage and maximum current in a typical rotor circuit will be the power-factor angle:

$$\theta_2 = \tan^{-1} \frac{X_R}{R_R} \tag{4.9}$$

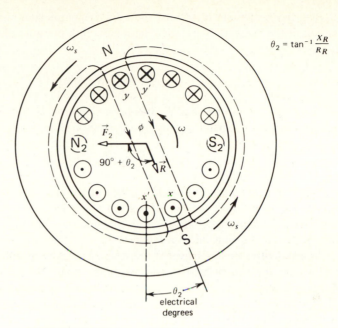

FIGURE 4.7 Shift in rotor MMF results from leakage reactance.

where R_R is the ac resistance of the same circuit. Peak current in a pair of bars occurs θ_2° in the cycle after peak induced voltage. During this interval, the rotor will have slipped θ_2 electrical degrees relative to the rotating air gap field. If maximum voltage is being induced in conductors x and y, maximum current will not flow in these conductors until they have slipped an additional θ_2 electrical degrees. Thus in Figure 4.7, maximum current occurs in conductors x' and y', which lag x and y in rotation by θ_2 electrical degrees. Under load, then, the entire rotor current pattern is shifted from the lightly loaded position, with the result that the rotor MMF vector $\vec{F}_2$ is now at an angle of $90° + \theta_2$ behind the flux, and therefore behind $\vec{R}$, the MMF responsible for the flux.

4.7 CIRCUIT MODEL OF THE INDUCTION MACHINE

The induction machine, in certain aspects, is like a synchronous machine. In others, it is like a transformer. Charles Steinmetz developed the most widely used circuit model of the polyphase induction machine. In developing his model, Steinmetz took a transformer approach. This model has proved very useful in understanding the characteristics of the induction machines. It is not too accurate in providing numerical data on their performance when the values of the model impedances are determined from laboratory measurements.

The Stator Circuit

As in the synchronous machine, the air gap flux is due to the resultant of the rotor and stator MMF vectors:

$$\vec{R} = \vec{F}_1 + \vec{F}_2 \tag{4.10}$$

If R is the magnitude of $\vec{R}$, and the effective permeance per pole of the magnetic circuit of the machine is $\mathcal{P}_p$, then the mutual flux per pole flowing across the air gap is given by

$$\phi = \mathcal{P}_p R \quad \text{Wb} \tag{4.11}$$

The angle between $\vec{R}$ and $\vec{F}_2$ has been shown to be $90° + \theta_2$. If $\vec{R}$ and $\vec{F}_2$ are known, $\vec{F}_1$ may be found by solving Equation 4.10:

$$\vec{F}_1 = \vec{R} - \vec{F}_2 \tag{4.12}$$

Furthermore, the discussion of three-phase stator windings in Chapter 2 pointed out that $\vec{F}_1$ is aligned with the axis of a given phase when the current is a maximum in that phase. The magnetic conditions in the motor at the instant the current in phase a is a maximum are illustrated in Figure 4.8.

In this figure, as in Chapter 2, positive I_1 is taken as entering the a' conductors and leaving the a (unprimed) conductors of the stator, thus producing MMF along the positive a-phase axis. In the discussion of synchronous machines, the direction of *current* was reversed for the *motor* mode of operation. In the present study of induction machines in the motor mode, it is expedient to reverse the *voltages* instead of the currents. This is equivalent to reversing connections to each phase winding before constructing the model. The relationships between the polarities assumed for Chapter 2 and those of the present chapter are shown in Figure 4.9. In Chapter 2, E_ϕ was a source voltage (a rise), driving the current out of the positive terminal a. For the induction motor, E_1, the negative of E_ϕ, is the *countervoltage* (a drop) in opposition to the flow of I_1, thus permitting electrical power to be fed *into* the machine. If e_1 and e_ϕ are the instantaneous values of E_1 and E_ϕ, respectively, then

$$e_\phi = \frac{d\lambda_\phi}{dt}$$

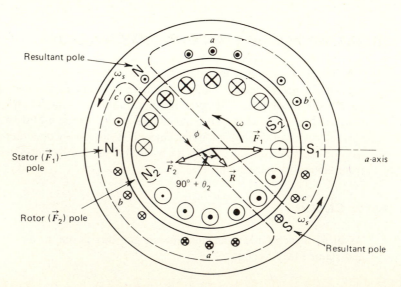

FIGURE 4.8 Current patterns and MMF resolution at the instant the stator current in phase a is maximum.

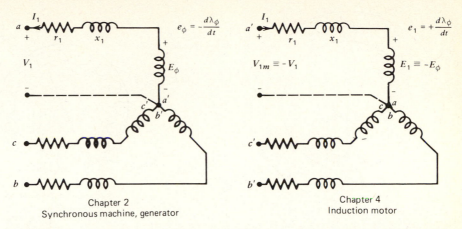

FIGURE 4.9 Relations between polarities assumed for synchronous and induction machine models.

and

$$e_1 = -e_\phi = +\frac{d\lambda_\phi}{dt}$$

where λ_ϕ is the instantaneous total linkage of the stator phase winding with the mutual flux between the rotor and stator. As in the synchronous machine, the peak voltage induced in each phase of the stator winding by the air gap flux is given by

$$|E_1|_{max} = |E_\phi|_{max} = N_{\phi 1}k_{w1}\omega_e\phi \qquad (4.13)$$

where

$$N_{\phi 1} = \text{number of turns in series in one phase winding of the stator}$$
$$k_{w1} = \textit{winding factor, which corrects for the distribution and pitch}$$
$$\text{of the stator coils}$$
$$\omega_e = 2\pi f_1, \text{the electrical angular line frequency}$$
$$\phi = \text{flux per pole of the field crossing the air gap}$$

Let N_{E1} be defined as in Chapter 2,

$$N_{E1} \overset{\Delta}{=} \frac{N_{\phi 1}k_{w1}}{\sqrt{2}} \qquad (4.14)$$

the *effective voltage turns per stator phase*. Then the rms voltage induced in each stator phase is given by Equation 4.13 divided by $\sqrt{2}$:

$$|E_1| = |E_\phi| = \omega_e N_{E1}\phi \qquad (4.15)$$

By Equation 4.11, this voltage is nonlinearly related to R:

$$|E_1| = |E_\phi| = \omega_e N_{E1}\mathscr{P}_p R \qquad (4.16)$$

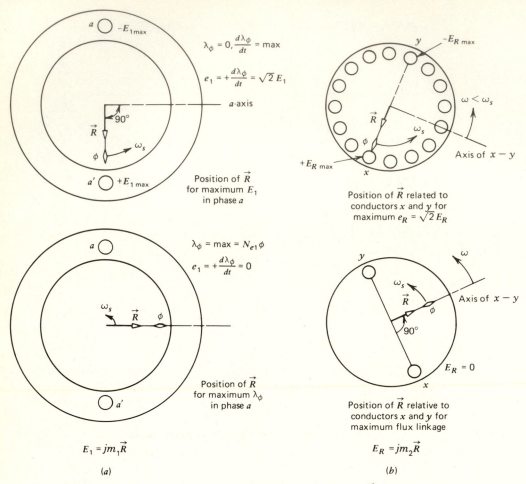

FIGURE 4.10 Phasor/vector relationships between the resultant MMF $\vec{R}$ and voltages induced in the stator and rotor windings (*a*) Stator. (*b*) Rotor.

The relationship is nonlinear because the magnetic circuit permeance $\mathcal{P}_p$ is subject to saturation. All the constants, together with the frequency and $\mathcal{P}_p$, in Equation 4.15 may be combined into a nonlinear multiplier m_1:

$$|E_1| = |E_\phi| = m_1 R \tag{4.17}$$

In Chapter 2 it was shown that

$$E_\phi = -jm_1\vec{R}$$

and a study of Figure 4.10*a* will show that maximum E_1 occurs in a given phase 90° *before* the flux due to $\vec{R}$ is a maximum in that phase. Thus the relationship between E_1 and $\vec{R}$ may be expressed as a phasor relationship:

$$E_1 = -E_\phi = +jm_1\vec{R} \tag{4.18}$$

Each phase winding of the stator has resistance and reactance due to leakage flux. An equivalent circuit of one phase of the stator is shown in Figure 4.11. In this circuit,

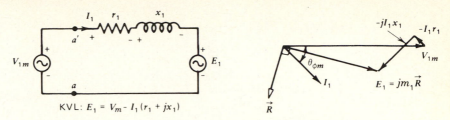

$$\text{KVL: } E_1 = V_m - I_1(r_1 + jx_1)$$

FIGURE 4.11 Preliminary stator model of an asynchronous machine.

r_1 represents the winding resistance and x_1 its leakage reactance. In the usual model of a *synchronous machine*, the resolution of magnetic field MMFs is modeled by a Kirchhoff's law phasor addition of *voltages*. In the Steinmetz *induction machine model*, however, a Kirchhoff's law addition of phasor *currents* at a node model the vector Equation 4.12. If by Equation 4.2, $\vec{F} = N_{I1}I_1$, this substitution may be made into Equation 4.12:

$$N_{I1} I_1 = \vec{R} - \vec{F}_2 \quad \text{A-turns} \tag{4.19}$$

Dividing by N_{I1}:

$$I_1 = \frac{\vec{R}}{N_{I1}} - \frac{\vec{F}_2}{N_{I1}} \quad \text{A} \tag{4.20}$$

Each term in Equation 4.20 is a phasor current that models a magnetomotive force vector. In addition, I_1 is the actual phase current in the stator winding. The two currents components on the right-hand side may be defined as follows:

$$I_\phi \overset{\Delta}{=} \frac{\vec{R}}{N_{I1}} \quad I_2 \overset{\Delta}{=} -\frac{\vec{F}_2}{N_{I1}} \tag{4.21}$$

The definition of I_ϕ may be modified by solving Equation 4.18 for $\vec{R}$:

$$I_\phi = \frac{\vec{R}}{N_{I1}} = \frac{E_1}{jm_1N_{I1}} \overset{\Delta}{=} \frac{E_1}{jX_m} \tag{4.22}$$

where, as in synchronous machine theory, the magnetizing reactance is defined by

$$X_m \overset{\Delta}{=} m_1 N_{I1}$$

Making these substitutions into Equation 4.20 gives

$$I_1 = I_\phi + I_2 \tag{4.23}$$

The circuit model of one phase of the asynchronous machine may now be drawn as in Figure 4.12. Note that Equation 4.23 expresses Kirchhoff's current law at node N and is the model of Equation 4.12: $\vec{F}_1 = \vec{R} - \vec{F}_2$.

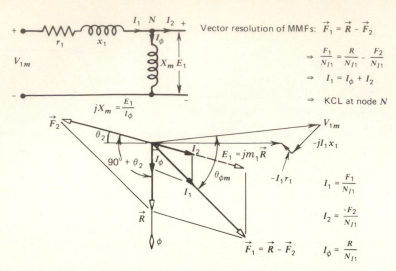

FIGURE 4.12 Stator-phase circuit model of an asynchronous machine, showing resolution of magnetic field vectors modeled by Kirchhoff's current law at a node.

The Rotor

The words "rotor circuit" have a clear and unambiguous meaning in the case of a wound rotor, where they mean one phase circuit of the polyphase winding. The voltage and MMF of such a winding have been fully discussed (Chapter 2). Since all rotor bars in a squirrel cage winding are interconnected, the precise meaning of a "rotor circuit" is not clear. For the present, however, a "circuit" may be considered to consist of two rotor bars approximately 180 electrical degrees apart, together with the corresponding current paths through the end rings.

Refer to Figure 4.10*b*. Considering the polarity chosen, the voltage induced in rotor circuit *x–y* leads the maximum flux linkage by 90°, and may be written as

$$E_R = jm_2\vec{R} \tag{4.24}$$

Since

$$|E_R| = \omega_{e2}N_{\phi2}k_{w2}\phi/\sqrt{2}$$

and

$$\phi = \mathcal{P}_pR$$

then

$$m_2 = \omega_{e2}N_{\phi2}k_{w2}\mathcal{P}_p/\sqrt{2} = 2\pi f_2N_{\phi2}k_{w2}\mathcal{P}_p/\sqrt{2} \tag{4.25}$$

In Equation 4.25, $N_{\phi2}$ is the turns per phase of the rotor winding and k_{w2} is its winding factor. Again, an effective *voltage turns per phase* may be defined:

$$N_{E2} \overset{\Delta}{=} \frac{N_{\phi2}k_{w2}}{\sqrt{2}} \tag{4.26}$$

a term that has definite meaning for a wound rotor but is rather vague for a squirrel cage. Since both E_1 and E_R are induced by ϕ, there is a definite relationship between the two:

$$E_R = j2\pi f_2 N_{E2} \mathscr{P}_p \vec{R} \qquad \text{(Equations 4.24, 4.25, 4.26)}$$

$$E_1 = j2\pi f_1 N_{E1} \mathscr{P}_p \vec{R} \qquad \text{(Equations 4.16, 4.18)}$$

Then

$$\frac{E_1}{E_R} = \frac{N_{E1} f_1}{N_{E2} f_2} \qquad (4.27)$$

But by Equation 4.7, $f_2 = sf_1$; so

$$\frac{E_1}{E_R} = \frac{N_{E1}}{sN_{E2}} \stackrel{\Delta}{=} \frac{1}{s} a_E \qquad (4.28)$$

where

$$a_E \stackrel{\Delta}{=} \frac{N_{E1}}{N_{E2}}$$

The symbol a_E represents the *voltage turns ratio* between the stator and rotor windings.

Each rotor circuit has a certain amount of resistance and leakage inductance. One such circuit may be modeled as in Figure 4.13. Let L_{IR} be the leakage inductance of one rotor circuit. Then the leakage reactance of the rotor is given by

$$X_R = 2\pi f_2 L_{IR} = s(2\pi f_1 L_{IR}) \qquad (4.29)$$

and is affected by the slip. Define

$$X_{RB} \stackrel{\Delta}{=} 2\pi f_1 L_{IR}$$

Then $\qquad\qquad\qquad\qquad\qquad\qquad\qquad\qquad\qquad\qquad\qquad\qquad\qquad$ (4.30)

$$X_R = sX_{RB}$$

where X_{RB} is the rotor leakage reactance at $s = 1$ (rotor "blocked" so that it cannot turn). The rotor circuit current is then

$$I_R = \frac{E_R}{R_R + jX_R} = \frac{1}{a_E} \frac{sE_1}{R_R + jsX_{RB}} \qquad (4.31)$$

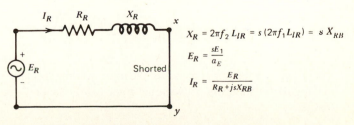

FIGURE 4.13 Model of one rotor circuit.

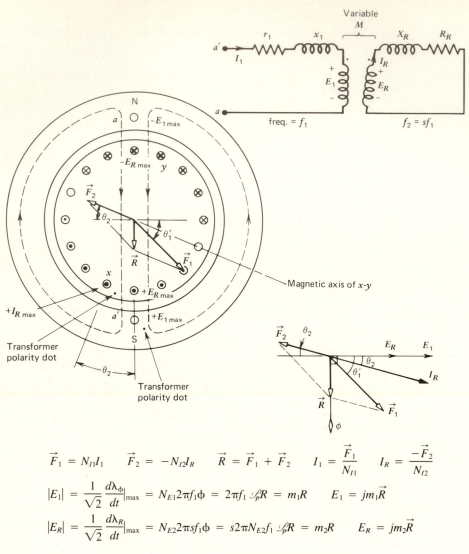

$$\vec{F}_1 = N_{I1}I_1 \quad \vec{F}_2 = -N_{I2}I_R \quad \vec{R} = \vec{F}_1 + \vec{F}_2 \quad I_1 = \frac{\vec{F}_1}{N_{I1}} \quad I_R = \frac{-\vec{F}_2}{N_{I2}}$$

$$|E_1| = \frac{1}{\sqrt{2}} \frac{d\lambda_\phi}{dt}\bigg|_{max} = N_{E1}2\pi f_1\phi = 2\pi f_1 \mathscr{L}_\rho R = m_1 R \qquad E_1 = jm_1\vec{R}$$

$$|E_R| = \frac{1}{\sqrt{2}} \frac{d\lambda_R}{dt}\bigg|_{max} = N_{E2}2\pi s f_1\phi = s2\pi N_{E2}f_1 \mathscr{L}_\rho R = m_2 R \qquad E_R = jm_2\vec{R}$$

FIGURE 4.14 Relative polarities of induction machine quantities.

To complete the model, the rotor circuit current must be related to the stator current component I_2. The magnitude of the rotor MMF F_2 is proportional to the currents in the rotor conductors and may be written

$$F_2 = N_{I2}|I_R| \tag{4.32}$$

(For a wound rotor, N_{I2} has a very definite value,

$$N_{I2} = \frac{q_2}{2} \frac{4}{\pi} \sqrt{2} \frac{N_{\phi2}}{p} k_{w2} \tag{4.33}$$

where q_2 is the number of phases of the rotor winding, usually 3.) Figure 4.14 illustrates the angular relationship between the rotor MMF vector $\vec{F}_2$ and the current phasor I_R; that is,

$$\vec{F}_2 = -N_{I2}I_R \tag{4.34}$$

Note that current into the dot at x and out of the paper at y would produce a magnetomotive force along the positive magnetic axis, but the position of $\vec{F}_2$ is along the negative axis. The situation is analogous to that of a transformer, where positive secondary voltage occurs at the dotted terminal and positive secondary current is taken as coming out of the dotted terminal, thus producing a negative MMF. By Equations 4.21 and 4.34,

$$I_2 = -\frac{F_2}{N_{I1}} = +\frac{N_{I2}}{N_{I1}} I_R \tag{4.35}$$

$$\stackrel{\Delta}{=} \frac{I_R}{a_I}$$

where $a_I = N_{I1}/N_{I2}$ is the *current turns ratio* between the stator and rotor windings. The load component of the stator current, I_2, may now be written in terms of rotor quantities. Applying Equation 4.35 to Equation 4.31,

$$I_2 = \frac{1}{a_I} I_R = \frac{1}{a_I a_E} \frac{sE_1}{R_R + jsX_{RB}} \tag{4.36}$$

Define

$$r_2 \stackrel{\Delta}{=} a_I a_E R_R$$
$$x_2 \stackrel{\Delta}{=} a_I a_E X_{RB} \tag{4.37}$$

Then

$$I_2 = \frac{sE_1}{r_2 + jsx_2} \tag{4.38}$$

With a stroke of genius, Steinmetz divided top and bottom of this expression for I_2 by s:

$$I_2 = \frac{E_1}{(r_2/s) + jx_2} \tag{4.39}$$

So the impedance that will give I_2 when connected across E_1 is

$$z_2 \stackrel{\Delta}{=} \frac{r_2}{s} + jx_2$$

The completed model is shown in Figure 4.15.

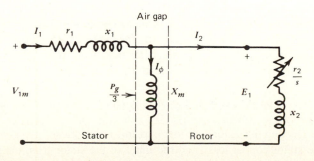

FIGURE 4.15 Steinmetz model of one phase of a three-phase induction machine.

4.8 LOSSES, POWER FLOW, AND EFFICIENCY OF INDUCTION MOTORS

As in all machines, the law of conservation of energy teaches that the output power of an induction motor, under steady-state conditions, is found by subtracting the losses from the input power. The input power is given by

$$P_{in} = \sqrt{3}V_L I_L \cos \theta_{\phi m} = 3|V_{1m}||I_1| \cos \theta_{\phi m} \qquad (4.40)$$

The losses are:

1. Stator copper loss:

$$\text{SCL} = 3I_1^2 r_1 \qquad (4.41)$$

2. Core losses $= P_{h+e}$.
3. Rotor copper loss. Since the model has referred the rotor resistances, regardless of the number of rotor phases, to the three-phase stator,

$$\text{RCL} = 3I_2^2 r_2 \qquad (4.42)$$

4. Friction and windage $= P_{fw}$
5. Stray load loss $= P_{LL}$, consisting of all losses not covered above, such as losses due to harmonic fields.

There is no general agreement as to how to treat core losses in the model. The core loss resulting from stator *leakage* flux is not negligible, as it is in a transformer. The rotor core loss varies with rotor frequency, and hence with the slip. Under running conditions the slip is only about 0.03, the rotor frequency is only about 2 Hz, and the rotor core loss is negligible. At starting, and during acceleration, however, the rotor core loss is high and decreasing, while the friction and windage start at zero and increase. As a result, the sum of the friction, windage, and core losses is roughly constant. This leads to the term *rotational losses*:

$$P_{rot} \stackrel{\Delta}{=} P_{fw} + P_{h+e} + P_{LL} \qquad (4.43)$$

While not correct in concept, this quantity has the questionable virtue of convenience!

It should be understood that different design groups and different authors have different ways of handling losses in induction machines which have proved to be satisfactory in their own work. In the present study, *performance of the motor will be calculated on the basis that only the copper losses are effective; then all other losses will be subtracted from the output power thus obtained.* (See Equation 4.46.)

A most important quantity in the analysis of induction machines is the power transferred by the air gap magnetic field from the stator windings to the rotor. This quantity will be given the symbol P_g:

$P_g \stackrel{\Delta}{=}$ power crossing the air gap from stator to rotor, called the *air gap power*

Neglecting the stator core losses at this point, the air gap power is given by subtracting the stator copper loss from the input to the stator from the electrical source:

$$P_g = P_{in} - \text{SCL} = \sqrt{3}V_L I_L \cos \theta_{\phi m} - 3I_1^2 r_1 \qquad (4.44)$$

the rotor power input P_g is partly consumed as rotor copper loss (RCL), and the remainder is available to develop mechanical power. Let DMP represent this developed mechanical power:

$$\text{DMP} \overset{\Delta}{=} \text{developed mechanical power}$$

Then

$$\text{DMP} = P_g - \text{RCL} = P_g - 3I_2^2 r_2 \qquad (4.45)$$

Having determined the *developed* mechanical power by subtracting only copper losses from the input, the net mechanical output power will be found by subtracting the rotational losses from the DMP. Let

$$P_{\text{out}} \overset{\Delta}{=} \text{mechanical power output}$$

Then

$$P_{\text{out}} = \text{DMP} - P_{\text{rot}} = \text{DMP} - P_{\text{fw}} - P_{h+e} - P_{\text{LL}} \qquad (4.46)$$

■ **Example 4.1**

An induction motor draws 25 A from a 460-V, three-phase line at a power factor of 0.85, lagging. The stator copper loss is 1000 W, and the rotor copper loss is 500 W. The "rotational" losses are friction and windage = 250 W, core loss = 800 W, and stray load loss = 200 W. Calculate (a) the air gap power P_g, (b) the developed mechanical power DMP, (c) the output horsepower, and (d) the efficiency.

Solution

(a) $P_g = P_{\text{in}} - \text{SCL}$

$$= \sqrt{3} \cdot 460 \cdot 25 \cdot 0.85 - 1000 = 16{,}931 - 1000 = 15{,}931 \text{ W}$$

(b) $\text{DMP} = P_g - \text{RCL} = 15{,}931 - 500 = 15{,}431 \text{ W}$

(c) $P_{\text{out}} = \text{DMP} - P_{\text{rot}}$

$$= \text{DMP} - (P_{h+e} + P_{fw} + P_{LL})$$

$$= 15{,}431 - 1250 = 14{,}181 \text{ W}$$

$$\text{Horsepower} = \frac{P_{\text{out}}}{746} = 19.0 \text{ hp}$$

(d) $\eta = \dfrac{P_{\text{out}}}{P_{\text{in}}} = \dfrac{14{,}181}{16{,}931}$, or 83.8% ■

4.9 AIR GAP POWER: THE MAGIC QUANTITY

Figure 4.15 shows the model of one phase of a three-phase induction motor. The rotor input power from one stator phase is shown as one third of the total air gap power, $P_g/3$. It is evident that this power must be consumed by the resistance r_2/s:

$$\frac{P_g}{3} = I_2^2 \frac{r_2}{s}$$

or

$$P_g = 3I_2^2 \frac{r_2}{s} \qquad (4.47)$$

From Equation 4.42, the rotor copper loss is given by

$$RCL = 3I_2^2 r_2 = sP_g \qquad (4.48)$$

by comparison with Equation 4.47.

The developed mechanical power is

$$DMP = P_g - RCL$$
$$= 3I_2^2\left(\frac{r_2}{s} - r_2\right) = 3I_2^2 r_2\left(\frac{1-s}{s}\right) \qquad (4.49)$$

or

$$DMP = 3I_2^2 \frac{r_2}{s}(1 - s) = P_g(1 - s) \qquad (4.50)$$

The torque developed by the electromagnetic energy conversion process is given by the developed mechanical power divided by the angular velocity of the rotor:

$$\tau_d = \frac{DMP}{\omega} \text{ N-m} \qquad (4.51)$$

But by Equation 4.6, $\omega = \omega_s (1 - s)$. Then

$$\tau_d = \frac{DMP}{\omega_s(1 - s)} = \frac{P_g(1 - s)}{\omega_s(1 - s)} = \frac{P_g}{\omega_s} \text{ N-m} \qquad (4.52)$$

Once the air gap power is determined, three quantities may be found from the slip and synchronous speed!

$$RCL = sP_g \text{ W (Equation 4.48)}$$

$$DMP = (1 - s)P_g \text{ W (Equation 4.50)}$$

$$\tau_d = \frac{P_g}{\omega_s} \text{ N-m (Equation 4.52)}$$

In the English system of units,

$$\tau_d = 7.04\frac{P_g}{n_s} \text{ lb-ft} \qquad (4.53)$$

where P_g is in watts and n_s is given by Equation 4.1 and is in rev/min. Since the developed torque is given by P_g/ω_s, the air gap power is often called the *torque in synchronous watts*.

Quite often the rotational losses are small, and τ_d, the developed torque, is sufficiently close to the output torque to be used for practical engineering purposes. The actual output torque is given by

$$\tau = \frac{P_{out}}{\omega} = \frac{DMP - P_{rot}}{\omega_s(1 - s)} \text{ N-m} \tag{4.54}$$

or in the English system,

$$\tau = 7.04\frac{P_{out}}{n} = \frac{5252 \times \text{output horsepower}}{n} \text{ lb-ft} \tag{4.55}$$

■ **Example 4.2**

If the frequency of the source in Example 4.1 is 60 Hz and the machine has four poles, find (a) the slip, (b) the operating speed, (c) the developed torque, and (d) the output torque.

Solution

(a) By Equation 4.48,

$$s = \frac{RCL}{P_g} = \frac{500}{15,931} = 0.0314$$

(b)

$$\omega_s = \frac{4\pi60}{p} = \frac{754}{4} = 188.5 \quad \text{rad/s}$$

$$n_s = \frac{120 \cdot 60}{4} = 1800 \quad \text{rev/min}$$

$$\omega = \omega_s(1 - s) = 188.5(1 - 0.0314) = 182.6 \quad \text{rad/s}$$

$$n = n_s(1 - s) = 1800 \cdot 0.9689 = 1744 \quad \text{rev/min}$$

(c)

$$\tau_d = \frac{DMP}{\omega} = \frac{15,431}{182.6} = 84.5 \text{ N-m}$$

(d)

$$\tau = \frac{P_{out}}{\omega} = \frac{14,181}{182.6} = 77.7 \quad \text{N-m}$$

Note: The output torque is 8 percent less than the developed torque. ■

4.10 SEPARATION OF MECHANICAL LOAD FROM ROTOR COPPER LOSS IN THE CIRCUIT MODEL

In the circuit model as shown in Figure 4.15, the resistance r_2/s absorbs the total rotor input from the stator phase; that is,

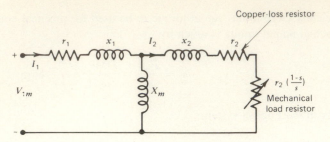

FIGURE 4.16 Circuit model with effect of mechanical load separated from rotor copper loss.

$$\frac{P_g}{3} = I_2^2 \frac{r_2}{s} = \frac{RCL}{3} + \frac{DMP}{3}$$

or by Equations 4.48 and 4.50,

$$I_2^2 \frac{r_2}{s} = I_2^2 r_2 + I_2^2 r_2 \left(\frac{1-s}{s}\right) \tag{4.56}$$

To check this last expression, divide the right-hand side of Equation 4.56 by I_2^2:

$$r_2 + r_2\left(\frac{1-s}{s}\right) = \frac{sr_2 + r_2 - sr_2}{s} = \frac{r_2}{s} \tag{4.57}$$

It is seen that r_2/s may be divided into two components: r_2, representing the rotor copper loss per phase, and $r_2(1-s)/s$, representing the developed mechanical power. The resulting circuit model is shown in Figure 4.16.

4.11 PERFORMANCE CALCULATIONS USING THE CIRCUIT MODEL

If the impedance elements of the model of a given motor are known, the steady-state performance of the motor at any speed may be calculated on the basis of relationships previously derived. The steps are:

1. Calculate synchronous speed:

$$\omega_s = \frac{4\pi f_1}{p} \quad \text{or} \quad n_s = \frac{120 f_1}{p}$$

2. For the desired speed, calculate the slip:

$$s = \frac{\omega_s - \omega}{\omega_s} = \frac{n_s - n}{n_s}$$

3. Calculate $z_2 = \dfrac{r_2}{s} + jx_2$ \hfill (4.58)

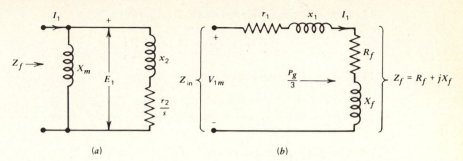

FIGURE 4.17 The field impedance concept in the induction machine model. (*a*) The field impedance. (*b*) The model employing Z_f.

4. Calculate the field impedance Z_f. This is the effect of resultant magnetic field on the impedance of one phase of the stator. It is the parallel combination of z_2 and jX_m (see Figure 4.17):

$$Z_f \equiv R_f + jX_f \overset{\Delta}{=} \frac{jX_m[(r_2/s) + jx_2]}{(r_2/s) + j(x_2 + X_m)} \quad \Omega \qquad (4.59)$$

5. Calculate the impedance looking in at the phase terminals:

$$Z_{\text{in}} = z_1 + Z_f = r_1 + jx_1 + R_f + jX_f \quad \Omega \qquad (4.60)$$

6. Calculate the phase current:

$$I_1 = \frac{V_{1m}\,\underline{|0°}}{Z_{\text{in}}} = |I_1|\,\underline{|\theta_{\phi m}} \quad \text{A} \qquad (4.61)$$

7. Calculate the power factor $= \cos \theta_{\phi m}$

8. Calculate the input power:

$$P_{\text{in}} = \sqrt{3}V_L|I_1| \cos \theta_{\phi m}$$

$$= 3|V_{1m}||I_1| \cos \theta_{\phi m} \quad \text{W}$$

9. Calculate the stator copper loss:

$$\text{SCL} = 3I_1^2 r_1 \quad \text{W} \qquad (4.62)$$

10. Calculate the air gap power:

$$P_g = 3I_1^2 R_f \quad \text{W} \qquad (4.63)$$

11. Calculate the rotor copper loss:

$$\text{RCL} = sP_g \quad \text{W}$$

12. Calculate the developed mechanical power:

$$DMP = (1 - s)P_g \ W$$

13. Calculate the developed torque:

$$\tau_d = \frac{P_g}{\omega_s} \ \text{N-m} \quad \text{or} \quad 7.04\frac{P_g}{n_s} \ \text{lb-ft}$$

14. Calculate the output power:

$$P_{out} = DMP - P_{rot} \ \ W$$

$$\text{Horsepower} = \frac{P_{out}}{746}$$

15. Calculate the output torque:

$$\tau = \frac{P_{out}}{\omega} \ \text{N-m} \quad \text{or} \quad 7.04\frac{P_{out}}{n} \ \text{lb-ft}$$

16. Calculate the efficiency:

$$\eta = \frac{P_{out}}{P_{in}}$$

■ **Example 4.3**

A four-pole, three-phase, 220-V, 60-Hz, 10-hp induction motor has the following model impedances:

$$r_1 = 0.39 \ \Omega \qquad x_1 = 0.35 \ \Omega \qquad X_m = 16.0 \ \Omega$$

$$r_2 = 0.14 \ \Omega \qquad x_2 = 0.35 \ \Omega$$

Rotational losses are 350 W.

(a) Calculate the performance of this motor for a speed of 1746 rev/min.

(b) Find the starting current and torque.

Solution

(a) $n_s = \dfrac{120 \cdot 60}{4} = 1800 \ \ \text{rev/min}$

$s = \dfrac{1800 - 1746}{1800} = 0.03$

$z_2 = \dfrac{r_2}{s} + jx_2 = \dfrac{0.14}{0.03} + j0.35 = 4.67 + j0.35 \ \Omega = 4.68 \ \underline{|4.29°}$

Note how r_2/s dominates this impedance.

$$Z_f = \frac{jX_m z_2}{z_2 + jX_m} = \frac{j16 \cdot 4.68\underline{|4.29°}}{4.67 + j16.35} = \frac{74.9\underline{|94.29°}}{17.00\underline{|74.06°}}$$

$$= 4.40\underline{|20.23°} \; \Omega$$

$$= 4.13 + j1.52 \; \Omega$$

$$\boldsymbol{R_f} = \mathbf{4.13} \; \Omega$$

$$Z_{in} = r_1 + jx_1 + Z_f = 0.39 + 4.13 + j(0.35 + 1.52)$$

$$= 4.52 + j1.87 = 4.89\underline{|22.48°} \; \Omega$$

Power factor $= \cos 22.48° = 0.924$

$$|I_1| = \frac{V_L}{\sqrt{3}|Z_{in}|} = \frac{220}{\sqrt{3} \; 4.89} = 26.0 \text{ A}$$

$$P_{in} = \sqrt{3} \cdot 220 \cdot 26 \cdot 0.924 = 9150 \text{ W}$$

$$P_g = 3I_1^2 R_f = 3 \cdot 26^2 \cdot 4.13 = 8380 \text{ W}$$

Developed power $= (1 - s)P_g = 0.97 \cdot 8380 = 8120 \text{ W}$

Output power $=$ DMP $- P_{rot}$

$$= 8120 - 350 = 7770 \text{ W}$$

Output horsepower $= \dfrac{7770}{746} = 10.4$

Developed torque $= 7.04\dfrac{P_g}{n_s} = 7.04\dfrac{8380}{1800} = 32.8$ lb-ft

Output torque $= \dfrac{7.04 \cdot P_{out}}{n} = \dfrac{7.04 \cdot 7770}{1746} = 31.3$ lb-ft

Efficiency $= \dfrac{7770}{9150} = 84.9\%$

(b) At starting, the machine is at standstill.

$$s = \frac{n_s - 0}{n_s} = 1$$

$$z_2 = \frac{r_2}{1} + jx_2 = 0.14 + j0.35$$

$$= 0.377\underline{|68.2°} \; \Omega$$

$$Z_f = \frac{16 \cdot 0.377\underline{|158.2°}}{0.14 + j16.35}$$

$$Z_f = \frac{6.03\lfloor 158.2°}{16.35\lfloor 89.5°} = 0.369\lfloor 68.7°$$

$$= 0.134 + j0.344 \ \Omega \qquad (R_f = 0.134 \ \Omega)$$

$$Z_{in} = 0.39 + 0.134 + j(0.35 + 0.344)$$

$$= 0.524 + j0.694 = 0.870\lfloor 52.9° \ \Omega$$

$$\text{Starting current} = \frac{220}{\sqrt{3} \cdot 0.870} = 146 \ \text{A}!$$

Note that the starting current is nearly six times the current for a slip of 0.03.

Rotational losses may be neglected at zero speed, and the starting torque is equal to the developed torque. At $s = 1$,

$$P_g = 3I_1^2 R_f$$

$$= 3 \cdot 146^2 \cdot 0.134 = 8570 \ \text{W}$$

$$\tau_d = 7.04\frac{P_g}{n_s} = 33.5 \ \text{lb-ft}$$

Note that, in spite of the high starting current, the starting torque is only slightly higher than that at 1746 rev/min. ∎

4.12 TORQUE/SPEED CHARACTERISTICS OF INDUCTION MACHINES

An expression for the torque of an induction machine as a function of its slip may be obtained by application of Thévenin's theorem to the circuit model. The rotor circuit, as referred to the stator, may be considered as being attached to an equivalent Thévenin generator, as shown in Figure 4.18. Looking to the left into terminals 1 and 2, the open-circuit voltage will be that fraction of V_{1m} appearing across jX_m:

$$V_{Th} = V_{1m}\left[\frac{jX_m}{r_1 + j(x_1 + X_m)}\right] \tag{4.64}$$

The Thévenin generator impedance will be the impedance appearing at terminals 1 and 2 with the generator V_{1m} replaced by a short (i.e., z_1 in parallel with jX_m):

$$Z_{Th} \stackrel{\Delta}{=} R_1 + jX_1 = \frac{jX_m(r_1 + jx_1)}{r_1 + j(x_1 + X_m)} \tag{4.65}$$

For approximate calculations, very little error is involved in

$$V_{Th} \cong V'_{1m} \stackrel{\Delta}{=} V_{1m}\left(\frac{X_m}{x_1 + X_m}\right) \tag{4.66}$$

since $r_1 \ll (x_1 + X_m)$. Also

$$X_1 \cong x_1 \tag{4.67}$$

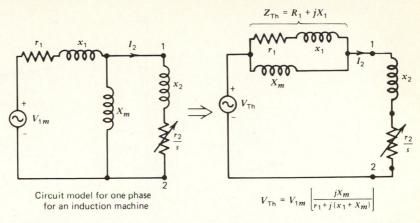

Circuit model for one phase
for an induction machine

$$V_{Th} = V_{1m} \left[\frac{jX_m}{r_1 + j(x_1 + X_m)} \right]$$

FIGURE 4.18 Application of Thévenin's theorem to the induction motor circuit model.

and

$$R_1 \cong r_1 \left(\frac{X_m}{x_1 + X_m} \right)^2 \qquad (4.68)^2$$

(In fact, letting $R_1 = r_1$ involves an error of only about 5 percent.)

An expression for I_2 may now be written involving the Thevenin generator quantities:

$$I_2 = \frac{V_{Th}}{R_1 + \dfrac{r_2}{s} + j(X_1 + x_2)} \qquad (4.69)$$

The developed torque is given by

$$\tau_d = \frac{P_g}{\omega_s} = \frac{1}{\omega_s} 3 I_2^2 \frac{r_2}{s}$$

Substituting the expression for I_2 from Equation 4.69,

$$\tau_d = \frac{3}{\omega_s} \frac{V_{Th}^2 (r_2/s)}{[R_1 + (r_2/s)]^2 + (X_1 + x_2)^2} \qquad (4.70)$$

Since by Equation 4.64 V_{Th} is proportional to the voltage at the motor terminals, *note that at any speed the torque is proportional to the square of the supply voltage.* When this expression for the developed torque is plotted as a function of the slip s, the resulting curve has the shape of Figure 4.19*a*. When plotted as a function of *speed*, the curve appears as in Figure 4.19*b*. This second figure shows:

1. There is a definite maximum torque τ_{max} in the motor range of operation, called the *breakdown torque*, or *pull-out torque*.
2. The full-load torque is about 0.4 to 0.5 of maximum torque. Thus the machine can handle short-time overloads without stalling.

---

[2]The real part of Z_{Th} (Equation 4.65) is $r_1\{X_m^2/[r_1^2 + (x_1 + X_m)^2]\}$, which is very nearly $r_1 X_m^2/(x_1 + X_m)^2$, since $r_1^2 \ll (x_2 + X_m)^2$.

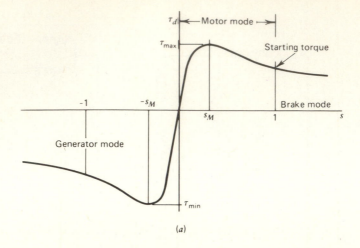

(a)

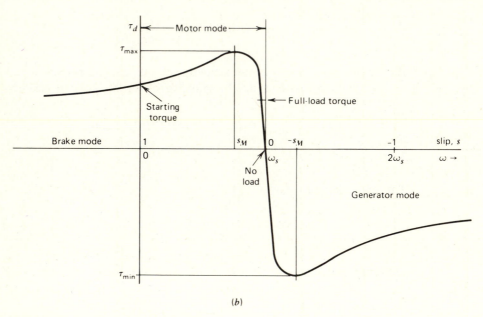

(b)

FIGURE 4.19 Torque of an induction machine. (a) Developed torque as a function of slip. (b) Developed torque as a function of speed.

3. Once the machine is running, only a small part of the curve is involved in going from no load to full load. *There is a linear variation in torque in this range.* At no load the speed is near synchronous speed ($s \cong 0$), and at full load the slip is still quite small.

4. The starting torque (i.e., torque at zero speed, or $s = 1$)[3] is slightly larger than the full-load torque. It may be made still larger by special design.

---

[3]The value of *starting torque* given by inserting $s = 1$ into Equation 4.70 does not take into account electrical transients that occur when voltage is applied suddenly to the motor terminals to start the motor. It is more correctly called *blocked-rotor torque*, but its value does not differ greatly from the effective average starting torque found when transients are considered.

5. The machine operates as a generator ($\tau_d < 0$, $\omega > 0$, negative power) when the machine is driven at speeds greater than synchronous speed (i.e., $s < 0$).

6. The machine acts as a brake when driven backward ($\tau_d > 0$, $\omega < 0$, mechanical power negative). In this region, both electrical and mechanical power are fed *into* the machine, and it becomes too hot very rapidly.

Breakdown Torque

Since ω_s is a constant when the frequency of the electrical supply is constant, the developed torque is proportional to P_g by Equation 4.52. The air gap power per phase, $P_g/3$, is all absorbed by the resistance r_2/s. Thus *breakdown torque τ_{max} corresponds to maximum power transfer into $r_2 s$.* Figure 4.18 shows that the terminals of r_2/s are attached to a generator having an internal impedance of $Z_{Th} + jx_2$. The laws of maximum power transfer show that, when the angle of a load impedance is fixed, maximum power is transferred into that impedance when its *magnitude* matches the *magnitude* of the internal impedance of the generator to which it is connected. Then for maximum P_g, and hence for maximum developed torque,

$$\frac{r_2}{s_M} = |Z_{Th} + jx_2|$$

$$= \sqrt{R_1^2 + (X_1 + x_2)^2} \qquad (4.71)$$

In this relation, s_M is the slip at which the motor develops maximum torque. Solving Equation 4.71,

$$s_M = \frac{r_2}{\sqrt{R_1^2 + (X_1 + x_2)^2}} \qquad (4.72)$$

Note that r_2 occurs in the *numerator,* and hence *the slip at which maximum torque occurs is proportional to the rotor resistance.*

The strength of the maximum torque may be found by inserting the expression for s_M (Equation 4.72) into the equation for the developed torque (Equation 4.70). Note that in Equation 4.70, s always occurs in terms of r_2/s and that r_2/s_M is, by Equation 4.71, independent of r_2! Therefore, *τ_{max} is independent of r_2:*

$$\tau_{max} = \frac{3}{\omega_s} \frac{V_{Th}^2}{2(R_1 + \sqrt{R_1^2 + (X_1 + x_2)^2})} \qquad (4.73)$$

Again, this torque is proportional to the square of the supply voltage. While R_1 is not negligible in comparison with $(X_1 + x_2)$ in the denominator of the τ_{max} expression, the reactance term plays a dominant role, and *maximum torque is limited by the rotor and stator leakage reactances.* For this reason a designer likes to keep the air gap as small as possible, so that as much as possible of the total flux of either the rotor or stator winding is *mutual* between the two, rather than linking only one winding, and thus becoming *leakage.*

The fact that τ_{max} is *independent* of rotor resistance while the slip at which it occurs is *proportional* to rotor resistance yields one of the principal advantages of the *wound-rotor machine.* Figure 4.20 shows the effect of r_2 on the torque/speed characteristic,

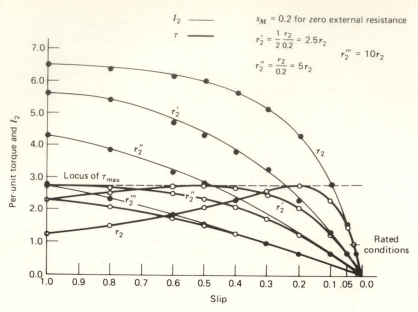

FIGURE 4.20 Effect of rotor resistance on developed torque and I_2 as functions of slip.

as well as the current speed/characteristic (derived from Equation 4.69), for constant reactances. It will be evident from this figure that, by inserting proper resistances into the rotor circuit, maximum torque may be obtained at starting (zero speed), with less starting current than would result if the resistance were not inserted. As the motor comes up to speed, the resistance of each rotor phase may be reduced to maintain high accelerating torque. For a "soft" start, the machine may be connected to the line with the rotor open-circuited. Relatively high resistances would then be connected to the slip rings and gradually reduced until sufficient torque was produced to start the load.

Example 4.4

For the motor of Example 4.3, find the pull-out torque and the slip at which it occurs.

Solution. *Pull-out* torque and maximum torque are synonymous. There are at least three procedures available for its calculation: (a) find V_{Th} and Z_1, then apply the formula of Equation 4.73; (b) find Z_1, then s_M by Equation 4.72, and then proceed as in Example 4.2 to find the torque; or (c) find s_M, use the Thevenin-modified model to find I_2, and then $\tau_{max} = (3/\omega_s)I_2^2(r_2/s_M)$. The first two of these methods will be applied for comparison.

(a) $$V_{Th} \cong V'_{1m} = V_{1m}\left(\frac{X_m}{x_1 + X_m}\right) = \frac{220}{\sqrt{3}}\left(\frac{16}{16.35}\right)$$

$$= 127.0 \cdot 0.9786 = 124.3 \quad V$$

$$X_1 \cong x_1 = 0.35 \ \Omega$$

$$R_1 \cong r_1\left(\frac{X_m}{x_1 + X_m}\right)^2 = 0.39 \cdot 0.9786^2 = 0.373 \ \Omega$$

$$\tau_{max} = \frac{3}{188.5}\left(\frac{124.3^2}{2(0.373 + \sqrt{0.373^2 + (0.35 + 0.35)^2})}\right)$$

$$= 105 \quad \text{N-m}$$

(b) $s_M = \dfrac{r_2}{\sqrt{R_1^2 + (X_1 + x_2)^2}} = \dfrac{0.14}{\sqrt{0.373^2 + (0.7)^2}} = 0.177$

$\dfrac{r_2}{s_M} = \dfrac{0.14}{0.177} = 0.791 \ \Omega$

$Z_f = \dfrac{jX_m[(r_2/s) + jx_2]}{(r_2/s) + j(x_2 + X_m)} = \dfrac{j16 \cdot (0.791 + j0.35)}{0.791 + j16.35}$

$\qquad = R_f + jX_f = 0.756 + j0.379$

$Z_{in} = z_1 + Z_f = 1.146 + j0.729 = 1.358\underline{|32.4°} \ \Omega$

$I_1 = \dfrac{220/\sqrt{3}}{1.358} = 93.5 \ A$

$P_g = 3I_1^2 R_f = 19{,}800 \ W$

$\tau_{max} = \dfrac{P_g}{\omega_s} = \dfrac{19.800}{188.5} = 105 \ \text{N-m}$

Converting to lb-ft, $\tau_{max} = 105 \cdot 0.738 = 77.5$ lb-ft, a little over twice the torque for a slip of 0.03 (about rated horsepower) found in Example 4.3. ∎

4.13 SOME USEFUL EXACT AND APPROXIMATE RELATIONSHIPS

To summarize relationships derived from the model:

For any fixed value of slip or speed:
(Frequency assumed constant.)
    Line current is proportional to voltage.
    Torque is proportional to voltage squared.
Maximum torque proportionalities:
    Maximum torque is proportional to voltage squared.
    Maximum torque is independent of r_2.
    The slip at which maximum torque occurs is proportional to r_2.

Some Approximate Relationships Near Rated Speed

Note in Figures 4.19 and 4.20 that both torque and current are linear functions of slip between synchronous speed and full-load speed, and even a little below full-load speed. Note also (Figure 4.20) that, as rotor resistance increases, the curves are linear over a greater range of slips. The reason for this linear behavior when the slip is small is that, in this limited range, the motor phase impedance is dominated by r_2/s. This fact was pointed out in Example 4.3. *As long as r_2/s is large compared to the other series reactances,* it may be stated that, *roughly,*

$$I_2 \cong \frac{V_{Th}}{r_2/s} = \frac{sV_{Th}}{r_2} \cong \frac{sV_{1m}}{r_2} \qquad \text{(A)}$$

Then

$$P_g = 3I_2^2(r_2/s) \cong \frac{3V_{1m}^2 s}{r_2},$$

and

$$\tau \cong \frac{3V_{1m}^2 s}{\omega_s r_2} = \frac{3pV_{1m}^2 s}{4\pi f_1 r_2} = C\frac{pV_{1m}^2 s}{f_1 r_2} \qquad \text{(B)}$$

where C is a constant of proportionality. The relationships A and B, above, will not be dignified by equation numbers, but they contain a great deal of information *applicable only to the linear range of operation near rated speed. Since both involve the ratio r_2/s, it is clear that as r_2 is increased, the s range over which they are valid also increases.*

Example 4.5

An induction motor, operating from a 60-Hz line, develops 10 hp at 1745 rev/min. At what speed will it run if the load torque is reduced to one half? What will be the horsepower at half torque?

Solution. The synchronous speed of a four-pole, 60-Hz induction motor is $120 \cdot 60/4 = 1800$ rev/min, so the machine very probably has four poles. Then at 10 hp,

$$s = \frac{1800 - 1745}{1800} = 0.0306$$

By Expression B, above, torque and slip are proportional; therefore at half torque, the slip will be 0.0153 and the speed will be

$$n = n_s(1 - s) = 1800(1 - 0.0153) = 1772 \quad \text{rev/min}$$

The torque at 10 hp was

$$\frac{10 \cdot 5252}{1745} \quad \text{lb-ft}$$

so the new horsepower output will be

$$\frac{1}{2}\frac{10 \cdot 1772}{1745} = 5.08 \text{ hp}$$

Example 4.6

A motor is operating at full load and is driving a mechanical load having a torque requirement independent of speed. For some reason the line voltage drops to 90 percent of rated value. Will the motor run hotter or cooler?

Solution. By Expression B,

$$s \cong \frac{\tau f_1 r_2}{C p V_{1m}^2}$$

and, since τ is constant,

$$\frac{s_2}{s_1} \cong \left(\frac{V_{1m(1)}}{V_{1m(2)}}\right)^2 = \left(\frac{1}{0.9}\right)^2 = 1.23$$

By Expression A,

$$\frac{I_{2(2)}}{I_{2(1)}} = \frac{s_2 V_{1m(2)}}{s_1 V_{1m(1)}} = 1.23 \cdot 0.9 = 1.107$$

Now the I^2R loss is nearly (I_1 is not quite equal to I_2) proportional to I_2^2, so

$$\frac{(\text{Copper loss})_2}{(\text{Copper loss})_1} \cong \left(\frac{I_{2(2)}}{I_{2(1)}}\right)^2 = 1.107^2 = 1.23$$

Thus a 10 percent reduction in voltage causes about a 23 percent increase in copper losses, for constant torque load. If the original speed were 1746 rev/min (i.e., $s = 0.03$ at 60 Hz), the speed at 90 percent voltage would be

$$1800(1 - 1.23 \cdot 0.03) = 1734 \quad \text{rev/min}$$

a reduction of less than 1 percent. Reduced line voltage hardly affects the mechanical performance of the motor but greatly increases the temperature. ∎

4.14 CONTROL OF PERFORMANCE CHARACTERISTICS BY ROTOR DESIGN

It has been assumed that the effective resistance of a squirrel cage rotor is constant. If this were the case, the designer of a squirrel cage motor would be limited to a torque/speed characteristic similar to *one* of those shown in Figure 4.20. That is, he or she could design a high-resistance rotor that would have a high starting torque and relatively low starting current; but this rotor would have poor speed regulation between no load and full load, and in addition would have a large rotor I^2R loss and thus low efficiency. On the other hand, a low-resistance rotor would have good running characteristics but would have relatively low starting torque and require a high starting current. Fortunately, the rotor frequency varies with speed, and the designer has some control over the leakage inductance of the rotor. The changes in rotor circuit leakage *reactance* with speed provide a means for controlling rotor *resistance*.

Figure 4.21 shows cross sections of several squirrel cage rotors to illustrate possible rotor bar shapes. The bars in rotor A are large and are near the rotor surface. They have low resistance and, since they are closely coupled to the stator winding, have relatively low leakage inductance. Such a rotor would have low slip at full load, high running efficiency, and high pull-out torque, but would have low starting torque and

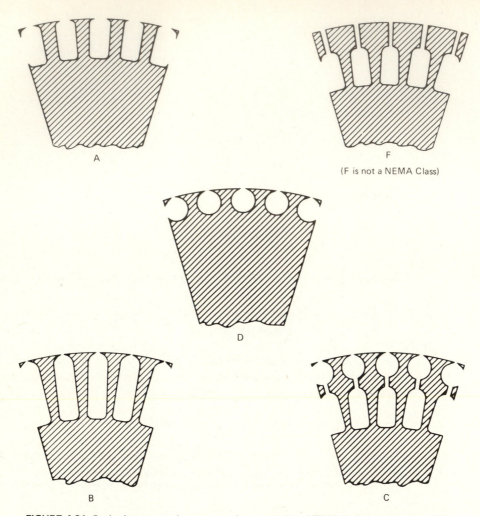

FIGURE 4.21 Squirrel cage rotor-bar cross sections to produce NEMA design-class characteristics.

would require a large starting current. This is called a "Design A" rotor by the National Electrical Manufacturers Association (NEMA, for short).

Compare the A rotor with the F rotor. The F rotor was once used in large motors designed for very easily started loads, such as industrial fans, where starting current had to be limited. It is no longer listed as a standard design. Here the rotor bars have been buried deep in the rotor iron, and the leakage inductance is large. The bars have a low resistance. A motor with such a rotor has low slip and good efficiency at full load, because r_2 is small. The high values of x_2 keeps the starting current low, but it also severely limits the pull-out torque and starting torque. (See Equations 4.70 and 4.73.)

Next examine rotor D in Figure 4.21. This is like A, in that the leakage reactance is low; but the bars are small and as a result have high resistance. A NEMA Design D rotor has characteristics rather like those for r_2'' in Figure 4.20; that is, high starting torque, low starting current, high slip, and low efficiency at full load.

Fortunately, the good characteristics of Designs F and D may be combined to obtain the best features of each. There are two ways of accomplishing this hybrid characteristic: by using two cages or by the use of "deep bars." A deep-bar design is shown as B in Figure 4.21, and a double-cage design is illustrated as C in Figure 4.21.

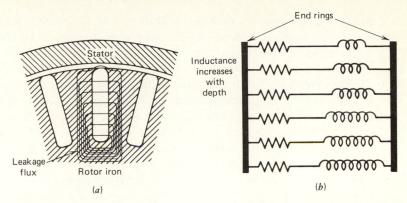

FIGURE 4.22 Deep-bar rotor effects. (*a*) Leakage flux paths. (*b*) Equivalent circuit.

Deep-Bar Rotors

The leakage flux paths for a deep-bar design are illustrated in Figure 4.22*a*. Since inductance is the number of flux linkages produced per ampere ($L = \lambda/I$), it is obvious that the parts of each bar extending deeply into the iron have higher leakage inductances than those parts of the bar cross section near the air gap, because more of the leakage flux links the deeper parts of the bar. One may think of each bar as being composed of several layers of equal cross section, and thus of equal resistance, but with inductances increasing with depth, as in Figure 4.22*b*.

Now under running conditions the slip is quite small, and by Equation 4.7 the frequency of the rotor currents is only 1 or 2 Hz. As a result, all of the leakage reactances are negligible, and the current distribution is essentially uniform throughout each rotor bar. The effective resistance of the rotor is that of all layers in parallel. This low resistance makes for low slip and high efficiency at full load. As the motor is overloaded, however, slip and rotor frequency increase. The reactances of the deeper levels of the bar come into play with two effects: (1) The value of x_2 in the circuit

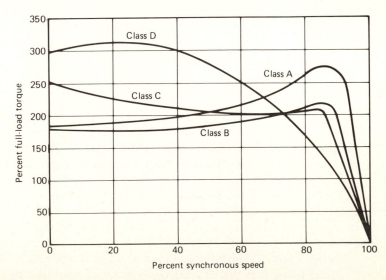

FIGURE 4.23 Typical torque speed characteristics of NEMA design-class squirrel cage motors.

model is larger than it would be for shallow bars like those of Design A, and the breakdown torque is reduced. (2) The rotor current begins to be crowded into those layers of the bars near the air gap, increasing the effective resistance of rotor circuits. Thus breakdown torque occurs at a somewhat greater slip than for Design A.

When the motor is first connected to the line for starting, the slip is unity and the frequency of the rotor currents equals the line frequency. Rotor-bar currents are concentrated near the air gap. Effective bar resistance and reactance are both high, with the result that starting current is only about 75 percent of that for Design A.

Thus a deep-bar rotor achieves good full-load efficiency at low slip, together with reduced starting current, at the expense of somewhat lower pull-out torque. A typical speed/torque curve for a deep-bar motor is shown as the Design Class B curve of Figure 4.23, although this same characteristic may also be obtained with a double-cage rotor.

Double-Cage Rotors

A double-cage rotor has two squirrel cages. The two cages may connect to the same end rings, or there may be separate end rings for each cage. The usual arrangement is that of NEMA Design C, as illustrated in Figure 4.21. The outer cage for Design C is composed of small, high-resistance bars, while the inner cage is made up of larger, low-resistance bars. Since the inner bars are almost totally surrounded by the iron of the rotor core, their leakage inductance is very great, indeed. For this reason they have little effect on motor performance except at low slips; that is, they come into play when the motor has reached running speed. At running speed the two cages are essentially in parallel, and r_2 in the circuit model is quite low. Full-load slip is low, almost as low as for Design A, and full-load efficiency is almost as high as for Design A.

At starting, the inner cage may nearly be disregarded, and the outer cage designed for optimum starting characteristics. The iron between the outer cage and the inner cage, together with the proximity of the outer cage with the stator windings, causes the leakage reactance to be low and permits a large maximum torque. If desired, the resistance of the outer cage may be chosen to realize this high torque at starting. The high outer-cage resistance keeps the starting current low.

Double-cage construction is the most costly of the squirrel cage designs, but it allows many of the advantages of the still more costly and less rugged wound rotor to be included in a squirrel cage machine.

NEMA DESIGN-CLASS SUMMARY

By proper shaping of the rotor bars, an extremely wide range of squirrel cage induction motor characteristics could be developed. To bring order out of the potential chaos, the National Electrical Manufacturers' Association has developed the broad "design classes" previously mentioned to meet widely recognized industrial needs. (*Do not confuse design class letters with code letters, both of which are found on induction motor nameplates.*) The torque, slip and current requirements to meet these design classifications are spelled out in NEMA Standard MG-1. They vary with horsepower rating, but the following general characteristics may be stated:

Design Class A. The full-load slip of a Design A motor must be less than 5 percent and is less than that of a Design B motor of equivalent rating. The breakdown torque is greater than that of Design B. Specified starting torques are the same

TABLE 4.1
Locked-Rotor Current of Three-Phase 60-Hz Integral-Horsepower Squirrel Cage Induction Motors Rated at 230 V

The locked-rotor current of single-speed, three-phase, constant-speed induction motors rated at 230 V, when measured with rated voltage and frequency impressed and with rotor locked, shall not exceed the following values:

Horsepower	Locked-Rotor Current-Amperes[a]	Design Letters
$\frac{1}{2}$	20	B,D
$\frac{3}{4}$	25	B,D
1	30	B,D
$1\frac{1}{2}$	40	B,D
2	50	B,D
3	64	B,C,D
5	92	B,C,D
$7\frac{1}{2}$	127	B,C,D
10	162	B,C,D
15	232	B,C,D
20	290	B,C,D
25	365	B,C,D
30	435	B,C,D
40	580	B,C,D
50	725	B,C,D
60	870	B,C,D
75	1085	B,C,D
100	1450	B,C,D
125	1815	B,C,D
150	2170	B,C,D
200	2900	B,C
250	3650	B
300	4400	B
350	5100	B
400	5800	B
450	6500	B
500	7250	B

[a]The locked-rotor current of motors designed for voltages other than 230 volts shall be inversely proportional to the voltages.

This table is reproduced by permission from NEMA Publication No. MG 1-1987, *Motors and Generators*, copyright 1988 by NEMA.

as those for Design B. The starting current is greater than that of Design B and may be more than eight times the rated full-load current. For this reason, Design A motors rated more than 7.5 hp usually require some device to limit the line current during starting. Design A motors have single-cage, low-resistance rotors.

Design Class B. This design provides a full-load slip of less than 5 percent (usually 0.02 to 0.03). The breakdown torque is at least 200 percent of full-load torque (175 percent for motors of 250 hp and above). The starting current usually does not exceed five times the rated current, but the upper limits are given by Table 4.1. Starting torque is at least as great as full-load torque for motors of 200 hp or less, and is over twice the rated torque for four-pole motors of less than 3 hp.

These characteristics may be achieved by either deep-bar or double-cage rotors. Design Class B has become the standard for most industrial applications, such as lathes, milling machines, boring machines, fans, and centrifugal pumps.

Design Class C. These motors have a full-load slip of less than 5 percent, usually slightly larger than Design B. The starting torque is larger than that of Design B, running 200 to 250 percent of rated torque. The breakdown torque is a few percent less than for Design B, and the starting current is about the same. These excellent overall characteristics are achieved by means of a double-cage rotor, which makes Design C the most expensive of the four design classes. Design Class C motors are used for hard-to-start loads, such as air compressors, reciprocating pumps, and conveyors that are fully loaded when started.

Design Class D. These motors are designed so that maximum torque occurs near zero speed. They have a very high starting torque (about 275 percent of full-load value) at relatively low starting current. (The starting current is limited to that of Design B.) The term *pull-out torque* does not apply to this design, since there may be no maximum in the torque speed curve between zero and synchronous speed. Full-load slip exceeds 5 percent. It is made large deliberately, and may be as much as 17 percent. For this reason, these are often called *high-slip* motors. Design D motors have a single, high-resistance rotor cage. Their full-load efficiency is less than that of the other designs.

These motors are used in applications in which it is important that the motor not stall under heavy overload, and in machines employing flywheel energy storage. A punch press, for example, often has a flywheel that delivers mechanical energy during the brief interval of the punching operation—usually a fraction of a second. The flywheel must slow down to release its kinetic energy, and the motor connected to it must permit this reduced speed without slowing below its pull-out speed. Design D motors have this characteristic. In the longer interval between punches, the motor gradually increases the speed of the flywheel, storing energy for the next punch. The average power required of the motor may be less than one horsepower to provide several hundred horsepower during the punch.

Typical torque/speed characteristics for the four design classes are shown in Figure 4.23.

4.15 TESTS TO DETERMINE CIRCUIT MODEL IMPEDANCES

The circuit model of a machine is useful in analyzing the device as part of a system, or in calculating machine performance on the assumption of constant ac terminal voltage and frequency, as in Section 4.11. The model, in essence, serves as a bookkeeping device to catagorize, organize, and assign numerical values to the complex phenomena of machine operation. The designer, for example, knows he or she must assign some value to x_2, the rotor leakage reactance, referred to the stator at stator frequency. To do this the designer must calculate slot-leakage, differential-leakage, and end-ring-leakage reactances for the rotor. Values must also be assigned to r_1, r_2, x_1, and X_m. To check a design, or in cases in which a motor is too large to be tested under full-load conditions in the laboratory, the impedances of the circuit model need to be evaluated by laboratory tests. Since Steinmetz's induction machine model is quite similar to his transformer model, the tests usually employed to evaluate its impedances are analogous to the open-circuit and short circuit tests on transformers. In addition, a dc measurement of r_1 is required.

Many factors must be taken into account in making these tests. Winding resistances vary with temperature, and rotor resistance is a function of frequency. Test procedures are presented in great detail in ANSI/IEEE Standard 112-1984. That standard should be consulted when accurate measurements are necessary. The following paragraphs, however, provide an understanding of the fundamental theory underlying these tests.

The No-Load Test

The no-load test corresponds to the open-circuit test on a transformer. However the test measures friction and windage losses in addition to core losses. Figure 4.24 shows the test circuit and the effect of zero mechanical load on the model. The mechanical load per phase, represented by $r_2(1 - s)/s$, is simply the internal windage and friction of the machine. Since this amounts to only a few percent of the rated mechanical load, the slip is very small, say, 0.001, and $r(1 - s)/s$ is very large compared to x_2 and r_2. It is thus essentially in parallel with R_c and jX_m, where R_c represents the core loss per phase. In Figure 4.24, $R_{fw\,c}$ represents the parallel combination of R_c and $r_2(1 - s)/s$, and absorbs a power equal to one third of the total friction, windage, and core loss. Since *under no-load conditions* r_2 is so much less than $r_2(1 - s)/s$, the no-load rotor copper loss is negligible. The no-load input power P_{nl}, then, may be taken to be the sum of the stator copper loss and P_{fwc}, the friction, windage, and core loss.

$$P_{nl} \equiv P_1 + P_2 = \text{SCL}_{nl} + P_{fwc} \tag{4.74}$$

where $\text{SCL}_{nl} = 3I_{nl}^2 r_1$. This permits the sum of the friction, windage, and core losses to be evaluated:

$$P_{fwc} = P_{nl} - 3I_{nl}^2 r_1 \tag{4.75}$$

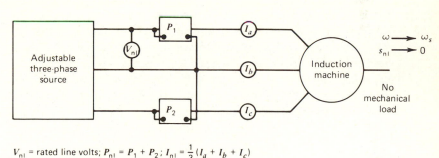

V_{nl} = rated line volts; $P_{nl} = P_1 + P_2$; $I_{nl} = \frac{1}{3}(I_a + I_b + I_c)$

$I_{h+e}^2 R_c$ represents core loss per phase = $\dfrac{P_{h+e}}{3}$ $R_{fw} = r_2 \left[\dfrac{1 - s_{nl}}{s_{nl}}\right] \gg r_2$ or x_2

$I_2^2 r_2 \left[\dfrac{1 - s_{nl}}{s_{nl}}\right]$ represents friction and windage per phase = $\dfrac{1}{3} P_{fw}$

$$P_{nl} = P_1 + P_2 = 3I_{nl}^2 r_1 + P_{h+e} + P_{fw}$$

FIGURE 4.24 No-load test on an induction machine.

where P_{nl} is the total three-phase power input to the machine at rated voltage and frequency, and I_{nl} is the average of the three line currents, with no mechanical load connected to the shaft.

Under no-load conditions, the power factor is quite low, 0.2 or less, so the circuit is essentially reactive. The current will be in the neighborhood of one third the rated current. These facts imply that r_1 is small compared to X_m and that R_{fwc} is much larger than X_m. The input impedance at no load is thus approximately jx_1 and jX_m in series:

$$|Z_{nl}| = \frac{V_{nl}}{\sqrt{3}I_{nl}} \cong x_1 + X_m \qquad (4.76)$$

Then

$$X_m = \frac{V_{nl}}{\sqrt{3}I_{nl}} - x_1 \qquad (4.77)$$

The values of r_1 and x_1 in Equations 4.75 and 4.77 have yet to be determined.

Direct-Current Test

In the case of a transformer, it was only necessary to obtain $R_{eq\ 1}$ or $R_{eq\ 2}$ to solve practical problems. A knowledge of the individual winding resistances was not essential. In the induction machine, however, r_2 plays a crucial role, and r_1 must be known to evaluate the stator copper loss.

With a steady direct current flowing in the stator windings, the stator magnetic field is fixed in space. If no external torque turns the shaft, the flux linkages with the rotor circuits are constant in time. As a result, no voltages are induced in the rotor, no rotor currents flow, and no torque is developed. There is no rotor MMF ($\vec{F}_2 = 0$), and the rotor circuits have no electrical effect on those of the stator. Thus the dc stator winding resistance may be measured independently of the rotor impedance.

The circuit is shown in Figure 4.25. The rheostat must be capable of handling the rated current of the motor. The current is adjusted to approximately rated value, so that the temperature of the winding approximates that of running conditions. (The

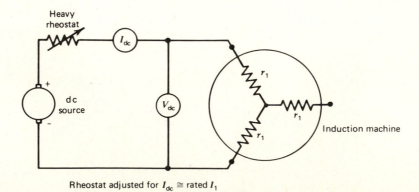

Rheostat adjusted for $I_{dc} \cong$ rated I_1

FIGURE 4.25 Direct-current Measurement of stator winding resistance.

winding temperature is often measured so that the test resistance may be corrected to actual operating temperature.) Applying Ohm's law to the circuit of Figure 4.25,

$$2r_1 = \frac{V_{dc}}{I_{dc}}$$

or

$$r_1 = \frac{V_{dc}}{2I_{dc}} \tag{4.78}$$

This value of r_1 may now be substituted into Equation 4.75 in obtaining the total of the friction, windage, and core losses of the machine.

The Blocked-Rotor Test

This test, which corresponds to the short-circuit test of a transformer, is *also called the locked-rotor test*. The shaft is clamped so that it cannot turn; thus $\omega = 0$ and $s = 1$ during this test. If full voltage at rated frequency were applied, the current would be five to eight or more times rated value, as indicated in the last section. For this reason, locked-rotor tests are not done at full voltage except for small motors. Instead, the applied voltage is usually adjusted upward from zero until the stator current is at rated value. Blocked-rotor test measurements are, then, taken at rated *current* rather than at rated voltage.

Under running conditions the frequency of the stator currents is line frequency f_1, while that of the rotor currents is sf_1, only a few hertz. In the blocked-rotor test, however, the slip is unity and the rotor and stator currents are at the same frequency.

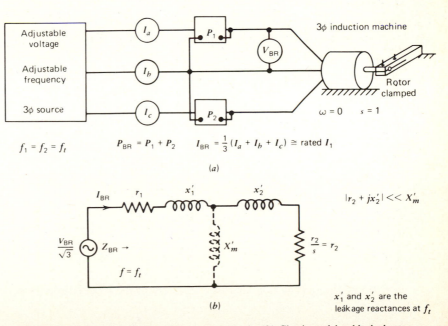

FIGURE 4.26 The blocked-rotor test. (*a*) Test circuit. (*b*) Circuit model at blocked rotor.

Since both rotor and stator quantities are to be evaluated, the frequency of the blocked-rotor test voltage f_t is usually a compromise between the normal stator and rotor frequencies.

The test circuit and the effect of blocking the rotor on the model are shown in Figure 4.26. The voltage appearing across one phase of the stator winding is $V_{BR}/\sqrt{3}$, and the magnitude of the impedance looking into the terminals of one phase of the machine is

$$|Z_{BR}| = \frac{V_{BR}}{\sqrt{3}I_{BR}} \tag{4.79}$$

where I_{BR} is the average of the three line currents and should be roughly equal to rated current.

With unity slip, $(r_2/s) = r_2$. The applied voltage is quite low and the iron core of the machine is unsaturated. As a result, $\mathscr{P}_p$ is greater than at rated voltage and X_m is larger than normal. Even at rated voltage, X_m is typically 25 times the magnitude of $r_2 + jx_2$. The effect of X_m may thus be neglected, and the assumption made that

$$Z_{BR} \equiv R_{BR} + jX'_{BR} = r_1 + r_2 + jX'_{BR} \tag{4.80}$$

where $X'_{BR} = x'_1 + x'_2$ is the blocked-rotor reactance at the test frequency f_t. The real part of Z_{BR} is given by the power per phase divided by the phase current squared:

$$R_{BR} = \frac{P_{BR}}{3I_{BR}^2} = r_1 + r_2$$

or

$$R_{BR} = |Z_{BR}| \cos \theta_{BR}, \quad \text{where } \cos \theta_{BR} = \frac{P_{BR}}{\sqrt{3}V_{BR}I_{BR}} \tag{4.81}$$

Now r_1 was found by the dc test (Equation 4.78). Then

$$r_2 = R_{BR} - r_1 = \frac{P_{BR}}{3I_{BR}^2} - \frac{V_{dc}}{2I_{dc}} \tag{4.82}$$

The imaginary part of Z_{BR} is given by

$$X'_{BR} = |Z_{BR}| \sin \theta_{BR}$$

or

$$X'_{BR} = \sqrt{Z_{BR}^2 - R_{BR}^2}$$
$$= \sqrt{\left(\frac{V_{BR}}{\sqrt{3}I_{BR}}\right)^2 - \left(\frac{P_{BR}}{3I_{BR}^2}\right)^2} \tag{4.83}$$

If rated frequency is f_B and the frequency of the blocked-rotor test is f_t, then let

$$X_{BR} \stackrel{\Delta}{=} \frac{f_B}{f_t}X'_{BR} \tag{4.84}$$

TABLE 4.2
Evaluating x_1 and x_2 From Blocked-Rotor
Reactance at Rated Frequency

Rotor	x_1 and x_2 as Fractions of X_{BR}
Wound	$x_1 = x_2 = 0.5X_{BR}$
Design A	$x_1 = x_2 = 0.5X_{BR}$
Design B	$x_1 = 0.4X_{BR}; x_2 = 0.6X_{BR}$
Design C	$x_1 = 0.3X_{BR}; x_2 = 0.7X_{BR}$
Design D	$x_1 = x_2 = 0.5X_{BR}$

Since the leakage reactances are proportional to frequency and are defined at $f_1 = f_B$,

$$x_1 + x_2 = X_{BR} = \frac{f_B}{f_t}X'_{BR} \tag{4.85}$$

The designer of the machine knows, on the basis of his or her calculations, how much of X_{BR} is x_1 and how much is x_2. There is no way to determine this ratio from tests on a squirrel cage motor, however. Experience with these machines has been condensed into Table 4.2, which shows approximately how X_{BR} divides between x_1 and x_2 for rotors of different designs. It has been pointed out that the sum $x_1 + x_2 = X_{BR}$ affects the breakdown torque. How X_{BR} is divided between x_1 and x_2 is relatively unimportant.

The *frequency used in the blocked-rotor test* requires consideration. At blocked rotor ($s = 1$), the rotor frequency is the same as that applied to the stator terminals. It has been noted that Design B and C rotors are designed deliberately so that their rotor resistances vary with frequency. The frequency of the test voltage thus will be critical in the determination of r_2 for motors of these two design classes, and most motors are of Design Class B. The variation of r_2 with frequency in wound-rotor motors and those of Designs A and D is not so great. The IEEE Standard 112-1984 recommends a blocked-rotor test frequency of not more than 25 percent of rated frequency. However, the rotor of a 60-Hz motor operates at about 2 Hz at full load and at 60 Hz at starting. A blocked-rotor test at 60 Hz may give a fairly reliable value of r_2 for calculating starting performance. Neither a 60-Hz test nor a 15-Hz test, however, gives very satisfactory values of X_{BR} and r_2 for calculating full-load performance. (Standard 112-1984 suggests alternative methods for the determination of r_2.)

In addition, the rotor heats very rapidly during the blocked-rotor test, and the value of r_2 is a function of temperature. The most consistent results are obtained if the machine is allowed to cool to room temperature before the blocked-rotor test is made and readings are taken immediately after voltage is applied.

The upshot of this discussion is that calculations based on the equivalent-circuit model cannot be expected to predict running performance of Design B and C motors very accurately if the model impedance elements are evaluated on the basis of blocked-rotor tests.

■ **Example 4.7**

The following test data were obtained on a 5-hp, four-pole, 220-V, three-phase, 60-Hz Design B induction motor having a rated current of 12.9 A.

dc Test	No-Load Test	Blocked-Rotor Test
$V_{dc} = 13.8V$	$V_{nl} = 220$ V	$V_{BR} = 23.5$ V
$I_{dc} = 13.0$ A	$f = 60$ Hz	$f_1 = 15$ Hz
	$I_a = 3.86$ A	$I_a = 12.8$ A
	$I_b = 3.86$ A	$I_b = 13.1$ A
	$I_c = 3.89$ A	$I_c = 12.9$ A
	$P_1 = 550$ W	$P_1 = 179$ W
	$P_2 = -350$ W	$P_2 = 290$ W

Calculate the performance of this motor at a slip of 0.03.

Solution. From the blocked rotor test,

$$I_{BR} = \frac{12.8 + 13.1 + 12.9}{3} = 12.93 \text{ A} \qquad (\text{rated } I_1 = 12.9 \text{ A})$$

$$|Z_{BR}| = \frac{V_{BR}}{\sqrt{3}I_{BR}} = \frac{23.5}{\sqrt{3} \cdot 12.93} = 1.049 \ \Omega$$

Note that, at 15 Hz, the voltage required to obtain rated current is only about 10 percent of rated volts. At 60 Hz, it will be about 15 percent.

$$P_{BR} = P_1 + P_2 = 179 + 290 = 469 \text{ W}$$

Then

$$R_{BR} = \frac{P_{BR}}{3I_{BR}^2} = 0.935 \ \Omega$$

and

$$\theta_{BR} = \cos^{-1} \frac{P_{BR}}{\sqrt{3} \cdot 23.5 \cdot 12.93} = 26.98°$$

(At 60 Hz, the blocked-rotor power factor angle is typically around 60°.)

$$X'_{BR} = |Z_{BR}| \sin \theta_{BR} = 0.476 \ \Omega$$

$$X_{BR} = \frac{f_B}{f_t}X'_{BR} = \frac{60}{15} \cdot 0.476 = 1.904 \ \Omega$$

By Table 4.2, for Design B,

$$x_1 = 0.4 \cdot 1.904 = 0.761 \ \Omega \qquad x_2 = 0.6 \cdot 1.904 = 1.142 \ \Omega$$

From the dc test

$$r_1 = \frac{1}{2} \frac{V_{dc}}{I_{dc}} = \frac{13.8}{2 \cdot 13.0} = 0.531 \ \Omega$$

Then

$$r_2 = R_{BR} - r_1 = 0.935 - 0.531 = 0.404 \ \Omega$$

From the no-load test,

$$I_{nl} = \frac{I_a + I_b + I_c}{3} = \frac{3.86 + 3.86 + 3.89}{3} = 3.87 \ A$$

$$Z_{nl} \cong x_1 + X_m = \frac{V_{nl}}{\sqrt{3}I_{nl}} = \frac{220}{\sqrt{3} \cdot 3.87} = 32.8 \ \Omega$$

$$X_m = Z_{nl} - x_1 = 32.8 - 0.761 = 32.1 \ \Omega$$

Friction, windage, and core losses:

$$P_{nl} = P_1 + P_2 = 550 - 350 = 200 \ W$$

$$P_{fwc} = P_{nl} - 3I_{nl}^2 r_1$$

$$= 200 - 3 \cdot 3.87^2 \cdot 0.531 = 176 \ W$$

It will be assumed that the stray load losses are negligible and that

$$P_{rot} = P_{fwc} = 176 \ W$$

The performance may now be calculated for $s = 0.03$:

$$\frac{r_2}{s} = \frac{0.404}{0.03} = 13.47 \ \Omega$$

$$z_2 = \frac{r_2}{s} + jx_2 = 13.47 + j1.142 = 13.52\underline{|4.85°}$$

$$Z_f = z_2 \| jX_m = \frac{j32.1 \cdot 13.52\underline{|4.85°}}{13.47 + j33.2} = \frac{434°\underline{|94.85°}}{35.8\underline{|67.91°}}$$

$$= 12.12\underline{|26.94°} = 10.81 + j5.49 = R_f + jX_f \ \Omega$$

$$Z_{in} = z_1 + Z_f = r_1 + R_f + j(x_1 + X_f)$$

$$= 11.34 + j6.25$$

$$= 12.95\underline{|28.86} \ \Omega$$

The power factor is $\cos 28.86° = 0.876$.

The line current is

$$|I_1| = \frac{220}{\sqrt{3}|Z_{in}|} = \frac{220}{\sqrt{3} \cdot 12.95} = 9.81 \ A$$

and the power drawn from the line is

$$\sqrt{3}V_L|I|\cos\theta_\phi = \sqrt{3} \cdot 220 \cdot 9.81 \cdot 0.876 = 3275 \text{ W}$$

$$P_g = 3I_1^2R_f = 3 \cdot 9.81^2 \cdot 10.81 = 3121 \text{ W}$$

The developed mechanical power is

$$\text{DMP} = 3121(1 - 0.03) = 3027 \text{ W}$$

The output horsepower is given by

$$\frac{\text{DMP} - P_{\text{rot}}}{746} = \frac{3027 - 176}{746} = 3.82 \text{ hp}$$

The efficiency of the motor at this slip is

$$\eta = \frac{P_{\text{out}}}{P_{\text{in}}} = \frac{3027 - 176}{3275} = 87.1\%$$

The motor would develop rated horsepower at a somewhat higher slip. ■

4.16 STARTING INDUCTION MOTORS

Wound-rotor motors present no great starting problem. The stator-winding terminals may be connected directly to the line with the rotor open-circuited, or with relatively high resistances connected to the slip ring brushes. The rotor circuit resistance may then be reduced until the motor starts, and then further reduced as it comes up to

TABLE 4.3
Table of NEMA Code Letters

Code Letter	Locked-Rotor kVA per Horsepower[a]	Code Letter	Locked-Rotor kVA per Horsepower[a]
A	0–3.15	K	8.0–9.0
B	3.15–3.55	L	9.0–10.0
C	3.55–4.0	M	10.0–11.2
D	4.0–4.5	N	11.2–12.5
E	4.5–5.0	P	12.5–14.0
F	5.0–5.6	R	14.0–16.0
G	5.6–6.3	S	16.0–18.0
H	6.3–7.1	T	18.0–20.0
J	7.1–8.0	U	20.0–22.4
		V	22.4 and up

[a]Locked-rotor kVA per horsepower range includes the lower figure up to, but not including, the higher figure. For example, 3.14 is letter A and 3.15 is letter B.
This table is reproduced by permission from NEMA Publication No. MG 1-1987, *Motors and Generators*, copyright 1988 by NEMA.

speed. At full operating speed, the brushes would be shorted together (zero external resistance in each phase).

Nearly all *squirrel cage* induction motors are capable of starting at full-rated voltage without being damaged. However the starting current is so high and the power factor at starting so low that the power supply may be adversely affected. The voltage may "dip" excessively when the motor is connected to the line. Nevertheless, induction motors up to several thousand horsepower are sometimes *"line started"; that is, they are started by simply connecting them to the line*. Design B motors were developed especially for line starting.

Estimating Starting Current from Code Letters

In addition to the design letter, squirrel cage motor nameplates are now required to be marked with a *code letter*. Be careful not to confuse the two. The code letter sets the limits of starting kVA per horsepower for the machine. See Table 4.3.

■ Example 4.8

A 10-hp, 230-V, three-phase induction motor is marked with code letter G. What is the upper limit of the starting current that may be expected at 230 V?

Solution. From Table 4.3, starting kVA per horsepower will be less than 6.3. The starting current will be less than

$$I = \frac{6.3 \cdot 10 \text{ hp} \cdot 1000}{\sqrt{3} \cdot 230} = 158 \text{ A} \qquad ■$$

In many cases it is necessary to limit the starting current to avoid excessive voltage dips. This is particularly true of Design A motors above 7.5 hp. Since the developed torque of a squirrel cage motor at any speed is proportional to the square of the current, starting torque always suffers from any scheme that limits starting current.

The student will immediately think of several ways in which starting current might be limited. A resistor or reactor may be placed in series with each line terminal of the motor, then shorted out as the machine nears rated speed. Resistors cost less than reactors but waste energy. Another method is to insert a step-down, three-phase transformer between the line and the motor during starting and then to remove it as the motor approaches operating speed. Such a starting transformer, together with the necessary switches or magnetic contactors and protective devices, is called a *starting compensator*.

Starting Compensators

A schematic diagram of such a device is shown in Figure 4.27. A starting compensator has a great advantage over a set of series impedances. The starting impedance of the motor is transformed by the square of the compensator turns ratio, thus line current at starting is reduced by the turns ratio squared. However the motor starting current is proportional to the voltage at the motor terminals, and thus is reduced by the first power of the turns ratio. The torque is then reduced by the square of the turns ratio, as is the line current. The reduction in starting line current by means of a compensator

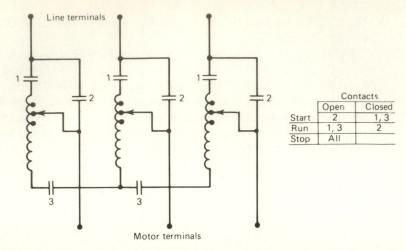

	Contacts	
	Open	Closed
Start	2	1, 3
Run	1, 3	2
Stop	All	

FIGURE 4.27 Starting compensator for induction motors or synchronous motors with damper windings.

is thus accompanied by a proportional reduction in starting torque. With a simple series impedance, the torque is reduced by the *square* of the line current reduction.

■ Example 4.9

Compensators are supplied with several secondary voltage taps to permit the user to select the turns ratio. A certain 100-hp, 1750-rev/min, 2300-V, three-phase motor draws 150 A and produces a starting torque of 120 percent full-load value when started at 2300 V. If a compensator is to be used, and the 80 percent voltage tap is selected, find the starting line current and torque.

Solution. The compensator turns ratio is

$$a = \frac{V_L}{0.8\ V_L} = \frac{1}{0.8} = 1.25$$

The voltage applied to the motor at starting is

$$\frac{2300}{a} = 0.8 \cdot 2300 = 1840 \text{ V}$$

and since the motor phase impedance is practically independent of voltage at $s = 1$,

$$I_{1\text{ start}} = \frac{1840}{\sqrt{3}\ |Z_{\text{in}}|} = \frac{1840}{2300} \cdot 150 \text{ A} = \frac{150}{a} = 0.8 \cdot 150 = 120 \text{ A}$$

Then by the transformer action, the line current is given by

$$I_L = \frac{I_{1\text{ start}}}{a} = \frac{120}{a} = \frac{150}{a^2} = (0.8)^2 \cdot 150 = 96 \text{ A}$$

In general,

$$I_L = \frac{\text{starting current at full voltage}}{a^2}$$

The starting voltage applied to the motor is V_L/a, or 1840 V, but the torque is proportional to the voltage squared. Then

$$\text{Starting torque with compensator} = \left(\frac{1}{a^2}\right) \cdot (\text{Starting torque at line voltage})$$

Full-load torque of the motor is given by

$$\tau_{fl} = \frac{\text{hp} \cdot 5252}{\text{rev/min}} = \frac{525,200}{1750} = 300 \text{ lb-ft}$$

Starting torque at rated voltage is then 120 percent of 300 or 360 lb-ft. Then

$$\tau_{st} = \frac{360}{a^2} = (0.8)^2 \cdot 360 = 230 \text{ lb-ft} \qquad \blacksquare$$

Two other ways of reducing starting current involve reconnecting the stator windings. In **Y–Δ starting,** the rated voltage of the machine with the stator phase windings in Δ must equal the nominal supply voltage. The machine is started with the windings connected in Y. They are switched to the Δ connection when the machine comes up to speed. The *phase* impedance at starting is independent of whether the phases are in Y or Δ. In the Y starting connection, the voltage across each phase is $1/\sqrt{3}$ times its normal value. The starting torque is one third of its Δ-connected value and the line current is also one third of what it would be if the machine were started at rated voltage in Δ.

Part-winding starting requires a motor designed with two windings per phase. An example would be a motor rated 230/460 V, in which the windings of each phase are connected in parallel for 230 V operation and in series for 460 V. For part-winding starting, the *lower* voltage must be used. The motor is started with only one set of phase windings connected to the line. The other set is connected in parallel as the motor approaches rated speed. The part-winding starting current and torque depend on the design of the motor and are not easy to predict.

4.17 SPEED CONTROL OF INDUCTION MOTORS

Induction motors are not good candidates for stepless speed control. It has been pointed out that they tend to overheat if the voltage is reduced. Speed control may be achieved successfully with wound-rotor motors. The motor slows down as more external resistance is inserted. Slip increases, and so does the rotor circuit copper loss ($RCL = sP_g$), but most of this loss is in the control resistors external to the motor. The efficiency of the system is poor at reduced speeds.

Many efforts have been made to obtain the smooth speed control provided by a wound rotor, without wasting energy dissipated in the external rotor resistors. These methods are now largely obsolete. They involve converting the energy, otherwise lost, into either mechanical energy to be supplied to the motor shaft or into electrical energy at line frequency to be returned to the line. An example is the Kramer system, which rectifies the output of the rotor slip rings and supplies this power to a dc motor connected to the shaft of the wound-rotor induction motor. Speed is controlled by controlling the field current of the dc motor. SCR or Ward–Leonard-controlled dc motors have replaced wound-rotor induction motor speed control systems for most industrial applications.

The *speed of an induction motor may be controlled effectively by changing the synchronous speed*. Since $\omega_s = 4\pi f_1/p$, there are only two ways to cause the synchronous speed to change.

Pole-Changing Methods

The number of poles of a squirrel cage motor may be changed by reconnecting the stator winding. The speed changes in steps. This method is not practical for wound-

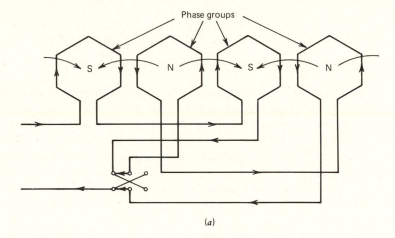

(a)

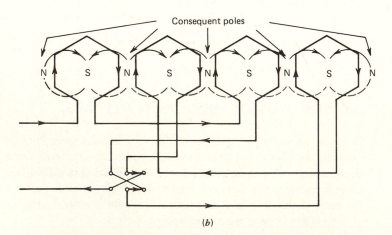

(b)

FIGURE 4.28 Obtaining two synchronous speeds by the method of consequent poles. (*a*) Interconnections of coil groups of one phase winding to obtain high-speed (four pole) operation. (*b*) Low-speed connection (eight poles), with four "consequent" poles.

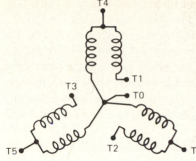

Variable Torque

Speed	L1	L2	L3	Insulate Separately	Tie Together
Low	T1	T2	T3	T4,T5,T6	—
High	T6	T4	T5	—	(T1,T2,T3)

τ_{max} at low speed is less than at high speed.

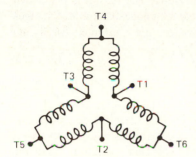

Constant Torque

Speed	L1	L2	L3	Insulate Separately	Tie Together
Low	T1	T2	T3	T4,T5,T6	—
High	T6	T4	T5	—	(T1,T2,T3)

Available horsepower at low speed is one-half that at high speed.

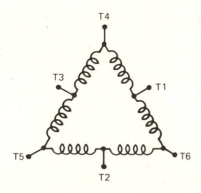

Constant Horsepower

Speed	L1	L2	L3	Insulate Separately	Tie Together
Low	T1	T2	T3	—	T4,T5,T6
High	T6	T4	T5	(T1,T2,T3)	—

τ_{max} at low speed is twice that at high speed.

FIGURE 4.29 Connection diagrams for consequent-pole, two-speed, squirrel cage induction motors.

rotor motors, because the rotor windings would also have to be reconnected to have the same number of poles as the stator. A squirrel cage rotor automatically develops a number of magnetic poles equal to those of the air gap field.

Two methods of pole changing are widely used: the method of *consequent poles* and *pole amplitude modulation* (PAM). The consequent-pole method usually provides only two synchronous speeds, one of which is half of the other. Of course, it is possible to install two separate stator windings and thus obtain four speeds. A three-speed system has been also worked out. The basic scheme is illustrated in Figure 4.28, which shows what happens in a typical phase of a two-speed polyphase winding. Each coil symbol represents a coil group. The coil pitch in pole-changing windings is usually made quite short (say, 60°) or quite long (say, 240°). A short pitch is shown in the figure. When the current to alternate pole groups is reversed, as shown in Figure 4.28*b*, all groups have the same magnetic polarity. The flux of the pole groups must then return in the spaces between the groups, thus inducing magnetic poles of opposite polarity on the surface of the stator core in these spaces. These induced poles are called *consequent poles*. The machine then has twice as many poles as before, and the synchronous speed is half of what it was.

In Figure 4.28 the two sets of pole groups are shown in series. They may also be connected in parallel, or they may be connected in series for one speed and in parallel for the other. Figure 4.29 shows standard connections to obtain different variations of torque and horsepower with speed. The stated characteristics, such as *constant horse-power,* are approximate. If the student applies the relationships of Section 4.13 to these circuits, he or she will be disappointed in the results. The reason is that when the numbers of poles are changed by consequent pole techniques, there are changes in leakage reactance, magnetizing reactance, and losses, not accounted for in the simple theory which forms the basis of Section 4.13.

Pole Amplitude Modulation

This rather glamourous and generally accepted name does not describe accurately the speed-changing method it represents. The result of its application is really suppressed-carrier amplitude modulation of the MMF pattern of the stator winding. Like the consequent-pole method, it is usually used to produce two-speed motors. *It has the advantage that the two speeds need not have a two-to-one ratio*.

Suppose, for example, that the stator field of a machine in sinusoidally distributed:

$$\mathcal{F}_1(\theta)_1 = F_{max} \sin \frac{p_1}{2} \theta_1$$

where p_1 is the number of poles of the winding. Let this be product modulated by another sinusoidal function of θ_1:

$$\mathcal{F}(\theta_1) = M \sin \frac{p_m}{2}\theta_1(F_{max} \sin \frac{p_1}{2} \theta_1)$$

$$= \frac{MF_{max}}{2} \left[\cos \left(\frac{p_1 - p_m}{2} \theta_1 \right) - \cos \left(\frac{p_1 + p_m}{2} \theta_1 \right) \right]$$

The result is that stator fields of two different numbers of poles are produced:

$$p_a = p_1 - k$$

$$p_b = p_1 + k$$

where $k = p_m$ in this case.

In practice, the modulating function is rather crude: a rectangular wave of amplitude $M = 1$; and this modulation is achieved by simply reversing connections to half of the pole groups in each phase winding.

A simple example will illustrate. Consider a three-phase, 8-pole winding. The synchronous speed is 900 rev/min at 60 Hz. The instantaneous MMF of the poles of one phase winding is shown in Figure 4.30a, together with the modulating function, a two-pole rectangular wave. The numbers of poles of the modulated field are thus 6 and 10. The "modulating function" is a fancy way of indicating that, to get a different

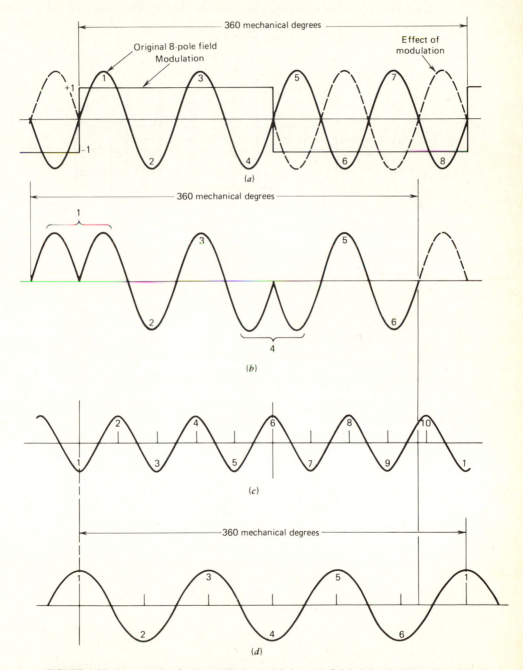

FIGURE 4.30 An example of pole amplitude modulation. (a) Original 8-pole field and modulation function. (b) Modulated field: a combination of 6- and 10-pole fields (6-pole field evident). (c) 10-pole component. (d) 6-pole component.

speed, the connections to pole groups 5, 6, 7, and 8 are reversed. The resulting MMF pattern for the same instantaneous phase current is shown in Figure 4.30b. It is evident that a 6-pole field is produced. The 10-pole field is not so obvious. Fourier analysis of the waveform of Figure 4.30b results in 10- and 6-pole fields of relative amplitudes shown in Figures 4.30c and 4.30d.

Which synchronous speed will be effective? To find out, the fields of the other two phase windings must be taken into account. The 6-pole field is a cosine function

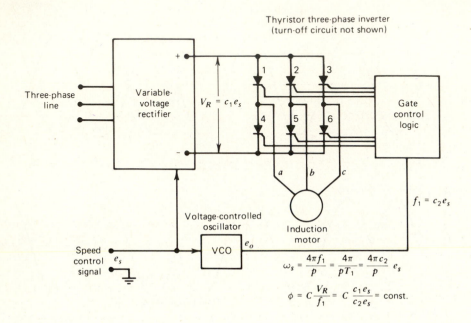

$$\omega_s = \frac{4\pi f_1}{p} = \frac{4\pi}{pT_1} = \frac{4\pi c_2}{p} e_s$$

$$\phi = C\frac{V_R}{f_1} = C\frac{c_1 e_s}{c_2 e_s} = \text{const.}$$

Interval	SCR's Fired
0–60°	1, 5, 3
60°–120°	1, 5, 6
120°–180°	1, 2, 6
180°–240°	4, 2, 6
240°–300°	4, 2, 3
300°–360°	4, 5, 3

SCRs not fired are OFF.

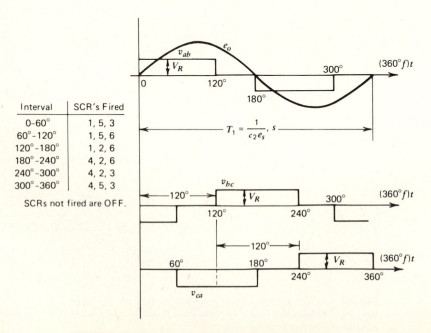

FIGURE 4.31 Induction motor speed controlled by frequency.

relative to the modulating function, and the 10-pole field is a negative cosine function. The modulating function spans 360 mechanical degrees. The modulating functions of the other two phases lead and lag this one by 120 mechanical degrees, respectively. For the 6-pole field, this displacement is 360 electrical degrees ($\frac{6}{2} \times 120°$); that is, the 6-pole components of the three phase-winding fields are all in *space* phase. But they are *not* in time phase. The three 6-pole fields are proportional to the currents in the three phase windings, and if these currents are balanced (or if the windings are Y connected, regardless of balance) their instantaneous sum is zero! Thus the addition of the 6-pole components of the three-phase MMFs results in zero net field. The 10-pole phase MMF patterns, on the other hand, are $\frac{10}{2} \cdot 120 = 600$ electrical degrees apart, or $360° + 240°$. Thus if the phase sequence for the original 8-pole field is *abc*, the 10-pole field will have a sequence *acb*. To prevent the machine from going backward when modulated to 10 poles, two of the line leads must be reversed at the same time the connections to the 4-pole groups in each phase are reversed. The 60-Hz synchronous speed of the modulated motor is 720 rev/min. Previous to the invention of PAM, it was necessary to have two separate windings in a machine to have two speeds which were not in a two-to-one ratio. PAM allows a great reduction in size and cost.

Controlling Speed by Frequency

The synchronous speed of an induction motor may be controlled by controlling the frequency of the voltage applied to its terminals. For three-phase motors, this requires a variable-frequency, three-phase source capable of supplying sufficient power to operate the motor. Such power supplies are generally complex and expensive. An example is shown schematically in Figure 4.31. A standard 60-Hz, squirrel cage motor will operate satisfactorily in such a system up to about 180 Hz. The voltage usually is controlled to be proportional to the frequency, to maintain constant air gap flux per pole. Solving Equation 4.15 for the flux per pole,

$$\phi = \frac{|E_1|}{\omega_e N_{E1}} = \left(\frac{1}{2\pi N_{E1}} \right) \frac{|E_1|}{f_1} \tag{4.86}$$

At any given slip, E_1 is proportional to V_{1m}, and it is observed that the flux is proportional to volts per hertz of the source. If the voltage were not varied along with the frequency, the core would saturate at low frequencies and there would be insufficient torque at high frequencies. With a constant volts/hertz supply, the available torque is the same at all speeds and the available horsepower is proportional to speed.

FOR FURTHER STUDY

THE CIRCLE DIAGRAM OF THE INDUCTION MACHINE

The circle diagram is very useful for predicting the performance of an induction machine in a qualitative way. It is helpful in organizing one's knowledge about how the machine will act in all possible modes of operation.

At one time, circle diagrams were drawn accurately to a large scale, and measurements taken from them were used for performance calculations. At least one of the early computer programs developed for induction machine analysis was based

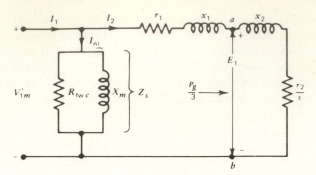

FIGURE 4.32 Approximate equivalent circuit for one phase of an asynchronous machine.

on the circle diagram. Although many complex diagrams have been devised, only the simplest will be discussed here.

The circle diagram is based on the *approximate equivalent circuit* of the induction motor shown in Figure 4.32. The shunt branch has been moved from points a, b to the input terminals and includes a resistor R_{fwc} to represent rotational losses. (If one wishes to make accurate calculations using the approximate circuit model, the phase terminal voltage should be reduced to a value which will make I_2 the same as it was before the shunt branch was moved. It can be shown that this reduced voltage is very nearly in phase with the actual terminal voltage V_{1m} and is given almost exactly by

$$V'_{1m} = V_{1m}\left(1 - \frac{x_1}{X_m}\right) \qquad (4.87)$$

Since all rotational losses are taken care of by R_{fwc}, the approximate circuit assumes zero slip at no load on the motor shaft. Then at no load, r_2/s is an open circuit, and the no-load line current I_{nl} is V'_{1m}/Z_s. By Kirchhoff's current law, the line current under load is

$$I_1 = I_{nl} + I_2 \qquad (4.88)$$

where

$$|I_2| = \frac{V'_{1m}}{|z_1 + z_2|} \qquad (4.89)$$

and

$$z_1 + z_2 = r_1 + \frac{r_2}{s} + j(x_1 + x_2) \qquad (4.90)$$

The current I_2 will lag V'_{1m} by the impedance angle

$$\theta_2 = \sin^{-1}\frac{x_1 + x_2}{|z_1 + z_2|} \qquad (4.91)$$

Equation 4.89 may be manipulated as follows:

$$|I_2| = \left(\frac{V'_{1m}}{x_1 + x_2}\right)\left(\frac{x_1 + x_2}{|z_1 + z_2|}\right) = \frac{V'_{1m}}{x_1 + x_2}\sin\theta_2 \qquad (4.92)$$

This will be recognized as the polar equation of a circle of the form $\rho = D\sin\theta$, where D is the circle diameter. Thus the locus of I_2 is a circle of diameter

$$D = \frac{V'_{1m}}{x_1 + x_2} \qquad (4.93)$$

By the way, $x_1 + x_2$ is the blocked-rotor reactance when the blocked-rotor test is made at rated frequency.

The equation for the circle diagram may be written by inserting

$$I_2 = |I_2|\,\underline{|-\theta_2} = \left(\frac{V'_{1m}}{x_1 + x_2}\sin\theta_2\right)\underline{|-\theta_2}$$

into Equation 4.88:

$$I_1 = I_{nl} + \left(\frac{V'_{1m}}{x_1 + x_2}\sin\theta_2\right)\underline{|-\theta_2} \qquad (4.94)$$

The resulting diagram with some additions is shown in Figure 4.33. Since I_1 is the sum of I_{nl} and I_2, the point of the I_1 phasor will coincide with that of I_2.

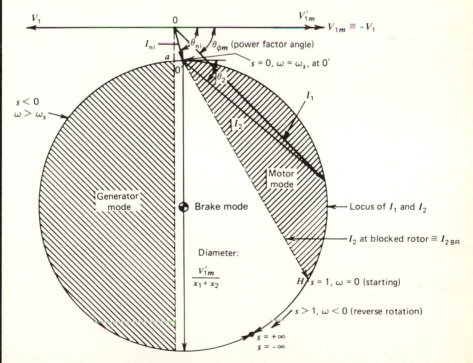

FIGURE 4.33 Current loci of an induction machine.

Thus the circle serves as the locus of both I_1 and I_2, though they radiate from different origins.

At starting, I_1 and I_2 will be at some point H. As the motor accelerates, the tips of the current phasors travel around the circle in a CCW direction until the output torque matches the load torque. If there is no shaft load, the motor will continue to accelerate to very near synchronous speed (according to the approximate model, synchronous speed actually will be reached). At this point, $I_2 = 0$ and $I_1 = I_{nl}$.

The circle diagram of Figure 4.33 represents any one phase of the machine. For a three-phase machine, the input power is given by

$$P_{in} = 3 V_{1m} I_1 \cos \theta_{\phi m} \qquad (4.95)$$

Note that V_{1m}, rather than V'_{1m}, appears in this expression for the power. The input voltage to the model was reduced to V'_{1m} to make the currents correct with the shunt branch displaced. Once the currents are determined, the voltage actually applied to the motor, V_{1m}, must be used in power calculations.

Conditions at Blocked Rotor

Refer to Figures 4.33 and 4.34. At blocked rotor, the output power is zero and the electrical power is all consumed as losses:

$$P_{in \ BR} = 3 V_{1m} I_{1 \ BR} \cos \theta_{\phi \ BR} = P_{rot} + SCL + RCL \qquad (4.96)$$

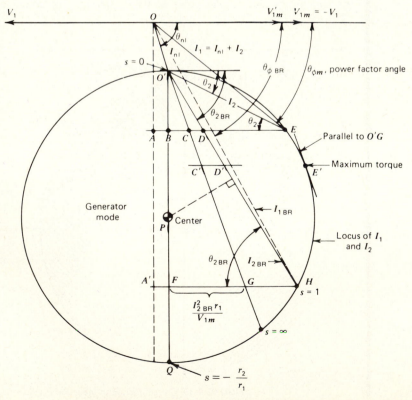

FIGURE 4.34 Circle diagram of an induction machine.

How can there be "rotational" losses at blocked rotor? Well, the approximate model assumed the rotational losses to include friction, windage, and *core losses*. Now at normal operating speeds, the *rotor* core losses are negligible. But when the slip is unity, as at blocked rotor, the rotor frequency equals the line frequency, and the losses in the rotor core are substantial. Therefore one is not too far off in assuming that, as speed decreases, rotor core losses increase to offset the decrease in friction and windage, so that the so-called "rotational" losses remain about constant. The amount of this constant loss is given by

$$P_{\text{rot}} = 3 \, |V_{1m}| \, |I_{\text{nl}}| \, \cos \theta_{\text{nl}} = 3 \, V_{1m} \, (A'F) \tag{4.97}$$

where $A'F = I_{\text{nl}} \cos \theta_{\text{nl}}$ is the distance from A' to F in Figure 4.34.

Since I_1 is the sum of I_2 and I_{nl}, and by examination of Figure 4.34, it will be seen that

$$|I_{1 \text{ BR}}| \cos \theta_{\phi \text{ BR}} = |I_{\text{nl}}| \cos \theta_{\text{nl}} + |I_{2 \text{ BR}}| \cos \theta_{2 \text{ BR}}$$

$$= A'F + FH = A'H$$

$$= AB + FH \tag{4.98}$$

Multiplying both sides of Equation 4.98 by $3V_{1m}$, and comparing with Equation 4.96,

$$P_{\text{in BR}} = 3 \, |V_{1m}| \, |I_{1 \text{ BR}}| \cos \theta_{\phi \text{ BR}} = 3 \, |V_{1m}| \, (A'H)$$

$$= 3 \, |V_{1m}| \, (A'F) + 3 \, |V_{1m}| \, (FH) \tag{4.99}$$

$$= P_{\text{rot}} + \text{SCL} + \text{RCL}$$

It is evident that the sum of the rotor and stator copper losses is given by a constant ($3V_{1m}$) times the distance FH:

$$3 \, |V_{1m}| \, (FH) = \text{SCL}_{\text{BR}} + \text{RCL}_{\text{BR}} = 3 \, |I_{2 \text{ BR}}|^2 \, (r_1 + r_2) \tag{4.100}$$

The distance *FH* can be divided into two segments to represent the two copper losses individually. The stator copper loss, per phase, according to the model of Figure 4.32 is given at blocked rotor by

$$\frac{\text{SCL}}{3} = |I_{2 \text{ BR}}|^2 \, r_1 \text{ W} \tag{4.101}$$

Let the distance *FG* be made equal to

$$FG = \frac{|I_{2 \text{ BR}}|^2 \, r_1}{|V_{1m}|} \text{ A} \tag{4.102}$$

Then multiplying this line segment by $3 \, |V_{1m}|$ gives the stator copper loss:

$$3 \, |V_{1m}| \, (FG) = 3 \, |V_{1m}| \, \frac{|I_{2 \text{ BR}}|^2 \, r_1}{V_{1m}} = 3 \, |I_{2 \text{ BR}}|^2 \, r_1 = \text{SCL}_{\text{BR}} \text{ W} \tag{4.103}$$

It follows from Equation 4.99 that the rotor copper loss at blocked rotor is given by

$$\text{RCL}_{BR} = 3 \, |V_{1m}| \, (FH - FG) = 3 \, |V_{1m}| \, (GH) \text{ W} \qquad (4.104)$$

Running Conditions

When running, the machine will be operating at some point on the current locus, such as E in Figures 4.34 and 4.35. Input power is given by

$$P_{in} = 3 \, |V_{1m}| \, |I_1| \, \cos \theta_{\phi m} = 3 \, V_{1m} \, (AE) \quad \text{W} \qquad (4.105)$$

and

$$\cos \theta_{\phi m} = \frac{AE}{OE} = \text{power factor}$$

Rotational losses are still $3 \, |V_{1m}| \, (AB)$. Thus the sum of the rotor and stator copper losses and the output power is given by $3 \, |V_{1m}| \, (BE)$. If the $I_{2\,BR}$ phasor is drawn from O' to H, and another straight line is drawn from O' through G and beyond, do their intersections with the line AE have any significance? Indeed they do! Compare the length FH with BD, where D is the intersection of $O'H$ with AE. By similar triangles,

$$\frac{BD}{FH} = \frac{O'B}{O'F} \qquad (4.106)$$

Now

$$O'B = |I_2| \, \sin \theta_2$$

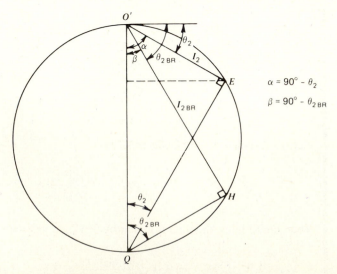

$$\alpha = 90° - \theta_2$$
$$\beta = 90° - \theta_{2\,BR}$$

FIGURE 4.35 Angular relationships in the circle diagram.

and

$$O'F = |I_{2BR}| \sin \theta_{2BR} \qquad (4.107)$$

Refer to Figure 4.35. Let the diameter of the circle passing through O' be $O'Q$, and draw lines from E to Q and H to Q. Since they are inscribed in a semicircle, the triangles $O'EQ$ and $O'HQ$ are right triangles. It will be seen that

$$\sin \theta_2 = \frac{|I_2|}{O'Q} \quad \text{and} \quad \sin \theta_{2\,BR} = \frac{|I_{2BR}|}{O'Q} \qquad (4.108)$$

Combining Equations 4.108, 4.107, and 4.106,

$$\frac{BD}{FH} = \frac{|I_2|^2}{|I_{2BR}|^2}$$

Multiplying numerator and denominator by $(r_1 + r_2)$ shows that

$$\frac{BD}{FH} = \frac{\text{copper loss at } I_2}{\text{copper loss at blocked rotor}} \qquad (4.109)$$

Since it has been shown that $3V_1(FH)$ is the total blocked-rotor copper loss, then

$$3|V_{1m}|(BD) = \text{SCL} + \text{RCL} \qquad (4.110)$$

under load (I_1 at point E). Since by similar triangles

$$\frac{BC}{BD} = \frac{FG}{FH} = \frac{r_1}{r_1 + r_2}$$

then

$$\text{SCL} = 3\,V_{1m}(BC) \quad \text{W}$$

and

$$\text{RCL} = 3V_{1m}(BD - BC) = 3V_{1m}(CD) \quad \text{W} \qquad (4.111)$$

Finally,

$$P_{\text{out}} = P_{\text{in}} - \text{losses}$$
$$= 3|V_{1m}|(AE) - 3|V_{1m}|(AB + BC + CD)$$

or

$$P_{\text{out}} = 3|V_{1m}|(DE) \quad \text{W} \qquad (4.112)$$

The efficiency is given by

$$\eta = \frac{P_{\text{out}}}{P_{\text{in}}} = \frac{DE}{AE} \qquad (4.113)$$

In the approximate model, the *air gap power*, or rotor input, is given by

$$P_g = P_{out} + RCL$$

$$= 3V_{1m}(CE) \quad W \tag{4.114}$$

Remember that $RCL = sP_g$. Then

$$s = \frac{RCL}{P_g} = \frac{CD}{CE}$$

and the speed is

$$\omega_s(1 - s) = \omega_s \left(\frac{DE}{CE}\right) \tag{4.115}$$

The torque is given by

$$\tau_{out} = \frac{P_g}{\omega_s} = \frac{3 \mid V_{1m} \mid (CE)}{\omega_s} \quad \text{N-m}$$

or

$$\tau_{out} = 7.04 \frac{3 \mid V_{1m} \mid (CE)}{n_s} \quad \text{lb-ft} \tag{4.116}$$

Often it is enough to know that the torque is proportional to the distance CE. Thus maximum torque corresponds to maximum CE, and it is found by drawing a tangent to the circle parallel to line $O'G$, locating E'. Then the slip for maximum torque is

$$s_M = \frac{C'D'}{C'E'} \tag{4.117}$$

and

$$\tau_{max} = \frac{3 \mid V_{1m} \mid (C'E')}{\omega_s} \quad \text{N-m} \tag{4.118}$$

Summarizing:

$$3 \mid V_{1m} \mid \times \left\{ \begin{array}{l} AE = \text{input power, watts} \\ AB = \text{rotational loss, watts} \\ BC = \text{stator copper loss, watts} \\ CD = \text{rotor copper loss, watts} \\ DE = \text{output power, watts} \\ \dfrac{DE}{746} = \text{output horsepower} \\ \dfrac{CE}{\omega_s} = \text{output torque, newton-meters} \\ \dfrac{GH}{\omega_s} = \text{starting torque, newton-meters} \end{array} \right.$$

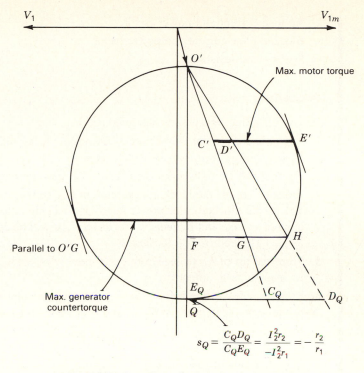

$$s_Q = \frac{C_Q D_Q}{C_Q E_Q} = \frac{I_2^2 r_2}{-I_2^2 r_1} = -\frac{r_2}{r_1}$$

FIGURE 4.36 Example applications of the circle diagram.

Also

$$\frac{CD}{CE} = \text{slip}$$

$$n_s \frac{DE}{CE} = \text{speed}$$

$$\frac{DE}{AE} = \text{efficiency}$$

$$\frac{AE}{OE} = \text{power factor}$$

It is usually sufficient to make a rough sketch of a circle diagram to satisfy oneself about some point regarding polyphase induction-machine performance, or to settle an argument. For example, Figure 4.36 makes it obvious that the maximum countertorque in the generator mode must be greater than the maximum developed torque in the motor mode. However as a basis for sketching, the method of accurate construction is as follows:

CONSTRUCTING THE CIRCLE DIAGRAM

Data Required:
    Rotor *phase* voltage

$$|V_{1m}| = \frac{V_L}{\sqrt{3}}$$

No-load current and power factor

Blocked-rotor current and power factor

Stator phase resistance

Procedure

1. Lay off a line to represent the reference V_{1m}.

2. Taking a convenient current scale and origin O, lay off I_{n1} lagging V_{1m} by the no-load power factor angle, thus locating O'.

3. From O and O' draw lines OAA' and $O'Q$, each perpendicular to V_{1m}.

4. From O, lay off the blocked-rotor current I_{1BR} to the same scale as I_{n1}, lagging V_{1m} by the blocked-rotor power factor angle (This is the blocked-rotor current at rated voltage and frequency. Normally, I_{1BR} will be 20 or more times as long as I_{n1}.) The end of I_{1BR} locates H.

5. From the tip of I_{n1} to the tip of I_{1BR}, draw $O'H \equiv I_{2BR}$. Measure I_{2BR} to find its magnitude in amperes.

6. From H, draw a line parallel to V_{1m} and perpendicular to OAA' and $O'Q$ until it intersects these lines. Mark A' and F at these intersections.

7. Having I_{2BR} and r_1, calculate $FG = |I_{2BR}|^2 r_1/|V_{1m}|$. Draw the line $O'G$.

8. Construct a perpendicular bisector to $O'H$. Since $O'H$ is a chord of the circle, this bisector will pass through the center of the circle at P. With radius $O'P$ or PH, draw the circle.

To analyze the performance under load, all that is necessary to know is one of the performance parameters, for example, the line current. If the line current is known, the point E may be found by scribing an arc of radius I_1, centered at O, with E being at the intersection of this arc with the circle. Next draw the line AE and locate B, C, and D.

4.18 THE ASYNCHRONOUS GENERATOR

The asynchronous, or induction, machine is sometimes used as a generator. Asynchronous generators are frequently used in wind power systems. They are often employed to supply additional power to a load in a remote area that is being served by a weak transmission line. Their advantages in this application are that they are rugged and relatively cheap, and that the engine or windmill driving the generator is not required to operate precisely at synchronous speed. The machine must be driven *above* synchronous speed, however, and the mechanical drive must be equipped with a control to cause the speed to increase as the electrical load increases.

In addition to supplying that part of the watt load not provided by the generator, the transmission line maintains the frequency constant and supplies the induction generator's magnetizing current. If the generator produces more power than the load requires, the excess is pumped back into the line. There must be an agreement between the power company and the owner of the generator as to who pays for the excess power. Also, care must be taken that the local generator does not energize the line when crews are working on it.

When constant frequency and constant voltage are not essential, an asynchronous generator may be operated unconnected to a power line. (For example, constant voltage and frequency are not necessary if the generated power is to be used to pump water or to provide heat.) This is called *stand-alone* operation, or the generator is said to be *isolated*. Under this condition the analysis of the performance of the machine becomes much more complex.

Induction Generator Connected to a Line

The characteristics of the asynchronous generator connected to a line are most easily understood in terms of the circle diagram. Refer to Figure 4.33. The machine will be

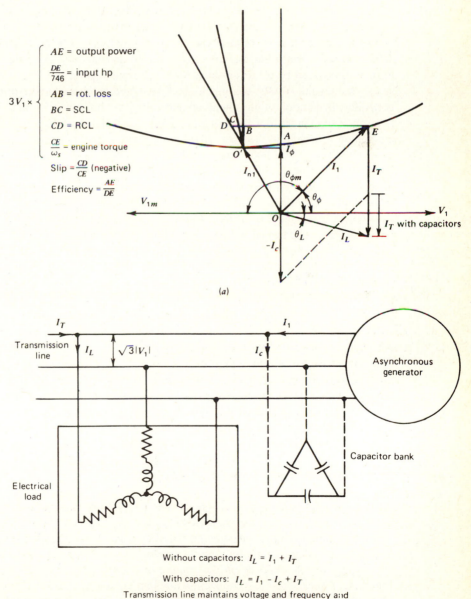

AE = output power

$\frac{DE}{746}$ = input hp

AB = rot. loss

BC = SCL

CD = RCL

$\frac{CE}{\omega_s}$ = engine torque

Slip = $\frac{CD}{CE}$ (negative)

Efficiency = $\frac{AE}{DE}$

$3V_1 \times$

(a)

Without capacitors: $I_L = I_1 + I_T$

With capacitors: $I_L = I_1 - I_c + I_T$

Transmission line maintains voltage and frequency and supplies lagging VARs to load and generator.

(b)

FIGURE 4.37 An application for an asynchronous generator. (*a*) Circle diagram of a loaded asynchronous generator. (*b*) System showing asynchronous generator supplying power to a local load.

generating when the power flow is reversed, relative to the motor mode, that is, when $\cos \theta_{\phi m}$ is negative, which means $\theta_{\phi m}$ is greater than 90°. The circle diagram indicates that this requires a negative slip, which means that the machine must be driven above synchronous speed by the mechanical power source. As the speed is gradually increased above synchronous speed, the current vectors first move from O' to a in Figure 4.33 without any generated output. In this region, the mechanical source is gradually taking over the job of supplying rotational losses from the electrical source. At a, all losses are being supplied mechanically, and the line merely supplies the reactive component of I_{n1}, that is, the magnetizing current flowing through X_m. A further increase in speed causes the machine to feed electrical power into the line, and the machine is operating in the generator mode. In this mode, $I_1 \cos \theta_{\phi m}$ lies along V_1, rather than V_{1m}.

Figure 4.37a shows a part of the circle diagram redrawn with V_1 rather than V_{1m} as the reference, that is, rotated 180° from Figure 4.33. Note that I_1 *must lead* V_1. This figure also illustrates the situation in which an induction generator is supplying a load also served by a transmission line. The transmission line is essential. It must supply the lagging VARs to compensate for the leading characteristic of the induction generator. In addition, it must hold the line voltage within suitable limits. Figure 4.37b is a schematic diagram of the system. The phasor diagram of Figure 4.37a illustrates a situation in which all the load power is supplied by the induction generator. In this case, the transmission line current is all at zero power factor, lagging; that is, it lags V_1 by 90°. Its magnitude must be sufficient to supply the reactive components of the generator and load currents combined.

It would be difficult for the transmission line to maintain voltage with a largely zero-power-factor lagging current. The situation may be improved by the installation of capacitors, as indicated by the dashed lines in the figure.

If there is an increase in load power requirements, either the transmission line must supply this power increment or the slip of the induction generator must increase. An increase in induction machine power requires a larger negative slip. Since the synchronous speed is held constant by the line frequency, the speed of the engine driving the induction machine must increase. That is why a throttle control is required that will increase the speed as the electrical power load increases. There need not be a perfect match between the induction machine power output and the load requirements. The transmission line can supply or carry away a small deficiency or surplus. This kind of auxiliary power source is relatively inexpensive. An ordinary squirrel cage motor may be used as the generator, and the engine controls need not be complicated. The induction machine does not have to be synchronized to the line as does a synchronous alternator.

The Isolated Asynchronous Generator

Analysis of the performance of an isolated asynchronous generator is much more complex than it is when the generator is connected to a line, because the frequency is not fixed and excitation must be provided by capacitors connected to the terminals. The frequency depends not only on the speed of the mechanical drive but also on the slip, which in turn depends on the amount of power required by the electrical load. Figure 4.38 shows both the connection diagram and circuit model of one phase of the three-phase system. The exciting capacitors normally would be connected in Δ, because this connection results in more nearly balanced line currents. However the model of one phase is on a line-to-neutral basis, so X_c in the model circuit is one third of the capacitive reactance connected line-to-line in the Δ arrangement. (Since the machine

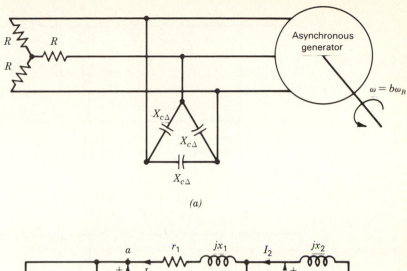

(a)

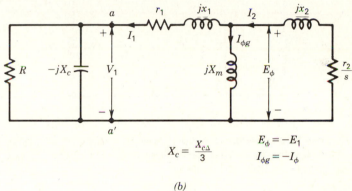

$$X_c = \frac{X_{c\Delta}}{3}$$

$$E_\phi = -E_1$$
$$I_{\phi g} = -I_\phi$$

(b)

FIGURE 4.38 Three-phase isolated asynchronous generator: (a) Connection diagram. (b) Model of one phase at rated frequency.

is operating in the generator mode, V_1 replaces V_{1m} as the terminal voltage, and the model has been inverted, with the unprimed terminal of the phase winding as the top, positive terminal. The voltage induced by the air gap flux is now E_ϕ, the negative of E_1; and E_ϕ/jX_m is defined as $I_{\phi g}$, which is the negative of I_ϕ.)

The load on the system is shown to be purely resistive. If it contains reactive elements, an equivalent parallel R–C branch must be calculated to permit using the following analysis. The combination of the exciting capacitors and the electrical load must present a leading power factor to the asynchronous generator, or the voltage will not build up.

EFFECT OF FREQUENCY VARIATIONS

Recall that the Steinmetz model of the induction machine is referred to the stator at stator frequency. Let the rated frequency of the machine be designated f_B and the corresponding synchronous speed be ω_B. Then let the ratio of driven speed of the generator to base synchronous speed be b:

$$b = \frac{\omega}{\omega_B} \tag{4.119}$$

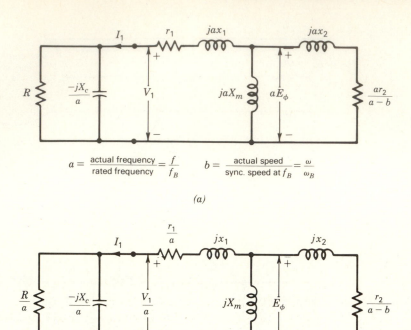

$$a = \frac{\text{actual frequency}}{\text{rated frequency}} = \frac{f}{f_B} \qquad b = \frac{\text{actual speed}}{\text{sync. speed at } f_B} = \frac{\omega}{\omega_B}$$

(a)

(b)

FIGURE 4.39 Circuit model of one phase of an asynchronous generator, its load and exciting capacitors, adjusted to include the effects of speed and frequency variations. (*a*) Effects of frequency variations. (*b*) The model reduced to rated frequency.

Similarly, define the ratio of generated frequency to rated frequency as *a*:

$$a = \frac{f}{f_B} \tag{4.120}$$

It will be assumed that all the model reactances of Figure 4.38 are measured at rated frequency. Then under operating conditions for the generator, the inductive reactances will be proportional to frequency and the capacitive reactance will vary inversly. Also, $|E_\phi|$, the voltage induced by the air gap flux will be proportional to frequency. These effects are shown in Figure 4.39*a*.

The synchronous speed varies with frequency also, and is given by $a\omega_B$. The slip under operating conditions is thus

$$s = \frac{\omega_s - \omega}{\omega_s} = \frac{a\omega_B - b\omega_B}{a\omega_B} = \frac{a - b}{a} \tag{4.121}$$

It should be noted that since the slip is negative in generator operation, the quantity $a - b$ is also negative. Dividing each impedance in the model *and* the voltage by a leaves the currents in the model unchanged, but allows the rated-frequency value of the voltage E_ϕ to be used. The result of this operation is shown in Figure 4.39*b*.

Factors Affecting Generated Frequency

Reference to the generator portions of the torque/slip curves of Figure 4.19 will show that the slip will be small and negative. The electrical load cannot be increased beyond the point at which maximum torque is developed. At that point, operation becomes unstable and generation collapses. The output frequency is given by

$$f = \frac{p\omega_s}{4\pi} = \frac{p}{4\pi}\frac{\omega}{1-s} \qquad\qquad (4.122)$$

Since the slip is negative, the quantity $(1 - s)$ will be greater than unity, but only slightly greater. The frequency, then, will depend primarily on the speed and will decrease slightly as the electrical power load is increased.

EXCITATION AND VOLTAGE BUILDUP

For a terminal voltage to be developed when an isolated asynchronous generator is brought up to speed, two conditions must exist. The rotor core must have some remanent magnetism, and sufficient exciting capacitance must be connected to the terminals. The rotor core need not be highly magnetized. Usually, previous operation of the machine will have left the rotor with sufficient magnetism. If not, connecting a storage battery to any two stator terminals (with the rotor stationary) will do the trick.

Figure 4.40 shows why the load on the machine must be leading. In Figure 4.40a, the N and S rotor magnetic poles are those due to remanent magnetism. No-load conditions are illustrated to avoid unnecessary complication. When the load resistance R is infinite, the slip is quite small and $r_2/(a - b)$ is essentially an open circuit. Then the magnetizing current $I_{\phi g} \approx -I_1$, as shown in Figure 4.40d. Figure 4.40a shows the remanent rotor poles in the position relative to the phase-a winding such that the flux linkages with that winding are decreasing at their maximum rate; and as a result, E_ϕ is at its peak. It is a very small peak because the remanent magnetism is weak. Figure 4.40b shows the situation 90 electrical degrees later. The rotor poles lie along the axis of the phase winding, and if the voltage is ever going to amount to anything, the a-phase current must be in such a direction as to increase the rotor's magnetization. Application of the right-hand rule shows that, ideally, I_1 should be at its negative maximum at this instant. Examination of the waveforms of Figure 4.40c demonstrates that this will occur if I_1 leads E_ϕ by 90 degrees, that is to say, if the impedance connected to the machine terminals is capacitive.

Figure 4.41 is a plot of $|E_\phi|$ as a function of $|I_\phi|$ for a 10-hp, two-pole, 208-V, three-phase machine. Saturation of the iron is evident. Obviously, X_m is not constant. However saturation is negligible in capacitors. At no load, $|I_{\phi g}| = |I_1|$, and since r_1 and x_1 are small, $I_1 \approx (E_\phi/-jX_c)$. So the capacitor characteristic will be a straight line when plotted on the same grid as that of the magnetization curve. The equation of the line will be

$$|E_\phi| = X_c|I_{\phi g}| \qquad\qquad (4.123)$$

Note that the slope of the line is X_c, the capacitance per phase of the exciting capacitors, assumed Y-connected. The curves are drawn for operation at rated frequency, 60 Hz. It is seen that for $X_c = 12.5\ \Omega$, the no-load voltage will be approximately 177 V,

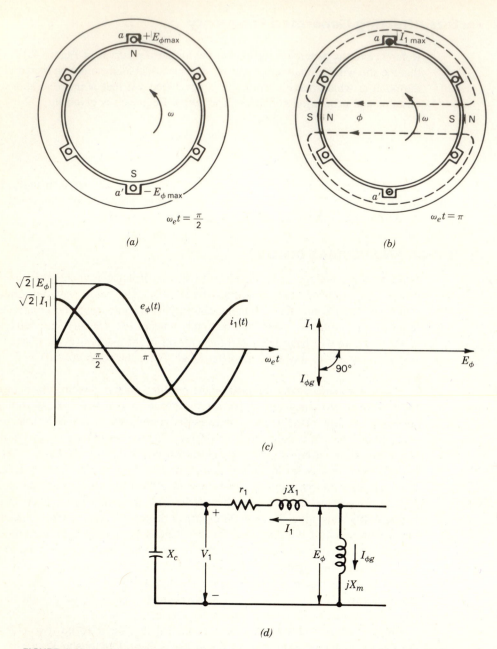

FIGURE 4.40 Conditions for the voltage buildup of an isolated induction generator. (*a*) Position of poles of rotor remanent magnetism at the instant of maximum induced voltage in phase *a*. (*b*) Reinforcement of rotor magnetism by leading current in the phase winding. (*c*) No-load waveforms and phasor diagram. (*d*) Circuit model at no load.

line-to-neutral, when the machine is driven at a speed that will result in an electrical frequency of 60 Hz. The critical capacitance is that which corresponds to the air gap line, in this case 114 μF per phase. For smaller capacitors, the voltage will not build up at all. The ordinates of the magnetization curve are proportional to frequency, and the slopes of the X_c lines are inversely proportional to frequency. Thus as the speed decreases, much larger capacitors are required to obtain voltage buildup.

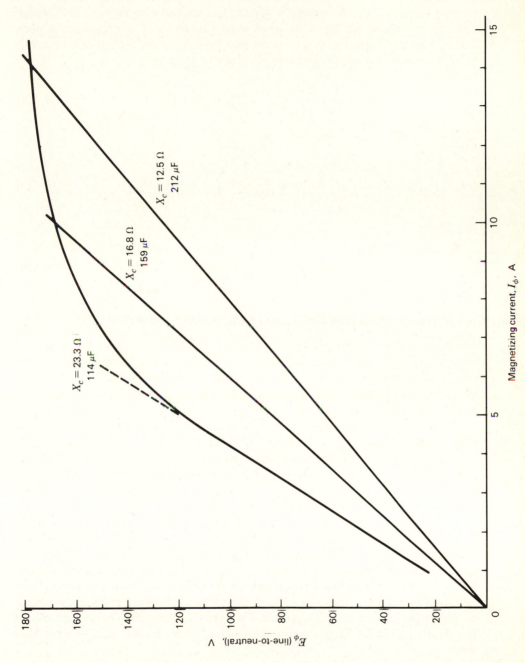

FIGURE 4.41 Magnetization curve of a 10-hp, two-pole, 208-V, three-phase, 60-Hz induction motor to be used as a generator, driven at 3600 rev/min.

MEASURING THE MAGNETIZATION CURVE

One way to obtain data for the magnetization curve of an asynchronous machine is to operate it as a motor with no shaft load from a three-phase, variable-voltage source of rated frequency. As the voltage is varied, data are taken of the input line voltage and current. Again, the slip is so small at no load that r_2/s is effectively an open circuit. Since X_m varies, it must be eliminated from equations used to determine E_ϕ.

The square of the input impedance per phase is given by

$$\frac{V_{n1}^2}{3I_{n1}^2} = r_1^2 + (x_1 + X_m)^2 \tag{4.124}$$

Then

$$X_m = \left(\frac{V_{n1}^2}{3I_{n1}^2} - r_1^2\right)^{1/2} - x_1 \tag{4.125}$$

and finally,

$$|E_\phi| = X_m I_{n1} = \left[\left(\frac{V_{n1}^2}{3I_{n1}^2} - r_1^2\right)^{1/2} - x_1\right]I_{n1} \tag{4.126}$$

DEFINITION OF THE BASIC ASYNCHRONOUS GENERATOR PROBLEM

The basic problem in determining asynchronous machine performance is to find the voltage and frequency supplied to a given electrical load when the speed and exciting-capacitor reactance are known. It is assumed that the circuit model impedances r_1, r_2, x_1, and x_2 have been determined at rated frequency, and that the magnetization curve has been plotted, also at rated frequency.

The electrical load must be stated in terms of resistance R ohms per Y-connected phase rather than so many watts, because the voltage is unknown at the outset. The reactance per phase of the exciting capacitors must be small enough to permit buildup and to maintain voltage over the expected range of loads and speeds. The correct reactance can only be determined by experiment or by working the problem several times until the results are suitable.

To solve the problem, equations are written in terms of the circuit impedances, the frequency ratio a, and the speed ratio b to represent conservation of power and conservation of VARs. Since the speed is a "given," the value of b is known. The power conservation equation turns out to be a fifth-order polynomial in a, having only one real root. This real value of a determines the frequency. The VAR relationship may be solved to determine the value of X_m at rated frequency. When this value is plotted on the magnetization curve, the generated voltage $|E_{1g}|$ is found. The circuit model is then easily solved to determine terminal voltage, power output, slip, or what have you?

SOME DETAILS OF THE SOLUTION

Refer to Figure 4.39*b*. The first step is to replace the reduced load at the terminals of the machine, consisting of $(R/a)\|(-jX_c/a^2)$, with the equivalent series impedance $R_L - jX_L$. Then the stator-winding reduced resistance and leakage reactance may be added to obtain the total strator-circuit impedance Z_s:

$$Z_s = R_L + \frac{r_1}{a} + j(x_1 - X_L) = \frac{X_c^2 R}{a(a^2 R^2 + X_c^2)} + \frac{r_1}{a} + j\left((x_1 - \frac{R^2 X_c}{a^2 R^2 + X_c^2}\right)$$

(4.127)

The reciprocal of Z_s gives the reduced stator-circuit admittance, here stated in terms of R_L and X_L for simplicity:

$$Y_s = \frac{R_L + (r_1/a)}{[R_L + (r_1/a)]^2 + (x_1 - X_L)^2} - j\frac{x_1 - X_L}{[R_L + (r_1/a)]^2 + (x_1 - X_L)^2}$$

(4.128)

$$= g_s - jb_s$$

Similarly for the rotor circuit,

$$Y_r = \frac{1}{[r_2/(a - b)] + jx_2}$$

$$= \frac{(a - b)r_2}{r_2^2 + (a - b)^2 x_2^2} - j\frac{(a - b)^2 x_2}{r_2^2 + (a - b)^2 x_2^2}$$

(4.129)

$$= g_r - jb_r$$

Let $E_\phi = |E_\phi|\underline{|0°}$. Then

$$Re|I_2| = -|E_\phi|g_r$$

and the generated power is

$$P_r = -|E_\phi|^2 g_r$$

(4.130)

The absorbed power is given by

$$P_s = |E_\phi|^2 g_s$$

The conservation of power equation is then

$$P_r = \frac{-|E_\phi|^2(a - b)r_2}{r_2^2 + (a - b)^2 x_2^2} = \frac{|E_\phi|^2[R_L + (r_1/a)]}{[R_L + (r_1/a)]^2 + (x_1 - X_L)^2} = P_s$$

or

$$\frac{R_L + (r_1/a)}{[R_L + (r_1/a)]^2 + (x_1 - X_L)^2} + \frac{(a - b)r_2}{r_2^2 + (a - b)^2 x_2^2} = 0$$

(4.131)

Substituting the expressions for R_L and X_L found in Equation 4.127 into the first term of the above equation gives

$$\frac{\dfrac{RX_c^2}{a(a^2 R^2 + X_c^2)} + \dfrac{r_1}{a}}{\left[\dfrac{RX_c^2}{a(a^2 R^2 + X_c^2)} + \dfrac{r_1}{a}\right]^2 + \left[x_1 - \dfrac{R^2 X_c}{a^2 R^2 + X_c^2}\right]^2}$$

which may be reduced to

$$\frac{a^3R^2r_1 + aX_c^2(r_1 + R)}{a^4R^2x_1^2 + a^2[R^2r_1^2 + x_1^2X_c^2 - 2x_1X_cR^2] + X_c^2(r_1 + R)^2}$$

When this form of the first term is used in Equation 4.131 and both terms are given a common denominator, the numerator of the combined terms may be set equal to zero:

$$(a - b)r_2\{a^4R^2x_1^2 + a^2[R^2r_1^2 + x_1X_c^2 - 2x_1X_cR^2] + (r_1 + R)^2X_c^2\}$$
$$+ [a^3R^2r_1 + a(r_1 + R)X_c^2](a^2x_2^2 - 2abx_2^2 + b^2x_2^2 + r_2^2) = 0 \quad (4.132)$$

After dividing Equation 4.132 by $R^2X_c^2$, terms may be collected in the following form:

$$Q_5a^5 + Q_4a^4 + Q_3a^3 + Q_2a^2 + Q_1a + Q_0 = 0 \quad\quad (4.133a)$$

Or alternatively,

$$a^5 + \frac{Q_4a^4 + Q_3a^3 + Q_2a^2 + Q_1a + Q_0}{Q_5} = 0 \quad\quad (4.133b)$$

where

$$Q_0 = -br_2A$$

$$Q_1 = r_2A + B(r_2^2 + b^2x_2^2)$$

$$Q_2 = -2bx_2^2B - br_2C$$

$$Q_3 = r_2C + (r_1 + R)(x_2/R)^2 + D[(r_2/x_2)^2 + b^2]$$

$$Q_4 = -b(Q_5 + D)$$

$$Q_5 = r_2(x_1/X_c)^2 + D$$

$$A = \left(\frac{r_1 + R}{R}\right)^2 \quad\quad B = \frac{r_1 + R}{R^2} = \frac{A}{r_1 + R}$$

$$C = (r_1/X_c)^2 + (x_1/R)^2 - 2(x_1/X_c)$$

$$D = r_1(x_2/X_c)^2$$

$$b = \frac{\text{given shaft speed}}{\text{synchronous speed at rated frequency}}$$

Each coefficient is a real number, and computer routines are available to determine the real root of Equation 4.123. Then

$$f = af_B \quad\quad (4.134)$$

where a is the real root of Equation 4.133.

     The next part of the problem is to find the generated voltage E_ϕ. The value of X_m under the given operating conditions must first be determined. This may be accomplished by setting up an equation for the conservation of VARs. Since there is no external VAR source for the system, the total VARs absorbed by all three branches of the circuit must be zero. (The stator circuit, containing the exciting capacitors, is a *source* of lagging VARs, so it will absorb a *negative* number of lagging VARs.) The conservation of VARs equation is

$$|E_\phi|^2 \left[\left(\frac{1}{X_m} \right) + b_s + b_r \right] = 0$$

or

$$\frac{1}{X_m} + b_s + b_r = 0 \tag{4.135}$$

Substituting from Equations 4.128 and 4.129,

$$\frac{1}{X_m} = -\frac{x_1 - X_L}{\left(R_L + \frac{r_1}{a} \right)^2 + (x_1 - X_L)^2} - \frac{(a - b)^2 x_2}{r_2^2 + (a - b)^2 x_2^2} \tag{4.136}$$

To simplify this equation, let

$$\Delta_s \overset{\Delta}{=} \left(R_L + \frac{r_1}{a} \right)^2 + (x_1 + X_L)^2$$

and

$$\Delta_r \overset{\Delta}{=} \frac{r_2^2}{(a - b)^2} + x_2^2 \tag{4.137}$$

Then Equation 4.136 becomes

$$\frac{1}{X_m} = -\frac{x_1 - X_L}{\Delta_s} - \frac{x_2}{\Delta_r}$$

$$= \frac{-(x_1 - X_L)\Delta_r - x_2\Delta_s}{\Delta_s\Delta_r}$$

Then

$$X_m = \frac{\Delta_s\Delta_r}{-(x_1 - X_L)\Delta_r - x_2\Delta_s} = \frac{\Delta_r}{-(x_1 - X_L)(\Delta_r/\Delta_s) - x_2} \tag{4.138}$$

From the conservation of power, Equation 4.131, it is observed that

$$\frac{R_L + (r_1/a)}{\Delta_s} + \frac{r_2/(a - b)}{\Delta_r} = 0$$

or

$$\frac{\Delta_r}{\Delta_s} = -\frac{r_2/(a - b)}{R_L + (r_1/a)}$$

Substituting this ratio in the denominator of Equation 4.138 yields a useful expression for finding the value of X_m which exists under the given conditions of speed, load, and exciting capacitive reactance:

$$X_m = \frac{[R_L + (r_1/a)]\{x_2^2 + [r_2/(a - b)]^2\}}{[r_2/(a - b)](x_1 - X_L) - [R_L + (r_1/a)]x_2}$$

(4.139)

where, from Equation 4.127,

$$R_L = \frac{X_c^2 R}{a(a^2R^2 + X_c^2)} \qquad \text{and} \qquad X_L = \frac{R^2 X_c}{a^2R^2 + X_c^2}$$

A line having a slope of X_m ohms may now be plotted on the magnetization curve grid. This line is the locus of all points in the field having this value. Where it intersects the magnetization curve tells the value of the internal voltage of the asynchronous generator, $E_{\phi 0}$, if it were operating at rated frequency. The actual value of this voltage is given by

$$E_\phi = a E_{\phi 0}$$

(4.140)

where a is the real root of the fifth-order Equation 4.131. Having found the frequency ratio a and the induced voltage converted to rated frequency, $E_{\phi 0}$, the actual induced voltage aE_ϕ may be used with the circuit of Figure 4.39a to solve for I_1, V_1, the output power per phase, and any other quantity of interest.

Approximate Methods

As in motor operation, the slip of an induction generator is quite small at full load—in the neighborhood of 0.03. That means that x_2 quite properly may be neglected in comparison to r_2/s. Solving Equation 4.121 for a yields

$$a = \frac{b}{1 - s}$$

(4.141)

With such a small slip, a is nearly equal to b. An approximate approach to finding a is justified, and is much easier than finding the root of a fifth-order polynomial. If x_1 is neglected also, the impedance of the stator circuit in Figure 4.39a is

$$Z_s = r_1 + \frac{-jR(X_c/a)}{R - j(X_c/a)}$$

(4.142)

and the real part of the corresponding admittance is

$$g_s = \text{Re}\left[\frac{1}{Z_s}\right] = \frac{R^2 r_1 + (X_c^2/a^2)(r_1 + R)}{r_1^2 R^2 + (X_c^2/a^2)(r_1 + R)^2}$$

(4.143)

With x_2 neglected, the admittance of the rotor circuit is simply

$$g_r = \frac{s}{r_2}$$

(4.144)

Solving the conservation of power equation $|E_\phi|^2 g_s + |E_\phi|^2 g_r = 0$, and substituting b for a (since they are nearly equal) into the expression for g_s, a close approximation for the slip is obtained:

$$s \approx -r_2 \left[\frac{R^2 r_1 + (X_c^2/b^2)(R + r_1)}{R^2 r_1^2 + (X_c^2/b^2)(R + r_1)^2} \right] \qquad (4.145)$$

This value of the slip now may be used in Equation 4.141 to find the value of a to be substituted into Equation 4.139 to obtain X_m. As before, a line of slope X_m is plotted to intersect the magnetization curve at $E_{\phi 0}$, the generated voltage referred to rated frequency.

Formulas for Terminal Voltage

Having found the frequency and E_ϕ by either the exact or approximate method, the solution of the circuit of Figure 4.39a is straightforward. However for the purpose of programming, a formula has been developed for the terminal voltage of the generator. An approximate relationship is also given:

$$V_1 = \frac{aE_{\phi 0}}{\{1 - a^2(x_1/X_c)]^2 + a^2[(x_1/R)^2 + (r_1/X_c)^2] + [1 + (r_1/R)]^2 - 1\}^{1/2}} \qquad (4.146)$$

$$\approx \frac{aE_{\phi 0}}{1 - a^2(x_1/X_c)} \qquad (4.147)$$

PROBLEMS

4.1. What are the advantages and disadvantages of induction machines as motors?

4.2. (a) What is the speed of the rotating field of a six-pole, three-phase motor connected to a 60-Hz line? Give your answer in rev/min and rad/s. (b) Repeat for 50 Hz.

4.3. In Problem 4.2(a), what are the slip speed and slip if the rotor speed is 1150 rev/min?

4.4. In Problem 4.2(b), what are the slip speed and slip if the rotor speed is 970 rev/min?

4.5. What are the frequencies of the rotor currents in Problems 4.3 and 4.4?

4.6. A six-pole, wound-rotor induction motor is used as a frequency changer. The stator is connected to a 60-Hz, three-phase line. The load is connected to the rotor slip rings. At what two speeds could the rotor be driven to supply 15 Hz to the load? How would the voltages at the load terminals compare for the two speeds?

4.7. In a certain three-phase induction motor, $N_{I1} = 350$ turns $[N_{I1} = (\frac{3}{2})(4/\pi)(N_{\phi 1}/p)\sqrt{2} \, k_{w1}]$. Under certain operating conditions, the peak rotor MMF F_2 is 1800 A-turns, the magnetizing current I_ϕ is 3.00 A, and the rotor circuit power factor is 0.940, lagging. Find I_1, the rms stator current.

4.8. A two-phase induction motor has a three-phase, Y-connected wound rotor. With the rotor open-circuited and stationary, the voltage between slip rings is 152.4 V when the stator phase voltage is 220 V. Find the voltage and current turns ratios a_E and a_I.

4.9. A six-pole, three-phase, 60-Hz induction motor is operating at a speed of 1164 rev/min. The input power at this speed is 33 kW and the stator copper loss is 1200 W. Find (a) the slip, (b) P_g, (c) rotor copper loss, (d) developed torque in newton-meters, and (e) developed horsepower. (f) If the rotational losses are 500 W, find the output torque and horsepower.

4.10. A two-pole induction motor draws 50 A from a 230-V, three-phase, 60-Hz line at 0.90 power factor. Stator and rotor copper losses are 1000 and 508 W, respectively. Friction and windage loss is 200 W, core loss is 750 W, and stray load loss is 180 W. Find (a) the efficiency, (b) speed, (c) output torque, and (d) horsepower output.

4.11. A two-pole, 60-Hz, 208–230-V, three-phase induction machine is rated at 10 horsepower at 3450 rev/min. The circuit model impedances are as follows:

$$x_1 = 0.423 \ \Omega \qquad x_2 = 0.252 \ \Omega \qquad X_m = 23.0 \ \Omega$$
$$r_1 = 0.148 \ \Omega \qquad r_2 = 0.144 \ \Omega$$

Rotational losses total 500 W. Calculate the actual output horsepower at rated speed when operated at 208 V, 60 Hz. Also calculate the line current, power factor, output torque, and efficiency.

4.12. Do Problem 4.11 with the line voltage at 230 V.

4.13. Calculate the performance of the machine of Problem 4.11 when operated at 208 V, 60 Hz at a slip of 0.03.

4.14. What are the starting current and torque in newton-meters of the machine of Problem 4.11, for a line frequency of 60 Hz when the applied voltage is (a) 208 V? (b) 230 V?

4.15. A six-pole, three-phase, 460-V, 60-Hz induction motor has the following model impedances:

$$r_1 = 0.800 \ \Omega \qquad x_1 = 0.700 \ \Omega \qquad X_m = 35.0 \ \Omega$$
$$r_2 = 0.300 \ \Omega \qquad x_2 = 0.700 \ \Omega$$

$$\text{Rotational loss} = 720 \ \text{W}$$

(a) Calculate the performance of this motor for a speed of 1164 rev/min.
(b) Calculate the developed torque at starting.

4.16. An eight-pole, three-phase, 230-V, 60-Hz induction motor has the following model impedances:

$$r_1 = 0.780 \ \Omega \qquad x_1 = 0.560 \ \Omega \qquad X_m = 32.0 \ \Omega$$
$$r_2 = 0.280 \ \Omega \qquad x_2 = 0.840 \ \Omega$$

$$\text{Rotational loss} = 224 \ \text{W}$$

      (a) Calculate the performance of this motor for a slip of 0.03.

      (b) Calculate the developed torque at starting.

4.17. Determine the speeds at which maximum torque occurs for the motors of Problems 4.15 and 4.16.

4.18. Determine the maximum torque at rated voltage for the motor of Problem 4.15. What would the maximum torque be at 80% rated line voltage?

4.19. A three-phase, wound-rotor induction motor has a Y-connected, three-phase rotor winding. The rotor winding resistance per phase is 0.1 Ω. With the slip rings shorted, this motor develops rated torque at a slip of 0.05 and a line current of 120 A. What external resistance must be inserted into each phase of the rotor circuit to limit the current to 120 A at starting? What fraction of full-load torque will be developed at starting with these resistors inserted?

4.20. What resistance per phase must be inserted into the rotor circuit of the machine of Problem 4.19 if rated torque is to be developed at half synchronous speed?

4.21. A four-pole, 60-Hz induction motor develops full-load torque at a speed of 1740 rev/min.

      (a) What will be its speed at half rated torque?

      (b) What will be its speed at half rated torque and half rated voltage?

      (c) What will be its speed at rated torque and voltage if the rotor resistance per phase is doubled?

4.22. The following data were obtained from tests on a 25-hp, four-pole, 230-V, three-phase, 60-Hz Design B induction motor:

    No load: 230 V, 20.1 A, 1200 W, 60 Hz
    Blocked rotor: 26.9 V, 62.7 A, 2350 W, 15 Hz
    dc Stator resistance test: 62.2 A, 14.4 V

Calculate the performance of this motor at a slip of 0.04.

4.23. A 40-hp, 2300-V, 60-Hz induction motor is marked with code letter F.

      (a) What is the range of possible starting currents if the machine is line-started?

      (b) What is the range of starting-line currents if the machine is started with a compensator and is connected to the 60% voltage tap?

4.24. In a 60-Hz PAM motor, the initial number of poles is 20 and the modulating function has four poles. At what two speeds will the motor run? How many poles has the field that is suppressed by having the poles of all three phases in space phase?

4.25. What voltage should be applied to a 5-hp, 220-V, 60-Hz motor, if operated at 180 Hz, to maintain the normal degree of iron saturation? Approximately what would be the rated horsepower at 180 Hz?

4.26. Sketch two circle diagrams to illustrate the effect of leakage reactance on maximum torque.

4.27. Use circle diagrams to show qualitatively the effect of X_m on the power factor of an induction motor at rated current.

4.28. The machine of Problem 4.11 is connected to a 230-V, 60-Hz, three-phase line and driven as a generator by a gasoline engine running at 3700 rev/min. Calculate the output power, current, and power factor. Calculate the input horsepower and engine torque.

4.29. The machine of Problem 4.11 is operated as an isolated induction generator. Assume that the magnetization curve of Figure 4.41 applies to this machine. The electrical load consists of three 10-Ω resistors in Y and three 63.3-μF capacitors in Δ. Find the terminal voltage and output power when the machine is being driven at 3400 rev/min.

5

DIRECT-CURRENT MACHINES

5.1 THE IMPORTANCE OF dc MACHINES

The first man-made sources of current electricity were batteries, and thus the first electromagnetic machines to be developed were dc machines. The first central power plant, developed by Thomas A. Edison to serve a part of New York City, had dc generators. When alternating current supplanted dc as the mode for the generation and transmission of electrical energy, induction and synchronous machines supplanted dc machine as motors and generators to a large extent. Still, today dc machines account for about 40 percent of the dollar volume in electrical-machine manufacturing and sales. Why is this so? First of all, most highway vehicles use lead–acid batteries for electrical energy storage. The starter motors, windshield wiper motors, fan motors, and motors to drive other accessories in vehicles must be dc motors. Literally millions of dc motors are built each year for these purposes. Second, in applications requiring accurate control of speed and/or torque, the dc motor is unsurpassed. In spite of their relatively high cost and maintenance requirements, dc motors are nearly the universal choice for driving power shovels, steel and aluminum rolling mills, electric elevators, railroad locomotives, and large earth-moving equipment. All of these applications require the precise control characteristics inherent in dc motors. A typical rolling mill motor is shown in Figure 5.1.

5.2 CONSTRUCTION OF dc MACHINES

The dc machine is something like a synchronous machine turned inside out, with a built-in, mechanical rectifier-inverter connected between the armature coils and the dc terminals. This rectifer-inverter is called a *commutator*. Figure 5.2 shows the basic elements.

The *armature winding* is composed of coils embedded in slots in the *rotor*. The rotor of a dc machine usually is simply called the *armature*. The armature has a cylindrical steel core consisting of a stack of slotted laminations. The slots in the

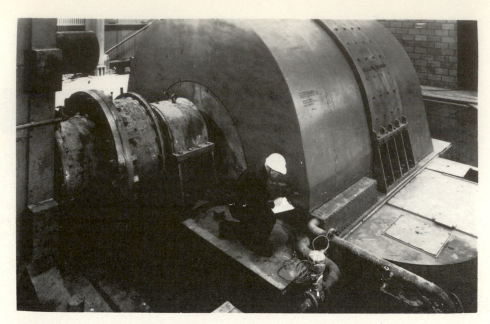

FIGURE 5.1 This 6000-hp, 40/80-rev/min reversing mill motor is representative of the motors used for continuous high-impact load duty on blooming and slabbing mills. (Courtesy of the General Electric Company.)

laminations are aligned to form axial slots in the outside surface of the core, in which the coil sides of the armature winding are placed. In small machines, the slots are sometimes skewed. Skewing makes winding a little more difficult, but it results in quieter operation and a slight reduction in losses. The laminated iron between the slots forms the "teeth" of the core. The armature coils are often held in place by wood or fiber "wedges" driven into the slots from one end of the core stack. Figure 5.3 shows a completed armature for a low-speed, 8000-hp machine in which the armature coils are held in the slots by means of glass fiber bands.

The *field poles* are located on the stator and project inward from the inside surface of the iron cylinder that forms the stator yoke. This yoke serves as a return path for the pole flux. Each iron pole consists of a narrower part, called the *pole core,* around which is placed the exciting winding, called a *field coil*. This "coil" may consist of two or more separate windings to provide for controling the strength of the field flux. A *pole shoe,* usually laminated, distributes the pole flux over the rotor surface. In permanent-magnet (PM) motors, the pole core is permanent-magnet material. In small PM machines there is no field pole winding. In larger PM machines, the pole core is surrounded by a magnetizing winding, to magnetize the core magnets after assembly and to demagnetize them for disassembly.

The surface of the pole shoe opposite the rotor is called the *pole face*. The space between the pole face and the rotor surface is called the *air gap*. As the teeth of the rotor core sweep past the pole faces, localized variations in the flux density within the pole shoe are produced. These flux density variations produce hysteresis and eddy current losses. Laminating the pole shoes greatly reduces the eddy current loss in the iron of the pole shoes.

The leads from the armature coils are connected to the *commutator*. The commutator consists of radial copper segments separated from each other by insulating material, usually mica. Current is conducted to the armature coils by carbon *brushes* that are

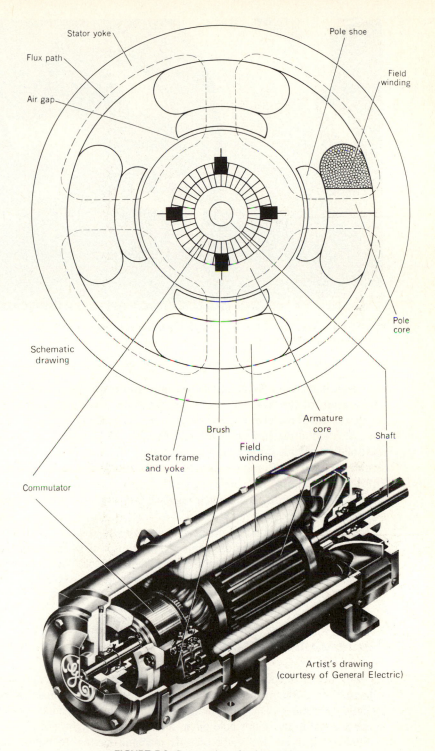

Stator yoke

Pole shoe

Flux path

Field winding

Air gap

Schematic drawing

Pole core

Brush

Armature core

Shaft

Stator frame and yoke

Field winding

Commutator

Artist's drawing
(courtesy of General Electric)

FIGURE 5.2 Construction of a dc machine.

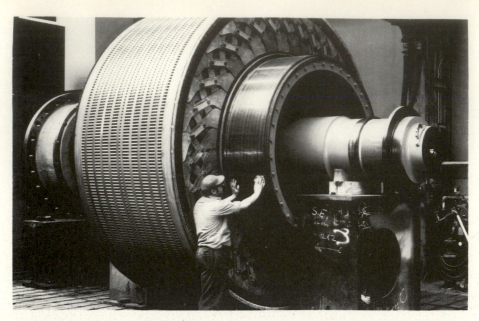

FIGURE 5.3 Armature of an 8000-hp, 700-V, 40-100-rev/min dc motor for a steel rolling mill. (Courtesy of the General Electric Company.)

held against the cylindrical surface of the commutator by the force of springs. The manufacture of brushes that will have the proper conductivity and sliding-friction characteristics for a given motor design is a very specialized scientific art. Brushes must be fitted to the surface of the commutator. They are held in *brush holders*, and must be free to slide radially in these holders to maintain contact with the commutator as they wear away. Current is carried into the brush by a "pigtail" or "brush shunt." This is a very flexible wire embedded in the brush material and connected to the brush holder. Brushes must be inspected regularly and replaced as necessary, as they are worn away.

The copper commutator segments are also subject to wear, leaving the harder mica insulation protruding above the copper. This causes the brushes to bounce, and the resulting arcs seriously damage the commutator surface. The commutator must then be resurfaced ("turned down") and the mica "undercut" to be below the new commutator surface.

5.3 HOW dc MACHINES WORK

As with all machines, the dc machine must be able to produce torque and generate voltage. The armature coils are diamond-shaped, just as those of synchronous or induction machines. The coil pitch, that is, the angular distance between coil sides, is about equal to the stator pole pitch, and is usually slightly less.

Figure 5.4 shows the stator of a dc machine having 16 poles. Suppose it were possible to put on X-ray spectacles and gaze down through the stator poles at the armature surface. Figure 5.5 shows shaded areas to represent those parts of the rotor surface covered by three adjacent stator poles. Seven typical coils are shown. (To make the drawing less complicated, these coils are shown as single-turn coils, and are

FIGURE 5.4 Stator for a large dc machine, showing brush rigging, main poles, interpoles, and inter-connections for compensating windings. (Courtesy of General Electric Company.)

shown on the armature core surface rather than buried in slots.) Motor operation is illustrated. In Figure 5.5a, coils 5, 6, and 7 are connected in series, as are coils 1, 2, and 3. The rotor is turning counterclockwise. The north stator poles produce inward flux, while that of the south pole is outward. As the rotor turns, coils 5, 6, and 7 are increasingly linked with inward flux while, 1, 2, and 3 are decreasingly linked with inward flux, resulting in the voltage polarities shown. Points a and d are at the same potential and have been connected together, as have points b and c. Now if the voltage generated between the terminals thus formed is considered the countervoltage of a motor, and if a battery of higher voltage is connected to these terminals to cause currents to flow as in Figure 5.5b, magnetic poles will be formed on the rotor surface, halfway between the stator poles. The polarities of these rotor poles are such as to interact with the stator field poles to produce torque in the direction of rotation. Thus motor action is confirmed.

In Figure 5.5c, the coil leads have been connected to segments of the commutator, and battery connections have been made through brushes. (Coil 4 has been connected to illustrate the symmetry of an actual winding. This coil is nongenerating and is shorted by the center brush.) Now the complete winding of the armature is symmetrical, as shown in Figure 5.6. So as the rotor of Figure 5.5c rotates, different coils will replace those shown, but the *pattern of coils remains stationary!* And even though the rotor rotates, its magnetic poles remain halfway between the stator poles and thus produce a constant torque. Coil 1 moves smoothly to the positions of coils 2 and 3 while generating a voltage of constant polarity. When it reaches the position of coil 4, $d\lambda/dt$ is zero (the flux linking the coil is equal to $+\phi$, the flux per pole), and the coil is now shorted by the $-$brush. As coil 1 moves on to the position of 5, the flux

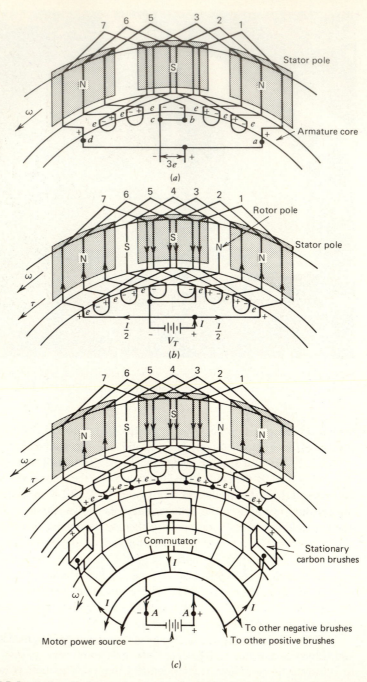

FIGURE 5.5 Production of voltage and torque in a dc machine (motor mode). (*a*) Voltage of six typical coils. (*b*) Magnetic poles on armature core surface produced by coil currents—motor action shown. (*c*) Conduction of current and potentials by the commutator.

linking the coil begins to decrease, inducing a voltage of reverse polarity. However note that the right-hand lead of the coil is now connected to the −brush, whereas in position 3 the left-hand lead was connected to that brush, with the result that the coil voltages always are directed to maintain constant brush polarity.

The voltage polarity of the brushes depends only on the direction of rotation and

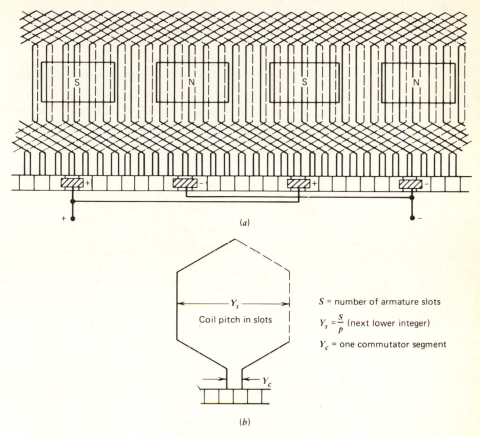

FIGURE 5.6 A completed lap winding. (*a*) Developed simplex lap winding. (*b*) Coil pitches.

the magnetic polarity of the field poles of the stator. It is independent of the direction of armature current flow. The armature current direction determines the polarity of the magnetic poles on the surface of the armature core and thus determines the direction of the torque. The following corollaries to the above discussion may be stated:

> *When current is flowing into the + brushes, the torque is in the direction of rotation and the machine is acting as a motor. The generated voltage is a "counter-voltage."*
>
> *When current is following out of the + brushes, the torque opposes the rotation. The machine is acting as a generator, and the voltage generated is a source voltage for the current.*

5.4 THE GENERATED VOLTAGE OF A dc MACHINE

Observe in Figures 5.5 and 5.6 that there are several armature coils in series between brushes. Note also that there are several of these series *paths* in parallel. In the windings shown in these figures, the number of parallel paths is equal to the number of brushes, and there is one brush for each stator pole. There are several other dc armature windings in common use in which the number of parallel paths is not equal to the number of poles, so let the symbol a be assigned to the number of parallel paths of current flow between the positive and negative terminals of the machine:

$$a \stackrel{\Delta}{=} \text{the number of parallel paths between armature terminals}$$

Direct-current machines are symmetrically wound, so each path is just like the others.

The voltage generated between terminals will depend on how many coils there are in series between terminals. Let C be the total number of coils in an armature winding

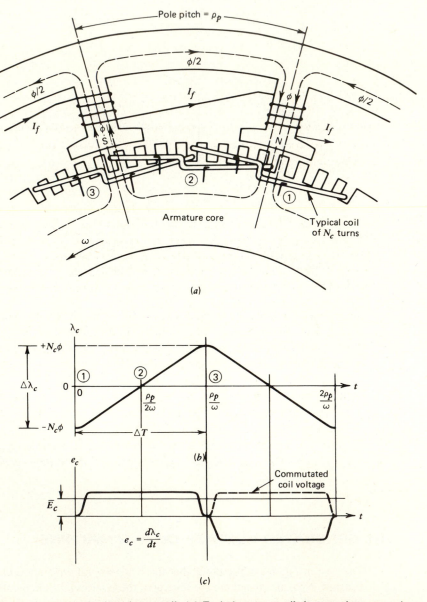

FIGURE 5.7 Generation of voltage in one coil. (a) Typical armature coil shown at three successive instants. (b) Coil flux linkage as a function of time. (c) Coil voltage.

and C_s the number shorted by brushes. Then the number of generating coils in series in each path between terminals is given by

$$C_p = \frac{C - C_s}{a} \cong \frac{C}{a} \tag{5.1}$$

since C_s is usually small compared to C.

The voltage induced in one coil may be derived in reference to Figure 5.7. A typical coil of N_c turns is shown in three positions as the armature rotates. In position 1 the coil encloses the total air gap flux of one north field pole. Let ϕ be the flux per pole, and *take positive flux as directed outward from the rotor*. At this instant, the coil flux linkage is $-N_c\phi$ Wb-turns and is constant for a very short interval (the coil is shorted by a brush at this instant). As the coil moves on, the flux linkage increases to zero at position 2, and as the armature continuously rotates at a speed of ω rad/s, the coil moves to position 3. At this point its flux linkage is at its positive maximum $+N_c\phi$. It is now enclosing the total air gap flux of a south field pole (and is again shorted by a brush). During this motion the voltage induced by Faraday's law is shown in Figure 5.7c.

As the coil moves beyond position 3, the coil flux begins to decrease, and the induced voltage reverses polarity. By this time the commutator segments to which the coil's leads are attached have moved to the other side of the brush. As far as the external circuit is concerned, the coil voltage has been "commutated," as indicated by the dashed curve of Figure 5.7c, and appears as a set of rather square, *unidirectional* pulses.

Since dc voltage is defined as the average voltage, the pole shoes are shaped to generate a voltage waveform having a high *average* value, that is, a flat-topped wave. This is accomplished by maintaining a reasonably uniform flux density over the pole face area. The average, or dc value, of the commutated voltage is shown as $\bar{E}_c$ in the figure. This voltage is easily calculated. It is given by the average rate of change of coil flux linkage over one half cycle:

$$\bar{E}_c = \frac{\Delta\lambda_c}{\Delta T} \tag{5.2}$$

Figure 5.7b shows that

$$\Delta\lambda_c = 2N_c\phi \tag{5.3}$$

The time required for this change in λ_c is the time required to move the coil one pole pitch at a speed of ω rad/s:

$$\Delta T = \frac{\rho_p}{\omega} \tag{5.4}$$

where ρ_p is the stator pole pitch in mechanical radians. Now

$$\rho_p = \frac{2\pi}{p} \tag{5.5}$$

where p is the number of stator poles, with the result that

$$\Delta T = \frac{2\pi}{p\omega} \tag{5.6}$$

Equation 5.2 for the dc value of the commutated voltage of one coil becomes

$$\bar{E}_c = \frac{2N_c p\phi\omega}{2\pi} \tag{5.7}$$

The generated voltage of the machine will be the dc volts per coil given by Equation 5.7, multiplied by the number of coils in series in each path between the terminals, a fact illustrated by Figure 5.8. Thus the internal dc voltage of the machine is given by the product of equations 5.1 and 5.7:

$$E_g = \frac{C}{a} \cdot \frac{2N_c p\phi\omega}{2\pi} \quad V \tag{5.8}$$

Designers of machines think in terms of the total number of "conductors" in the armature winding before deciding how these conductors are to be collected into coils. The armature coils are quite similar to stator coils of three-phase machines. Reference to Figure 2.5 will show that a four-turn coil will have four "conductors" in each coil "side" for a total of eight conductors. Thus the number of conductors per coil is twice the number of turns. Let Z_c be the number of conductors per coil:

$$Z_c = 2N_c \text{ conductors} \tag{5.9}$$

Then the total number of conductors in an armature winding is

$$Z = C \cdot 2N_c \tag{5.10}$$

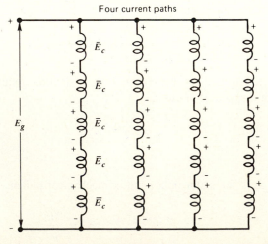

FIGURE 5.8 Generated voltage of a dc machine related to nonshorted coils per path.

It will be noted that this quantity appears in the equation for the generated voltage, Equation 5.8, which may be rewritten

$$E_g = \frac{Zp\phi\omega}{2\pi a} \quad \text{V} \tag{5.11}$$

Equation 5.11 assumes the flux per pole, ϕ, to be in webers and ω to be in mechanical radians per second. All of the symbols in this expression represent constants, except ϕ and ω. Thus a simplified expression for the generated voltage may be written

$$E_g = K_a\phi\omega \quad \text{V} \qquad K_a \overset{\Delta}{=} \frac{Zp}{2\pi a} \tag{5.12}$$

where K_a is called the *armature constant*. The voltage E_g is the internal source voltage when the machine is generating, and it is the countervoltage in the motor mode.

In industry, shaft speeds are still most often measured in revolutions per minute (rev/min). Let n represent the speed in these units. Then

$$E_g = K'_a\phi n \tag{5.13}$$

where

$$K'_a \overset{\Delta}{=} \frac{Zp}{60a}$$

5.5 CIRCUIT MODEL OF A dc MACHINE

The circuit model of the basic dc machine is quite simple, as illustrated in Figure 5.9. The armature symbol includes brushes, to indicate the source voltage to be direct rather than alternating. Thevenin's theorem permits the internal winding resistance r_a to be considered a circuit element external to the generating process. The inductance of the armature is quite small and may be neglected for nearly any consideration, except the commutation of thyristors in solid-state control systems. The Kirchhoff's-law voltage equation for the armature circuit, in the motor mode, for motor operation and out of the positive terminal, is

$$E_g = V_T - I_a r_a \tag{5.14}$$

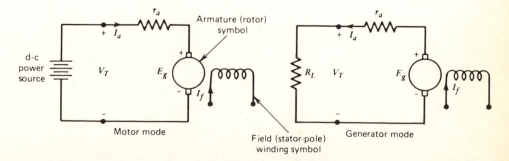

FIGURE 5.9 Circuit model of a dc machine.

where E_g is the generated countervoltage given by Equation 5.12 or 5.13, and V_T is the voltage af the terminals. The *armature current* is given the symbol I_a. The armature current is assumed to be flowing into the positive terminal for motor operation and out of the positive terminal in the generator mode. The Kirchhoff's-voltage-law relation for the generator mode is

$$V_T = E_g - I_a r_a \tag{5.15}$$

By Equation 5.12, the generated voltage E_g depends on the shaft speed ω and the magnetic flux per pole, ϕ. The strength of the pole flux will be practically constant in permanent-magnet machines. In machines with field windings, the flux will depend on the ampere-turns per pole of those windings. Each field pole winding has the same number of turns as those of the other poles. If there is only one field winding per pole, as is often the case, the MMF per pole is given by

$$\mathcal{F}_f = N_f I_f \quad \text{A-turns} \tag{5.16}$$

where N_f is the turns per pole of the single winding and I_f is the dc current supplied to that winding. As a result of saturation of the iron parts of the magnetic circuit of a dc machine, the flux per pole is a nonlinear function of the pole MMF:

$$\phi = \mathcal{P}_p \mathcal{F}_f \tag{5.17}$$

where the effective permeance per pole decreases as $\mathcal{F}_f$ increases. The same thing happens in synchronous machines, and a typical plot of ϕ as a function of $\mathcal{F}_f$ is quite similar to the plot of ϕ as a function of R in Figure 2.29 of Chapter 2. Equation 5.12 may now be written

$$E_g = K_a \omega \mathcal{P}_p \mathcal{F}_f \tag{5.18}$$

and it will be observed that the generated voltage is proportional to the rotor speed and is nonlinearly related to the field MMF per pole. A typical plot of E_g as a function $\mathcal{F}_f$ is shown in Figure 5.10 for two speeds. The rated or "base" speed of the machine

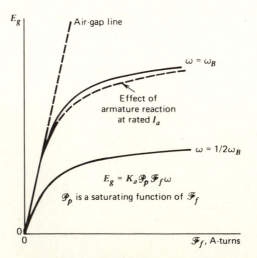

FIGURE 5.10 Magnetization curves of a dc machine.

is given the symbol ω_B. The curve at rated speed is given several names in the literature and is alternatively called the *saturation curve*, the *magnetization curve*, or the *open-circuit characteristic*. The straight-line or unsaturated part of the curve and its extension are called the *air gap line*. *The saturation curve is usually plotted as a function of I_f rather than $\mathcal{F}_f$*.

The effect of the magnetic field of the armature has been neglected in the above discussion. The student will note in Figure 5.5 that the magnetic poles of the armature are half pole pitch, or 90 electrical degrees from the field poles. This places the armature poles opposite the large air gap between the salient field poles. The armature MMF thus has a relatively minor effect on the field pole flux ϕ. This effect is called the *armature reaction*. It results in a reduction in ϕ which may not be negligible when the machine is being started or reversed, or when it is heavily loaded. The effects of armature MMF on machine performance are discussed in Section 5.12.

5.6 DEVELOPED TORQUE

Power flowing into the armature terminals of a dc motor is given by

$$P_{in} = V_T I_a \quad W \tag{5.19}$$

This is partly consumed as *armature copper loss*, given by

$$ACL = I_a^2 r_a \quad W \tag{5.20}$$

The remaining power is available for conversion to mechanical energy. This will be called the *developed power P_d*:

$$P_d = P_{in} - ACL = V_T I_a - I_a^2 r_a \quad \text{(motor mode)} \tag{5.21}$$

Reference to Figure 5.9 and Equations 5.14 and 5.21 will show that

$$P_d = I_a(V_T - I_a r_a) = E_g I_a \quad W \tag{5.22}$$

In a generator the output power is the developed power minus the armature copper loss:

$$P_{out} = E_g I_a - ACL$$
$$= E_g I_a - I_a^2 r_a = P_d - ACL \tag{5.23}$$

The developed torque of a dc motor or the countertorque of a dc generator will be called the *developed torque*, and is given by the developed power divided by the shaft angular velocity:

$$\tau_d = \frac{P_d}{\omega} = \frac{E_g I_a}{\omega} \quad \text{N-m} \tag{5.24}$$

In pound-feet,

$$\tau_d = \frac{7.04 P_d}{n} = \frac{7.04 E_g I_a}{n} \quad \text{lb-ft} \tag{5.25}$$

where n is the shaft speed in rev/min. By equation 5.12, $E_g = K_a \phi \omega$, and so

$$\tau_d = K_a \phi I_a \text{ N-m} \tag{5.26}$$

In the English system of units,

$$\tau_d = 7.04 K_a' \phi I_a \text{ lb-ft} \tag{5.27}$$

where K_a' is defined by Equation 5.13.

■ Example 5.1

A certain dc machine is rotating at 1200 rev/min. The cross-sectional area of the face of each of its four poles is 200 cm^2 and the average flux density in the air gap under the pole faces is 0.78 T. The armature winding has 95 coils of two turns each. The machine is wave wound, with $a = 2$ paths. The resistance of the winding between terminals is 0.20 Ω. If the voltage of the terminals is 250 V, is the machine operating as a motor or a generator? What power is being developed? Describe the torque.

Solution. The flux per pole is given by $B \cdot A$, where A is in m^2:

$$\phi = 200 \cdot 10^{-4} \cdot 0.78 = 0.0156 \text{ Wb}$$

$$Z = 2CN_c = 2 \cdot 95 \cdot 2 = 380 \text{ conductors}$$

$$\omega = \frac{1200}{60} 2\pi = 125.7 \text{ rad/s}$$

$$K_a = \frac{Zp}{2\pi a} = \frac{380 \cdot 4}{4\pi} = 121.0 \text{ V-s/Wb}$$

$$E_g = K_a \phi \omega = 121.0 \cdot 0.0156 \cdot 125.7 = 237.2 \text{ V}$$

The armature circuit looks like this:

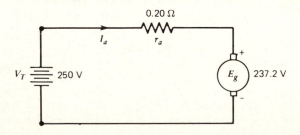

Since the terminal voltage is greater than the internal voltage, current is flowing into the positive terminal. The machine is absorbing power from the dc source and hence acts as a motor. The armature current is given by

$$I_a = \frac{V_T - E_g}{r_a} = \frac{12.8 \text{ V}}{0.20 \text{ Ω}} = 64.0 \text{ A}$$

The input power is

$$P_{\text{in}} = V_T I_a = 250 \cdot 64.0 = 16,000 \text{ W}$$

The armature I^2R loss is

$$I_a^2 r_a = 64.0^2 \cdot 0.20 = 819 \text{ W}$$

The power converted into mechanical power P_d is the armature input minus the armature copper loss:

$$P_d = 16,000 - 819 = 15,181 \text{ W}$$

or 20.3 hp. To obtain the actual output horsepower, one would have to subtract rotational losses, which are unknown in this problem. The *developed* torque, however, can be obtained by dividing the developed mechanical power by the rotor's angular velocity:

$$\tau_d = \frac{P_d}{\omega} = \frac{15,181}{2\pi \ (1200/60)} = 121 \text{ N-m}$$

or

$$\frac{20.3 \cdot 5252}{1200} = 88.8 \text{ lb-ft}$$

Since the machine is in the motor mode, this torque is in the direction of rotation.

∎

5.7 BRUSHLESS dc MACHINES

A brushless dc machine is one in which the commutation is done by means of electronics. The concept is not new, but improvements in magnet materials and the reduction of costs of integrated circuits and electronic power switches have made brushless dc machines marketable recently. Their advantages are low noise and radio interference, no requirement for commutator maintenance and brush replacement, and no dust due to brush wear. Their big disadvantage at this writing is cost.

The electronic part of a brushless dc motor performs the functions of a commutator: maintaining a 90 electrical-degree displacement between the rotor and stator magnetic poles, and acting as a rectifier/inverter between the dc terminals and the alternating currents and voltages of the armature winding. These functions are accomplished more easily if the armature winding is located on the stator. The field poles are then on the rotor. To reduce the number of electronic power switches, the stator is wound like a two-phase or, more often, a three-phase ac machine. This is simpler than providing a switch for each coil, as a commutator does. The electronic inverter converts its dc input into polyphase ac, which is fed to the armature winding on the stator. So far, what has been described is an inverter-fed synchronous machine. However there is one more feature that makes all the difference. In a brushless dc machine, the switching elements in the inverter are controlled by shaft position sensors, so that the stator magnetic poles run ahead of the rotor poles, maintaining a constant angular distance from them. Thus the torque angle is constant rather than being a variable, as it is in a synchronous machine.

The brushless dc machine has torque, speed, and voltage equations like those of a shunt dc machine. It has the excellent control characteristics of a dc machine. However it is essentially a shaft position-controlled *synchronous* machine, and it is best understood from the standpoint of synchronous machine theory. For that reason, a more extended discussion of these interesting systems is given in Chapter 2, Section 2.22.

5.8 FIELD EXCITATION OF WOUND-POLE MACHINES

It has been pointed out that the field poles of the dc machine stator may be excited by means of blocks of permanent-magnet material or by windings carrying direct current. Field windings allow more flexibility of operation at the expense of field winding I^2R or *copper* losses. At the present time, it is impractical to design machines larger than about 200 hp with PM fields, so large dc machines all have field windings.

Energy is stored in the magnetic field of the machine when the field magnets in a PM machine are magnetized, or during the current buildup after the field windings in a non-PM machine are connected to the source of magnetizing current. As the field current builds up, a voltage is induced in the winding:

$$e_f = L_{ff}\frac{di_f}{dt} \tag{5.28}$$

and energy is stored in the magnetic field at a rate given by

$$P_{\text{mag}} = e_f i_f = \left(L_{ff}\frac{di_f}{dt}\right)i_f \tag{5.29}$$

In these expressions, L_{ff} is the self-inductance of the field circuit and i_f is the instantaneous value of the field current. The energy stored in the field is

$$W_\phi = \int_0^\infty P_{\text{mag}}\,dt = \tfrac{1}{2}L_{ff}I_f^2 \tag{5.30}$$

where I_f is the steady-state value of the field current. The time constant of the field circuit is usually a few seconds. Once a steady state is reached, no more power is required to provide field energy, but power must be supplied to the winding equal to the $I_f^2 r_f$ loss, called *field copper loss*. Thus after the initial transient, all the power input to the field winding is lost as heat. Field circuits may be designed so that this loss is only about 1 percent of rated power. (In small machines the percentage field copper loss will be considerably higher.)

When the field current is reduced to zero, the flux collapses to a very low residual value. Nearly all of the stored energy must be dissipated somehow. If the field circuit is opened by means of a switch, much of this energy is dissipated in an arc across the opening contacts. The $L(di/dt)$ voltage at the field terminals is quite high when the circuit is suddenly opened, and the energy stored in the field is enough to be lethal. Often a resistor is shorted across the shunt field terminals just before the current is switched off, in order to reduce this voltage. Such a resistor is called a *field discharge resistor*. Sometimes resistors made of nonlinear material are connected permanently across the field windings of dc and synchronous machines. Such resistors have a very high resistance at normal excitation voltages but switch to a very low resistance when the voltage exceeds a specified level.

Ways of Exciting Field Windings

The field windings may be connected to a separate dc source, such as a dc generator or rectifier. Such a source is called an *exciter*, as it is for synchronous machines. A dc machine provided with a special field current source is said to be *separately excited*.

The windings may be designed to be connected across the armature terminals. These windings have many turns per pole (hundreds of turns in large machines, thousands of turns in machines of, say, 1 hp or less) and have sufficient resistance (50 to 100 per unit) to limit the field current to 1 or 2 percent of full-load armature

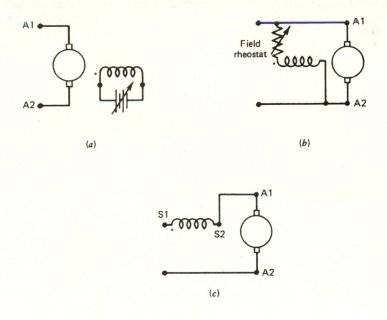

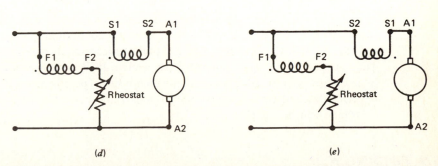

FIGURE 5.11 Typical field excitation methods for a dc machine. (*a*) Separately excited. (*b*) Shunt. (*c*) Series. (*d*) Cumulative compound motor or differential compound generator. (*e*) Cumulative compound generator or differential compound motor.

current. Such field windings are called *shunt fields*, because they are designed to be connected in parallel with the armature and *shunt* is an early word that designated a parallel connection. A machine with only one field winding, and that winding connected across the armature terminals, is said to be *shunt connected*, or sometimes *shunt wound*. (See Figure 5.11*b*.) The line supplies the excitation for a shunt motor. In the generator mode, the armature supplies the field current in addition to the load current.

The field windings may be connected in series with the armature. The winding is then called a *series field*. Such windings are designed to carry the full armature current. The cross sectional area of the conductors in this winding must then be approximately *a* times the area of the armature conductors, because each armature conductor carries I_a/a A. Since the current in a series field is so heavy, the resistance of the winding must be kept low in order to have reasonably low field copper losses. Resistances of tens of milliohms are typical, or more correctly in the neighborhood of 0.01 per unit or less. A machine having *only* a series winding is said to be a *series machine*, or *series wound*.

A machine may have both series and shunt field windings. If so, it is said to be *compound wound*. If the series and shunt field windings are so connected that their MMFs aid each other on each pole, the machine is said to be *cumulative compound*. If they oppose each other, the machine is said to be *differential compound*. It is sometimes economical to design a basically shunt dc motor so that there is an appreciable reduction in field flux under load, as a result of armature reaction, and then to provide a weak cumulative series field to counteract the armature reaction effect. Such a motor has the characteristics of a shunt machine, although it is really cumulative compound. For this reason it is called a *stabilized shunt motor*.

Figure 5.11 illustrates the various connections schematically. In this figure, the field windings are dotted like transformer windings to indicate relative magnetic polarity. Current into the dotted terminals produces MMF in the assumed positive direction. According to NEMA standards, field terminals are marked to indicate relative polarity of the windings. Shunt field windings are labeled F1 and F2; series field windings are labeled S1 and S2. Currents *into* F1 and S1 produce a cumulative compound effect. Currents into F1 and out of S1 result in a differential effect. The armature terminals are labeled A1 and A2, such that for generator operation, and shaft rotation counterclockwise when viewed from the end opposite the drive, A1 will be positive for current flowing into F1 or S1. The student may be interested in writing rules for rotation in the opposite direction, motor operation, and other conditions.

Miscellaneous Information

In a compound wound machine, the series field coil is usually on the outside of the shunt field coil on each pole. The two compound connections shown in Figure 5.11 have the shunt field connected on the end of the series winding away from the armature. This is called a *long-shunt* compound connection. If F1 were instead connected to A1 in each case, a *short-shunt* connection would result. There is practically no difference in performance between long- and short-shunt connections. Convenience alone should determine which to use.

5.9 FINDING $K_a\phi$

In most calculations of speed and/or torque it is not necessary to know K_a, but it *is* necessary to determine the quantity $K_a\phi$ or $K_a'\phi$. This quantity may be obtained from a specific set of data. For example, suppose the armature circuit resistance of a motor is known, and it is further known that at a given field excitation and armature terminal voltage, the machine operates at a certain speed at a certain armature current. Then by Equations 5.12 and 5.14,

$$K_a\phi = \frac{E_g}{\omega} = \frac{V_T^* - I_a^* r_a}{\omega^*} \quad \text{V-s/rad}$$

or by Equation 5.13,

$$K_a'\phi = \frac{V_T^* - I_a^* r_a}{n^*} \quad \text{V-min/rev} \tag{5.31}$$

where the quantities marked with asterisks are the values known for a particular condition.

■ **Example 5.2**

A dc motor is operating at 1200 rev/min. It draws 100 A from the line at a terminal voltage of 230 V. The armature resistance between terminals is 0.07 Ω.

(a) Find the torque being developed by this motor.
(b) Find the speed and armature current of this motor when the torque is 300 N-m, for the same excitation.

Solution

(a)
$$E_g^* = V_T^* - I_a^* r_a = 230 - (100 \cdot 0.07) = 223 \text{ V}$$

$$\omega^* = \frac{1200}{60} 2\pi = 40\pi \text{ rad/s}$$

$$K_a\phi = \frac{E_g^*}{\omega^*} = 1.77 \text{ V-s/rad}$$

$$\tau_d = K_a\phi I_a = 1.77 \cdot 100 = 177 \text{ N-m}$$

(b)
$$I_a = \frac{\tau_d}{K_a\phi} = \frac{300}{1.77} = 169 \text{ A}$$

$$\omega = \frac{E_g}{K_a\phi} = \frac{V_T - I_a r_a}{K_a\phi} = \frac{230 - (169 \cdot 0.07)}{1.77}$$

$$= 123 \text{ rad/s or } n = 1177 \text{ rev/min} \quad ■$$

$K_a\phi$ From the Open-Circuit Characteristic

It is often desirable to find $K_a\phi$ corresponding to a known value of field excitation. This is made possible by obtaining the open-circuit characteristic of the machine. The machine is driven at rated speed by another machine, engine, or what have you? With no electrical load connected to the armature terminals, the armature terminal voltage

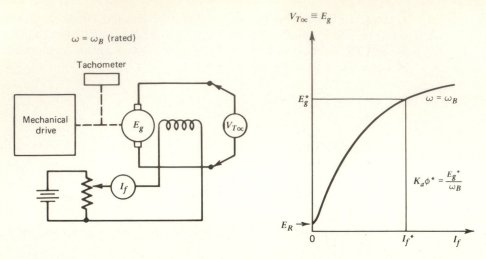

FIGURE 5.12 The open-circuit characteristic of a dc machine (also called the saturation, or magnetization, curve).

is recorded as the field current is increased from zero to a value sufficient to produce a voltage substantially greater than rated voltage. (If the machine is compound wound, the shunt field alone is excited to obtain data for this curve.) Under open-circuit conditions, $I_a = 0$, and $V_T = E_g$. A plot of open-circuit voltage as a function of field current produces the magnetization, or *saturation*, curve at rated speed, such as that of Figure 5.12. (Compare with Figure 5.10 for $\omega = \omega_B$.)

Since $K_a\phi = E_g/\omega$, it is simple to find the value of $K_a\phi$ corresponding to a given field current I_f^*. The open-circuit voltage corresponding to this field current is simply divided by the speed at which the open-circuit data were taken. Referring to Figure 5.12, for $I_f = I_f^*$,

$$K_a\phi^* = \frac{E_g^*}{\omega_B} \tag{5.32}$$

■ Example 5.3

The magnetization curve of a dc machine and other data are given in Figure 5.13. This machine is connected as a shunt motor to a 125-V dc line. The field current is adjusted for 0.70 A. Find the speed of this machine when it develops a torque of 30 N-m. Assume the windings to be at a temperature of 75°C.

Solution. From the magnetization curve at 1200 rev/min and a shunt field current of 0.70 A,

$$E_g^* = 90.0 \text{ V}$$

and

$$\omega_B = \frac{1200}{60}2\pi = 40\pi \quad \text{rad/s}$$

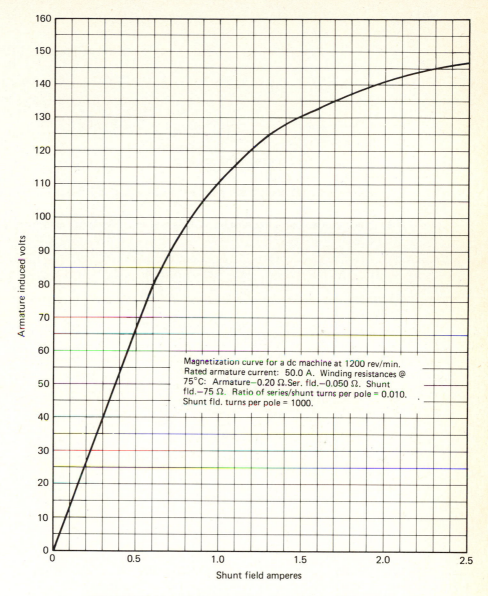

FIGURE 5.13 Magnetization curve for a dc machine at 1200 rev/min.

Then

$$K_a\phi^* = \frac{E_g^*}{\omega_B} = \frac{90.0}{40\pi} = 0.716 \quad \text{V-s/rad}$$

The speed is found from $\omega = E_g/K_a\phi$. The terminal voltage is known, and to find E_g the armature current must be calculated to determine the $I_a r_a$ drop. A knowledge of the $K_a\phi$ permits calculation of I_a from the developed torque. The developed torque is given as 30 N-m. Then

$$\tau_d = 30 = K_a\phi^* I_a = 0.716 I_a$$

$$I_a = \frac{30}{0.716} = 41.9 \quad \text{A}$$

and

$$E_g = V_T - I_a r_a = 125 - (41.9 \cdot 0.20) = 116.6 \text{ V}$$

Finally,

$$\omega = \frac{E_g}{K_a \phi^*} = \frac{116.6}{0.716} = 162.8 \text{ rad/s}$$

$$n = 1555 \text{ rev/min}$$

Two other techniques could have been used in solving this problem:

1. Since the excitation is constant, $K_a \phi$ is constant (assuming no armature reaction). Then

$$\frac{\omega}{\omega_B} = \frac{n}{n_B} = \frac{E_g/K_a \phi}{E_g^*/K_a \phi} = \frac{E_g}{E_g^*}$$

E_g and E_g^* are found as before. Then $n = n_B(E_g/E_g^*) = (116.6/90.0) \cdot 1200 = 1555$ rev/min.

2. The torque could be converted to lb-ft and the English-units equations employed:

$$\tau_d = 30.0 \cdot 0.738 = 22.1 \quad \text{lb-ft}$$

$$K_a' \phi = \frac{E_g^*}{n_B} = \frac{90.0}{1200} = 0.0750 \quad \text{V-min/rev}$$

$$\tau_d = 7.04 K_a' \phi I_a = 7.04 \cdot 0.0750 I_a = 22.1 \text{ lb-ft}$$

$$I_a = 41.9 \text{ A}$$

As before,

$$E_g = 116.6 \text{ V} = K_a' \phi n$$

$$n = \frac{116.6}{0.0750} = 1555 \text{ rev/min} \qquad \blacksquare$$

COMPOUND MACHINES

The total MMF per pole of a compound wound machine is the sum of the ampere-turns per pole of the shunt and series windings:

$$\mathcal{F}_{\text{pole}} = \mathcal{F}_f + \mathcal{F}_{\text{se}} = N_f I_f \pm N_{\text{se}} I_{\text{se}} \tag{5.33}$$

where N_{se} is the number of series field turns per pole and I_{se} is the series field current. In Figure 5.11, $I_{\text{se}} \equiv I_a$. However a very low resistance is sometimes shunted across S_1-S_2 to adjust the strength of the series field; in this case I_{se} is less than I_a. Such a resistor is called a *diverter*, and may be a short piece of heavy nichrome ribbon. The positive sign is used in Equation 5.33 for cumulative MMFs, and the negative sign when they are differential.

When a machine is connected for compound operation, the *effective* value of the shunt field current is used to find $K_a\phi$ on the open-circuit characteristic. The effective field current is that value of *shunt* field current that would produce the same MMF as the shunt and series windings acting together:

$$N_f I_f^* = \mathcal{F}_{\text{pole}} = N_f I_f \pm N_{se} I_{se} \tag{5.34}$$

Solving,

$$I_f^* = I_f \pm \frac{N_{se}}{N_f} I_{se} \tag{5.35}$$

(The ratio N_{se}/N_f may be measured by driving the machine at no load with the series and shunt fields independently excited at relatively low levels, and in opposition to each other. The currents are adjusted so that the open-circuit voltage is reduced to its residual-magnetism level, that is, so that simultaneous opening of the two field circuits produces no change in the armature terminal voltage. Then $N_{se} I_{se} = N_f I_f$ or $N_{se}/N_f = I_f/I_{se}$.)

Example 5.4

The machine of Figure 5.13 is driven as a cumulative-compound generator at 1140 rev/min. The series field is shunted by a diverter so that $I_{se} = 0.600 I_a$. The shunt field is separately excited at 1.30 A. Find the terminal voltage at full load and at no load, and the voltage regulation.

Solution. At full load,

$$I_a = I_B = 50.0 \text{ A}$$

$$I_{se} = 0.600 \cdot 50.0 = 30.0 \text{ A}$$

The N_{se}/N_f ratio is given as 0.010. Then

$$I_f^* = I_f + (N_{se}/N_f)I_{se}$$

$$= 1.30 + (0.01 \cdot 30.0) = 1.60 \text{ A}$$

From Figure 5.13, E_g^* corresponding to $I_f^* = 1.60$ A is 132.5 V, and

$$K_a'\phi = \frac{E_g^*}{n_B} = \frac{132.5}{1200} = 0.1104 \quad \text{V-min/rev}$$

The generated voltage is then

$$E_g = K_a'\phi n = 0.1104 \cdot 1140 = 125.9 \text{ V}$$

By Kirchhoff's law,

$$V_T = E_g - I_a R_a$$

where R_a is the total armature circuit resistance including r_a and the series field shunted by its diverter. Now the voltage drop across the series field is hardly worth worrying about, but it is possible to calculate it. The fraction of armature current flowing through the series field depends on the relative resistance of the winding resistance and that of the diverter. If r_{se} is the series field winding resistance and R_d the diverter resistance,

$$\frac{I_{se}}{I_a} = \frac{1/r_{se}}{(1/r_{se}) + (1/R_d)} = \frac{R_d}{r_{se} + R_d}$$

The resistance of the parallel combination is of interest:

$$\frac{r_{se}R_d}{r_{se} + R_d} = r_{se}\frac{I_{se}}{I_a} = 0.050 \cdot 0.600 = 0.030 \ \Omega$$

Then

$$R_a = 0.030 + r_a = 0.230 \ \Omega$$

and at full load,

$$V_{Tfl} = 125.9 - 50.0 \cdot 0.230 = 114.4 \ \text{V}$$

At no load, $I_a = 0$ and $V_T = E_g$. However there is no series field excitation at no load:

$$I_f^* = I_f + 0 = 1.30 \ \text{A}$$

From the magnetization curve, $E_g^* = 125$ V, and

$$V_{Tnl} = E_g = E_g^*\frac{n}{n_B} = 125\frac{1140}{1200} = 118.8 \ \text{V}$$

$$\text{Voltage regulation} = \frac{V_{Tnl} - V_{Tfl}}{V_{Tfl}} = \frac{4.4}{114.4}, \text{ or } 3.8\% \qquad \blacksquare$$

5.10 SPEED CONTROL OF SHUNT AND PERMANENT-MAGNET MOTORS

Equations 5.12 and 5.14 are expressions for the voltage induced by rotation of the armature windings through the field flux. The first of these is from the point of view of Faraday's law, and the second from Kirchhoff's voltage law. When these are equated,

$$E_g = K_a\phi\omega = V_T - I_a r_a$$

and solved for the shaft speed ω, the following **magic equation** results:

$$\omega = \frac{V_T - I_a r_a}{K_a\phi} \qquad (5.36)$$

This amazing relationship tells everything about the speed of a dc motor. The armature current is usually load dependent and is thus not usually available as a design parameter to the engineer of speed control systems. The remaining variables that may be used for speed control are, obviously, the field flux ϕ, the armature terminal voltage V_T, and the armature circuit resistance r_a. Of course, ϕ cannot be used as a control variable in PM motors. Inspection of Equation 5.36 will show that increasing the terminal voltage will increase the speed if the flux per pole is constant. That is, speed will vary in the same direction as V_T for a PM or separately excited machine. For a given armature current, the speed will decrease as the armature circuit resistance increases (at a cost of greater I^2R loss). And finally, speed is *inversely* proportional to the flux per pole. *Increased* field current *decreases* the speed. *If the field circuit is opened accidentally,* ϕ drops to its residual value, which is quite small. *The machine under these conditions "runs away" and will destroy itself in a few seconds if the armature circuit is not opened in time.* The armature current becomes quite high during runaway, because the drop in ϕ reduces the countervoltage to a very low value before the machine reaches a high speed. Thus a circuit breaker in the armature circuit may protect the machine from damage.

Shunt Field Control

The most economical and efficient method of controlling the speed of a wound-pole dc motor is by controlling the current in the field winding and in that way causing the field flux to vary. A rheostat in series with the shunt field winding will do the job. See Figure 5.11b. (The series field requires a very large current to produce much pole magnetization and thus is not suitable as a speed control winding.) The shunt field current is only a percent or two of the armature current, and only a *relatively* inexpensive rheostat is required.

The right-hand side of Equation 5.36 may be written in intercept-slope form:

$$\omega = \frac{V_T}{K_a\phi} - \frac{r_a}{K_a\phi}I_a \tag{5.37}$$

When there is no mechanical load connected to the shaft of the motor, the armature current is quite small, in the order of 5 percent of rated current. If there were no rotational losses, the no-load current would be zero. The no-load speed is given very nearly by

$$\omega_{nl} = \frac{V_T}{K_a\phi} \tag{5.38}$$

Curves of speed as a function of I_a with $K_a\phi$ as the parameter are shown in Figure 5.14. In this figure, the rated, or *base*, value of armature current has been given the symbol I_B. The *base case* assumes the terminal voltage to be the rated voltage of the machine, $V_T = V_B$, and that the shunt field is connected directly across V_T; that is, that the rheostat resistance is zero and I_f and ϕ are at their maximum values. This value of flux per pole is designated ϕ_B.

Speed regulation is defined by

$$\text{Regulation} \overset{\Delta}{=} \frac{\omega_{nl} - \omega_{fl}}{\omega_{fl}} \tag{5.39}$$

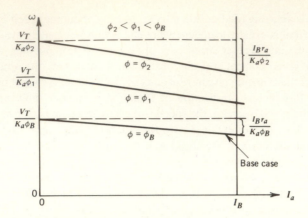

FIGURE 5.14 Field control of dc motor speed.

where ω_{fl} is the speed at rated current. A big regulation is a bad regulation from the control point of view. Shunt motors are classed as *constant speed* motors; that is, their speed regulation is less than 8 percent. In larger machines it will be less than 5 percent. With field current speed control, the speed increases as field current decreases. The slope of the speed-versus-I_a curves also increases proportionately, and the speed regulation remains constant.

Field rheostat control allows for controlling speed at or above base speed. The upper speed limit is determined by two factors. There is a mechanical limit—the highest speed at which the machine may be operated safely. There is also a stability limit. When the field MMF is weak, armature reaction may further weaken the field as I_a increases, resulting in speed instability. A two-to-one speed range is possible with a standard motor. Special motors may be purchased at higher cost, which permit a wider range of control by this method.

Armature Rheostat Speed Control

According to Equation 5.36, speed may be controlled by changing r_a. The *internal* resistance of the machine cannot be changed conveniently. However the total resistance of the armature circuit may be varied by means of a rheostat in series with the armature. This rheostat must be capable of carrying the heavy armature current. It is thus more expensive than a field rheostat. A part of the cost is offset by the fact that the armature speed control rheostat also serves as a starting resistor. The armature resistances of dc machines are so low (about 0.05 per unit) that motors rated more than a few horsepower require that a resistor be inserted in series with the armature when first connected to the line. This starting resistor is shorted out in steps as the countervoltage builds up with the speed. Such starting resistors are required with field rheostat-controlled motors, but the short duty cycle allows them to be much smaller and less expensive than the resistors of armature speed control rheostats.

Figure 5.15 shows the schematic diagram for this type of control, and speed-control characteristics obtained. If a PM motor is used, the field circuit is eliminated. It will be observed that the speeds actually obtained depend very much on the speed/torque characteristic of the mechanical load on the motor. The speed regulation is poor at the lower speeds. This makes the speed hard to control unless the load characteristic is unchanging. The efficiency is also poor at low speeds. At half the no-load speed, for example, $E_g = V_T/2 = I_a(r_a + R_{ex})$, and half of the power flowing into the ar-

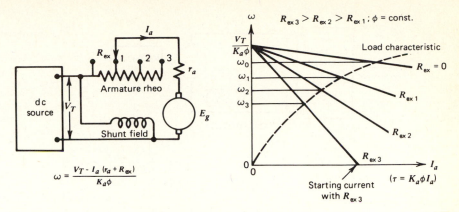

FIGURE 5.15 Armature rheostat speed control.

mature circuit is consumed as I^2R loss. Thus at half speed the system can be no more than 50 percent efficient.

The advantages of the armature rheostat control are (1) starting and speed control functions may be combined in one rheostat, (2) the speed range begins at zero speed, and (3) the cost is much less than that of other systems that permit control down to zero speed.

This type of control is used in applications where operation is at top speed most of the time, with the lower, inefficient speeds required only occasionally. Direct-current-operated electric cranes would be an example. Small battery-driven electric vehicles, such as golf carts, also use this type of speed control with permanent-magnet or series motors, because more efficient systems have been driven off the market by price competition.

Armature Terminal-Voltage Speed Control

The great versatility of this type of control is the fundamental reason for the continued existence of large dc machines. The basic circuit is shown in Figure 5.16. The motor is separately excited or has a PM field. The speed is easily controlled from zero to maximum safe speed in either forward or reverse directions. By means of inverse current feedback, it is possible to control torque rather than speed.

The controlled-voltage source may be a dc machine, or it may be a solid-state controlled rectifier. If a dc machine is used, the system is called a *A Ward-Leonard system* in honor of its inventor. Figure 5.17 is a simple schematic diagram of such a system, with a synchronous motor to supply mechanical input to the generator. Ward–Leonard systems are more expensive than those involving solid-state controlled rectifiers, and their response may be slower. However they have advantages in some applications. In a diesel-electric locomotive, for example, the prime power source is mechanical. It may be less costly to have the engine drive a dc generator than to employ an ac generator followed by a controlled rectifier.

Power shovels are usually found at the end of a long, weak power line. To maintain adequate voltage levels, the power factor must be kept high or made leading. A synchronous motor is ideal for this purpose, whereas a rectifier has a very poor input power factor at low loads. Consequently, most large power shovels and drag lines employ Ward–Leonard control systems, with the generators driven by synchronous motors. Figure 5.18 is a photograph of the motor–generator set ("M–G set") for a

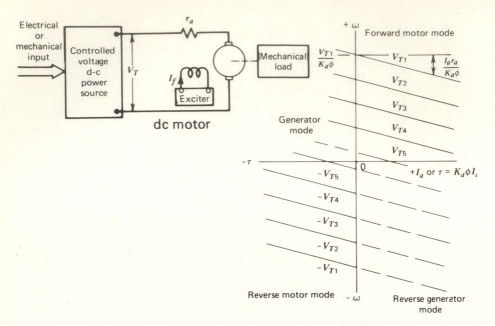

FIGURE 5.16 Variable armature-terminal-voltage speed control.

large power shovel. This M–G set provides the variable voltages for not only the shovel itself but for boom adjustment, platform rotation, and forward and reverse motion.

Direct-current motors are noisy when operated from rectifier sources. When quiet operation is important, electrical elevator drives are usually Ward–Leonard systems.

Another advantage of Ward–Leonard systems is their ability to "regenerate." This means the ability to return stored energy to the supply line. For example, when a power shovel lifts a heavy load, the synchronous machine of Figure 5.17 is operating as a motor, the *generator* is generating and the *motor* is in the motor mode, and potential energy is being stored in the mass of the shovel and the load. When the load is being lowered, the rate of descent may be controlled by operating the "motor" in

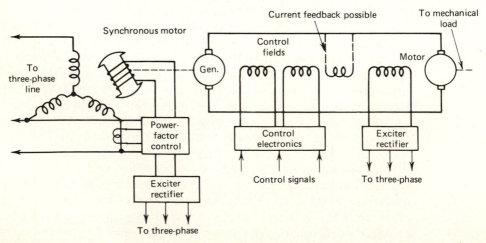

FIGURE 5.17 A Ward-Leonard system.

FIGURE 5.18 Shovel motor–generator set. (Courtesy of the General Electric Company.)

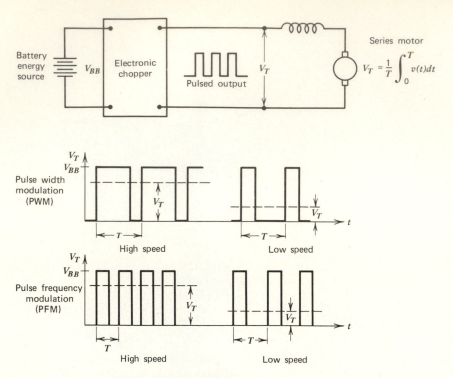

FIGURE 5.19 Chopper variable-V_T speed control.

the generator mode, using its countertorque as a brake. The "generator," under these conditions, goes into the motor mode, driving the synchronous machine as a generator and thus returning energy to the power line.

If regeneration is to take place when a solid-state package is used as the controlled-voltage source, it must include an inverter as well as a rectifier. Without the inverter the motor may still be used as a brake by dumping the reverse energy into a resistor. For many, if not most industrial applications requiring critical speed or torque control, the solid-state controlled rectifier in combination with a dc motor is receiving wide acceptance. Steel and aluminum rolling mills are typical applications.

In applications in which the power source is a battery and efficiency is an important consideration, various "chopper" drives provide variable armature terminal voltage to dc motors as a means of speed control. Examples are battery-driven automobiles. These choppers may employ thyristors or power transistors. It has been pointed out that the dc value of a voltage or current is its *average* value. The chopper is essentially a switch that turns on the battery for short time intervals. It may vary the average or dc value of the terminal voltage by varying the pulse width (pulse width modulation, or PWM), or the pulse frequency (PFM), or both. Figure 5.19 illustrates the general idea. Series motors or permanent-magnet motors are often used in these systems.

5.11 PER-UNIT VALUES OF dc MACHINE RESISTANCES AND LOSSES

The per-unit system offers a convenient basis for comparing the resistance of the various windings of a dc machine. Let rated terminal voltage be defined as the base voltage V_B and rated *armature* current be defined as base current I_B. The base resistance is defined as the ratio of these two quantities:

$$R_B \triangleq \frac{V_B}{I_B} \tag{5.40}$$

One may think of R_B as a load that will draw rated current from the armature of a generator at rated voltage. Base power may be defined as

$$P_B \triangleq V_B I_B = I_B^2 R_B \tag{5.41}$$

The chief losses in a dc machine are as follows:

LOSS		APPROXIMATE PER-UNIT VALUE AT FULL LOAD
Armature copper loss		0.05
Shunt field copper loss		0.01
Series field copper loss		0.005
Rotational or "stray *power*" loss $=$	$\left\{ \begin{array}{l} \text{Friction } + \\ \text{windage } + \\ \text{core loss} \end{array} \right\}$	0.04
Brush drop loss		Variable
Stray *load* loss		0.01

Armature Copper Loss

The calculation of this loss is based on a measurement of the dc resistance of the armature at standstill. For some purposes the resistance of the brushes is included, in which case brush drop loss is not considered a separate loss. The armature copper loss is given by

$$\text{ACL} = I_a^2 r_a \tag{5.42}$$

At full load,

$$\text{ACL}_{\text{fl}} = I_B^2 r_a \tag{5.43}$$

and the full-load armature copper loss in per-unit value is given by

$$\frac{\text{ACL}_{\text{fl}}}{P_B} = \frac{\text{ACL}_{\text{fl}}}{V_B I_B} = \frac{I_B^2 r_a}{V_B I_B} = \frac{r_a}{R_B} \tag{5.44}$$

It is evident that *the per-unit full-load armature copper loss equals the per-unit armature resistance*, about 0.05. Similarly, *the per-unit series field resistance R_{se}/R_B is in the neighborhood of* 0.005 *per unit.*

Shunt Field Copper Loss

In machines of 5-kW rating or larger, the shunt field current under normal operating conditions is only about 1 percent of rated armature current; that is,

$$\frac{I_f}{I_B} = \frac{V_B/R_f}{I_B} \cong 0.01 \qquad (5.45)$$

where R_f includes the field rheostat resistance:

$$R_f = R_{\text{rheo}} + r_f$$

The $I_f^2 R_{\text{rheo}}$ loss in the field rheostat is not counted as a machine loss. If r_f is the shunt field winding resistance, then the shunt field copper loss is given by

$$\text{FCL} = I_f^2 r_f \qquad (5.46)$$

From the rough approximation of 5.45,

$$\frac{R_B}{R_f} \cong 0.01$$

which implies that the per-unit resistance of the shunt field circuit is around 100, or about 2000 times the armature circuit resistance. To allow for an adequate range of field current control, the winding resistance is often made only about half to two thirds of the expected total field circuit resistance; so the per-unit r_f may range from 50 to 100.

Rotational Losses

In a dc machine, the core loss is a true rotational loss. Except in the pole faces, the flux in the stator remains constant in steady-state operation. The flux in the armature core is constantly alternating as the machine rotates, and at zero speed the core loss is zero. The rotational losses, then, consist of the sum of the friction, windage, and core losses. In a generator or constant-speed motor, these are considered part of the *constant* losses. The rotational losses are also called *stray power losses*, a term that is easily confused with another loss, called the *stray load loss*, soon to be discussed.

The two constant losses under constant-speed conditions are the rotational losses P_{rot} and the shunt field copper loss. Then at constant speed, maximum efficiency occurs when

$$P_{\text{rot}} + I_f^2 r_f = I_a^2 r_a \qquad (5.47)$$

where r_a includes the series field resistance in a compound machine.

Brush Drop Loss

The mechanism by which current is transferred from the carbon brushes to the commutator is very complex. Essentially, an arc exists between the two surfaces, the volt-ampere characteristic above a few amperes has the constant-voltage characteristic of an arc. The brush drop voltage V_{BD} consists of the sum of the voltage drops across the positive brush set and across the negative brush set. This voltage always acts as a voltage *drop* with respect to the armature current, and thus reverses when the armature current reverses. The brush drop loss is given by

$$\text{Brush drop loss} = V_{BD}I_a \tag{5.48}$$

The brush drop is about 2 V but may be higher in some machines. For efficiency calculations, it is assumed to be 2 V for machines with carbon-graphite brushes having pigtails:

$$P_{BD} = 2I_a \tag{5.49}$$

Stray Load Loss

When the efficiency of a dc machine is calculated on the basis of the above losses, it is found to be too high by about 1 percent. Obviously, not all losses have been accounted for. One major reason for this discrepancy is that the armature copper loss is calculated on the basis of the dc resistance of the armature winding, while the current in this winding is actually alternating. An incremental loss should be included to account for skin effects, eddy currents in the conductors, and core loss due to leakage flux in the iron surrounding the armature conductors. The hysteresis and eddy current losses in the pole faces are also somewhat increased at high armature currents. There is no commonly accepted method by which to measure these "stray load" losses. They are often considered negligible in small dc machines. *IEEE Standard 113–1973 suggests that the total stray load loss may be taken as 1 percent of the output power.*

■ **Example 5.5**

A 125-kW, 250-V, 1800-rev/min cumulative-compound dc generator has the following winding resistances:

$$r_a = 0.025 \ \Omega \qquad r_{se} = 0.010 \ \Omega \qquad r_f = 30 \ \Omega$$

The machine is long-shunt connected. Its stray power loss at rated voltage and 1800 rev/min is 5000 W. When operated at rated speed, load, and terminal voltage, the shunt field current is 5 A. Find the efficiency and input horsepower requirements under these conditions.

Solution

$$P_{\text{out}} = 125,000 \ \text{W}$$

$$\text{Shunt field copper loss} = I_f^2 r_f = 250 \cdot 30 = 750 \ \text{W}$$

$$I_a = I_{se} = I_{\text{load}} + I_f = \frac{125,000}{250} + 5 = 505 \ \text{A}$$

$$\text{Series field copper loss} = I_{se}^2 r_{se} = 505^2 \cdot 0.01 = 2550 \ \text{W}$$

$$\text{ACL} = I_a^2 r_a = 505^2 \cdot 0.025 = 6380 \ \text{W}$$

$$\text{Brush drop loss} = 2I_a = 1010 \text{ W}$$

$$\text{Stray load loss} = 1\% \text{ of } 125 \text{ kW} = 1250 \text{ W}$$

$$P_{\text{rot}} \equiv \text{stray power loss} = 5000 \text{ W}$$

$$\text{Total losses} = 16,940 \text{ W}$$

$$\text{Efficiency} = \frac{P_{\text{out}}}{P_{\text{out}} + \text{losses}} = \frac{125,000}{141,940} = 88.1\%$$

It's a minor point, but the drive motor must supply the field rheostat power as well as those losses involved in the efficiency calculation. The total shunt field circuit power is 5A · 250 V, or 1250 W. Subtracting the field copper loss gives 500 W lost in the rheostat. The input power is, then,

$$P_{\text{in}} = P_{\text{out}} + \text{efficiency losses} + \text{rheostat losses}$$

$$= 125,000 + 16,490 + 500 = 141,900 \text{ W, or } 190.3 \text{ hp}$$

For this machine, maximum efficiency would result with an armature current given by Equation 5.47:

$$5000 + 750 = I_a^2(r_a + r_{\text{se}}) = I_a^2 \cdot 0.035$$

$$I_a = \sqrt{164,286} = 405 \text{ A}$$

or at about 80 percent of full load.   ■

5.12 DIRECT-CURRENT MOTOR STARTING AND DYNAMIC BRAKING

Direct-current motors have an armature resistance of only about 0.05 per unit. If an attempt were made to start a dc motor by simply connecting it to a line of rated voltage, the armature current would be damaging:

$$I_a \text{ line start} = \frac{V_B}{r_a} = \frac{V_B}{r_{a \text{ pu}} R_B} = \frac{I_B}{r_{a \text{ pu}}} \cong 20I_B$$

or, in other words, 20 times rated current. Small machines have higher per-unit armature resistances. Their low rotor inertias allow them to accelerate rapidly and build up countervoltage quickly. For these reasons, dc machines up to, say, 3 hp may often be line-started safely. Larger machines require that a resistor be inserted in series with the *armature* when the machine is first connected to the line, to limit starting current to about two or three times the rated current. (The shunt field should receive full line voltage during starting procedure, so the starting resistor must *not* be connected so as to reduce the shunt field voltage.) As the motor accelerates, the countervoltage E_g builds up and causes a reduction in the armature current. The starting resistor may then be cut out in steps as E_g increases. When E_g has reached a sufficiently high value, the starting resistor may be shortened completely, and the motor is then connected directly to the line.

Figure 5.20 shows a "simplified" diagram of an automatic starter for a dc motor. The "simplified" diagram is one in which the component symbols are placed for

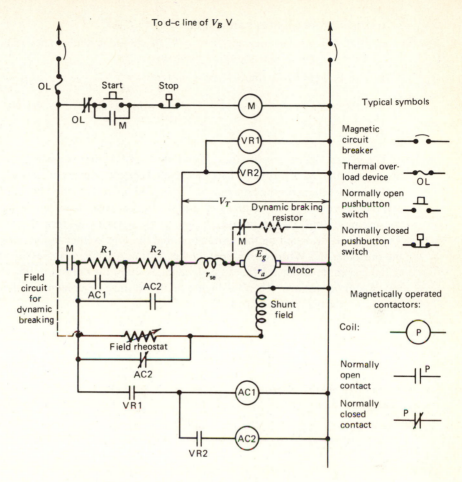

FIGURE 5.20 Simplified diagram of a *counter-emf* starter for a dc motor.

convenience in drawing the circuit. This is contrasted with the *wiring* diagram, in which the relative physical locations of the components are shown, with their terminals numbered and wiring interconnections detailed. The particular circuit shown in Figure 5.20 uses the terminal voltage of the armature circuit to determine when to close the contacts that short out the starting resistors R_1 and R_2. For this reason, it is called a *counter-emf* starter. Other starters simply use time delay relays to determine the sequence of events, while others are *current*-sensitive relays.

Many starters now use SCRs or other electronic power switches instead of magnetic contractors. Electronic logic replaces relay logic. However initial design usually begins with drawing a circuit such as Figure 5.20 as a basis for planning the logic. The procedure for designing the starting resistors is the same, unless current limits or internal voltage drops of the SCRs must be taken into account.

In the circuit of Figure 5.20, a contact is provided to short out the field rheostat during starting. At the same instant that the final section of the starting resistor is shorted, the field rheostat is reinserted into the shunt field circuit, and the machine assumes the speed for which the rheostat was set. Starters provided with a field-rheostat-shorting contact are called *variable-speed starters*. If there is no field rheostat, this contact need not be included. The starter is then called a *constant-speed* starter.

It is best to have full voltage on the shunt field during starting for several reasons:

1. High flux maximizes starting torque: $\tau_d = K_a\phi I_a$.
2. High flux prevents overspeed: $\omega = E_g/K_a\phi$, and ϕ tends to be weakened by the severe armature reaction during starting.
3. Fast buildup of E_g limits starting current:

$$I_a = \frac{V_T - E_g}{R_a} = \frac{V_T - K_a\phi\omega}{R_a}$$

The system operates as follows. When the "start" button is pushed, the coil of the main contactor M is excited. This closes all normally open contacts labeled M. A heavy contact connects the armature and shunt field circuits to the line. A small control contract shorts the "start" button, so that it may be released without interrupting the starting procedure. At the instant M closes, the motor is not turning and E_g is zero. The starter is designed on the assumption that the line voltage is the rated voltage of the machine, V_B. With two sections in the starting resistor, it is possible to limit the initial current to about 2.5 times rated value. To limit the current to twice rated value requires three sections. Then for Figure 5.20, we must be content to allow the initial current to reach 2.5 times rated:

$$I_a = \frac{V_B - \overset{0}{\cancel{E}_g}}{R_a + R_1 + R_2} = \frac{V_B}{R_a + R_1 + R_2} = 2.5I_B \qquad (5.50)$$

In per-unit, divide by I_B:

$$I_{a\ pu} = \frac{I_a}{I_B} = \frac{V_B/I_B}{R_a + R_1 + R_2} = \frac{R_B}{R_a + R1 + R_2}$$

$$= \frac{1}{R_{a\ pu} + (R_1 + R_2)_{pu}} = 2.5 \qquad (5.51)$$

In these expressions, $R_a = r_a + r_{se}$. It is fair to assume that per-unit R_a is about 0.06. Taking the reciprocal of Equation 5.51,

$$0.06 + (R_1 + R_2)_{pu} = 0.40 \qquad (5.52)$$

and

$$R_1 + R_2 = 0.34 \text{ per unit}$$

The motor will now begin to accelerate, and $E_g = K_a\phi\omega$ will rise. The voltage V_T will also rise across the voltage-sensitive relay coils VR1 and VR2, and the armature current decreases according of the relationship

$$I_a = \frac{V_B - E_g}{R_a + R_1 + R_2} \qquad (5.53)$$

Relay VR1 is usually adjusted so that the first section of the starting resistor, R_1, is shorted by the accelerating contactor AC1 when I_a drops to rated value I_B. At this current, Equation 5.53 is solved for E_g:

$$E_g = V_B - I_B(R_a + R_1 + R_2) \qquad (5.54)$$

and

$$V_T = E_g + I_B R_a = V_B - I_B(R_1 + R_2) \tag{5.55}$$

In per-unit,

$$V_{Tpu} = 1 - 0.34 = 0.66$$

So VR1 is adjusted to close at 66 percent of rated voltage. Figure 5.21 shows what has happened to this point. When VR1 closes, the coil of accelerating contactor AC1 is energized. Voltage-sensitive relays are usually too delicate to operate contacts heavy enough to handle the armature current. Hence the need for separate accelerating con- tactors. Resistor section R_1 is chosen such that the current again rises to 2.5 per unit when R_1 is shorted by AC1. Mechanical inertia prevents E_g from changing during the few milliseconds AC1 takes to close. Then

$$I_a = \frac{V_B - E_g}{R_a + R_2} = 2.5 \text{ per unit} \tag{5.56}$$

By Equation 5.54, with $R_a + R_1 + R_2 = 0.40$ per unit, $E_g = 0.60$ per unit. Inserting this value into Equation 5.56,

$$1 - 0.60 = 2.5(0.06 + R_2)$$

gives

$$R_2 = 0.10 \text{ per unit}$$

Then

$$R_1 = R_1 + R_2 - R_2 = 0.24 \text{ per unit}$$

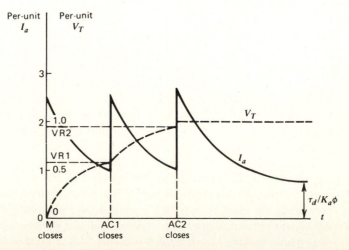

FIGURE 5.21 Armature current and terminal voltage during acceleration for the starter of Figure 5.20.

The motor continues to accelerate and I_a again decreases as E_g and V_t continue to rise. Relay VR2 is set to close when the armature current again drops to rated value I_B:

$$I_a = \frac{V_B - E_g}{R_a + R_2} = I_B$$

In per-unit:

$$\frac{1 - E_g}{0.16} = 1$$

or

$$E_g = 0.84 \text{ per unit}$$

Then

$$V_T = E_g + I_B R_a = 0.84 + 0.06 = 0.90 \text{ per unit}$$

so VR2 is adjusted to close at 90 percent of rated voltage. When AC2 closes,

$$I_a = \frac{V_B - E_g}{R_a} = \frac{0.16}{0.06} = 2.67 \text{ per unit}$$

This is an acceptable value, although slightly above the original 2.5 limit. As the motor reaches its final speed, I_a will be determined by the mechanical load. The final results are

$$R_1 = 0.24 \text{ per unit}$$

$$R_2 = 0.16 - R_a = 0.10 \text{ per unit}$$

Dynamic Braking

If a particular application requires that a motor stop quickly after the stop button is pressed, a mechanical or eddy current brake may be provided, or *dynamic braking* may be used. The mechanical brake may be operated by a magnetic solenoid. An eddy current brake consists of a disc of conducting material acted on by the magnetic field of a coil. The disc rotates with the motor shaft. The coil is energized at the same time the motor is turned off, and eddy currents are generated in the disc as long as the shaft continues to rotate. These eddy currents interact with the coil's field to produce torque in opposition to the rotation, thus slowing the motor rapidly.

In *dynamic braking,* the shunt field of the motor is left connected to the line after the armature is disconnected by the opening of the main (M) contactor. When M opens, a resistor is connected across the armature terminals. With the shunt field energized, the machine acts as a generator, and a countertorque is developed which rapidly slows the machine. The resistor is called a *dynamic braking resistor*. A simple of dynamic-braking circuit is shown by the broken lines in Figure 5.20.

FOR FURTHER STUDY

REVERSING, JOGGING, AND PLUGGING

"Jogging" means running a machine for just a few revolutions, without going through the entire starting sequence. Providing for this type of operation is important in applications requiring accurate positioning, such as in lathes and cranes.

"Plugging" means a sudden reversal from full speed in one direction to full speed in the opposite direction. If the armature connections of a motor are suddenly reversed, the countervoltage becomes a source voltage, which, when added to the line voltage, causes a very high current to flow, unless additional resistance is inserted in the circuit (see Figure 5.22). Figure 5.23 is a simplified diagram of a starter for a dc machine that also provides for jogging and plugging. Some typical operating sequences for the circuit of Figure 5.23 follow. References in parentheses are to line numbers on the diagram.

1. *Jog Forward.* All relays initially deenergized. Depress Jog Fwd (13). F2 energized (16) via N.C. R2 (15). Armature connected to line, forward polarity (3, 6). Main contactor is energized (21). The plugging resistor R_p and accelerating resistors R_1 and R_2 remain in series with the armature.

2. *Forward Acceleration.* All relays initially deenergized. Depress Forward. F1 energized via N.C. contacts of Reverse and Jog Fwd buttons (13) and of R2 (15). F2 is energized (16). M is energized (20). The armature, with R_1, R_2 and R_p in series, is connected to the line with forward polarity (3 and 6). The forward circuit is locked in by F1 contacts across the Forward button (15). The P_f coil is connected from the $+$ side of R_p to the negative line (2), and P_f closes immediately, energizing P (10, 11), shorting out the plugging resistor R_p (5).

   The voltage-sensitive relay 1A is connected across the armature via a forward-biased diode and contacts of F1 and M (8). When the counter-emf of the armature has risen sufficiently, 1A closes, energizing a second voltage-sensitive relay 2A (9). When the motor is nearly at full speed, 2A closes (5) and the armature is connected directly the line through the series field.

3. *Plugged Reverse.* Assume forward acceleration has been completed. The Reverse button is depressed. The N.C. contacts of F2 (18) prevent R1 from being

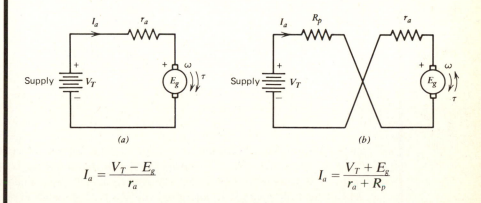

$$I_a = \frac{V_T - E_g}{r_a} \qquad\qquad\qquad I_a = \frac{V_T + E_g}{r_a + R_p}$$

FIGURE 5.22 Plugging as a means of braking or reversal. (*a*) Effective circuit before plugging. (*b*) After plugging.

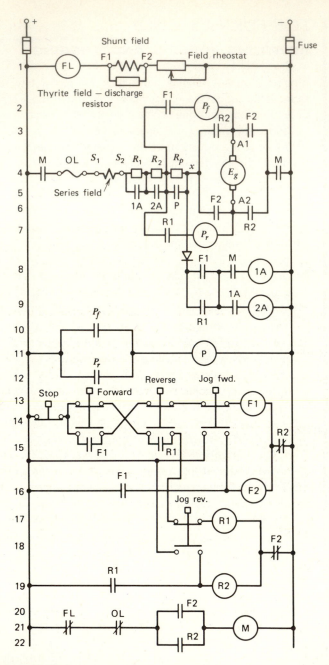

FIGURE 5.23 Direct-current motor starter and control, with two starting resistors, forward and reverse plugging, and forward and reverse jogging.

energized immediately, however, F1 is deenergized. 1A and 2A are deenergized (8). P_f is deenergized (2). P is deenergized (10). F2 is deenergized (16) as F1 opens. The armature forward contacts open (3 and 5). M is deenergized (20). When F2 is fully open, R1 is energized via F2's N.C. contacts (18). The reverse circuit is locked in by R1 contacts across the Reverse button (14). R2 is energized (19), connecting the armature for reverse operation and M is energized (22). P_r is connected across R_p and the armature in series (7).

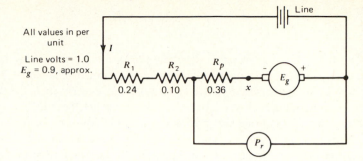

FIGURE 5.24 Armature circuit when plugging is initiated but before rotation has reversed.

The above process takes place in a fraction of a second. Mechanical inertia keeps the motor going forward at nearly full speed during this interval, and thus E_g is only slightly less than line voltage when M closes. At this instant, the armature circuit is as in Figure 5.24.

Resistance values shown are per unit for a starter that limits starting and plugging currents to 2.5 times rated current ($r_a = 0.06$). Calculations will show that the voltage across P_r is essentially zero. As the motor decelerates and begins to accelerate in reverse, the voltage across P_r rises, and eventually its contacts close (12), energizing P and shorting out the plugging resistor R_p (5).

While the motor is still going forward, point x in the figure is negative, and accelerating contactors are not energized because the diode connecting them to the armature is reverse-biased. Once the motor reverses, acceleration proceeds normally.

Starters that do not provide for plugging must have some means of prohibiting the reversal of the motor at high speed. Plugging may also be used as a means of fast braking, if a *zero speed switch* is provided that disconnects the machine's armature from the line as the speed approaches zero.

5.13 ARMATURE REACTION AND COMMUTATION

Armature reaction is the effect of the armature MMF field upon the pole flux ϕ. Since the armature poles are usually halfway between the stator field poles, the interaction is indirect. However the brushes are sometimes shifted from their normal positions to improve commutation. When this is done, the armature MMF has a more direct effect on ϕ; that is, the armature reaction is increased. It is evident that the subjects of armature reaction and commutation are related.

The Nature of the Armature MMF Field

Figure 5.25 shows a "developed" or flattened-out view of a cross section of part of the armature and stator of a dc machine. The circles in the armature slots contain crosses or dots to indicate current directions. The current pattern is held stationary by the action of the commutator. If S is the number of armature slots, the current in each slot is the number of conductors per slot times the current in each conductor, I_a/a:

$$I_s \stackrel{\Delta}{=} \text{current per slot} = \frac{Z}{S}\frac{I_a}{a} \qquad (5.57)$$

The MMF that the armature develops across the air gap at any point is determined easily by means of Ampère's law in circuital form, which states: *The magnetomotive force around any closed path is equal to the net current enclosed.* It is seen in Figure 5.25 that path a encloses zero net armature current, because the number of amperes out of the paper inside this path is equal to the number of amperes flowing in. This shows that the points of zero *armature* MMF are on the pole axes. Paths b and c, respectively, enclose currents of I_s and $2I_s$. If the MMF drop in the iron is negligible compared to that of the air gap, and if the armature MMF at the pole axes is zero, it is seen that $\mathcal{F}_{gb} = I_s$ and $\mathcal{F}_{gc} = 2I_s$, where $\mathcal{F}_{gb}$ and $\mathcal{F}_{gc}$ are measured where paths b and c cross the air gap, other than at the pole centers. Continuing this process results

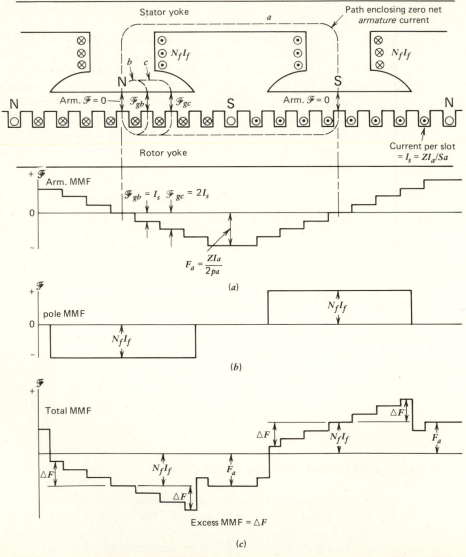

FIGURE 5.25 MMF of a dc machine without interpoles or compensating windings. (*a*) Armature MMF alone. (*b*) Pole MMF alone. (*c*) Total MMF.

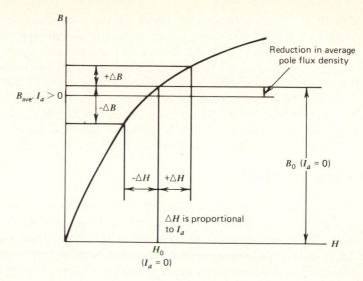

FIGURE 5.26 *B–H* curve of pole shoe iron, showing the effect of armature reaction on the average flux density under a pole.

in a plot of armature MMF as a function of distance on the armature surface which looks like that shown in Figure 5.24*a*. The peak MMF is halfway between the poles, and is equal to the number of armature conductors in one half pole pitch:

$$F_a = \frac{I_a}{a} \cdot Z \cdot \frac{1}{2p} = \frac{ZI_a}{2pa} \quad \text{A} \tag{5.58}$$

Effect of Armature Reaction on Pole Flux and Commutation

When the pole magnetomotive force is added to that of the armature, the total MMF pattern is obtained; as shown in Figure 5.25*c*. Note that the field is strengthened at one pole tip and weakened at the other. The result is that the field intensity in the iron at one side of the pole is increased and that on the other side of the pole is decreased by the same amount. The effect on the flux densities at the two pole tips is illustrated in Figure 5.26. Saturation prevents the ΔH on the strengthened pole tip from increasing the flux density under that pole tip very much. At the other pole tip, the subtraction of ΔH due to the armature field pulls the iron out of saturation and reduces the flux density in the air gap significantly. The net result is that the *average* flux density in the air gap under the pole shoe is reduced. Since ϕ is the surface integral of B over the pole face area, the flux per pole is reduced. This is one of the effects called *armature reaction.*

Flashover

When a motor or generator is subjected to severe overloads, the armature MMF causes a heavy concentration of the pole flux at one tip. As a result, the voltages in the coils passing under the poles are unequal. Comparison of Figures 5.25 and 5.6 will show that the armature coils influenced by the high flux densities are connected to those commutator segments near the brushes. The air near the brushes is highly ionized as a result of the arc between each brush and the commutator surface. Under heavy overload, it is possible to induce enough voltage in the coils experiencing the increased flux density to break down the ionized air between commutator segments near the brushes. The resulting arc further ionizes the air around the commutator to the extent that it becomes sufficiently conductive to allow an arc to be established from brush to brush. This phenomenon is called *flashover*. The arc is sufficiently hot to melt the commutator segments if it is not extinguished quickly. Since a spare armature for a large mill motor may cost nearly a quarter of a million dollars, it is important that flashover be prevented. The preventive for flashover is compensating windings.

Compensating Windings

Compensating windings located in slots in the pole faces eliminate the problems of armature reaction and flashover by balancing out the armature MMF under the pole faces. All of the currents in the armature slots under a given pole flow in the same direction. If (1) the currents in the compensating windings are made to flow in opposite direction, (2) if the current in the compensating windings is proportional to the armature current, and (3) if the number of ampere conductors in the pole face is equal to the number of ampere conductors in that part of the armature winding directly below the pole face, the two magnetomotive forces will exactly cancel in the air gap under the pole. The pole flux is then undisturbed by the armature field.

Figure 5.27 shows schematically and pictorially the installation of compensating windings. Figure 5.37 is a photograph of a PM dc motor stator with compensating windings, showing the heavy conductors which connect them into the armature circuit. The compensating conductors are all in series, and the total compensating winding is connected in series with the armature.

Compensating windings are extremely expensive. They are not used unnecessarily. They will be found only in the following cases:

1. In large machines subject to heavy overloads or plugging
2. In smaller motors subject to sudden reversal and high acceleration
3. In metadynes (q.v.)

The number of conductors required in each pole face is easily calculated. In Figure 5.27a, ρ is the pole pitch and α the pole arc, in degrees or radians. The number of ampere conductors per pole in the armature winding is

$$\frac{Z}{p} \frac{I_a}{a}$$

since each conductor carries I_a/a A. The fraction of this total under a pole face is then

$$\text{Armature MMF/pole face} = \frac{\alpha}{\rho} \frac{ZI_a}{pa} \qquad (5.59)$$

The current in the compensating conductors is I_a. Let N_c be the number of compensating-winding conductors per pole face. For exact compensation (called *100 percent compensation*),

$$N_c I_a = \frac{\alpha}{\rho} \frac{ZI_a}{pa}$$

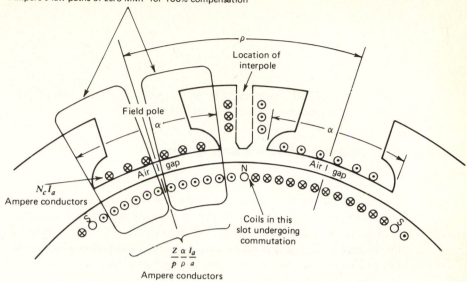

Ampere's law-paths of zero MMF for 100% compensation

Location of interpole

Field pole

$N_c I_a$
Ampere conductors

Air I gap

Air I gap

Coils in this slot undergoing commutation

$$\frac{Z}{p}\frac{\alpha}{\rho}\frac{I_a}{a}$$

Ampere conductors

(a)

(b)

FIGURE 5.27 Features of compensating windings. (a) Geometry and current patterns. (b) Detail of the stator of a 1620-hp, 900-rev/min dc motor, showing compensating winding ($N_c = 9$) and interpoles. (Figure 5.27b courtesy of General Electric Canada, Inc.)

or

$$N_c = \frac{\alpha Z}{\rho p a} \qquad (5.60)$$

This is not likely to come out to be an integer, and a lesser integer number is used, because 60 percent compensation is usually sufficient.

Commutator Sparking Causes and Cures

Examination of Figure 5.5c will show that the current in an armature coil must reverse during the short interval during which the coil is short-circuited by a brush. When the coil is shorted, its current flows from one commutator segment to the next through the brush, rather than in the external circuit (see Figure 5.28). Figure 5.29 shows how the current is shorted coil varies during the brief interval of short circuit. As soon as the coil is shorted, its current i_c begins to decay. But the resulting $L(di/dt)$ voltage, by Lenz's law, tries to prevent this, with the result that, if nothing is done to oppose the $L(di/dt)$ voltage, the generated voltage of the machine forces a current reversal just as the coil becomes unshorted. The large voltage appearing between the commutator segments to which the coil is connected causes sparks at the trailing edge of the brush. Each spark acts as a tiny explosion, removing a minute piece of the commutator copper. This phenomenon is capable of ruining the commutator surface in a few hours of operation.

The cure of this destructive sparking is to induce a voltage in the coil undergoing commutation that will overcome the $L(di/dt)$ voltage and assist in the reversal of the current. To accomplish this, **nearly all dc machines of more than 1 hp are provided with interpoles.** These are narrow poles attached to the stator yoke and located halfway between the main field poles. They can be seen between the main poles in Figures 5.27, 5.31 and 5.37. Since, from Figure 5.29, the $L(di/dt)$ voltage will depend on I_a, the strength of the interpoles must also be proportional to I_a. For this reason, their windings are connected in series with the armature. Once their strength is correctly adjusted at the factory, there is no need for further adjustment. The same may be said for compensating windings. Consequently, leads to these *commutating windings* are not brought out to the terminal box. These windings remain a permanent part of the armature circuit, as shown in Figure 5.30.

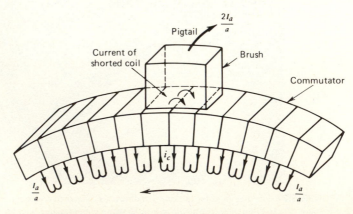

FIGURE 5.28 Current path for coil being commutated.

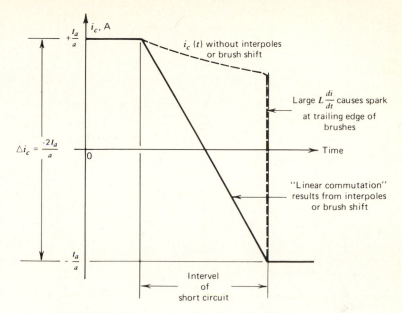

FIGURE 5.29 Coil current as a function of time during commutation.

The polarity of the voltage induced by the interpoles in a coil undergoing commutation must be such as to assist in the current reversal in that coil. That is, the voltage should oppose the coil current flow in the direction it had just before the coil was shorted by a brush and encourage it to flow in the reversed direction it will have immediately afterward. The right-hand drawings in Figure 5.31 illustrate the correct voltage polarities for motor and generator operation and the magnetic polarities of the interpoles required to induce these voltages. Note that in both cases the MMF of the interpoles is opposed by the armature MMF at its strongest point—midway between the poles. Since the interpoles must force flux into the armature winding in spite of the opposition of the peak armature MMF, the number of interpole turns is chosen to make the MMF of each one about 25 percent greater than F_a. Compensating windings buck out part of the armature MMF, so interpoles do not need to be so strong in machines equipped with compensating windings.

Figure 5.31 also shows how to determine the polarity of interpoles. First, to determine the polarities of the armature poles remember that the developed torque is in the direction of rotation in a motor and is opposed to rotation in a generator. That

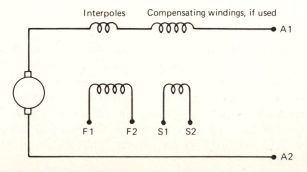

FIGURE 5.30 Permanent connection of commutating windings into the armature circuit.

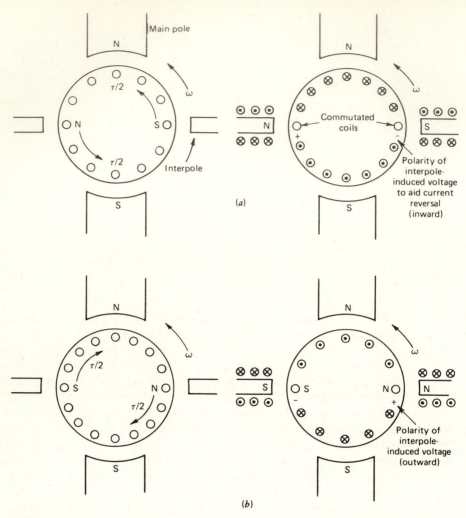

FIGURE 5.31 Interpole polarities. (*a*) Motor. (*b*) Generator. Note: Interpole MMF always opposes armature MMF.

means that the armature poles in a motor will have the same polarities as the field poles behind them; in a generator, the opposite polarities. The interpoles must oppose these armature poles, north against north, south against south.

NUMBER OF INTERPOLES

The purpose of interpoles is to provide flux so that the commutating coil's $d\lambda/dt$ due to rotation is enough to overcome $L(di/dt)$. It makes no difference to the coil whether this flux is provided by one pole or two. Consequently, the number of interpoles may be made equal to the number of field poles, or half as many, as in Figure 1.9*b* in Chapter 1. In smaller machines, one often finds one interpole for every two main poles. The difficulty with this plan is that interpole fluxes must return through the main field poles of opposite polarity. Thus the fluxes of the main field poles are unbalanced.

BRUSH SHIFT

In very small machines, economics prohibits the use of interpoles. If the machine is to be used in an application requiring rotation in both directions, the commutator must be made oversize and brush materials chosen to minimize sparking. If the motor only runs in one direction, the positions of the brushes may by shifted so that the coils undergoing commutation have current reversal aided by flux from the main poles. See Figure 5.32. Remember that to aid commutation, flux must enter the armature in opposition to the armature MMF. This means that when the brushes are shifted in the proper direction to reduce sparking, armature poles are brought under the influence of field poles of like polarity. Now, like poles of the armature and field repel each other to produce the developed torque. *Thus to reduce sparking, brushes must be shifted in a direction opposite to the direction of the torque*.

This means of improving commutation is subject to two difficulties:

1. The voltage induced to aid commutation is not proportional to armature current.
2. The armature MMF is no longer 90 electrical degrees from the field poles and has a component opposing the field MMF. Thus armature reaction is increased.

The first of these difficulties means that a certain degree of brush shift provides correct spark elimination at only one load. Before interpoles were invented, the brush rigging of dc machines was provided with a handle to permit adjustment of brush shift as the load on the machine varied.

Figure 5.32 shows that when brushes are shifted, certain armature ampere turns are directly opposed to those of the main field. In a generator, this causes poor voltage regulation. In a motor, the speed tends to rise with load, a less stable condition. The

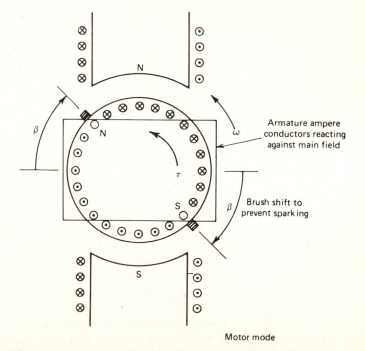

Motor mode

FIGURE 5.32 Brush shift to prevent sparking causes increased armature reaction.

effect on torque is discussed in Chapter 8, Section 8.9. In a PM machine, the demagnetizing effects on the main poles may be intolerable. In spite of these difficulties, it is often advisable to shift brushes in fractional-horsepower machines for unidirectional rotation applications, in order to prolong commutator life.

The Effects of Armature Reaction on Speed and Torque

When the armature is carrying a heavy current, the armature MMF field tends to reduce the flux of each stator pole, even when there is no brush shift. This effect is illustrated in Figure 5.26. When the brushes are shifted in the proper direction to reduce sparking, there is an even greater reduction in pole flux as the armature current increases. The effect of these armature-reaction phenomena on the magnetization curve is shown in Figure 5.10. If the no-load magnetization curve is being used to find $K_a\varphi$ or $K'_a\varphi$, the effect of armature reaction may be approximated by a modification of Equation 5.35: $I^* = I_f \pm (N_{se}/N_f)I_{se} - AI_a$, where the coefficient A is evaluated empirically at full load armature current.

Equations 5.26 and 5.27 show that the developed torque of a dc machine is proportional to the flux per pole. Thus the presence armature reaction causes the torque per ampere to be somewhat less at higher armature currents. Equation 5.36, the magic equation, shows motor speed to be inversely proportional to the pole flux. Armature reaction results in higher speeds under load than would be predicted on the assumption of constant φ. These effects on the speed and torque of a shunt motor are shown by the broken line curves of Figure 5.33 (a). In series and compound motors, the series fields are made strong enough to overcome the effects of armature reaction. The destabilizing effects of armature reaction are noted in the discussion of the *stabilized shunt* motor in Section 5.7. When a machine has both interpoles and adequate compensating windings, the effects of armature reaction on speed and torque are nullified.

5.14 CHARACTERISTICS OF SERIES AND COMPOUND dc MOTORS

Combinations of series and shunt windings on the field poles of dc machines can provide a wide range of torque/speed characteristics for motor operation or can tailor the volt/ampere characteristics of a generator to a particular load requirement. However ac motors have been developed to meet most industrial needs for motors with fixed torque/speed characteristics. Direct-current *generators* are rapidly being replaced by controlled, solid-state rectifiers. Yet there are situations, such as in mines, where the ready availability of dc supplies makes the use of dc *motors* attractive. In certain speed control applications, particularly in electric transportation, series or compound motors have characteristics superior to those of separately excited shunt machines.

Schematic diagrams for the shunt, series, and compound connections are shown in Figure 5.11. Methods of calculating speed and torque are outlined in Sections 5.6 through 5.8. With this background, it is easy to deduce the overall performance characteristics of dc machines connected in various ways.

The *series motor* has a very high starting torque. As its speed increases, its torque decreases, resulting in an essentially constant-horsepower characteristic. For this reason, it is ideal for traction applications (locomotives, street cars), electric automobiles, and the like, because the need for different gear ratios for different speed ranges is eliminated.

In Figure 5.33, the series dc motor is compared to the shunt machine. The speed

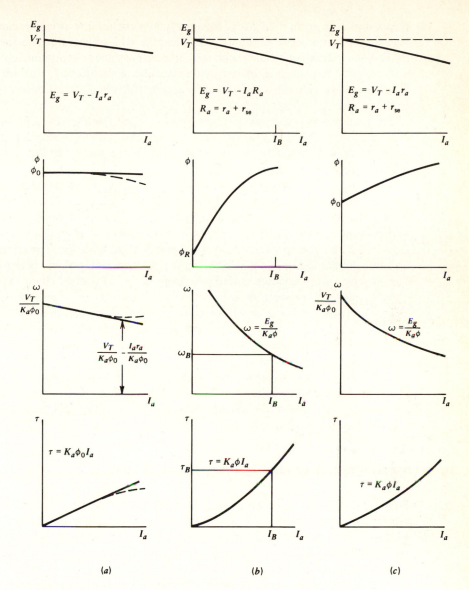

FIGURE 5.33 Direct-current motor characteristics. (*a*) Shunt. Effects of armature reaction shown by broken-line curves. (*b*) Series. (*c*) Cumulative compound.

is determined by the magic equation, Equation 5.36. The generated voltage decreases as load on the motor increases, as a result of the drop in the armature circuit resistance $I_a R_a$. This resistance is the sum of the armature resistance r_a and the series field resistance r_{se}:

$$R_a = r_a + r_{se} \qquad (5.61)$$

There is no shunt field winding, so that

$$\mathscr{F}_{\text{pole}} = N_{se} I_a \qquad \text{(series motor)} \qquad (5.62)$$

As a result, the flux per pole, ϕ, has the usual saturation curve relationship with the armature current I_a, as shown in the second curve of Figure 5.33*b*. The speed as a function of I_a is given by $E_g/K_a\phi$. It is nearly hyperbolic. The armature current at no load is only that required to supply losses, and the no-load pole flux is essentially ϕ_R, the residual magnetism of the stator core. In machines above approximately 1 hp, the no-load speed is so high as to destroy the machine. Consequently, *a series dc motor should never be connected to its load by a belt*. If the belt should break, a runaway would result.

Since the developed torque is given by $\tau_d = K_a\phi I_a$, and ϕ is nearly proportional to I_a, the torque of a series motor at low to moderate currents increases as the square of the armature current. As the core of the machine saturates, the increase in torque with I_a becomes more linear.

The *cumulative compound dc motor* has speed/torque characteristics intermediate between those of the shunt and series motors. These characteristics are shown in Figure 5.33c. The no-load flux is determined by the shunt field. This value is shown in ϕ_0 in the second curve of Figure 5.33c. The no-load speed is thus limited to a safe value, $\omega_{nl} = V_T/K_a\phi_0$. The speed drops off considerably under load. For a given torque, less armature current is required than if the shunt field were acting alone, because the pole flux increases with armature current. A shunt motor, of course, could have its field current adjusted to produce the same torque as a compound or series motor at a given armature current. However it will be noted that the armature current of a compound motor does not rise as rapidly with torque increases as does that of a shunt machine. This ability to "lug," that is, to slow down with increased torque load without too great an increase in line current, is very valuable in many industrial applications.

Differential compound motors are almost never used. Most mechanical loads require more torque as the speed increases. The rising speed/torque characteristic of differential compound motors results in an unstable situation when they are connected to such loads. Speed increases torque, which increases speed, etc.

5.15 PERMANENT-MAGNET dc (PMDC) MACHINES

Permanent-magnet dc machines employ magnetized materials to supply pole flux, rather than current-carrying coils. There are literally millions of PMDC machines produced each year. They are used extensively in automobiles to operate windshield wipers and washers, drive blowers for heaters and air conditioners, to raise and lower windows, and so on. These motors power slot cars and electric tooth brushes. They serve as starter motors for outboard motors and lawn mowers, and are often employed in equipment using batteries as a source of power. Permanent-magnet motors up to 200 hp have been developed for industry.

One major advantage of these motors is that they require no field current. This fact means that there will be a considerable saving in energy over equivalent wound-pole machines during a typical machine lifetime. Although permanent-magnet materials tend to be expensive, the size of a permanent-magnet field pole may be much less than that of the equivalent wound pole. This means that the overall size of the machine is reduced. The reduction in the cost of other materials compensates, at least in part, for the magnet cost. In small machines there is a definite cost advantage in permanent-magnet field poles.

The flux per pole, ϕ, is not adjustable in PMDC machines. This makes them undesirable as generators, in many cases, because there is no simple way to control the output voltage. Constant flux also means that speed and torque must be controlled by the high-current armature circuit. In wound-pole machines, these quantities may be controlled by the low-current field circuit. However over most of the control range, it is advantageous to use armature circuit control in preference to field circuit control, even with wound-pole machines, so the adoption of PMDC motors for industrial applications requiring precise control involves no great sacrifice. The elimination of a separate field current supply is often a great advantage, in fact.

The effect of constant flux on the performance characteristics of dc machines is quite simple. In fact the performance PMDC motors is quite similar to that of shunt machines. From Equations 5.12 and 5.26,

$$E_g = K_a \phi \omega \tag{5.63}$$

and

$$\tau = K_a \phi I_a \tag{5.64}$$

With ϕ constant, these become

$$E_g = k\omega \quad \text{V} \tag{5.65}$$

$$\tau = kI_a \text{ N-m} \tag{5.66}$$

where

$$k = K_a \phi \tag{5.67}$$

In the English system of units,

$$E_g = k_V n, \text{V} \tag{5.68}$$

$$\tau = 7.04 k_V I_a \tag{5.69}$$

$$= k_\tau I_a \text{ lb-ft} \tag{5.70}$$

In these last expressions, k_V is the *voltage constant* of the machine, and $k_\tau \equiv 7.04 k_V$ is its *torque constant*. The magic speed equation, Equation (5.36), becomes

$$\omega = \frac{V_T - I_a R_a}{k} \quad \text{rad/s} \tag{5.71}$$

or

$$n = \frac{V_T - I_a R_a}{k_V} \quad \text{rev/min} \tag{5.72}$$

These equations are basically the same as for shunt motors with constant field current. Consequently, the armature circuit methods of speed control, discussed in Section 5.10, apply directly to PMDC motors. These include armature rheostat control, variable terminal-voltage systems, and chopper controls.

Magnet Materials and Applications

There are three classes of permanent-magnet materials being used for PMDC motors: Alnicos, ceramics (ferrites), and rare-earth materials. Alnico magnets are used in motors having ratings in the range of 1 to 200 horsepower. Ceramic magnets are most economical in fractional horsepower motors, and may show an economic advantage over Alnico up to about 10 hp. The rare-earth magnet materials are very costly, but are the best economic choice in very small motors. There have been calculations to show that they might have a cost advantage in very large motors as well.

Magnet materials are chosen largely on the basis of their residual flux density B_r, their maximum energy product $(BH)_{max}$, their coercive force H_c, and their *intrinsic* coercive force H_{ci}. Temperature affects these characteristics, and the susceptibility of a material to temperature variations must be considered. Temperatures in PMDC motors tend to be higher than in wound-pole motors, because they must be made totally enclosed to prevent the magnets from collecting magnetic junk from their environment.

To understand the significance of these magnetic paramenters, consider a specimen of magnetic material to be placed between the poles of an electromagnet as in Figure 5.34a. Furthermore, assume that the reluctance of the flux path through the electro-

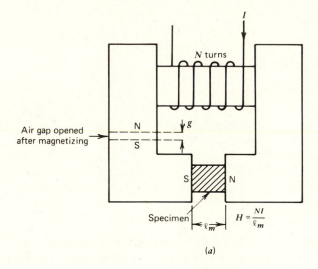

(a)

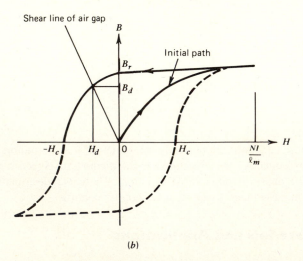

(b)

FIGURE 5.34 Example to define PM parameters. (a) Specimen in magnetizing jig. (b) B–H loop of specimen material.

magnet is negligible. If a sufficient current is passed through the coil, the magnetic material will be driven far into saturation as indicated on the hysteresis loop of the material in Figure 5.34b. If the current is now reduced to zero, the flux density in the material will return to the residual value B_r. Now if the specimen were absent when the current was turned on, the flux density between the poles of the electromagnet would have been

$$B_{air} = \mu_0 H \tag{5.73}$$

With the specimen in place, its internal flux density may be considered as the sum of B_{air} and that due to the "magnetization" of the material:

$$B = B_{air} + B_m = \mu_0(H + M) \tag{5.74}$$

where M is called the *magnetization* of the specimen. Note that the units of M are ampere-turns per meter ("amperes per meter" in the SI system). When the current is reduced to zero, $H = 0$, $B = B_r$, and

$$B_r = \mu_0 M_r \tag{5.75}$$

defines the relationship between the residual flux density and the magnetization of the material in the specimen.

The upper, left-hand quadrant (second quadrant) of the B–H loop is the part of interest to the designer. As long as the magnetic circuit remains closed, the permanent-magnet specimen can perform no useful function. *It must supply flux to an air gap to produce force on another magnet or induce voltage in a moving coil.* Suppose an air gap of permeance $\mathcal{P}_g$ is cut into the magnetic circuit after the specimen is magnetized, as indicated in Figure 5.34a. Magnetic poles will appear across the gap, having the polarity indicated. Note that these poles will not only drive flux across the gap but also that their polarity is such as to oppose the magnetization of the specimen. With no current in the winding, the MMF across the gap is

$$\mathcal{F}_g = \frac{\phi_g}{\mathcal{P}_g} \tag{5.76}$$

where ϕ is equal to the flux density in the specimen, B_d, times the specimen area A_m:

$$\mathcal{F}_g = \frac{B_d A_m}{\mathcal{P}_g} \tag{5.77}$$

The reverse field intensity applied to the specimen is then

$$H_d = -\frac{\mathcal{F}_g}{\ell_m} = -\frac{B_d A_m}{\mathcal{P}_g \ell_m} \tag{5.78}$$

The values of H_d and B_d may be found from the dimensions of the specimen, A_m and ℓ_m, and the permeance of the air gap. Note that

$$B_d = -\left(\frac{\mathcal{P}_g \ell_m}{A_m}\right) H_d \tag{5.79}$$

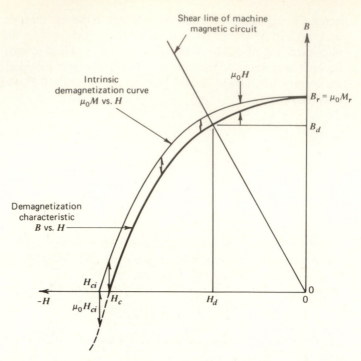

FIGURE 5.35 Demagnetization and intrinsic demangetization curves for an Alnicolike material.

so a line may be drawn from the origin having a slope $-\mathscr{P}_g \ell_m/A_m$, until it intersects the *B–H* curve for the material in the second quadrant. *This construction line is called a "shear line."* The intersection will be the operating point for the permanent magnet. In a motor, the shear line is determined by the dimensions of a pole magnet and the permeance of the path of that magnet's flux. The maximum possible value of the product $B_d H_d$ for a given material is called the *energy product* of that material. The units of this quantity are tesla-amperes per meter, or weber-amperes per cubic meter. By integration of Faraday's law, we learn that webers are equivalent to volt-seconds, so the units of $B_d H_d$ are watt-seconds per cubic meter, or joules per cubic meter. Examination of Figure 5.34 will show that the energy product is the area of the largest rectangle that will fit under the demagnetization curve of the material.

An important consideration is the *permanence* of the "permanent" magnet. Under what conditions will it become completely demagnetized? This will not occur unless the magnetization *M* is reduced to zero. Solving Equation 5.74 for $\mu_0 M$,

$$\mu_0 M = B - \mu_0 H \tag{5.80}$$

In the second quadrant, *H* is negative. It will be seen, then, that $\mu_0 M$ exceeds the flux density *B* by an amount $\mu_0 H$. The situation is illustrated by Figure 5.35, which shows a *demagnetization characteristic* (i.e., the second quadrant of the *B–H* loop) similar to that of an Alnico. If a reverse field of H_{ci} is applied to the magnet, it will be completely demagnetized. For an Alnico material, H_{ci} does not differ a great deal from H_c.

Figure 5.36 shows demagnetization characteristics for an Alnico and a ceramic material for comparison. Note how much more flux density is available from an Alnico. This is the reason they are chosen for large motors. Alnico magnets are much more sensitive to demagnetization, however. Suppose the magnetic circuit of a motor with

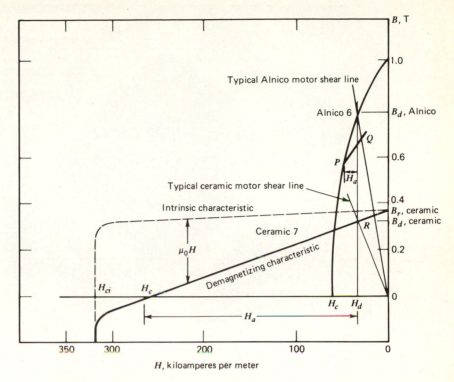

FIGURE 5.36 Demagnetization characteristics of Alnico 6 and Ceramic 7.

Alnico magnets is designed so that the shear line is as indicated in the figure. The magnet operating point is at B_d, H_d. If a demagnetizing MMF is applied to the magnetic circuit such that H_d is reduced by an amount H_a, the operating point will move to point P. When the demagnetizing MMF is removed, the flux density moves up along a so-called *subsidiary B–H loop*, approximately parallel to the top of the demagnetization characteristic, to point Q. We see that there is a considerable loss in flux density; that is, the magnet has been partially demagnetized.

Now the chief cause of demagnetization in PMDC machines is armature reaction. Reference to Figure 5.25 will show that when armature current flows, one pole tip is demagnetized by an amount ΔF, which is easily calculated. At starting, when the armature current is several times the rated current, ΔF becomes quite large. If the pole shoe were made of Alnico, it would become demagnetized. Therefore when Alnico magnets are used, an iron pole shoe must be provided to protect the magnet from demagnetization. Figure 5.37 shows the structure of a typical Alnico motor.

Alnico magnets are so powerful that it is impractical to dismantle an Alnico PMDC motor for repair without first demagnetizing the magnets. For this reason, these machines are provided with windings around the PM poles to permit magnetization and demagnetization in situ. Once the machine is reassembled, the magnets are remagnetized. Plants using many Alnico PMDC motors keep on hand a magnetizer–demagnetizer such as that shown in Figure 5.38. These devices provide a current pulse to the magnetizing windings lasting about 10 ms.

Returning to Figure 5.36, it will be seen that ceramic magnets are much more difficult to demagnetize. Suppose the normal operating point of the magnet is at R. Note how great H_a can be without leaving the linear part of the demagnetizing characteristic. When the demagnetizing MMF is removed, the operating point moves back

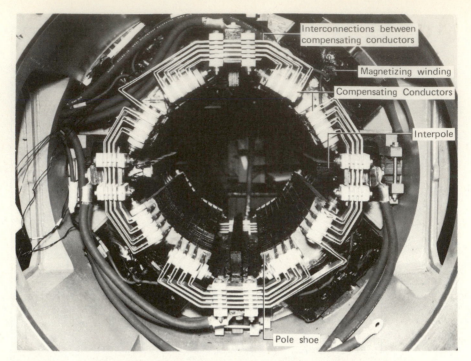

FIGURE 5.37 Construction of an Alnico PMDC motor. (Courtesy of the General Electric Company.)

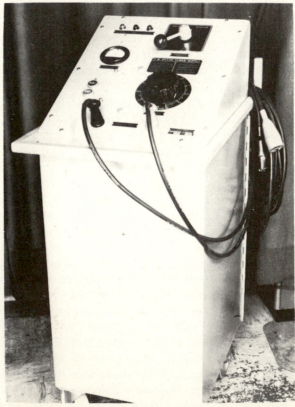

FIGURE 5.38 Typical magnetizer–demagnetizer for Alnico PMDC motors. (Courtesy of the R. B. Annis Company.)

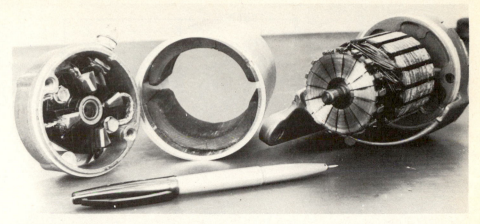

FIGURE 5.39 Construction of a ceramic pole starter for an outboard motor. This is a four-pole motor. Each arc of ceramic material provides two magnetic poles.

to R, and the magnet functions normally. Thus ceramic pole magnets may be designed without iron pole shoes. Figure 5.39 shows a typical construction.

Another interesting fact about ceramic magnets is that their permeability is nearly equal to μ_0. In calculating the shear line, the length of the *air gap* includes the magnetic path length of the magnet itself. Even removal of the armature does not cause enough change in the operating point to demagnetize the magnet. Ceramic motors may thus be disassembled and reassembled without affecting the performance of the motor. This is fortunate in that it would be impractical to install magnetizing windings in motors of the small sizes which usually employ ceramic magnets.

5.16 SELF-EXCITED dc GENERATORS

It has been mentioned that rectifiers are rapidly replacing dc generators as sources of dc power for most applications. Those dc generators that are being installed at the time of this writing are in most cases separately excited or, if self-excited, are provided with automatic voltage regulators. The characteristics of self-excited generators without voltage regulators are of interest to those who must deal with older installations, and to students who wish to round out their knowledge of capabilities of dc machines in the generator mode.

Generator Buildup

"Self-excited" means that a generator supplies its own field excitation. If a dc generator that is to be self-excited is driven at an adequate speed by some prime mover, it must generate some small voltage, even when the field current is zero. Remanent magnetism in the stator core usually provides sufficient voltage. If it does not, no current will flow in the field winding when it is connected to the armature circuit, and thus it cannot be self-excited. If the core has become demagnetized, one must "flash the field," that is, briefly apply a dc voltage to the field winding to induce some residual magnetism into the core. Not much voltage is required, and a car battery will often do the job. The polarity of the terminal voltage of the generator will be the same as that of the residual-flux voltage, and this is determined by the polarity of the residual flux and by the direction of rotation. If the direction of rotation is fixed and the polarity

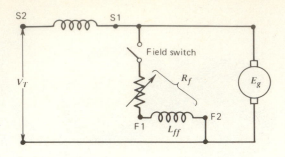

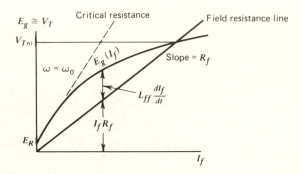

$$E_g(I_f) = (K_a\omega_0 N_f \mathcal{P}_p)I_f \qquad V_T \cong E_g = I_f R_f + L_{ff}\frac{dI_f}{dt}$$

Equation of field resistance line: $V_T = I_f R_f$

FIGURE 5.40 Voltage buildup of a dc generator.

of the residual voltage is incorrect, the field may be "flashed" to reverse the flux polarity.

The field connections to the armature must be such that increased field current increases the flux. The residual voltage at rated speed, measured at the armature terminals, will amount to only a few percent of the machine's rated voltage. If this drops even slightly when the field is connected, a reversed connection is indicated. Interchanging the two field leads will cause the voltage to rise to a level determined by the speed and the resistance of the field circuit.

The circuit for a short-shunt compound generator at no load is given in Figure 5.40, together with its open-circuit characteristic. At no load, $I_a = I_f$ and the internal $I_a r_a$ drop of the armature is only about 0.0005 per unit—entirely negligible. The two governing equations for the system during voltage buildup are, for the armature,

$$V_T \cong E_g = K_a\phi\omega = K_a N_f I_f \mathcal{P}_p \omega_0 \tag{5.81}$$

and for the shunt field,

$$V_T = I_f R_f + L_{ff}\frac{dI_f}{dt} \tag{5.82}$$

where R_f includes the resistance of the field winding and that of the field rheostat. Equation 5.81 is the nonlinear relationship represented by the open-circuit characteristic. The speed at which the generator is being driven is ω_0. When the field switch is first closed, $V_T = E_R = L_{ff}(dI_f/dt)$, and because dI_f/dt is positive, the field current begins to increase. As it does so, V_T increases.

Solving Equation 5.82 for the inductive voltage,

$$L_{ff}\frac{dI_f}{dt} = V_T - I_f R_f \tag{5.83}$$

The straight line

$$V_T = I_f R_f \tag{5.84}$$

also shown in Figure 5.40, is called the *field resistance line*. Comparing Equations 5.83 and 5.84, it is seen that the $L(dI_f/dt)$ voltage, as I_f increases, is the difference between the field resistance line and the saturation curve. When the field current reaches a value corresponding to the intersection of the field resistance line with the open-circuit characteristic, $L(dI_f/dt)$ is zero, and there will be no further increase in I_f. Buildup has ceased, and the point of intersection determines the no-load voltage of the generator. Since R_f is the slope of the field resistance line, the point of intersection, and hence the no-load voltage, may be adjusted by the field rheostat. If the rheostat is set so that the slope is above the *critical resistance* shown in the figure, buildup will not occur. *In the case of a series generator*, the *load* resistance must be below the critical value before buildup takes place.

Generator Characteristics

As the load on a *shunt generator* increases, the armature $I_a r_a$ drop increases. This has a double effect. The terminal voltage drops in accord with Kirchhoff's law,

$$V_T = E_g - I_a r_a$$

and E_g is reduced as I_a increases, because the field current depends on V_T:

$$I_f = \frac{V_T}{R_f} \tag{5.85}$$

(in the steady state, $dI_f/dt = 0$). Figure 5.41 shows a typical *external characteristic* of a self-excited shunt generator. The voltage regulation is too poor for most applications. Of course, the field rheostat may be adjusted manually to obtain the desired terminal voltage at a given load.

Connections for compound operation of generators are shown in Figure 5.11. A *cumulative-compound generator* is capable of much better voltage regulation. Since the load current flows through the series field, the pole MMF actually increases with load, and so does the generated voltage E_g.

If the increase in E_g is not sufficient to overcome the *IR* drops in the armature and series field resistances at full load, there will be a positive voltage regulation and the machine is said to be *undercompounded*.

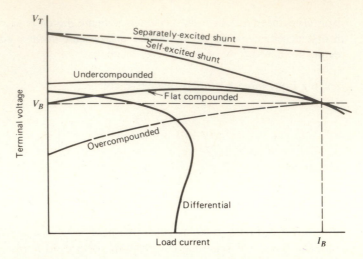

FIGURE 5.41 External volt/ampere characteristics of separately excited dc generators.

If the increase in E_g exactly compensates for the armature and series field IR drops at full load, the voltage regulation will be zero, and the generator is said to be *flat-compounded*.

If the terminal voltage at full load is greater than that at no load (negative regulation), the generator is *overcompounded*.

These characteristics are illustrated in Figure 5.41. An overcompounded generator is useful in mining applications, where the load may be at some distance from the generator. The rising voltage characteristic of the generator may be made to compensate for the IR drop in the line, so that the voltage at the load may be virtually constant.

A *differential-compound* generator is often used in power shovel drives. It has a nearly constant-current output over a part of its characteristic, also shown in Figure 5.41. When connected to a shunt motor in a Ward–Leonard system, a constant torque system results. This is an example of negative current feedback.

5.17 ARMATURE WINDINGS FOR COMMUTATOR MACHINES

The armature windings used as illustrations in this chapter have been *simplex lap* windings, because they are easily understood. There are, of course, other kinds of armature windings for use with commutators. The ones most widely used are:

Winding	Number of Paths
Simplex lap	$a = p$
Multiplex lap	$a = mp$
Simplex wave	$a = 2$
Frog leg	$a = 2pm_{\text{lap}}$

In the above table, m is the so-called "plex" of the winding. A duplex winding has a plex of 2. A triplex winding has a plex of 3, and so on.

Experience over many years has shown that, all things considered, it is impractical to design an armature winding capable of handling more than about 250 A per path. It will be shown that wave windings have many advantages over lap windings, but they are limited to two paths. They cannot be used in machines rated above 500 A, *but they are almost always used in machines of more than two poles having less than that current rating*.

The number of paths through a simplex lap winding is equal to the number of poles. A 10-pole, lap-wound machine may be designed to handle up to 2500 A, but might be limited to low speeds by voltage considerations. For design flexibility at high current ratings, multiplex lap windings are employed. The basic idea of duplex lap windings is that one complete simplex lap winding is installed, using every other commutator segment. Then another simplex winding is connected to the segments not used previously. For triplex windings every third segment is used for the first winding, and so on. A triplex lap winding for a ten-pole machine may be rated up to 6000 A. Multiplex wave windings have commutation difficulties and are not used.

Equalization Requirement in Lap Windings

Each path of a simplex lap winding is influenced by a different pair of poles. This will be evident by examining Figures 5.5 and 5.6. It is practically impossible to make the fluxes of all poles identical. If they were identical in a particular machine at the time of manufacture, bearing wear will eventually allow the armature to sag toward the poles in the lower half of the machine, shortening the air gaps of these poles and increasing their strengths relative to those of the upper half. The result of uneven pole strength is that the voltages induced in the parallel paths are unequal. Since the paths are in parallel, circulating currents flow in the winding. The internal resistances of the paths are so low that a 4 percent difference in voltage between two paths will cause a circulating current approximately equal to the rated current of the winding! With this degree of unbalance, any load connected to the machine would cause it to overheat. This intolerable situation is corrected in lap windings by making heavy connections between those coils that are $n \cdot 360$ electrical degrees apart (n an interger) and thus should be at the same potential. These connections are called *equalizers*. They are expensive, but necessary, in lap-wound machines. Not every coil needs to be connected to all other coils $n \cdot 360°$ away, but a large percentage must be so connected. These equalizing connections may be made at the commutator or may be made to leads brought out of the coils at the other end of the rotor.

The effect of equalizers is to strengthen weak poles and weaken strong ones, thus reducing the imbalance in voltages in the various paths. This action is illustrated in Figure 5.42. When coil sides 380 electrical degrees apart are moving under the centers of poles of unequal strength, currents circulate through the winding and equalizers that produce magnetic poles on the rotor surface in addition to the rotor poles that produce torque. Two typical coils involved in this process are shown in the figure. The magnetic effects of the equalizer component of the currents in the coil sides shown under the south field pole cancel. The net magnetic effect of the equalizer current in these two coils is like that of a single coil spanning 360 electrical degrees. At the instant illustrated, the axis of this 360° coil lies under the center of the south field pole. If the two north field poles were of equal flux density, the voltages induced in the two coils would be equal; and since they oppose each other in the equalizer circuit, the equalizer current would be zero. However since the left-hand north pole is stronger, $|d\lambda/dt|$ in the left coil is greater than that of the coil being influenced by the weak

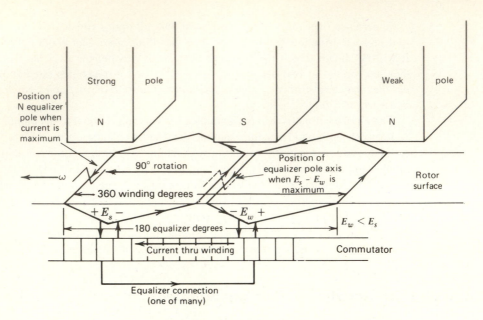

NOTE: Current in equalizer circuit does not reach its maximum until 90
        equalizer-circuit electrical degrees later than the time of the above
        picture, that is, when the N pole of the equalizer current pattern
        is in opposition to the strong N field pole.

FIGURE 5.42 Action of equalizers in compensating for unequal field pole strengths.

north pole. In other words, E_s, the voltage of the left coil, is greater than E_w, the voltage being induced in the right coil. The voltage applied to the equalizer circuit is the difference between these two. It is an alternating voltage of half the frequency of the normal coil voltages. However it *is* an ac voltage, and the circuits through which the equalizer currents flow are highly inductive. For this reason, the currents, and thus the equalizer pole MMFs, are strongest about 90 equalizer circuit degrees after the unbalance voltage is maximum. This brings the magnetic poles due to equalizer currents directly beneath the field poles of unequal strength. In the figure, a north equalizer pole is at its peak strength when opposite a strong north field pole. At the same time, a south equalizer pole will appear under the weak north field pole, strengthening its flux. The result is that the imbalance between the poles is reduced to that necessary to cause sufficient magnetizing current to flow in the equalizer circuits.

Wave Windings

Figure 5.43 shows the features of a simplex wave winding for a four-pole machine. Note that there are two coils in series between adjacent commutator segments and that all four poles are involved in inducing the voltage between segments. There can be no voltage unbalance between paths, because the voltage of each path is the sum of voltages induced by all poles. If there are more than four poles, the number of coils between segments is $p/2$.

The lead from the second coil of the two-coil series circuit may be connected to the commutator segment ahead of or the segment behind the one to which the lead of the first coil is attached. If it is connected to the segment ahead (leads cross), the winding is said to be *progressive*; if to the one behind, it is *retrogressive*. Figure 5.43 shows a progressive winding. The choice between these two connections determines

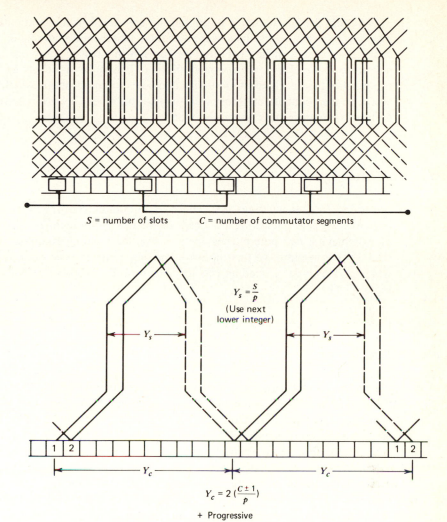

S = number of slots C = number of commutator segments

$$Y_s = \frac{S}{p}$$
(Use next
lower integer)

$$Y_c = 2\left(\frac{C \pm 1}{p}\right)$$

+ Progressive
− Retrogressive

FIGURE 5.43 Features of a simplex wave winding.

the relative polarity of the brushes, but the performance of the machine is not otherwise affected.

A simplex wave winding has only two parallel paths, regardless of the number of poles. Let C be the number of commutator segments. Since there are two leads to each coil, and two leads connected to each segment, the number of coils in the winding equals the number of segments. The brushes are one pole pitch apart. Then the number of segments between brush centers is C/p. It has been pointed out that the number of coils in series between segments is $p/2$. Thus the number of coils used in connecting that part of the winding from one brush to the next of opposite polarity is

$$\frac{C}{p} \cdot \frac{p}{2} = \frac{C}{2} \text{ coils} \tag{5.86}$$

or half the coils. In going from the second brush to the next brush (which would have the same polarity as the first brush), the other half of the coils would be used. Thus there can be only two paths.

Note that the leads from a given coil are connected to the commutator at points two brush pitches apart. In a wave winding, those coils undergoing commutation connect all brushes of like polarity together. For this reason, only one brush of each polarity is required in a wave-wound machine. It may be an advantage to have only two brushes in a machine of more than two poles, in applications which prevent easy access to the commutator. Most wave-wound machines, however, have a brush for each pole to reduce the current per brush.

Frog Leg Windings

Equalizers for lap windings are connected to commutator segments 360 electrical degrees (two pole pitches) apart. Leads of wave-winding coils are also connected to the commutator two pole pitches apart. A frog leg winding combines a lap winding with a wave winding, so that the wave-winding coils serve as equalizers for the lap winding. This type of winding gets its name from the shape of its coils illustrated in Figure 5.44.

5.18 METADYNES

Metadynes form a class of dc machines that use the armature MMF as a magnetizing field. There are a very great many possible metadynes. Most serve as control generators with special characteristics to meet particular needs. A wide variety of metadynes has been used in Europe, but only two have found widespread use in the United States. These are the Rosenburg generator and the amplidyne.

The Rosenburg generator, now obsolete, was used at one time in railroad passenger cars. It was driven by the car wheels and provided constant voltage for lighting and charging batteries over a wide range of speeds. The polarity of the voltage generated was independent of the direction of shaft rotation. This allowed the cars to be coupled into the train with either end toward the engine.

Amplidynes are two-stage rotary amplifiers. At this time, they are being displaced by solid-state power amplifiers, although thousands are still in service and a few continue to be manufactured. Power gains as high as 10^5 are possible, but the response

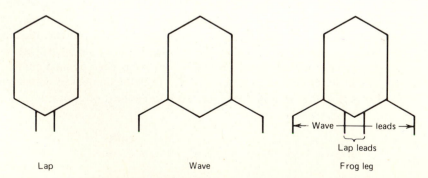

Lap   Wave   Frog leg

FIGURE 5.44 Coil shapes for dc armature windings.

time of high-gain machines may be several seconds. A more practical gain is 10^4 (80 db), with a bandwidth extending from zero to 3 Hz.

The Amplidyne

Suppose the brushes of a conventional dc shunt-wound generator were shorted together and the machine driven at constant speed. Then a very small field current would cause rated current to flow in the armature.[1] The peak armature MMF field F_{Aq} (see Figure 5.25a) would lie halfway between the field poles (i.e., along the q axis 90 electrical degrees from the stator field). Now suppose another set of stator poles to be installed between the original poles, so that the armature MMF could set up a large flux. (The interpoles would have to be removed, and since the stator field flux would be quite small, the original poles could be narrowed to admit large new poles between them.) The result of shorting the brushes and installing these additional poles would be that a very weak stator field magnetomotive force would produce a very strong flux along an axis displaced by 90 electrical degrees from the stator field axis, that is, the q axis.

This armature-induced flux would induce a voltage in the armature, much larger than that induced by the stator field, and this voltage can be utilized by placing another set of brushes on the commutator, halfway (90 electrical degrees) from the shorted set. However if a load is connected to these brushes, the load currents in the armature conductors cause a second armature reaction field halfway between the new poles, that is, along the axes of the original field poles. As luck would have it, this field is in opposition to the weak field set up by the shunt field windings, with the result that connecting even a very small load to the new brushes would cause the terminal voltage to drop nearly to zero. If the new poles are provided with compensating windings, however, this second armature field may be nullified, so that the stator field can regain control. The resulting machine is an *amplidyne*. It has two stages of amplification:

1. From the weak stator field current to the moderately strong current in the shorted brushes.
2. From the moderate current and low voltage in the shorted brush axis to the much higher voltage and higher current in the load brush axis.

Figure 5.45 shows the basic magnetic geometry of the amplidyne. This figure requires some comment. Turn back to Figures 5.5 and 5.6, and note the position of the brushes. Their mechanical location is on the axes of the stator poles. But note that they are connected to conductors halfway between the poles. Their correct mechanical position depends on how the leads are brought down from the coils, but their *electrical* position is halfway between the poles. In Figure 5.45, the stator field pole (i.e., the control field pole) axis is defined as the direct, of d axis, while the axis halfway between these poles is called the quadrature, or q axis. In this figure, the brushes are shown in their *electrical* positions. This practice simplifies analysis, because the *electrical brush axis coincides with the axis of the armature MMF pattern*. But remember

---

[1]The armature resistance is about 0.05 per unit, so E_g would only have to be 0.05 per unit to cause rated current to flow. The field current for rated E_g is about 0.02 per unit. Thus for rated current in the shorted brushes, the field current would be about 0.001 per unit.

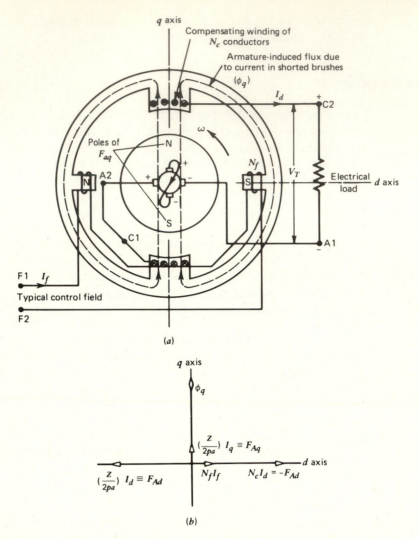

FIGURE 5.45 Amplidyne diagrams, part I. (a) Magnetic geometry. (b) Magnetic fields as vectors.

when examining an actual machine that the brushes are usually half a pole pitch away from where they often appear in schematic diagrams.

The control fields (see Figure 5.46a) may have differing numbers of turns to match the impedance characteristics of the control signal sources. The total d-axis MMF F_d is the sum of the armature field F_{Ad} due to the load current $(Z/2pa)I_d$, the compensating-winding MMF N_cI_d, and the control field MMFs, as shown in Figure 5.45b. The compensating field MMF is equal and opposite to F_{Ad}. The d-axis MMF relations are

$$F_{Ad} = \frac{Z}{2pa} I_d \text{ (see Equation 5.58)} \tag{5.87}$$

$$N_cI_d = -F_{Ad} \tag{5.88}$$

$$F_d = F_{Ad} + N_cI_d + N_{f1}I_{f1} + N_{f2}I_{f2} + \cdots$$

$$= N_{f1}I_{f1} + N_{f2}I_{f2} + \cdots \tag{5.89}$$

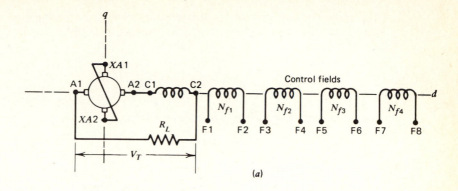

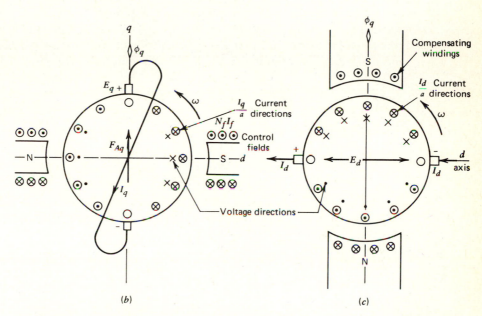

FIGURE 5.46 Amplidyne diagrams, part II. (*a*) Schematic diagram. (*b*) Control field MMF controls I_q and q-axis armature MMF F_{Aq}. (*c*) F_{Aq} induces q-axis flux ϕ_q, ϕ_q induces E_d. Armature d-axis MMF F_{Ad} is balanced by compensating winding.

So the net d-axis MMF is due to the control fields alone. This net d-axis MMF acts on the permeance of the d-axis magnetic circuit to produce the d-axis flux:

$$\phi_d = \mathscr{P}_d F_d \tag{5.90}$$

This flux induces the q-axis voltage in the circuit of the shorted brushes:

$$E_q = K_a \phi_d \omega \tag{5.91}$$

The only resistance in the q-axis circuit is the armature winding and brush resistance r_a. Then the current in the shorting lead is

$$I_q = \frac{E_q}{r_a} \tag{5.92}$$

The corresponding currents in the armature conductors produce an armature MMF field along the *electrical* brush axis; that is the q axis, designated F_{Aq}:

$$F_{Aq} = \frac{Z}{2pa} I_q \qquad (5.93)$$

which induces *the major flux in the machine*, ϕ_q:

$$\phi_q = \mathscr{P}_q F_{Aq} \qquad (5.94)$$

The production of this q-axis flux is illustrated by Figures 5.46b and 5.46c. The large q-axis poles make $\mathscr{P}_q$ large. The q-axis flux induces a set of voltages in the armature coils displaced 90 electrical degrees in space from those in which E_q was induced, as shown in Figure 5.46c. (Actually, most of the coils at any instant will be involved in phenomena in both axes, and d and q quantities will add in some quadrants of the armature and will subtract in others). This d-axis voltage is given by

$$E_d = K_q \phi_q \omega \qquad (5.95)$$

and the terminal voltage is

$$V_T = E_d - I_d r_{ad} \qquad (5.96)$$

where r_{ad} must include the resistance of the compensating windings. The student may wish to combine the above equations to obtain an equation for the transfer resistance for a typical control field:

$$R_{T1} = \frac{E_d}{I_{f1}}$$

If the resistance of a control field winding is known (r_{f1}), the voltage gain may be calculated:

$$\mu_1 = \frac{E_d}{r_{f1} I_{f1}} \qquad (5.97)$$

and the power gain may be calculated for a given load resistor:

$$G = \frac{V_T^2/R_L}{I_{f1}^2 r_{f1}} \qquad (5.98)$$

where

$$V_T = E_d \left(\frac{R_L}{R_L + r_{ad}} \right) \qquad (5.99)$$

The amplidyne does not give something for nothing, but merely controls the power flow from the mechanical power source coupled to its shaft to the electrical terminals A1 and C2. The q-axis current in the shorted brushes requires a small torque to be produced by the mechanical drive:

$$\tau_{dq} = K_a \phi_d I_q = \frac{E_q I_q}{\omega} \qquad (5.100)$$

This is small because ϕ_d is small, and $\omega \tau_{dq}$ is just enough to provide for the $I_q^2 r_a$ losses. The load current, however, produces the major countertorque:

$$\tau_{dd} = K_a \phi_q I_d = \frac{E_d I_d}{\omega} \qquad (5.101)$$

and $\omega \tau_{dd}$ supplies the load power plus the d-axis copper losses.

A Constant-Current Metadyne

As an example of another possible metadyne, consider the circuit shown in Figure 5.47. The current flowing in the q-axis circuit is given by:

$$-I_q = \frac{V_T - E_q}{r_{aq}} \qquad (5.102)$$

where r_{aq} is the armature winding resistance between q-axis brushes and E_q is the q-axis voltage induced by the d-axis flux. Since there are no windings on the stator, the d-axis flux is induced by the d-axis armature MMF. Let $\mathcal{P}_d$ be the permeance of the metadyne's magnetic circuit along the d axis. The armature MMF along this axis is given by

$$F_d = \frac{Z}{2pa} I_d \qquad (5.103)$$

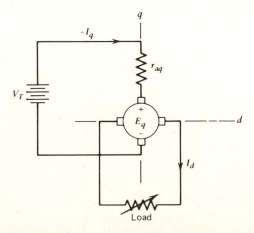

FIGURE 5.47 Constant-current metadyne with no stator windings.

and the d-axis flux is

$$\phi_d = F_d \mathcal{P}_d = \frac{Z\mathcal{P}_d}{2pa} I_d$$

$$= \frac{\pi}{p^2} K_a \mathcal{P}_d I_d \tag{5.104}$$

The voltage induced by this flux into the q axis is

$$E_q = K_a \phi_d \omega = \frac{\pi}{p^2} K_a^2 \omega \mathcal{P}_d I_d$$

$$\overset{\Delta}{=} K_m \mathcal{P}_d \omega I_d \tag{5.105}$$

where

$$K_m = \frac{\pi K_a^2}{p^2} \tag{5.106}$$

Substituting this expression for E_g into Equation 5.102 gives

$$-I_q = \frac{V_T - K_m \omega \mathcal{P}_d I_d}{r_{aq}} \tag{5.107}$$

Rearranging,

$$I_d = \frac{V_T + I_q r_{aq}}{K_m \omega \mathcal{P}_d} \tag{5.108}$$

If the metadyne is driven at constant speed and the machine remains unsaturated so that $\mathcal{P}_d$ is constant, the denominator of Equation 5.108 will be constant. Then if r_{aq} is quite small,

$$I_d \cong \frac{V_T}{K_m \omega \mathcal{P}_d} \tag{5.109}$$

which shows that the load current will be nearly constant, and its magnitude will depend on the voltage applied to the q-axis brushes.

PROBLEMS

5.1.   List the parts of the magnetic circuit of a dc machine through which the pole flux flows in completing its path.

5.2.   State the purposes of the following:

(a) Armature winding
(b) Armature core
(c) Field winding
(d) Pole shoes

(e) Pole core

(f) Commutator

5.3. What things distinguish motor action from generator action in a dc machine?

5.4. The coils in a certain dc machine have three turns each. The stator poles each produce a flux of 0.01 Wb. When a coil encloses the flux of a south stator pole, what is its flux linkage.?

5.5. (a) In Problem 5.4, what is the change in flux linkage of the coil, $\Delta\lambda_c$, as the armature rotates an angle equal to one pole pitch?

(b) What is the sign of this change if a south stator pole is taken as producing positive flux in the air gap?

(c) In what length of time must this rotation take place if the dc value of the commutated coil voltage is to be 5 V?

(d) At what speed, in rev/min, must the machine rotate if it has four poles?

5.6. A dc machine has eight poles. What is the pole pitch ρ_p in mechanical degrees? At what speed, in rev/min, must this machine be driven so that its coils move one pole pitch in 4.17 ms?

5.7. A certain six-pole dc machine has 95 commutator segments and has a simplex lap winding such as that of Figure 5.6. (a) How many coils are there in the winding? (b) How many current paths are there?

5.8. A four-pole dc machine has a flux per pole of 0.04 Wb. Each coil has one turn. The number of parallel paths through the winding is 2. If the speed of rotation is 125 rad/s, what is the average commutated voltage generated by each coil?

5.9. In the machine of Problem 5.8 there are four brushes, and each brush shorts two coils. How many coils must there be in the winding if the machine is to generate 250 V?

5.10. A four-pole dc machine has a normal flux per pole of 0.01 Wb. It has 75 coils in its armature of four turns each. The armature is wave-wound ($a = 2$). What voltage is generated at a speed of 1200 rev/min?

5.11. What are the armature constants K_a and K'_a for the machine of Problem 5.10?

5.12. A 1000-hp, 500-V dc motor has an armature resistance of 0.007 Ω. The armature current is 1550 A when delivering 1000 hp at 750 rev/min. (a) Find E_g and the developed torque. (b) What is the armature copper loss?

5.13. What are the rotational losses in the machine of Problem 5.12 under rated conditions?

5.14. A 750-kW, 500-V dc generator has a terminal voltage of 500 V at rated current. It is being driven at a speed of 450 rev/min. The armature resistance is 0.007 Ω. (a) What is E_g under rated conditions? (b) What torque is developed? What is the armature copper loss?

5.15. The generator of Problem 5.14 has a rotational loss of 12,180 W. How many watts of mechanical power are required to drive the generator? How many horsepower? What is the shaft torque?

5.16. Find $K'_a\phi$ for the machine of Problem 5.14.

5.17. A 7.5-hp dc motor has a shunt field inductance of 100 H. How much energy is stored in its field when the field current is 0.80 A? This energy is sufficient to operate a 6-W lamp for how long?

5.18. The field winding of the machine of Problem 5.17 has a resistance of 75 Ω. What is the time constant of the winding?

5.19. What are the purposes of a field discharge resistor?

5.20. Draw the circuit diagram for a long-shunt, stabilized-shunt motor. Why are stabilized-shunt motors made?

5.21. If the machine of Problem 5.12 is delivering 1000 hp at rated voltage and current at a speed of 750 rev/min, find $K_a\phi$ and $K'_a\phi$.

5.22. The saturation curve for a dc machine is given in Figure 5.13. Plot a curve of $K_a\phi$ as a function of shunt field amperes, assuming zero series field current.

5.23. For the machine of Figure 5.13, find $K_a\phi$ for zero series field current if $I_f = 1.60$ A.

5.24. In the machine of Figure 5.13, what is $K_a\phi$ if the series and shunt fields aid each other, and $I_f = 1.20$ A and $I_{se} = 50.0$ A?

5.25. What is the MMF per stator pole in Problem 5.24?

5.26. A machine is operating under conditions such that $K_a\phi$ is 0.96. The developed torque is 80 N-m. The armature circuit resistance is 0.12 Ω. If the terminal voltage is 250 V, find the speed of this machine when operating (a) as a motor, (b) as a generator.

5.27. Find the speed of the machine of Example 5.3 for a field current of 0.5 A, for the same developed torque.

5.28. The machine of Figure 5.13 is operated as a long-shunt, cumulative-compound motor connected to a 115-V dc line. The field rheostat is adjusted so that the speed of the motor is 1200 rev/min at an armature current of 50 A. There is no diverter. (a) Calculate E_g. (b) From the saturation curve, find the effective field current. (c) What is $(N_{se}/N_f)I_a$? (d) What is I_f? (e) What rheostat resistance is required?

5.29. In Problem 5.28, assume the no-load current to be 5.0 A. (a) With the motor connected to the 115-V line, what is E_g at no load? (b) What is the *effective* field current at no load if I_f remains the same as in the previous problem? (c) From the saturation curve, Figure 5.13, what would E_g be at 1200 rev/min with this excitation? (d) What is $K'_a\phi$ at no load? (e) What is the no-load speed? (f) What is the speed regulation?

5.30. Assume that the accompanying saturation curve is for a 25-hp, 230-V dc motor having a rated armature current of 90.0 A. It is cumulative-compound, with a N_{se}/N_f ratio of 0.002. The armature circuit resistance is 0.15 Ω, including the series field. What field current is required to make the speed 1800 rev/min at an armature current of 90.0 A? What are the developed torque and horsepower under these conditions? Approximately what would be the no-load speed? (Assume $I_a = 0$ at no load.)

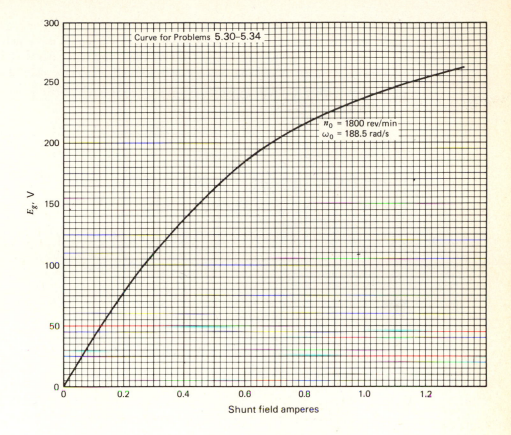

Curve for Problems 5.30–5.34

n_0 = 1800 rev/min
ω_0 = 188.5 rad/s

E_g, V

Shunt field amperes

5.31. A 25-hp, 230-V dc shunt motor has an armature circuit resistance of 0.20 Ω and a shunt field winding resistance of 216 Ω. The saturation curve for this machine is shown above. If the load is such that the armature current is essentially constant at 60 A, find the speed range possible with a 500-Ω field rheostat.

5.32. For the motor of Problem 5.31, what resistance must be inserted in series with the armature to obtain a speed of 1200 rev/min at an armature current of 60 A and zero field rheostat resistance? Assume the line voltage to be 230 V.

5.33. In Problem 5.32, calculate the power lost in the speed-reducing resistor. If the stray power loss is 450 W and the stray load loss is 150 W, calculate the overall efficiency of the system.

5.34. The motor of Problem 5.31 has 175 V supplied to the shunt field while its armature is fed by a chopper. The amplitude of the pulses applied to the armature terminals is 288 V. The pulse frequency is 60 Hz. Find the speed of the machine when the pulse width is 10 ms and the average armature current is 40 A.

5.35. The Ward-Leonard system shown below employs two dc machines described by Figure 5.13, one used as the motor and the other as a generator. The field current of the motor is supplied by a rectifier and is maintained constant at 0.600 A. The mechanical load characteristic is such that the armature current is 50.0 A at speeds of ±1800 rev/min. The induction motor drives the generator at 1150 rev/min. What is the range of control voltages that must be supplied to the generator shunt field winding to drive the load at any speed over the range of ± 1800 rev/min? The series fields of the machines are not used. What power is developed by the motor at 1800 rev/min and what is the input power to the control field at this speed?

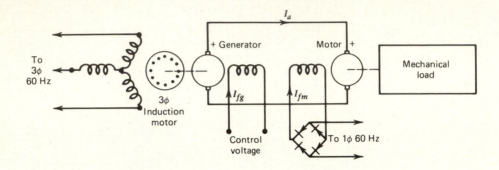

5.36. Calculate R_1 and R_2 for the dc motor starter of Figure 5.20 if this starter is to be used with the motor of Figure 5.13. (a) Assume $R_a = 0.06$ per unit. (b) Assume the actual per-unit value of R_a. (Recall that $R_a = r_a + r_{se}$). The rated terminal voltage of this machine is 115 V.

6
SINGLE-PHASE MACHINES

6.1 WHY SINGLE-PHASE MOTORS ARE DIFFERENT

Most homes and rural areas are supplied with single-phase ac electrical power. Consequently, most electrical home appliances and electrically driven farm machines employ single-phase motors. They drive washing machines, refrigeration and air-conditioning compressors, grain dryers, fans, pumps, sewing machines, vacuum cleaners, clocks, phonographs, hand tools, and so on. Literally millions of single-phase motors are manufactured each year.

The difficulty with single-phase as a power source for motors is that it does not lend itself to producing a rotating magnetic field. Figure 2.17 shows the magnetic field set up by a single phase winding. It "breathes"; that is, it does not move around the air gap but remains stationary as it oscillates in magnitude and polarity. This pulsating field, acting alone, does not produce starting torque.

Several schemes have been developed to circumvent this difficulty. Each results in a motor with specific characteristics suitable to a certain range of uses. This chapter discusses the most important of these motor types.

6.2 THE UNIVERSAL MOTOR

Perhaps the most obvious solution to the single-phase problem is to design a dc motor so that it will run well on ac. The direction of the torque developed by a dc machine depends on the direction of current flow in the armature conductors and the polarity of the field magnets. If a dc machine is designed so that (1) when the line current reverses direction, the field and armature currents reverse simultaneously, and (2) the core loss with alternating flux is relatively low, then a successful single-phase machine results.

The first of these criteria is met by connecting the armature and field windings in series. The second is achieved by laminating the stator core. (A shunt machine does not perform well on ac because the field circuit inductance is larger than that of the armature. This causes the field pole reversals to be out of phase with the current reversals in the armature. The result is that the torque is backward during part of each half cycle, lowering average torque and reducing the efficiency.)

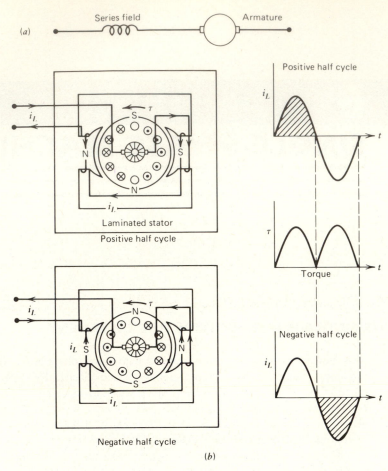

FIGURE 6.1 The universal motor. (*a*) Circuit diagram. (*b*) Principle of operation.

A series dc motor designed to operate also on ac is called a *universal motor,* because it will run successfully on any frequency from dc up to its design frequency. The top design frequency for universal motors is 60 Hz. Figure 6.1 shows the principle of operation. A photograph of a universal motor for a vacuum cleaner is shown in Figure 6.2. When operated on ac, this kind of motor develops its torque in pulses—one pulse each half cycle.

The speed/torque characteristic of a universal motor is shown in Figure 6.3. It is typical of a series motor. (Compare with Figure 5.33*b*.) No-load speed is quite high, often in the range of 20,000 rev/min. It is limited by windage and friction. Having high speed capability, universal motors of a given horsepower rating are significantly smaller than other kinds of ac motors operating at the same frequency. Their starting torque is relatively high. These characteristics make universal motors ideal for devices such as hand drills, hand grinders, food mixers, routers, vacuum cleaners, and the like, which require compact motors operating at speeds greater than 3000/3600 rev/min. (i.e., greater than the maximum synchronous speed available at 50/60 Hz.)

6.3 SINGLE-PHASE INDUCTION MOTORS

The numerous varieties of single-phase induction motors make it impossible to provide a general treatment that applies to all types. This section introduces the broad class

(a)

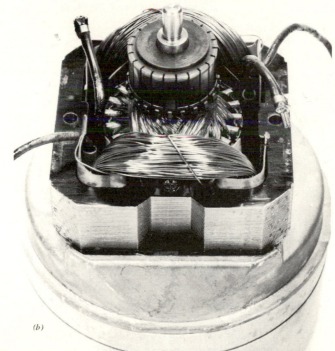

(b)

FIGURE 6.2 A universal motor from a vacuum cleaner. (Courtesy of AMETEK, Lamb Electric Division.) (a) Cover removed to show brushes. (b) Brush assembly removed to show field poles and armature.

of single-phase motors that develop starting torque by means of two stator windings. Included are the split-phase, capacitor start, two-value capacitor, and permanent-split capacitor motors. Following the sections on these motors, material is provided on shaded-pole and repulsion induction motors.

The operation of single-phase induction motors is quite similar to that of two-phase

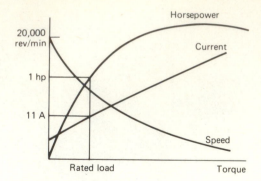

FIGURE 6.3 Characteristics of a 115-V universal motor at 60 Hz.

motors. Two-phase motors have two phase windings located in slots in the stator core, which are displaced from each other by 90 electrical degrees. These windings are supplied by a source that provides two equal phase currents 90° out of phase with each other. Such an arrangement produces a rotating magnetic field of constant ampere-turn amplitude. This is evident when one considers that the MMF of one phase would be

$$F_a = N_a i_a \cos \theta_1 = N_a \sqrt{2}I_\phi \cos \omega t \cos \theta_1 \tag{6.1}$$

while that of the other would be

$$F_b = N_a i_b \cos(\theta_1 + 90°) = N_a \sqrt{2}I_\phi \cos(\omega t + 90°) \cos(\theta_1 + 90°)$$

$$= -N_a \sqrt{2}I_\phi \sin \omega t(-\sin \theta) = N_a \sqrt{2}I_\phi \sin \omega t \sin \theta \tag{6.2}$$

The total field of a two-phase motor is, then,

$$F_a + F_b = N_a \sqrt{2}I_\phi(\cos \omega t \cos \theta_1 + \sin \omega t \sin \theta_1)$$

$$= N_a \sqrt{2}I_\phi \cdot \tfrac{1}{2}[\cos(\omega t + \theta_1) + \cos(\omega t - \theta_1) + \cos(\omega t - \theta_1)$$

$$- \cos(\omega t + \theta_1)] \tag{6.3}$$

$$= N_a \sqrt{2}I_\phi \cos(\omega t - \theta_1)$$

Equation 6.3 is the expression for a traveling MMF wave having angular velocity of ω electrical radians per second and a constant amplitude of $\sqrt{2}I_\phi$ A-turns.

Single-phase motors are provided with two phase windings, called the *main* and *auxiliary* windings, which have coil groups displaced 90 electrical degrees from each other. The auxiliary winding is often called the *aux*, or *phase*, winding. The main winding is also called the *run*, or *running*, winding. In a single-phase motor, these two windings generally do not have the same number of turns. The currents in the two windings are unequal and are not necessarily 90° out of phase. Thus the performance is that of an *unbalanced* two-phase motor.

Figure 6.4 shows how the main and auxiliary windings are placed in the slots of the stator core of a four-pole, single-phase motor. In the foreground are the coils of the main winding. The winding for each pole consists of one or more (in this case, three) concentric coils. Such an arrangement is called a spiral, or *concentric,* winding. The coils for this particular motor are wound of enameled aluminum wire. The coils

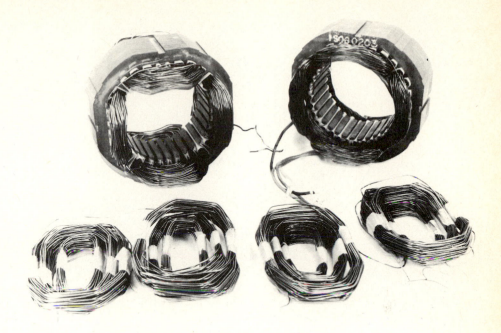

FIGURE 6.4 Stator windings for a four-pole, single-phase induction motor. (Photo by the author. Components courtesy of Emerson Electric Company.)

are wound so that the magnetic polarity of the coil groups is alternately N, S, N, S for a given instantaneous current direction.

In the background of Figure 6.4 are two laminated stator cores. The one on the left has been partially wound and contains a main winding exactly like that of the coils in the foreground. The right-hand stator is complete. The aux winding has been added. Note that this winding occupies less volume, and is displaced $\frac{1}{2}$ pole pitch (90° electrical) from the main winding.

Once the motor gets started, it will run with only one winding connected the line. There are two theories as to how this happens. These are called the *cross-field theory* and the *counterrotating-field theory*. According to the cross-field theory, the *rotor* provides a magnetic field, displaced approximately 90 electrical degrees in both space and time from that of the remaining stator winding. The rotor thus substitutes for the missing stator winding in the development of a rotating magnetic field. Figure 6.5 illustrates this part of the theory.[1] It is assumed that the auxiliary winding has been disconnected and that only the main winding is connected to the single-phase line. When the main-winding current is at its peak, those rotor conductors directly beneath it experience maximum flux, and their velocity relative to that flux causes maximum voltage to be induced in them. However the high permeance of the magnetic circuit embraced by the rotor bar pairs that happen to be under the main winding at any given time cause the currents in these bars to lag the speed voltages by nearly 90°. Thus the currents in these bars are near their maximum when the current in the main winding is zero (time $= t_2$ in Figure 6.5). This produces a field 90° ahead of that of the main winding in the direction of rotation, as shown in Figure 6.5*b*. At time t_3, the main-

---

[1] The cross-field theory involves both speed and transformer voltages and has been worked out in great detail. Only one aspect is described here.

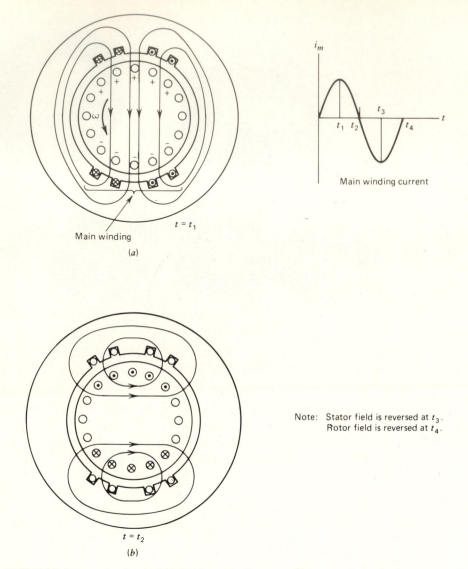

FIGURE 6.5 Rotor and main winding combine to produce a rotating magnetic field in a single-phase induction motor. (*a*) Rotor voltage induced by rotation through main-winding flux at t_1 (maximum main-winding current). (*b*) Rotor currents lag speed voltages by nearly 90°. Rotor cross flux is near maximum when main-winding MMF is zero.

winding field has reversed, and the rotor field is near zero. At t_4, the stator field is again zero and the rotor field has reversed. Note that the direction of rotation of the field is the same as that of rotor rotation.

Thus a single-phase induction motor will run in the direction in which it is started. The machine is started with both main and auxiliary windings connected. The auxiliary winding *may* be disconnected automatically as the operating speed is approached. The circuit of the auxiliary winding is designed so that its current leads that of the main winding. Thus the auxiliary-winding field builds up first. Ideally, the auxiliary current should lead the main-winding current by 90 degrees. Reversing the connections to either the main or the auxiliary winding will reverse the direction of *starting*.

In summation, single-phase induction motors will not start on their main windings alone but must be started by an auxiliary winding or other means. Once started, they run in the direction of starting. *Single-phase motors are classified according to starting method.*

6.4 THE SPLIT-PHASE MOTOR

Figure 6.6 shows the circuit diagram of the resistance split-phase motor, often simply called a *split-phase motor*. Figure 6.7 is a photograph of a cutaway motor of this type.

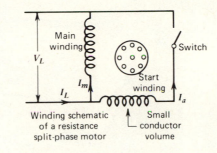

Winding schematic
of a resistance
split-phase motor

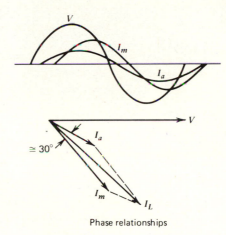

Phase relationships

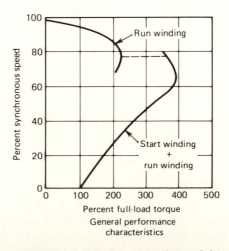

General performance
characteristics

FIGURE 6.6 Split-phase motor characteristics.

FIGURE 6.7 Cutaway split-phase motor. (Courtesy of General Electric Company.)

In this particular motor, as in most general-purpose, single-phase induction motors, disconnection of the auxiliary winding after starting is accomplished by a centrifugal switch. Occasionally a current-sensitive relay will be used, particularly with "hermetic" motors, which are sealed into refrigeration units, or with submersible-pump motors.

The auxiliary winding of split-phase motors has fewer turns than the main winding, and is wound of smaller wires. As a result, the X/R ratio is less, and the auxiliary-winding current is more nearly in phase with the line voltage than that of the main winding. The usual phase difference is 30° or less. This is not even close to the ideal of 90°, so the starting torque is relatively low. The small phase difference between the two winding currents makes the starting line current, which is the sum of the two winding currents, quite high. The *running* characteristics, however, are as good as those of any single-phase motor that operates on its main winding alone.

Split-phase motors are relatively inexpensive and are used to drive easily started loads, such as fans, saws, and grinders.

6.5 CAPACITOR MOTORS

Capacitor motors are single-phase induction motors that employ a capacitor in the auxiliary-winding circuit to cause a greater phase split between the currents in the main and auxiliary windings.

Capacitor Start Motors

Capacitor start motors have a capacitor connected in series with the auxiliary winding. As in the case of the resistance split-phase motor, this winding is disconnected as the motor comes up to speed.

The VAR rating required of the capacitor is such that, until the development of ac electrolytic capacitors in the 1930s, the cost of starting capacitors was prohibitive. Figure 6.8 shows the diagram and characteristics of the capacitor start motor. Note that the starting torque is much higher and the starting current much lower than those of the resistance split-phase motor. These desirable characteristics both result from the fact that the capacitor permits the auxiliary winding circuit to be designed for the ideal 90° current shift ahead of the main winding. The penalties are the cost of the capacitor and the slightly reduced reliability that results from the inclusion of another component.

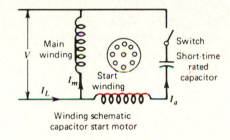

Winding schematic
capacitor start motor

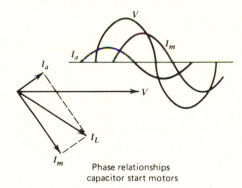

Phase relationships
capacitor start motors

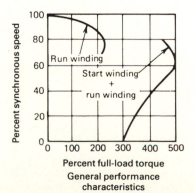

General performance
characteristics

FIGURE 6.8 The capacitor start motor.

Capacitor start motors are used for hard-starting loads, such as compressors, conveyors, pumps, and some machine tools.

Two-Value Capacitor Motors

The impedances of both the main and auxiliary windings vary with motor speed. By the use of two capacitors, however, it is possible to achieve balanced, two-phase operation of the motor at starting and at one other speed, if the auxiliary winding remains connected. The speed chosen would be close to that at which rated horsepower is developed, say, at about 80 percent of rated horsepower. The auxiliary winding must be designed for continuous operation.

The circuit for a two-value capacitor motor is shown in Figure 6.9. The two capacitors are connected in parallel at starting. The starting capacitor is almost always electrolytic. The running capacitor must handle alternating current continuously, and is usually of oil-filled paper construction. Running capacitors contribute considerably to the cost of these motors.

Two-value capacitor motors are quiet and smooth running. They have a higher efficiency than motors that run on the main winding alone. A still higher efficiency is possible if the capacitor is connected in series with the main winding rather than the auxiliary winding, but the capacitor required is very large and prohibitively expensive.

PSC Motors

The letters PSC stand for *permanent-split capacitor*. PSC motors have no starting switch. The capacitor is permanently wired in series with the auxiliary winding. The winding design and capacitor size are a compromise between cost, starting torque, and running characteristics. They are fairly quiet and have a relatively high efficiency and power factor. The elimination of the starting switch reduces the size and cost and improves reliability.

Figure 6.10 shows the circuit diagram and operating characteristics of a PSC motor. The breakdown torque is much lower than that of other motors. A high-resistance rotor will increase the starting torque but lower the running efficiency. The speed of PSC motors with high-resistance rotors may be controlled by taps on the windings,

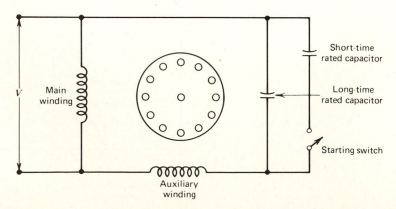

FIGURE 6.9 Schematic diagram of the two-value capacitor single-phase induction motor.

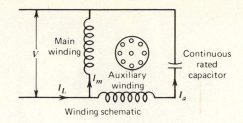

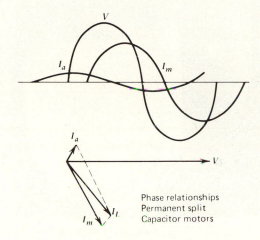

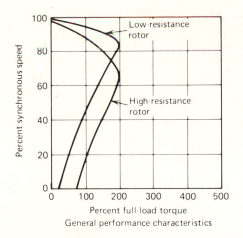

FIGURE 6.10 The PSC motor.

in a manner similar to that used with shaded-pole motors. PSC motors are used for fans and blowers in heaters and air conditioners, and to drive refrigerator compressors.

6.6 THE SHADED-POLE INDUCTION MOTOR

The shaded-pole motor is a special type of single-phase induction motor. Its operation differs from that of most single-phase motors in that the magnetic field produced by

FIGURE 6.11 A shaded-pole motor.

the stator *does* have a rotating component, without employing a centrifugal switch or external components. A typical two-pole, shaded-pole motor is shown in Figure 6.11. This motor was taken from an electric can opener. Note that the motor has a squirrel cage rotor and that parts of each salient stator pole are enclosed by heavy shorted, single-turn copper coils. These are called *shading coils*. Figure 6.12 shows steps in the construction of a shaded-pole motor having several poles. A typical application would be to drive a cooling fan in a computer.

The concept from which these motors derive their name is that the shorted turns that surround portions of each pole "shade" those parts from increasing flux, thus forcing the core flux to flow through the unshaded areas. The process is illustrated in Figure 6.13. If the coil resistance and leakage flux are neglected, the voltage induced in the exciting coil by the core flux is equal to the applied voltage. Taking the applied voltage to be a cosine wave,

$$\sqrt{2}V_T \cos \omega t = N \frac{d\phi}{dt} \qquad \text{(Faraday's law)}$$

Integrating,

$$\phi = \frac{\sqrt{2}V_T}{N\omega} \sin \omega t \qquad (6.4)$$

The flux variation over one cycle is divided into four periods of one-quarter cycle each, in Figure 6.13. By Lenz's law, currents flowing in the shading coils will always oppose changes in flux. During period 1, the flux is circulating counterclockwise and increasing. The MMFs of the shading coils oppose the flux, causing most of the flux to flow through the unshaded parts of the poles. During period 2, the flux is in the same direction, but is decreasing. Lenz's law currents in the shading coils reverse to try to prevent the collapse of flux. As a result, the shading-coil MMFs aid the coil MMF, and the major portion of the flux is drawn through the shaded parts of the poles. Note that the field passing through the rotor has rotated slightly.

During periods 3 and 4, the flux has reversed, and again rotates slightly. Figure 6.13b indicates that a nonuniformly rotating magnetic field passes through the rotor, producing a weak induction motor action. Note that rotation is in the direction from the unshaded toward the shaded part of the poles. Thus a shaded-pole motor can be reversed only by providing two sets of shading coils which may be opened and closed; or it may be reversed permanently by inverting the core.

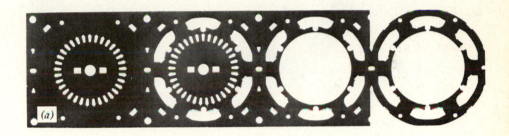

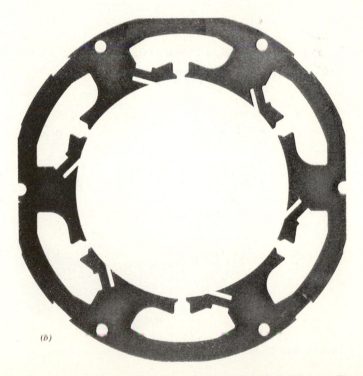

FIGURE 6.12 Construction of a multipole, shaded-pole motor. (Courtesy of Emerson Electric Company.(*a*) Stamping sequence for rotor and stator laminations. (*b*) Stator showing slots for shading coils.

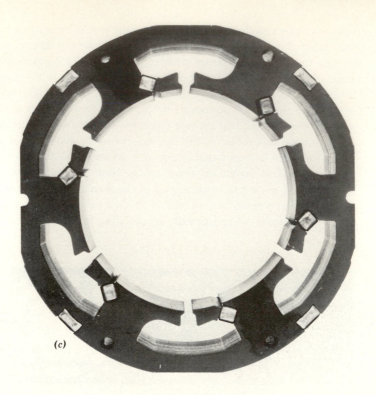

(c)

(d)

FIGURE 6.12 Continued. (c) Stator with shading coils installed. (d) Completed stator with rotor inserted.

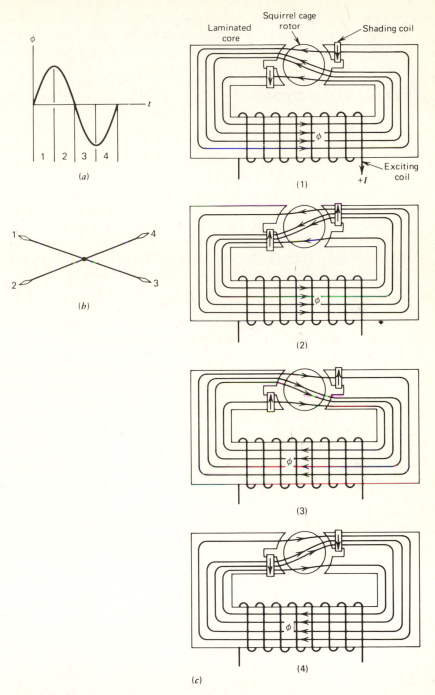

FIGURE 6.13 Operation of a shaded-pole motor. (*a*) Flux as a function of time. (*b*) Angles of flux passing through the rotor. (*c*) Flux pattern during the four parts of the flux cycle.

The speed of shaded-pole motors may be varied by changing the volts per turn, and thus changing the maximum flux. This is most economically accomplished by providing taps on the exciting coil. For constant supply voltage, Equation 6.4 shows that the *more* turns there are, the *less* will be the flux and the lower the speed.

Shaded-pole motors are rather weak and inefficient, but there is little to go wrong with them, and they are quite inexpensive. They are used to drive devices that require little starting torque, such as fans and phonographs. The shaded-pole principle is used in starting electric clocks and other single-phase synchronous timing motors.

6.7 REPULSION MOTORS

The repulsion principle is similar to that by which universal motors operate, the chief distinction being that the stator MMF induces the current in the repulsion rotor winding. Repulsion motors have excellent characteristics, but they are very expensive to manufacture. They have commutators and thus require more attention and maintenance than single-phase induction motors. They have been displaced by two-value capacitor motors (*q.v.*) for nearly all applications. A few continue to be made to drive hard-starting farm equipment, such as grain elevators and augers.

The rotor of a repulsion motor is just like that of a dc machine. The stator core is laminated to reduce eddy current loss, and its poles are excited by a winding connected to the single-phase line. Voltages are induced in the rotor coils by transformer action. Figure 6.14 shows the repulsion principle, using the polarity conventions adopted for metadynes in Chapter 5. The brushes are shifted from the position they would have in a dc machine. The repulsion motor runs in the direction opposite to that of the brush shift. With the brushes shifted, some rotor conductors act as a transformer secondary, the stator winding serving as the primary. The brushes are shorted together and, as a result, a heavy current flows in the rotor windings. The direction of the currents is such that the MMF of those conductors acting as the secondary opposes the MMF of the stator winding, as in a transformer. The rotor currents produce magnetic poles on the surface of the rotor on the electrical brush axis, as shown in the figure. Repulsion between the rotor poles and the stator poles produces the torque. As the line current I_1 alternates, the rotor and stator currents reverse almost simultaneously. As a result, the rotor and stator magnetic poles reverse every half cycle, producing torque pulses that are always in the same direction.

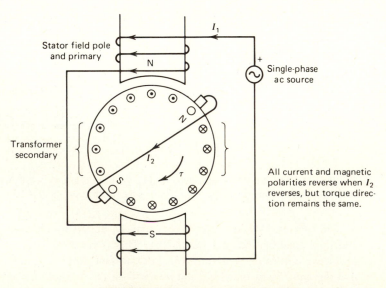

FIGURE 6.14 The repulsion motor principle.

The unmodified repulsion motor has characteristics of a series dc motor. However the repulsion principle is often used simply as a means of providing high starting torque at relatively low starting current. When acceleration is nearly complete, provision is often made to convert the machine to a single-phase induction motor to provide essentially constant-speed running characteristics. This may be done by installing a squirrel cage winding in the rotor in addition to the commutator winding (*repulsion–induction motor*), or by shorting out the commutator by a centrifugal-force-operated device (*repulsion-start induction-run motor*). Such motors are ideal for use in the rural areas that may be served by long, single-phase lines, because their high starting torque is delivered without excessive current, and their running speed is nearly constant over a wide torque range.

6.8 THE CIRCUIT MODEL OF SINGLE-PHASE INDUCTION MOTORS

The main and auxiliary windings have, in general, unequal numbers of turns and carry unequal currents. Their magnetic axes are, however, at right angles to each other, electrically speaking. They thus produce an unbalanced two-phase magnetic field. Figure 6.15 shows a schematic version of the situation. The terms N_m and N_a include pitch, distribution, and all other factors required to convert the rms current into the MMFs of the windings. The symbol I_m is a phasor representing $\sqrt{2}|I_m|\cos(\omega t + \theta_m)$, and I_a is a phasor representing $\sqrt{2}|I_a|\cos(\omega t + \theta_a)$. Thus if $\theta_m = 0$ and $\theta_a = 90°$, $|I_m| = |I_a| = I_\phi$, and $N_a = N_m$, then the conditions of Equations 6.1, 6.2, and 6.3 are obtained and a forward rotating field of uniform magnitude results. It is not necessary that the two currents be equal in magnitude, as long as $N_aI_a = jN_mI_m$; that is, as long as both windings have equal ampere-turns and the current in the auxiliary winding leads the main-winding current by 90°. Similarly, if $N_aI_a = -jN_mI_m$, a uniform backward-rotating field is obtained. Note that if F_a, the auxiliary winding MMF, lies 90° behind F_m, the main winding MMF, in *space*, a forward-rotating field is produced when F_a peaks first, that is, when F_a *leads* F_m in *time*.

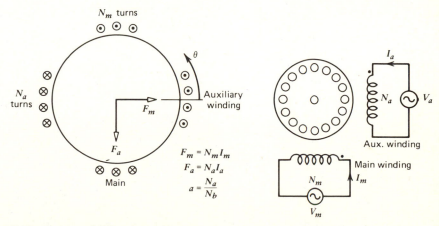

Note: F_a must peak before F_m to produce
a forward-rotating field.

FIGURE 6.15 Schematic representation of a single-phase or unbalanced two-phase motor.

Let the effective turns ratio of the two windings be defined as

$$a \triangleq \frac{N_a}{N_m} \qquad (6.5)$$

Then for a uniform forward field,

$$N_a I_a = a N_m I_a = j N_m I_m$$

or

$$I_a = j \frac{I_m}{a} \qquad (6.6)$$

Similarly, for a uniform backward field,

$$I_a = -j \frac{I_m}{a} \qquad (6.7)$$

An unbalanced two-phase set of quantities may be expressed as the sum of two balanced sets of opposite phase sequence, that is, by the method of symmetrical components. Thus as in Figure 6.16, the unbalanced fields of the main and auxiliary windings may be expressed in terms of the components of a forward-rotating field and of a backward-rotating field:

$$F_m = N_m I_m = F_f + F_b \qquad (6.8)$$

$$F_a = N_a I_a = j F_f - j F_b = a N_m I_a \qquad (6.9)$$

These equations permit the *currents* in the two windings to be analyzed into symmetrical components. Equations 6.8 and 6.9 may be written

$$N_m I_m = N_m I_{mf} + N_m I_{mb}$$

$$a N_m I_a = j N_m I_{mf} - j N_m I_{mb} \qquad (6.10)$$

or

$$I_m = I_{mf} + I_{mb} \qquad (6.11)$$

$$I_a = j \frac{I_{mf}}{a} - j \frac{I_{mb}}{a} \qquad (6.12)$$

FIGURE 6.16 Time relationships in the symmetrical component sets that represent unbalanced two-phase operation.

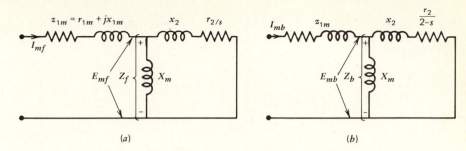

FIGURE 6.17 Circuit models for the forward and backward components of a single-phase motor. (a) One of two phases of the forward motor model, referred to the main winding. (b) One of two phases of the backward motor model, referred to the main winding.

These equations may be used to determine the forward and backward components of the winding currents. Multiplying Equation 6.12 by ja, and adding to Equation 6.11,

$$I_m + jaI_a = 2I_{mb}$$

or

$$I_{mb} = \tfrac{1}{2}(I_m + jaI_a) \tag{6.13}$$

Similarly,

$$I_m = \tfrac{1}{2}(I_m - jaI_a) \tag{6.14}$$

These considerations lead to the concept that a single-phase motor may be treated as the superposition of two balanced two-phase motors, one going forward and the other backward. First, consider the forward motor. The circuit model for one phase would be as shown in Figure 6.17a. The total power input to the rotor of the forward motor would be, for two phases,

$$P_{gf} = 2I_{mf}^2 R_f \tag{6.15}$$

where R_f is the real part of Z_f and

$$Z_f = \frac{jX_m[(r_2/s) + jx_2]}{(r_2/s) + j(X_m + x_2)} \tag{6.16}$$

The speed of the backward motor is minus the speed of the forward motor. Then the slip of the backward motor is given by

$$s_b = \frac{\omega_s - (-\omega)}{\omega_s} = \frac{\omega_s + \omega_s(1 - s)}{\omega_s} = 2 - s \tag{6.17}$$

This leads to the circuit model for one phase of the backward motor, given in Figure 6.17b. These models permit the calculation of the voltages induced in the main winding by the forward- and backward-rotating fields, E_{mf} and E_{mb}:

$$E_{mf} = I_{mf}Z_f$$

$$E_{mb} = I_{mb}Z_b \tag{6.18}$$

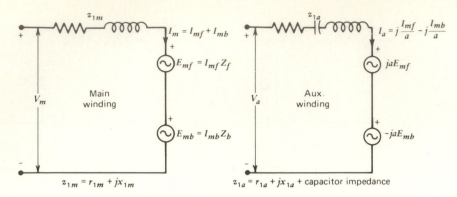

Voltage-source symbols represent voltages induced by forward and backward-rotating fields.

FIGURE 6.18 Equivalent circuits of main and auxiliary windings.

The field-induced voltages in the auxiliary winding will be the turns ratio times those induced in the main winding. The forward field reaches the auxiliary winding 90 electrical degrees before it gets to the main winding. For the backward field, the sequence is reversed. Thus the voltages induced in the auxiliary winding by the forward and backward fields will be

$$E_{af} = jaE_m$$
$$E_{ab} = -jaE_{mb} \tag{6.19}$$

These considerations lead to the equivalent circuits for the main and auxiliary windings in Figure 6.18. Note that the series impedance of the auxiliary winding has been defined to include the impedance of the series capacitor, if any. *Normally* the main and auxiliary winding are both connected to the single-phase line, so that $V_a = V_m$.

To construct a circuit model, let the main- and auxiliary-winding terminal voltages also be expressed in terms of symmetrical components. This is possible because *any* two phasor voltages may be represented thus. Let

$$V_m = V_{mf} + V_{mb}$$
$$V_a = jaV_{mf} - jaV_{mb} \tag{6.20}$$

Then by solving Equations, 6.20,

$$V_{mf} = \frac{1}{2}\left(V_m - j\frac{V_a}{a} \right)$$
$$V_{mb} = \frac{1}{2}\left(V_m + j\frac{V_a}{a} \right) \tag{6.21}$$

Note from Figure 6.18 that

$$V_m = I_m z_{1m} + E_{mf} + E_{mb}$$

or by Equations 6.11 and 6.18,

$$V_m = (I_{mf} + I_{mb})z_{1m} + I_{mf}Z_f + I_{mb}Z_b \tag{6.22}$$

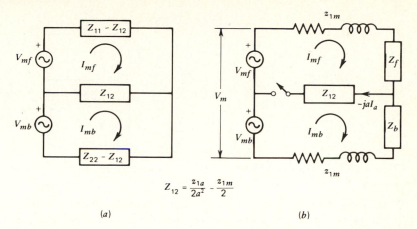

$$Z_{12} = \frac{z_{1a}}{2a^2} - \frac{z_{1m}}{2}$$

(a)   (b)

FIGURE 6.19 Circuit model of a single-phase motor. (a) Circuit form. (b) Circuit model.

Similarly,

$$V_a = I_a z_{1a} + jaE_{mf} - jaE_{mb}$$

or   (6.23)

$$V_a = I_a z_{1a} + jaI_{mf}Z_f - jaI_{mb}Z_b$$

In these expressions, z_{1a} includes the impedance of the series capacitor, if any. Substituting Equations 6.12, 6.22, and 6.23 into Equations 6.21 and reducing results in

$$V_{mf} = I_{mf}\left(\frac{z_{1m}}{2} + \frac{z_{1a}}{2a^2} + Z_f\right) - I_{mb} \cdot \frac{1}{2}\left(\frac{z_{1a}}{a^2} - z_{1m}\right)$$

$$V_{mb} = I_{mb}\left(\frac{z_{1m}}{2} + \frac{z_{1a}}{2a^2} + Z_b\right) - I_{mf} \cdot \frac{1}{2}\left(\frac{z_{1a}}{a^2} - z_{1m}\right)$$

(6.24)

These equations are of the form

$$V_{mf} = Z_{11}I_{mf} - Z_{12}I_{mb}$$

$$V_{mb} = -Z_{12}I_{mf} + Z_{22}I_{mb}.$$

(6.25)

which represent a circuit such as that of Figure 6.19a. In this figure,

$$Z_{11} - Z_{12} = \frac{1}{2}\left(z_{1m} + \frac{z_{1a}}{a^2}\right) + Z_f - \frac{1}{2}\left(\frac{z_{1a}}{a^2} - z_{1m}\right) = z_{1m} + Z_f \quad (6.26)$$

$$Z_{22} - Z_{12} = \frac{1}{2}\left(z_{1m} + \frac{z_{1a}}{a^2}\right) + Z_b - \frac{1}{2}\left(\frac{z_{1a}}{a^2} - z_{1m}\right) = z_{1m} + Z_b \quad (6.27)$$

The equivalent circuit model of the single-phase motor is, then, that of Figure 6.19b. From Equations 6.13 and 6.14, the current through Z_{12} is

$$I_m - I_{mb} = -jaI_a$$

(6.28)

When the starting switch opens, I_a goes to zero, so the switch is properly shown in series with Z_{12}.

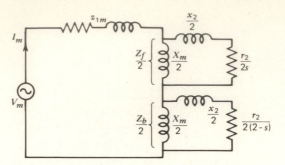

FIGURE 6.20 Model of a single-phase motor with starting switch open.

Once the starting switch is open, $I_{mf} = I_{mb}$; and since $I_m = I_{mf} + I_{mb}$, the current in the circuit is $I_m/2$, and the voltage applied to the circuit is $V_{mf} + V_{mb} = V_m$. The circuit may be redrawn to represent the main winding, operating alone. By reducing each impedance to half its value, the current will become equal to the main-winding current. The model of the single-phase motor after the starting switch opens is shown in Figure 6.20.

6.9. USING THE SINGLE-PHASE INDUCTION MOTOR MODEL

If the model impedances and turns ratio are known, the procedure for calculating the performance of a motor is as follows:

CASE I. AUXILIARY WINDING CONNECTED

(This case includes PSC motors, two-value capacitor motors, and capacitor start motors and resistance split-phase motors with the starting switch closed.)

1. Select the slip at which the performance is to be calculated.
2. Calculate Z_f, Z_b, z_{1m}, and z_{1a}/a^2, where z_{1a} includes the capacitor impedance, if any. Calculate $Z_{12} = \frac{1}{2}[(z_{1a}/a^2) - z_{1m}]$.
3. Calculate $V_{mf} = \frac{1}{2}[V_m - j(V_a/a)]$; $V_{mb} = \frac{1}{2}[V_m + j(V_a/a)]$. Usually both windings are connected to the same line, so that $V_m = V_a$ and

$$V_{mf} = \frac{V_L}{2}\left(1 - \frac{j}{a}\right)$$

$$V_{mb} = \frac{V_L}{2}\left(1 + \frac{j}{a}\right)$$

(6.29)

Note that these will be complex numbers. They must be treated as complex numbers in the following calculations.

4. Solution of the model allows the calculation of I_{mf} and I_{mb} as follows:

$$I_{mf} = \frac{V_{mf}(z_{1m} + Z_b + Z_{12}) + V_{mb}Z_{12}}{(z_{1m} + Z_f + Z_{12})(z_{1m} + Z_b + Z_{12}) - Z_{12}^2}$$

(6.30)

$$I_{mb} = \frac{V_{mb}(z_{1m} + Z_f + Z_{12}) + V_{mf}Z_{12}}{(z_{1m} + Z_f + Z_{12})(z_{1m} + Z_b + Z_{12}) - Z_{12}^2}$$

(6.31)

5. Calculate the winding currents, the line current, power factor, and input power:

$$I_m = I_{mf} + I_{mb} \tag{6.32}$$

$$I_a = j\frac{I_{mf}}{a} - j\frac{I_{mb}}{a} \tag{6.33}$$

$$I_L = I_m + I_a = |I_L| \underline{/\theta_L} \tag{6.34}$$

$$\text{Power factor} = \cos\theta_L \tag{6.35}$$

$$P_{\text{in}} = |V_L||I_L| \cos\theta_L \tag{6.36}$$

6. Calculate the developed torque and developed mechanical power:

$$\tau_d = \frac{2}{\omega_s}(I_{mf}^2 R_f - I_{mb}^2 R_b) \text{ N-m} \tag{6.37}$$

$$\text{DMP} = 2(I_{mf}^2 R_f - I_{mb}^2 R_b)(1 - s) \text{ W} \tag{6.38}$$

If rotational losses are known, the output power is given by

$$P_{\text{out}} = \text{DMP} - P_{\text{rot}} \text{ W} \tag{6.39}$$

$$\text{Horsepower} = \frac{P_{\text{out}}}{746}$$

$$\text{Efficiency} = \frac{P_{\text{out}}}{P_{\text{in}}} \tag{6.40}$$

CASE II. RUNNING WITH AUXILIARY WINDING OPEN

(Capacitor start motors, and resistance split-phase motors after starting switch opens.)
An open auxiliary winding is equivalent to infinite z_{1a}, which means $Z_{12} = \infty$. Examination of the circuit model for this situation. Figure 6.20, shows that

$$I_m = \frac{V_m}{z_{1m} + \frac{1}{2}Z_f + \frac{1}{2}Z_b} = |I_L| \underline{/\theta_L} \tag{6.41}$$

and the input power is given by

$$P_{\text{in}} = |V_L||I_m| \cos\theta_L \tag{6.42}$$

Then

$$P_{gf} = 2I_{mf}^2 R_f = I_m^2 \frac{R_f}{2}$$

$$P_{gb} = 2I_{mb}^2 R_b = I_m^2 \frac{R_b}{2} \tag{6.43}$$

$$\tau_d = \frac{P_{gf} - P_{gb}}{\omega_s} \text{ N-m} \tag{6.44}$$

$$DMP = (P_{gf} - P_{gb})(1 - s) \tag{6.45}$$

$$P_{out} = DMP - P_{rot} \tag{6.46}$$

CASE III. STARTING TORQUE

At starting $s = 1$ and $2 - s = 1$. Then $Z_f = Z_b \triangleq Z_{st}$. Let the real part of Z_{st} be defined as R_{st}:

$$R_{st} = \text{Re}(Z_{st}) = \text{Re}\left[\frac{jX_m(r_2 + jx_2)}{r_2 + j(X_m + x_2)}\right] \tag{6.47}$$

The starting torque is given by

$$\tau_{st} = \frac{2}{\omega_s}(I_{mf}^2 R_f - I_{mb}^2 R_b) = \frac{2R_{st}}{\omega_s}(I_{mf}^2 - I_{mb}^2) \tag{6.48}$$

It can be shown that

$$I_{mf}^2 - I_{mb}^2 = aI_m I_a \sin \alpha \tag{6.49}$$

where α is the phase displacement between I_m and I_a, the currents in the two windings at starting. Then

$$\tau_{st} = \frac{2aI_m I_a R_{st}}{\omega_s} \sin \alpha \tag{6.50}$$

CASE IV. CAPACITOR IMPEDANCE FOR BALANCED TWO-PHASE OPERATION AT A SPECIFIC SPEED

Balanced operation results when the backward field is eliminated, that is, when $I_{mb} = 0$. Equation 6.31 shows that this is accomplished when

$$V_{mb}(z_{1m} + Z_f + Z_{12}) + V_{mf}Z_{12} = 0 \tag{6.51}$$

Then

$$Z_{12} = \frac{-V_{mb}(z_{1m} + Z_f)}{V_{mf} + V_{mb}} \tag{6.52}$$

But by definition,

$$V_{mf} + V_{mb} = V_m = V_L \tag{6.53}$$

where V_L is the line voltage, assumed to be applied to both the main-winding circuit and to the auxiliary-winding circuit. Also by Equation 6.21,

$$V_{mb} = \frac{V_L}{2}\left(1 + \frac{j}{a}\right)$$

Making these substitutions into equation 6.52 results in

$$Z_{12} = -\frac{1}{2}\left(1 + \frac{j}{a}\right)(z_{1m} + Z_f) \tag{6.54}$$

Now, from Equations 6.24 and 6.25,

$$Z_{12} = \frac{1}{2}\left(\frac{z_{1a}}{a^2} - z_{1m}\right) \tag{6.55}$$

from which

$$\begin{aligned} z_{1a} &= a^2(2Z_{12} + z_{1m}) \\ &= R_c - jX_c + r_{1a} + jx_{1a} \end{aligned} \tag{6.56}$$

or

$$R_c - jX_c = z_{1a} - (r_{1a} + jx_{1a}) \tag{6.57}$$

Equation 6.54 may be used to find Z_{12}, which is then substituted into Equation 6.56 to obtain z_{1a}. Subtracting the auxiliary-winding leakage impedance gives the desired capacitor impedance.

■ **Example 6.1**

Predict the performance of a $\frac{1}{4}$-horsepower, four-pole, capacitor start, single-phase induction motor at a slip of 0.05; $r_{1m} = 1.9$, $x_{1m} = 2.6$, $r_2 = 3.6$, $x_2 = 2.6$, and $X_m = 56$ ohms. Assume it is operated on a 115-V, 60-Hz line and that the auxiliary winding has been disconnected by the centrifugal switch. Rotational losses total 25 W.

Solution. Computing the impedances of Figure 6.20,

$$z_{1m} = 1.9 + j2.6 \ \Omega$$

$$\frac{Z_f}{2} = j28\|\left(\frac{1.8}{0.05} + j1.3\right) = 13.1 + j17.3 \ \Omega$$

$$\frac{Z_b}{2} = j28\|\left(\frac{1.8}{1.95} + j1.3\right) = 0.84 + j1.27 \ \Omega$$

With $V_m = 115 \ \underline{|0°}$,

$$I_m = I_{mf} = I_{mb} = 4.34 \ \underline{|-53.24°} \ \text{A}$$

Thus

$$P_{\text{in}} = (115)(4.34) \cos(53.24°) = 299.3 \ \text{W}$$

$$P_g = P_{gf} - P_{gb} = 4.34^2(13.1 - 0.84) = 231.2 \ \text{W}$$

$$\tau_d = \frac{231.1}{2\pi 60/2} = 1.23 \text{ N-m}$$

$$\text{DMP} = 231.2(0.95) = 219.7 \text{ W}$$

$$P_{\text{out}} = 219.7 - 25 = 194.7 \text{ W} \qquad \text{or} \qquad 0.261 \text{ hp}$$

$$\text{Efficiency} = \frac{194.7}{299.3} = 0.65 \qquad \blacksquare$$

■ **Example 6.2**

Calculate the starting torque for the motor of the previous example if $r_{1a} = 12$, $x_{1a} = 6.5$, and $X_c = -20 \ \Omega$. Assume $a = 1.6$.

Solution. Since $s = 2 - s = 1.0$ at starting,

$$Z_{st} = Z_f = Z_b = 3.28 + j2.69 \ \Omega$$

$$z_{1a} = 12 + j6.5 - j20 = 12 - j13.5 \ \Omega$$

$$Z_{12} = \frac{1}{2}\left(\frac{12 - j13.5}{1.6^2}\right) - (1.9 + j2.6) = 1.39 + j3.94 \ \Omega$$

$$V_{mf} = \frac{115}{2}\left(1 - \frac{j}{1.6}\right) = 67.81\underline{|-32.0°} \ \text{V}$$

$$V_{mb} = \frac{115}{2}\left(1 + \frac{j}{1.6}\right) = 67.81\underline{|+32.0°} \ \text{V}$$

$$I_{mf} = 11.77\underline{|-54.93°} \ \text{A}$$

$$I_{mb} = 4.37\underline{|-19.70°} \ \text{A}$$

$$\tau_{st} = \frac{2}{60\pi}(3.28)(11.77^2 - 4.37^2) = 4.15 \text{ N-m}$$

Also note that

$$I_m = I_{mf} + I_{mb} = 15.55\underline{|-45.61°} \ \text{A}$$

$$I_a = \frac{j}{1.6}(I_{mf} - I_{mb}) = 5.37\underline{|+18.00°} \ \text{A}$$

and the line current is

$$I = I_m + I_a = 18.56\underline{|-30.60°} \ \text{A} \qquad \blacksquare$$

PROBLEMS

6.1. Describe how to reverse the following single-phase motors. State the speed conditions under which reversal may be accomplished. (a) Universal motor.

(b) Resistance split phase. (c) Capacitor start. (d) PSC. (e) Two-value capacitor. (f) Shaded-pole motor.

6.2. Which single-phase motor would you choose for the following applications? (a) Low-cost phonograph turntable. (b) $\frac{1}{20}$-hp fan. (c) Conveyor that must be started fully loaded. (d) Variable-speed sewing machine. (e) Bench grinder operating at about 3500 rev/min. (f) Hand-held grinder operating at 15,000 rev/min. (g) Water pump.

6.3. A certain two-pole universal motor draws 3.0 A from a 115-V, 60-Hz line as it rotates at 4200 rev/min under rated load. Each of the two field windings has 85 turns. The total series field resistance and reactance are 4 and 20 Ω, respectively. The armature resistance is 5 Ω and the armature reactance is 10 ohms.

   (a) Calculate the counter-emf E_g. (*Hint:* Assume it is in phase with the current.)
   (b) Find the voltage that would be measured between the brushes. Note that this is the armature voltage, neglecting brush drop.
   (c) Calculate the motor power factor.
   (d) Find the average developed torque.
   (e) Draw a phasor diagram showing line voltage (reference), line current, field voltage, and armature voltage.

6.4. If the shaft of the motor in Problem 6.3 were blocked, calculate the resulting current and input power.

6.5. Calculate the peak flux per pole at rated load for the motor of Problem 6.3.

6.6. The shaft load on the motor of problem 6.3 is reduced so that the current drops to 2.2 A. Estimate the new speed. Repeat the calculations.

6.7. For a certain universal motor, $R_f = 3.2$ Ω and $R_a = 3.6$ Ω. When the motors draws 4.0 A, the field voltage is 24 V, the armature voltage is 90 V, and the input power is 140 W.

   (a) Find the field reactance.
   (b) Find the armature reactance.
   (c) Calculate the core loss.

6.8. A $\frac{1}{4}$-hp, 115-V, 60-Hz four-pole split-phase motor has the following equivalent-circuit impedances:

$$r_{1m} = 2.54 \ \Omega \qquad x_{1m} = 2.90 \ \Omega \qquad r_2 = 2.36 \ \Omega$$
$$r_{1a} = 10.70 \ \Omega \qquad x_{1a} = 2.43 \ \Omega \qquad x_2 = 1.73 \ \Omega$$
$$X_m = 59.13 \ \Omega \qquad a = 0.916$$

Friction, windage, and core loss = 44.4 W
Calculate the performance of this motor at a slip of 0.0328.

6.9. Calculate the starting current and torque of the motor of Problem 6.8.

6.10. A $\frac{1}{4}$-hp, 115-V, 60-Hz four-pole capacitor start motor has the following circuit model impedances:

$$r_{1m} = 2.09 \; \Omega \qquad x_{1m} = 2.44 \; \Omega \qquad r_2 = 2.14 \; \Omega$$
$$r_{1a} = 11.02 \; \Omega \qquad x_{1a} = 6.41 \; \Omega \qquad x_2 = 1.33 \; \Omega$$
$$X_m = 40.64 \; \Omega \qquad a = 1.621 \qquad Z_c = -j18.55 \; \Omega$$

Friction, windage, and core loss = 71.3 W
Calculate the starting current and torque.

6.11. Calculate the performance of the motor of problem 6.10 at a slip of 0.0317. Assume the starting switch has opened.

6.12. A $\frac{3}{4}$-hp, 208-V, 60-Hz, six-pole PSC motor has the following circuit model impedances:

$$r_{1m} = 5.70 \; \Omega \qquad x_{1m} = 6.50 \; \Omega \qquad r_2 = 8.86 \; \Omega$$
$$r_{1a} = 9.62 \; \Omega \qquad x_{1a} = 10.06 \; \Omega \qquad x_2 = 5.08 \; \Omega$$
$$Z_c = -j133 \; \Omega \qquad X_m = 85.65 \; \Omega \qquad a = 1.244$$

Friction, windage, and core losses = 41.9 W

(a) Calculate the starting torque and current.

(b) Calculate the performance at a slip of 0.100.

6.13. A special-purpose, four-pole, two-value capacitor motor is rated at $\frac{1}{2}$-hp, 115 V, 60 Hz. The circuit model impedances are:

$$r_{1m} = 1.04 \; \Omega \qquad x_{1m} = 2.83 \; \Omega \qquad r_2 = 1.87 \; \Omega$$
$$r_{1a} = 1.29 \; \Omega \qquad x_{1a} = 3.01 \; \Omega \qquad x_2 = 1.75 \; \Omega$$
$$X_m = 72.09 \; \Omega \qquad a = 1.032$$

Friction, windage, and core losses = 32.1 W
Starting capacitor 282 μF (two capacitors in parallel)

(a) Calculate the starting torque and current.

(b) Calculate the running-capacitor impedance for zero backward field at a slip of 0.035. Discuss the resistance value.

(c) Calculate the performance of the motor for a slip of 0.035 for the capacitor reactance found in (b). Assume zero R_c.

7
MACHINES FOR SPECIAL JOBS

7.1 INTRODUCTION

Many applications exist that do not lend themselves to the machines previously discussed, for a variety of reasons. Some loads require constant torque over a wide speed range. Some require synchronous operation with smooth acceleration from standstill. Others require high acceleration/deceleration rates. In some applications, linear motion may be preferred over rotary motion. The requirements of certain servomechanisms (systems for controlling position or speed) may dictate characteristics of the servomotor. Increased use of computer control requires motors that may be easily and accurately controlled digitally. Also advances in materials and power semiconductor devices have opened new possibilities, especially variable-speed machine/drive packages competing for traditional motor applications.

This chapter includes qualitative information on principles of operations and performance characteristics for several special machines. Certainly, it is not exhaustive—varieties of special motors are nearly as numerous as the loads which they drive.

7.2 THE HYSTERESIS MOTOR

The hysteresis motor is an interesting example of beneficially utilizing a phenomenon that is often considered undesirable. Hysteresis motors may be classified as self-starting, synchronous ac motors. As the name implies, the operation of this motor depends upon the hysteresis of the iron. Specifically, the rotor is constructed of high-hysteresis iron. It has no slots, windings, or salient poles. Although it is possible to construct such a rotor as a solid iron cylinder, most rotors are formed from a core covered by a shell of the hysteresis material. The core may be of laminated iron or some nonmagnetic material. An iron core will result in a radial flux pattern through the hysteresis shell, in contrast to the circumferential pattern produced by a nonmagnetic core.

FIGURE 7.1 "Inside-out" hysteresis synchronous motor with permanent-split-capacitor stator. (Photo courtesy of Hurst Manufacturing Division, Emerson Electric Co.)

The stator of a hysteresis motor may be single-phase or polyphase, so long as it produces a revolving magnetic field. It is highly desirable for the spatial distribution of the stator MMF to be as nearly sinusoidal as possible, to minimize rotor losses due to minor hysteresis loops. Therefore single-phase stators are often of the permanent-split-capacitor design. Figure 7.1 shows an *inside-out* construction where the hysteresis shell rotates around a permanent-split-capacitor stator. The added inertia helps smooth torque ripples produced by unbalance in the stator MMF. For low-power applications where efficiency is not of great concern (such as ac clock motors), shaded-pole stators have been utilized to produce reliable and inexpensive hysteresis motors.

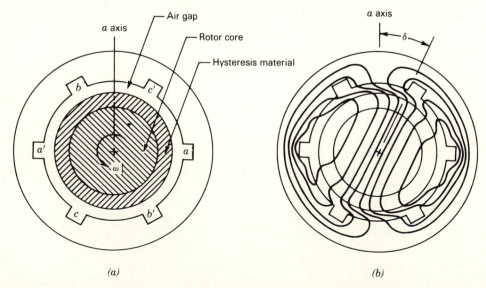

FIGURE 7.2 Two-pole hysteresis motor. (*a*) Cross section illustrating construction features. (*b*) Typical magnetic flux distribution for iron core rotor.

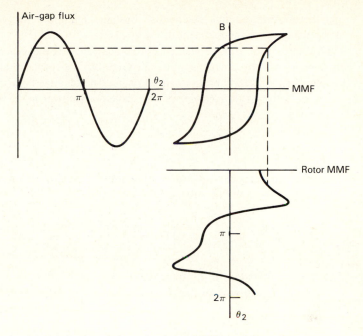

FIGURE 7.3 Graphical derivation of one cycle of rotor MMF, assuming a sinusoidal flux distribution and typical hysteresis characteristic.

Figure 7.2 illustrates the cross section of a typical hysteresis motor. A three-phase, two-pole winding is assumed, and the axis of the a-phase coil group is used as the angular reference. Note the symmetrical properties of the rotor. Its (laminated) core is surrounded by a uniformly thick layer of magnetic material with a B–H curve characterized by substantial hysteresis losses (the hysteresis loop has a large area).

A simplified explanation of motor operation will assume infinite rotor resistance, that is, no induced rotor currents. The relationship between magnetic flux and rotor MMF is illustrated by Figure 7.3. Taken as a whole, this figure depicts one complete cycle of flux and the corresponding excursion around the hysteresis loop. The stator voltage and the spatial flux distribution are nearly sinusoidal, since the applied voltage is sinusoidal. The crucial point is that an angle exists between the flux and the rotor MMF, and thus torque is developed. Significantly, this angle is determined by the hysteresis loop of the rotor material, and it therefore remains constant throughout the acceleration period. Also the flux per pole is primarily a function of the rotor material,

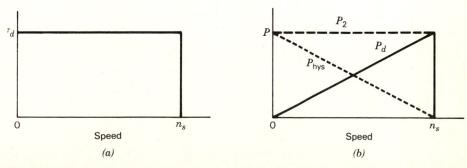

FIGURE 7.4 Idealized characteristics of a hysteresis motor. (*a*) Torque versus speed. (*b*) Developed power and hysteresis power loss versus speed.

and therefore it remains essentially constant. Therefore the voltage induced in the stator windings is independent of rotor speed.

Because of the unique characteristics described in the preceding paragraph, both the developed torque and the power delivered to the rotor are approximately constant throughout the subsynchronous speed range. The idealized curves are shown in Figure 7.4. Neglecting other rotor losses, the developed power is the rotor power minus the power lost to hysteresis:

$$P_d = P_2 - P_{\text{hys}} \tag{7.1}$$

Since the hysteresis curve is traversed at slip frequency, the time required for the rotor material to complete one full cycle of the loop is

$$\Delta T = \frac{2}{p} \frac{60}{(n_s - n)} \quad \text{s} \tag{7.2}$$

The power dissipated in the rotor due to hysteresis equals the area of the hysteresis curve times the volume of the material divided by ΔT. The developed power is thus

$$P_d = P_2 - A_{\text{hys}}(\text{Vol}) \frac{p(n_s - n)}{120} \quad \text{W} \tag{7.3}$$

Since P_d must be zero at starting, it is evident that

$$P_2 = A_{\text{hys}} (\text{Vol}) \frac{pn_s}{120}$$

and

$$P_d = A_{\text{hys}} (\text{Vol}) \frac{pn}{120} \tag{7.4}$$

The developed torque for subsynchronous operation may then be calculated as

$$\tau_d = \frac{P_d}{\omega} = \frac{A_{\text{hys}} (\text{Vol}) \, p}{4\pi} \text{ N-m} \tag{7.5}$$

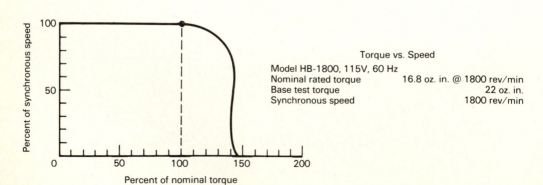

FIGURE 7.5 The actual characteristic of the hysteresis motor pictured in Figure 7.1. (Courtesy of Hurst Manufacturing Division, Emerson Electric Co.)

The interesting result is that hysteresis torque does not depend on speed, but rather is constant from zero to n_s. It should be noted that the previous development assumes infinite rotor resistance and therefore ignores the effects of eddy currents. Below synchronous speed, eddy currents produce a rotor MMF which interacts with the air gap flux to produce torque. This torque may be thought of as an induction torque superimposed upon the hysteresis torque. The result is well illustrated by Figure 7.5, which is the actual speed vs. torque characteristic for the motor pictured in Figure 7.1. When comparing this curve with the idealized characterized of Figure 7.4a, it is necessary to interchange the axes, a practice common in industry.

Even allowing for nonideal behavior, the motor exhibits a relatively flat torque characteristic. As a result, the hysteresis motor is able to smoothly accelerate and synchronize any mechanical load it is capable of starting. Once synchronous operation is attained, the torque will be that required by the load, the rotor material will cease to traverse the hysteresis loop, and the angle δ will be constant at some reduced value. In short, the machine will have become a permanent-magnet synchronous motor.

Hysteresis motors typically are built in fractional horsepower sizes and exhibit low efficiency and poor power factor. They are selected for applications requiring smooth acceleration of hard-to-start loads which must run at synchronous speed. Clock motors, tape drives, and phonograph turntables are some examples. Some published studies suggest that motors as large as 250 horsepower could be designed with 90+ percent efficiencies and 70+ percent power factors.

7.3 SYNCHRONOUS RELUCTANCE MOTORS

Any motor with a revolving stator field may be operated as a synchronous motor of the variable-reluctance type if the rotor is suitably designed and some other means of starting the motor is provided. The rotor construction must result in variation of reluctance with the rotor angle θ_2, such that salient poles are created, that is, regions of differing reluctance separated by 90 electrical degrees.

The centers of the low and high reluctance regions are referred to as the *d* axis (direct) and *q* axis (quadrature), respectively. Under steady-state synchronous opera-

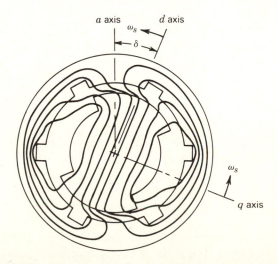

FIGURE 7.6 A typical flux pattern in a reluctance motor under load.

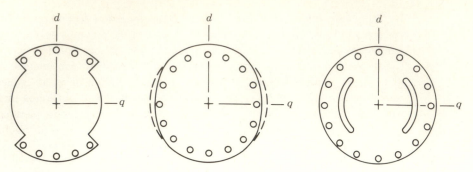

FIGURE 7.7 Three possible rotors for a two-pole induction-start, reluctance-run synchronous motor.

tion, this will produce an angular displacement between the stator MMF and the d axis. Such a condition is illustrated in Figure 7.6. The result is reluctance torque. This torque has a nonzero average value only at synchronous speed. Therefore a reluctance motor must develop starting torque by some other means. At subsynchronous speeds, variations in d–q axis reluctances will result in slip frequency ripple in the torque.

Reluctance torque is present in many ac machines. For example, a salient-pole synchronous machine may develop significant reluctance torque. At light loads, it will be sufficient to maintain synchronism, even with zero field excitation and negligible residual magnetism.

The reluctance principle may be incorporated into the design of the hysteresis motor to produce a hysteresis-reluctance synchronous motor. This is usually accomplished by removing some hysteresis material from the shell (outer or inner surface), or slotting the rotor stack to produce the required variation in reluctance. The resulting motor starts by means of the hysteresis principle, but has higher synchronous pullout torque. Efficiency and power factor are also improved.

Another example is the induction-start, reluctance-run synchronous motor, in which the rotor cage is left intact but the rotor core is modified to alter the reluctance. Figure 7.7 displays some possible rotor designs. This type of motor starts as an ordinary squirrel cage induction motor, except that starting torque is a function of rotor position and torque during run-up will exhibit the slip-frequency ripple described above. Each electrical cycle will be divided into equal acceleration/deceleration periods produced by the reluctance torque. Induction torque will accelerate the load to near synchronous speed. Proper synchronous operations depends upon the reluctance torque being sufficient greater than the load torque so that the rotor will "snap" into synchronism during an acceleration half cycle.

7.4 LINEAR MOTORS

A linear motor is a machine that converts electrical power into translational mechanical power. Of course, linear motion can be obtained for many applications from conventional rotating motors by employing worm gears, rack and pinion gears, pulleys, and so on. For other applications, vibration, noise, and/or economic constraints may suggest a linear motor. Linear motors have been used in applications varying from precision position control to mass transit traction. Others include conveyor systems, loom shuttle drives, conductive fluid pumps (especially for liquid metals), and sliding door openers.

Many types of linear motors exist. Conceptually, any rotary motor may be "unrolled" to create a linear one of the same type. Thus dc, induction, and synchronous (with

PM, wound field, and reluctance "rotors") are all possible. Linear motors have also been built for which construction of the rotary antitype is not feasible (though it is theoretically possible). Figures 7.8*a–c* illustrate the relationship between a squirrel cage induction motor and a single-sided linear induction motor with a ladder-type secondary. As suggested by the remaining parts of Figure 7.8, the magnetic material corresponding to the rotor stack could be separated from the ladder, and the ladder could be replaced by a conducting sheet. Also a second primary winding could be added to form a double-sided motor.

Three significant deviations from rotary machines immediately present themselves. First, the terms *stator* and *rotor* have significance only as they relate to the corresponding rotating machine, since either may be the stationary member. When the primary or the armature is the moving member, some means of connecting the power source must be devised. If the distance traveled is short, a flexible cable may be suitable. Applications such as mass transit require a power distribution rail and brush pickup.

A second major difference is that forces normal to the direction of travel do not automatically cancel in linear motors, as they do in rotating machines. Normal forces may degrade performance by creating a magnetic pull several times the gravitational pull on the moving member. In many applications it is possible to avoid this problem by properly designed geometries. For example, a two-sided, moving primary could

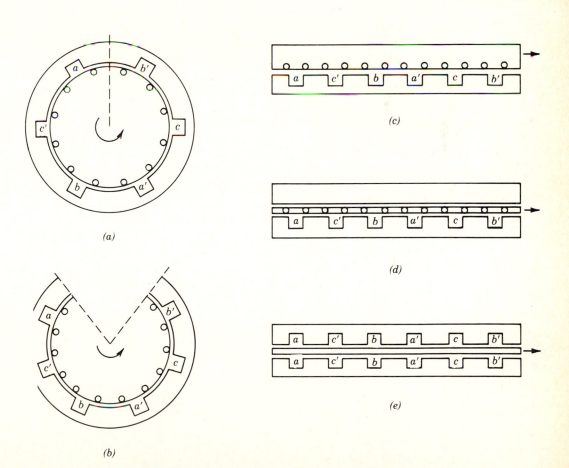

FIGURE 7.8 The conceptual development of a two-pole linear induction motor.

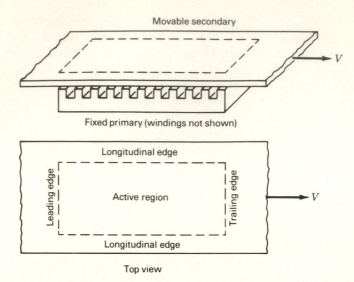

FIGURE 7.9 A moving secondary, linear induction motor, with the secondary regions identified.

be used with a vertical fixed secondary. In some cases, a clever design may even utilize the normal force for levitation of large loads, though separate magnets are typically provided for this purpose.

The third feature distinguishing linear motors from their rotary counterparts is the open-ended air gap. Unless either the *stator* or *rotor* is extended, the magnetic circuit changes as motion occurs. Even then, the active area of the motor (corresponding to the air gap of a rotating machine) has definite ends and associated end effects.

For the purpose of discussion, a single-sided linear induction motor with a short, stationary primary will be considered. (See Figure 7.9.) The secondary always extends beyond the width of the primary, and the resulting longitudinal edges may be treated much like end rings of squirrel cage rotors. However the end effects of leading and trailing edges have no counterpart in the rotary machine. Even at synchronous speed (zero slip), secondary current will flow to oppose the sudden application of magnetic flux as the secondary moves over the leading end of the primary. Current also flows when the flux returns to zero as the secondary exits the active region.

The equivalent circuit analysis utilized for rotary induction motors is inadequate to predict the nature of these end effects and their influence on motor performance. Rather, electromagnetic field theory must be employed to examine entry-end and exit-end traveling waves and their effect on air gap flux. The analysis is beyond the scope of this book, but some general observations can be made. The exit end wave attenuates quickly and has little effect on motor performance. The entry end wave may usually

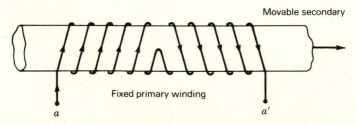

FIGURE 7.10 A tubular linear motor with one phase winding shown.

be neglected for low-speed applications when the number of poles is six or greater. Obviously, a transition region exists between *low-speed* and *high-speed* operation. End effects may typically be neglected if the velocity satisfies the following inequality:

$$v < \left[\frac{4\omega\rho_s g}{\mu_0} \right]^{\frac{1}{2}}$$

(7.6)

where

$$v = \text{translational velocity, in m/s}$$

$$\omega = \text{electrical radian frequency}$$

$$\rho_s = \text{secondary surface resistivity}$$

$$g = \text{effective air gap}$$

For high-speed applications (velocities significantly greater than calculated by Equation 7.6), the entry end wave may extend the entire length of the primary. The typical result is a reduction in air gap flux, which reduces thrust. Power factor and efficiency are also adversely affected. To make matters worse, these degradations in performance are much more pronounced in the low-slip region, where normal operation occurs. Properly designed compensating windings in the primary may be used to minimize the adverse effects of the entry end wave.

A tubular configuration of the linear motor may be formed by rerolling the primary windings about the axis of motion. One phase of such a winding is shown in Figure 7.10. The other two windings would be symmetrically placed at 120° electrical spacing. Note that, unlike flat linear motors, the tubular designs results in cancellation of the normal forces. Also eliminated are the longitudinal edge effects. Such a motor could be used for transporting a conducting rod or producing a reciprocating short-stroke motion. An interesting modification with robotic applications is the use of helical windings to produce both linear motion and rotation.

7.5 STEPPER MOTORS

A stepper motor may be thought of as a polyphase synchronous motor having salient stator poles The name *stepper* derives from the most common application for these machines, that is, rotating a fixed angular step in response to each input pulse received by their controller. When this type of motor is supplied from an electronic drive, accurate position control and precise rotational speeds are the natural consequences. The reader should be aware that various electromechanical devices exist which achieve stepper characteristics by employing solenoids and ratchets. These are not under consideration here.

The characteristics of stepper motors make them ideally suited to many applications requiring incremental motion, especially where digital control is used. A few common ones are printer head and paper feed drives in computer printers, disc drives, digital plotters, medical equipment that dispenses precise dosages, computer-controlled tools, and process control valves.

Though construction details vary widely, stepper motors may be divided into three main categories based on principal of operation. These are variable reluctance, permanent magnet, and hybrid types (which utilize both principles for torque develop-

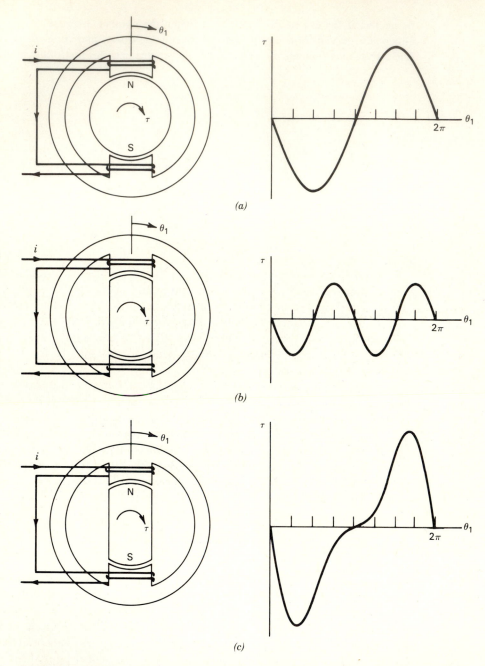

FIGURE 7.11 Torque development in the three basic types of stepper motors. (*a*) Permanent-magnet rotor. (*b*) Variable reluctance rotor. (*c*) Hybrid rotor.

ment). Figure 7.11 illustrates the three rotor types and their respective torque/displacement characteristics.

Permanent-Magnet Type

Consider the simplified two-phase stator with two-pole permanent-magnet rotor shown in Figure 7.12*a*. The rotor is shown in a stable equilibrium position for the case when $i_a > 0$ and $i_b = 0$. Stepping action is produced by switching currents i_a and i_b, supplied

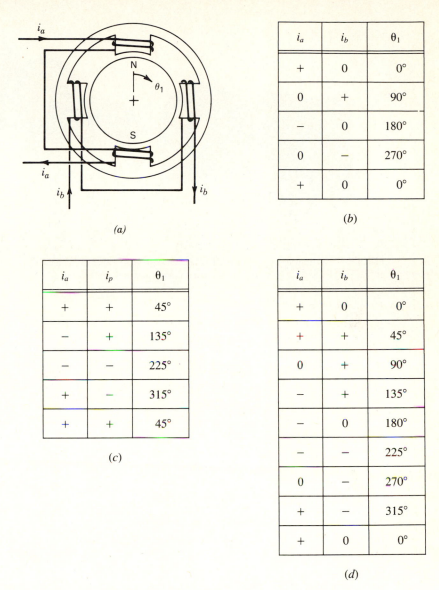

i_a	i_b	θ_1
+	0	0°
0	+	90°
−	0	180°
0	−	270°
+	0	0°

(b)

(a)

i_a	i_p	θ_1
+	+	45°
−	+	135°
−	−	225°
+	−	315°
+	+	45°

(c)

i_a	i_b	θ_1
+	0	0°
+	+	45°
0	+	90°
−	+	135°
−	0	180°
−	−	225°
0	−	270°
+	−	315°
+	0	0°

(d)

FIGURE 7.12 Two-phase permanent-magnet stepper motor. (a) Cross section. (b) Full-step, single-winding sequence. (c) Full-step, two-winding sequence. (d) Half-step sequence.

from a bipolar source. Three possible sequences for producing clockwise rotation are given. The first energizes only one phase winding at a time. The resulting step size is one half the pole pitch. The second method energizes both windings simultaneously and would thus require a larger power supply. The resulting steps are the same size, but the effective stator pole positions are located midway between the pole pieces. Advantages of this scheme are increased torque and damping. The final method, known as *half-stepping,* alternates between one and two phases. The resulting step size is one fourth the pole pitch.

Various techniques have been used to produce smaller step sizes. Perhaps the most obvious is to increase the number of phases and/or the number of poles. Figure 7.13 illustrates a two-phase, 24-pole, stacked-can construction with permanent-magnet rotor (the ceramic rotor has 24 radially oriented poles). The stator stacks are displaced by

FIGURE 7.13 Exploded view of a 24-pole, stacked-can, permanent-magnet stepper motor. (Photo courtesy of Hurst Manufacturing Division, Emerson Electric Co.)

one half pole pitch. As with the two-pole example, each full step results in a rotation of one half pole pitch, giving this motor 48 full steps per revolution, or a step size of 7.5°. Again half-stepping could be employed to double the number of steps. Carrying this concept a little further, if the source is capable of supplying current at several discrete levels, the effective stator poles may be shifted in even smaller steps. The possible resolution is a property of the drive rather than the motor. This technique is known as *microstepping* and requires a relatively complicated electronic supply.

Variable Reluctance Type

A variable reluctance rotor is constructed of ferromagnetic material with salient poles. Figure 7.14 illustrates a three-phase, two-pole motor, which has a 60° step. Since the rotor has no magnetic polarization, the three-phase supply may be unipolar. Note that the rotor advances one tooth pitch (180° in this case) during the complete electrical cycle of three steps.

Many variations within the general classification of *doubly salient* machines have been utilized to obtain the smaller step sizes. For example, a variable reluctance rotor can be produced with teeth at a pitch slightly less than that of teeth on the stator poles, producing a vernier effect, so that a complete electrical cycle advances the rotor only one tooth pitch. The step size may be computed as follows:

$$\text{Step size} = \frac{\text{rotor tooth pitch}}{\text{steps/cycle}} \tag{7.7}$$

Figure 7.15 shows a three-phase, variable reluctance stepper motor with 32 rotor teeth. Careful examination of this photo reveals that though stator teeth on a given pole have the same pitch as the rotor teeth, the spacing between teeth on adjacent poles is $\frac{4}{3}$ times this pitch. Deenergizing phase *a* and energizing phase *b* results in a rotation of only one third the tooth pitch. Thus this motor has 96 steps per revolution, or a step angle of 3.75°, and the direction is determined by the phase sequence.

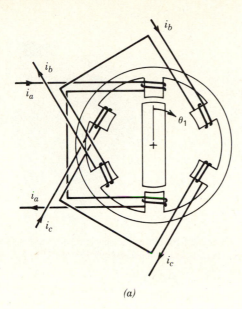

(a)

i_a	i_b	i_c	θ_1
+	0	0	0°
0	+	0	60°
0	0	+	120°
+	0	0	180°

(b)

i_a	i_b	i_c	θ_1
+	+	0	30°
0	+	+	90°
+	0	+	150°
+	+	0	210°

(c)

i_a	i_b	i_c	θ_1
+	0	0	0°
+	+	0	30°
0	+	0	60°
0	+	+	90°
0	0	+	120°
+	0	+	150°
+	0	0	180°

(d)

FIGURE 7.14 Three-phase, variable reluctance stepper motor. (*a*) Cross section. (*b*) Full-step, single-winding sequence. (*c*) Full-step, two-winding sequence. (*d*) Half-step sequence.

Hybrid Type

Typical construction for the hybrid rotor uses two variable reluctance stacks rotated by one half the tooth pitch and separated by an axial permanent magnet. The axial magnet causes all teeth on a given stack to be the same magnetic polarity. The rotation permits rotor teeth of opposite polarity (which will be on opposite ends of the rotor) to simultaneously assume positions of minimum reluctance. The step size may again be calculated using Equation 7.7. Figure 7.16 presents a simple hybrid rotor which could be used with the two-phase, bipolar stator of the previous PM example. The resulting motor would have a step size of 18°, or 20 steps per revolution.

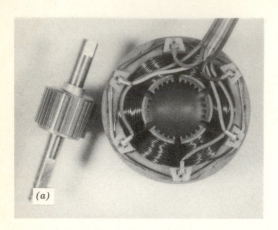

FIGURE 7.15 Three-phase, 96 steps/revolution, variable reluctance stepper motor. (*a*) Complete rotor and stator removed. (*b*) Assembled motor with end cap. (Components courtesy of Hurst Manufacturing Division of Emerson Electric Co.)

As with variable reluctance types, smaller step sizes may be obtained by vernier interaction between stator and rotor teeth. One common hybrid motor has 50 rotor teeth and a two-phase, four-pole/phase stator with teeth pitched at 48 per full circle. This design results in 1.8° per step, or 200 steps per revolution.

Steady-State and Dynamic Characteristics

An idealized torque/speed characteristic for a stepper motor is shown in Figure 7.17. Note that two modes of operation are possible. The normal mode could be called *locked step*. In this mode the combined load plus the rotor inertia is less than the developed torque, with the result that the rotor may come to rest (or at least experience a reversal of momentum) between steps. Starts, stops, and instantaneous reversals are possible. The second mode, called *slewing,* maintains synchronism but does not permit stops or reversals. Some applications utilize both modes. For example, a magnetic tape drive may read/write at a step rate in the normal mode but rewind/fast-forward in the slewing mode. The transitions to and from slewing must be properly controlled to avoid loss of synchronism.

Since developed torque is directly related to the stator current, the stator-winding time constant L/R is a most significant parameter affecting performance at high step rates. External resistance is often provided for the purpose of reducing this time constant. Another common technique is to construct the motor with two windings per phase and energize them in opposite directions from a unipolar supply (sometimes referred to as a "four-phase, bifilar" motor). This achieves equivalent control with a less expensive unipolar electronic drive. And since the two windings must fit in the same space, smaller wire is used and the resulting increase in winding resistance improves performance. The trade-off is that the magnitude of the current must be reduced, which reduces the torque at low step rates.

Another factor that may limit stepper motor performance is mechanical resonance. The natural resonant frequency typically occurs at relatively low speeds (below 200 steps per second). If the motor must operate near its resonant frequency, unacceptable increases in overshoot and settling time or a decrease in torque may result. These problems can often be reduced by adding inertia (decreasing the resonant frequency) or providing some type of mechanical or electronic damping.

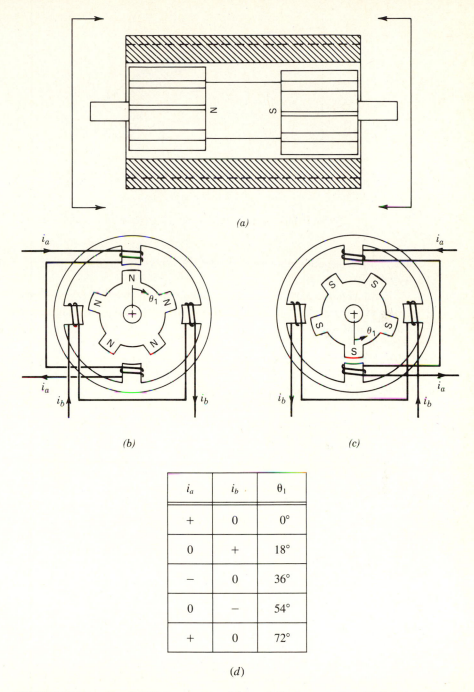

FIGURE 7.16 Two-phase hybrid stepper motor. (*a*) Longitudinal cross section. (*b*) End view of north rotor end. (*c*) End view of south rotor end. (*d*) Full-step, single-winding sequence.

Comparison of Stepper Types

Stepper motors are available with step sizes ranging from 90° to 0.72°. Maximum step rates vary from about 100 to as many as 10,000 steps per second. Torque capacities vary over a wide range. Selection of a stepper motor for a particular application must consider the overall performance of the motor with electronic drive. There are, how-

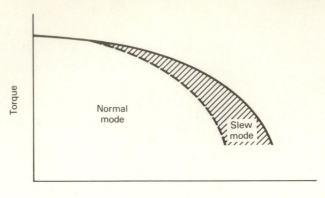

FIGURE 7.17 Typical torque/speed characteristic for stepper motors.

ever, a few general qualitative comparisons that can be made between the three basic types.

Permanent-magnet steppers are available in the larger step sizes, usually above 5°. They exhibit excellent torque/volume ratios but typically have high inertia rotors. The rotor provides braking and holding torque even with the stator windings deenergized. The variable reluctance motor is ideally suited to the smaller step sizes. By comparison, it develops less torque. However, since it may be constructed with a low-inertia rotor, its torque/inertia ratio may be superior. The rotor is capable of "free-wheeling," since braking and detent torque must be provided by the stator. The hybrid stepper combines the smaller step size of the variable reluctance rotor with the higher torque of the permanent magnet. Some braking and holding torque is also present. As might be expected, it is the most expensive of the three types to manufacture.

7.6 SWITCHED RELUCTANCE MOTORS

Switched reluctance motors are relatively recent, compared to most other types of electrical machines. Though the operational principles have long been understood, cost-effective application of the device had to await advances in power switching semiconductors.

The switched reluctance motor is essentially a single-stack, variable reluctance stepper motor. However the motor and required electronic drive are normally designed for efficient power conversion for loads requiring variable-speed operation rather than typical stepper applications. The number of rotor poles is usually small, with the result that the rotation per step is large. Also steppers are rarely constructed in the midintegral horsepower range, where switched reluctance motors appear to have much potential. Recent studies, some based on actual prototype testing, have concluded that switched reluctance motor efficiency compares well with that of induction motors in the integral horsepower sizes. Economic comparisons favor the switched reluctance motor when variable speed is needed, requiring the induction motor to be fed by an inverter.

Figure 7.18 shows a typical cross section similar to the simplified variable reluctance stepper of Figure 7.14, but with a four-pole rotor. One full cycle of the three-phase supply would result in a rotation of 90°. Direction is of course determined by the phase sequence, as is speed by the frequency. Four-phase configurations are also common.

Perhaps the most outstanding features of the switched reluctance motor are the

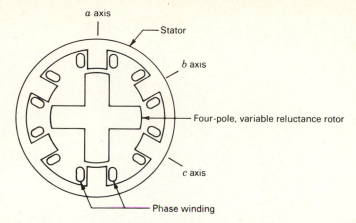

FIGURE 7.18 Cross section of a three-phase, switched reluctance motor with a 30° step angle. Rotation is counterclockwise for positive phase sequence.

simplicity and ruggedness of its construction. The salient-pole rotor is a laminated stack with no windings, commutator, slip rings, or permanent magnets. The stator also has salient poles, each wrapped with a concentrated winding resembling that used for dc machine field poles. Replacement of these windings, should it be required, is relatively easy. Even the electronic drive is simplified, being unipolar and requiring only one active switch per phase, which is in series with the motor winding. This type of drive inherently avoids the "shoot-through" problem found in many power convertors, which must supply alternating current.

Switched reluctance drives are inherently variable speed and are normally considered to compete with induction motor/inverter drive systems. The motor alone has comparable efficiency and torque/weight figures and superior torque/inertia ratio. Comparison of performance is complicated by the variety of possible inverter types, since overall system performance is the issue. Having emphasized this disclaimer, it may be generally stated that there is little difference between overall efficiencies of the two. The induction system typically enjoys a slightly higher power factor or, more precisely, a lower ratio of input kVA (peak or average) to output power. If switched reluctance motors gain an increasing market share, their popularity will be attributable to the low-cost, high-reliability construction.

7.7 LOW-INERTIA DC MOTORS

Certain applications require higher acceleration and deceleration rates than can be obtained by conventional motor designs. Examples are high-speed tape readers, oscillographs and x–y recorders, layer winders, point-to-point tool positioners, and many others. One option is to employ mechanical clutches and brakes. However the mechanical assemblies are expensive and add significantly to system complexity and maintenance.

The motors discussed in this section have been developed specifically to provide high performance characteristics when used in direct-drive control applications. They are dc motors redesigned to eliminate iron from the rotor. As such, they retain the desirable speed and torque control properties discussed in Chapter 5. Removing iron from the rotor achieves higher acceleration in two ways. The most obvious is that rotor mass and inertia are greatly reduced. Also, the armature time constant is drastically reduced, since it is directly proportional to the inductance of the armature winding. As a side benefit, the lower inductance reduces sparking and significantly increases

FIGURE 7.19 Exploded view of a low-inertia, printed-circuit dc motor. (Photo courtesy of PMI Motion Technologies).

brush life. Also eliminating the rotor stack eliminates "cogging" associated with the variable reluctance of its teeth and slots.

One popular design is the "pancake" motor, so-called because of its flat, round rotor. The stator is usually of the permanent-magnet type. As shown in Figure 7.19, the rotor (armature) is actually a multisided printed circuit board. The armature conductors and contact pads for the brushes (there is no distinct commutator) are embossed on the rotor surface with metallic foil. One advantage of this design is that the number

FIGURE 7.20 Low-inertia, "basket-weave" dc motor. (Photo courtesy of Pacific Scientific.)

of armature "coils" may be quite large, producing uniform torque, even at low speeds. The magnetic fields produced in this machine are axially oriented rather than radially. Typical sizes for these motors are in the fractional and subfractional horsepower ranges, but smaller integral sizes are also available. Depending on the specific application, acceleration from zero to a few thousand revolutions per minute may be accomplished within 10 milliseconds.

Rotor inertia is reduced even more by the hollow "basket weave" rotor design. Wire is wound to form a cylindrical shell, then reinforced by glass fibers and epoxy resin. The final assembly is cup-shaped and fits around a stationary core. This core may be a permanent magnet, with the outer stator housing providing the return flux path. Alternately, such a motor could also be constructed with a stator similar to a conventional PM dc motor with an added iron sleeve extending inside the armature basket, as shown in Figure 7.20. As in conventional motors, the magnetic fields are oriented radially. This design is practical only to about 0.5 horsepower. Mechanical time constants on the order of 1 millisecond are common, typically hundreds of times faster than ordinary dc motors of the same rating and perhaps ten times faster than the pancake motor. Current density must be high to achieve the large torques, and adequate ventilation is essential to prevent rotor overheating.

7.8 TORQUE MOTORS

Certainly, all electric motors develop torque. The term *torque motor,* however, is normally reserved for a machine designed specifically to maintain constant torque or tension for low-speed operation (perhaps including stalling). Both dc and induction motors are utilized for these applications, so "torque motor" often refers to the intended use rather than to the principle of operation or type of construction.

Three types of motion are common in torque motor applications. The motor may be required to rotate at speeds much below what is considered normal, and the rotation could even be backward (braking) during at least part of its operational cycle. A second possibility is for the rotation to be restricted to a few revolutions at most. Various types of mechanical actuators require this type of motion. The third possibility is for the motor to produce torque with little or no movement, much as a spring. Proper control would permit either tension or compression.

Due to the low speeds, power efficiency is not a meaningful measure for torque motors. Performance is better expressed by the ratio of torque to input power and may be given at locked-rotor. Maximum permissible stall duty cycle is also a significant parameter. Depending on environmental and spatial constraints, auxiliary cooling often provides an economical alternative to oversizing torque motors for thermal reasons.

As suggested previously, the desired output from a torque motor may be either constant torque or constant tension. A common application requiring the latter is maintaining constant tension on some material while the takeup spool fills. Assuming the linear velocity of the material does not change as the spool fills, the rotational speed will gradually decrease as the radius (and circumference) increase. The torque must increase if constant tension is to be maintained. Therefore the torque/speed curve must have a negative slope in the first quadrant.

Alternating-Current Types

Alternating-current torque motors are polyphase induction motors with the rotor designed to produce the desired torque/speed characteristic. Where constant torque is the

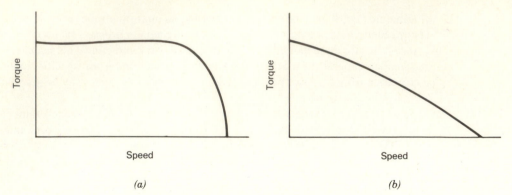

FIGURE 7.21 Typical torque/speed characteristic for (*a*) an induction motor designed for constant torque, and (*b*) a high-slip induction motor designed for constant-tension applications.

goal, it can be approximated from locked-rotor to speeds as high as 70 percent of synchronous. For constant-tension applications, a rotor exhibiting extremely high slip characteristics is utilized ($s_M > 1$). Figure 7.21 represents typical curves for ac torque motors.

Direct-Current Types

The excellent control characteristics of the dc motor, discussed in Chapter 5, make torque motor applications possible with standard machines. For separately excited or shunt machines with constant field excitation and for permanent-magnet machines, torque is directly proportional to the armature current. Therefore controlling the torque is simply a matter of controlling the current. Figure 7.22 shows torque/speed curves for such a machine as r_a varies. Constant tension can be approximated by designing the machine with the proper value of armature resistance, or setting the effective armature resistance by means of a rheostat and operating it on a constant voltage bus.

Sometimes the term *torque motor* is used to refer to a direct-drive, dc servomotor specifically designed for some low-speed or limited-travel load. Such motors are often

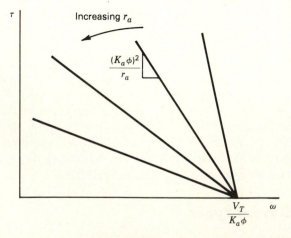

FIGURE 7.22 Typical torque/speed characteristics for a dc motor for various values of r_a and constant field excitation.

used as operators for valves, dampers, and many positioning systems. They are characterized by short axial length, fast response (low armature inductance), and high accuracy and resolution (no gear trains with dead bands and backlash). So-called *frameless* designs eliminate a separate motor housing and incorporate the motor with the mechanical device to be controlled.

7.9 SYNCHROS

The terms *synchro* and *selsyn* (contraction for "self-synchronous") are both used to refer to special wound-rotor induction motors which are applied in pairs to provide shaft position control and/or synchronism of two remote machines. Applications range from controlling valves to indicating the position of some remote device. Most commonly, the primary winding is single-phase and located on the rotor, while the three-phase stator winding acts as the secondary. While standard wound-rotor induction motors could be used, synchros normally have low-friction bearings to reduce error and mechanical dampers to improve dynamic performance. Equivalent control may be obtained electronically through shaft position sensors and feedback control schemes. Historically, selsyns proved less costly and more reliable, but improvements in electronics and less expensive alternate motors are decreasing selsyn applications.

The configuration shown in Figure 7.23 may be used for applications with low torque requirements. As shown, both rotors are energized from the same ac supply. An unbalanced set of three single-phase voltages (in time phase with the rotor voltage) will be induced in the stator phase windings of both machines. These voltages will be equal if, and only if, the two rotors are at the same relative positions. Assuming one machine is the transmitter, its rotor position will be fixed by the controlling mechanical input. If the rotor of the other machine (receiver) is not at the same position, unequal phase voltages will exist. This will result in stator current, thus MMF, and thus torque seeking to align the receiver's rotor position with that of the transmitter. Equilibrium is reached when the output torque decreases to equal the torque required

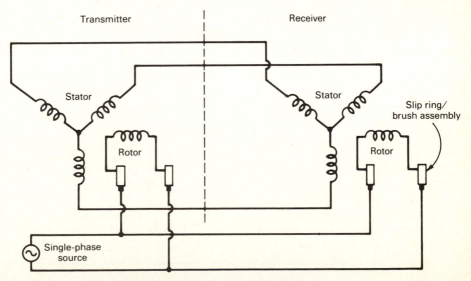

FIGURE 7.23 Low-torque synchro system.

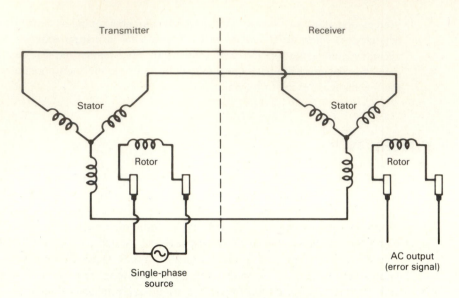

FIGURE 7.24 Synchro system with control transformer.

by the transmitter's load. Therefore it is essential that the load torque be small and that friction be minimized in the synchros.

Loads requiring higher torques may be controlled using the scheme shown in Figure 7.24. Assuming that stator leakage reactances and winding resistances may be neglected, the receiver's stator voltage will equal those induced in the transmitter. Therefore by Faraday's law, the magnetic flux distributions in the two machines must be the same. Analysis will show that the voltage induced in the receiver's rotor winding is approximately the product of the primary excitation voltage and the cosine of the angular displacement between the two rotors. Thus when the receiver's rotor position matches that of the transmitter's, its induced voltage will equal the primary excitation voltage. If, however, it is displaced by 90°, the flux linkage and induced voltage will be zero. If this 90° shift is mechanically compensated, the voltage induced in the receiver's

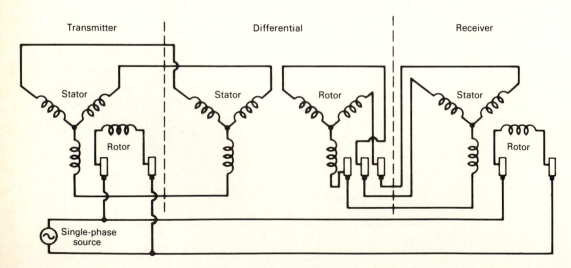

FIGURE 7.25 Low-torque synchro system with differential.

rotor is an error signal, which may be amplified and used to drive the load. For small angular errors, the signal approximates a linear function of the error. The receiver in this configuration is sometimes called a *control transformer*.

Synchros are also constructed with three-phase rotor windings. For low-torque applications, it is common to supply both transmitter and receiver stator windings from the same three-phase source. The rotor windings are then connected in parallel. Again by Faraday's law, flux distributions must be the same in both machines. Equilibrium will exist in the rotor circuit only when the rotors are synchronized at the same relative positions. Otherwise, a torque will be produced which tends to align the two rotors.

A three-phase synchro with a 1 : 1 turns ratio between stator and rotor windings may also be utilized to produce a rotation equal to the sum or difference of the rotation of two shafts. When such a machine is connected as shown in Figure 7.25, it operates as a *differential synchro*. That is, its rotor turns the difference between the rotor movements of the transmitter and the receiver. The stator winding flux linkage must match that of the transmitter, just as the rotor-winding flux linkage must match the stator of the receiver. Since both single-phase rotors are powered from the same source, if both transmitter and receiver rotate in phase synchronism the differential stator and rotor flux linkages will vary in unison (with no motion). Any difference in rotation between the transmitter and receiver will be compensated by rotation of the differential rotor. Interchanging any two leads on either its stator or rotor would cause the differential to rotate an amount equal to the sum of transmitter and receiver rotation.

PROBLEMS

7.1. Operated at 60 Hz, what is the maximum developed torque for a hysteresis motor constructed with a rotor containing 20 cm^3 of hysteresis material having a B–H curve area of 16 kJ/m^3?

7.2. A certain hysteresis motor with a three-phase, four-pole stator having $r_1 = 40$ Ω draws locked-rotor current of 0.35 A at 0.40 lagging power factor from the 120–208-V, 60-Hz line. The 5-cm-long rotor has hysteresis material 5 mm thick with an outside diameter of 6 cm. Assume stator core losses are constant at 5% and the rotor exhibits ideal hysteresis characteristics. (a) Calculate the area inside the B–H curve. (b) Find the power dissipated in the rotor under locked-rotor conditions. (c) Find the developed torque during acceleration.

7.3. A three-phase, six-pole linear induction motor has a stationary primary 1.0 m long. The secondary surface resistivity is $3.4 \cdot 10^{-6}$ Ω and the air gap is 1 cm. If the steady-state slip is 5%, find the secondary speed. Is the low-speed approximation valid for this linear motor?

7.4. A three-phase, eight-pole linear induction motor has a stationary secondary constructed of 1-cm-thick aluminum. (The surface resistivity is the material's volume resistivity divided by its thickness.) The primary is 10 m long and the air gap is 2.5 cm. If the steady-state slip is 4%, find the secondary speed. Is the low-speed approximation valid for this linear motor?

7.5. For the stacked-can, permanent-magnet stepper motor of Figure 7.13, i_a and i_b represent the currents in the two stator cans. Taking the equilibrium point when $i_a > 0$ and $i_b = 0$ as the zero-angle reference, construct the full-step, single-winding table for one complete electrical cycle.

7.6. A variable reluctance stepper motor is constructed with the stator of Figure 7.14 and a four-pole rotor similar to that shown for the switched reluctance motor of Figure 7.18. (a) Construct a table similar to Figure 7.14*b* for a complete electrical cycle of the full-step, single-winding sequence. (b) Repeat for the full-step, two-winding sequence. (c) Repeat for the half-step sequence.

7.7. For the variable reluctance stepper motor of Figure 7.15, construct the full-step, single-winding table for one complete electrical cycle. Take the equilibrium point when $i_a > 0$ and $i_b = i_c = 0$ as the zero-angle reference.

7.8. Repeat Problem 7.5 for the half-step sequence.

8
FORCES AND TORQUES IN ELECTROMAGNETIC SYSTEMS

8.1 THE ENERGY CONVERSION APPROACH TO FORCE OR TORQUE CALCULATION

It is often desirable to know the forces being exerted on the various parts of some system that contains stored energy. One way to find the force on a given part is to imagine that that part moves a small distance Δx. The change in stored energy which results from this motion is taken to be the work done on the part during the motion, ΔW, in joules. If Δx is in meters, the force, in newtons, is given by

$$ f = \frac{\Delta W}{\Delta x} $$

This is the *principle of virtual work*. Before this method can be applied with confidence to electromagnetic systems, its validity must be proved and the limits of its applicability understood.

8.2 ENERGY STORED IN A MAGNETIC FIELD

Electrical engineers always seem to approach problems from the circuit point of view; so to be comfortable, the energy in a magnetic field will be discussed in terms of an inductive circuit. Consider the electromagnetic system of Figure 8.1 consisting of a magnetic circuit excited by a coil of N turns. By Kirchhoff's law, the voltage applied to the electrical circuit terminals is given by

$$ v = ir + e \tag{8.1} $$

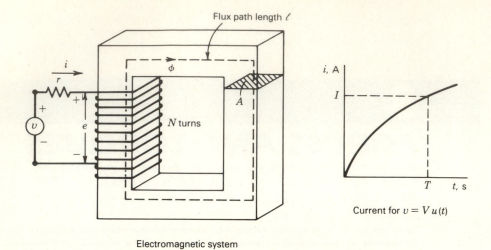

Electromagnetic system

FIGURE 8.1 Electromagnetic system to illustrate energy storage.

where e is the voltage induced in the coil by the magnetic flux. The instantaneous power input to the system is given by

$$p = vi = i^2r + ei \quad \text{W} \tag{8.2}$$

Suppose that a dc voltage is applied to the circuit at time $t = 0$ and that at the end of T seconds the current has attained a value of I amperes. The energy input to the system during this interval is

$$W_{\text{in}} = \int_0^T p \, dt = \int_0^T i^2r \, dt + \int_0^T ei \, dt \quad \text{J} \tag{8.3}$$

It is seen that the energy expression has been split into two parts: the portion dissipated as resistance loss in the winding and the remainder, which is stored in the magnetic field:

$$W_\phi = \int_0^T ei \, dt \quad \text{J} \tag{8.4}$$

Now if λ is the magnetic-field flux linkage with the N-turn coil,

$$\lambda = N\overline{\phi} \tag{8.5}$$

(See Chapter 3, Equations 3.5–3.13, for fuller explanations of these terms.) Then by Faraday's law,

$$e = \frac{d\lambda}{dt} = N\frac{d\overline{\phi}}{dt} \quad \text{V} \tag{8.6}$$

Then

$$W_\phi = \int_0^T \frac{d\lambda}{dt}i \, dt = \int_0^{\lambda_T} i \, d\lambda \quad \text{J} \tag{8.7}$$

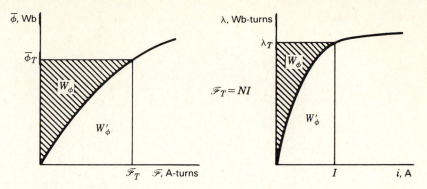

FIGURE 8.2 Alternative magnetization curves showing magnetic-field energy and coenergy.

or alternatively,

$$W_\phi = \int_0^T N\frac{d\overline{\phi}}{dt} i\, dt = \int_0^T Ni\frac{d\overline{\phi}}{dt}\, dt = \int_0^{\phi_T} \overline{\mathscr{F}}\, d\overline{\phi}\quad \text{J} \tag{8.8}$$

These equations show, as illustrated in Figure 8.2, that the energy stored in the magnetic field is equal to the area between the magnetization curve for the system and the flux axis or the flux linkage axis, depending on whether Equation 8.7 or Equation 8.8 is used.

Coenergy

The area below the magnetization curve, that is, the area between the curve and the current or MMF axis, is called the *coenergy*. (Should the units be "cojoules?") This quantity is sometimes useful in determining forces in a system. The symbol for coenergy is W_ϕ', and the coenergy is given by

$$W_\phi' = \phi_T \mathscr{F}_T - W_\phi = \lambda_T I - W_\phi \tag{8.9}$$

If the magnetization curve is assumed to have no hysteresis, then for each value of i there would be only one value of λ, and with this assumption,

$$W_\phi' = \mathscr{F}\phi - W_\phi = i\lambda - W_\phi \tag{8.10}$$

where W_ϕ is determined for the same value of i. It should be noted that, although the approach to the magnetic field was through the electric circuit, the energy and coenergy of the field have been found in terms of magnetic quantities—flux and MMF.

Field Energy in Magnetically Linear Systems

In the nonlinear system which has been discussed, the magnetization curves may be expressed by the nonlinear equations

$$\phi = \mathscr{F}[\mathscr{P}(\phi)] \qquad \text{or} \qquad \lambda = i[\mathscr{L}(\lambda)]$$

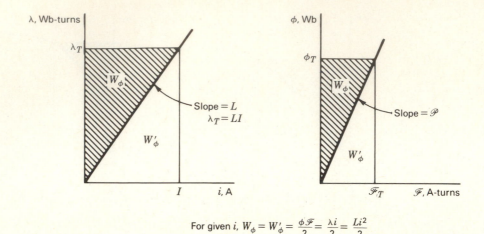

For given i, $W_\phi = W'_\phi = \dfrac{\phi \mathscr{F}}{2} = \dfrac{\lambda i}{2} = \dfrac{Li^2}{2}$

FIGURE 8.3 Alternate magnetization curves, field energy, and coenergy in a linear electromagnetic system.

where $\mathscr{P}(\phi)$ is a nonlinear permeance and $\mathscr{L}(\lambda)$ is a nonlinear inductance. In a linear system, the permeance and inductance are constant:

$$\phi = \mathscr{P}\mathscr{F} \qquad \mathscr{P} \text{ constant}$$

$$\lambda = Li \qquad L \text{ constant}$$

If the same experiment as before is performed on the system of Figure 8.1, now assumed to be linear, the voltage induced by the field flux may also be written in terms of the inductance:

$$e = N\frac{d\overline{\phi}}{dt} = \frac{d\lambda}{dt} = L\frac{di}{dt} \tag{8.11}$$

The expressions for the field energy in Equations 8.7 and 8.8 are still valid, and in addition one may write

$$W_\phi = \int_0^T ei\, dt = \int_0^T L\frac{di}{dt}i\, dt = \int_0^I Li\, di = \frac{1}{2}LI^2 \tag{8.12}$$

Figure 8.3 shows the alternate magnetization curves and field energy for linear systems. It is seen that the energy and coenergy are represented by triangles of equal area in this case. Again, since for each value of i or $\mathscr{F}$ there is only one corresponding value of λ or ϕ, the expressions for the field energy and coenergy may be written for any current as

$$W_\phi = W'_\phi = \frac{\phi\mathscr{F}}{2} = \frac{\lambda i}{2} = \frac{Li^2}{2} \qquad \text{LINEAR ONLY} \tag{8.13}$$

Energy Density in a Linear Medium

Consider a volume of material or space within a magnetic field having constant permeability μ. The cross-sectional area, measured perpendicular to the flux, is A m^2, and the length ℓ m is measured parallel to the applied H field, such that the MMF applied to the volume is $\mathcal{F} = H\ell$. Then the energy density of the field within that volume is

$$w_{\text{fld}} = \frac{\text{energy in volume}}{\text{volume}} = \frac{\phi\mathcal{F}/2}{A\ell} = \frac{1}{2}\left(\frac{\phi}{A}\frac{\mathcal{F}}{\ell}\right) = \frac{BH}{2} \quad \text{J/m}^3 \qquad (8.14)$$

or, since $B = \mu H$,

$$w_{\text{fld}} = \frac{\mu H^2}{2} = \frac{B^2}{2\mu} \quad \text{J/m}^3 \qquad (8.15)$$

8.3 FORCES OR TORQUES IN LINEAR SYSTEMS WITH ONE ELECTRICAL INPUT

The work done on an object in a mechanical system is given by

$$W_{\text{mech}} = \int f\, dx \quad \text{J} \qquad \text{for linear motion}$$

and $\qquad (8.16)$

$$W_{\text{mech}} = \int \tau\, d\theta \quad \text{J} \qquad \text{for rotation}$$

The energy conversion approach to determining force or torque is to find first an expression for the differential mechanical output dW_{mech}. Then from Equations 8.16,

$$dW_{\text{mech}} = f\, dx \qquad \text{or} \qquad dW_{\text{mech}} = \tau\, d\theta \qquad (8.17)$$

and

$$f = \frac{dW_{\text{mech}}}{dx}\, \text{N} \qquad \text{or} \qquad \tau = \frac{dW_{\text{mech}}}{d\theta}\, \text{N-m} \qquad (8.18)$$

The problem in determining dW_{mech} changes, depending on whether the system is linear or nonlinear and on whether the currents are known or the fluxes (or flux linkages) are known. In all cases, however, conservation of energy is the basis for calculating dW_{mech}. This is a shortcoming of the method, because a conservative magnetic system must be assumed and the effects of core loss on the developed forces are not accounted for. However the method is no worse than others in this regard. One of its beauties is that its application to a few general systems allows one to develop considerable insight into the directions of forces on parts of a system and the ability to judge whether a system will develop any force at all.

Consider a magnetically operated plunger, or *solenoid*, such as is shown in Figure 8.4. The magnetic circuit will be assumed to be linear. The problem is to find the force exerted on the plunger by the magnetic field as a function of x for two cases:

Case I: The coil current or MMF is known.
Case II: The flux or flux linkage is known.

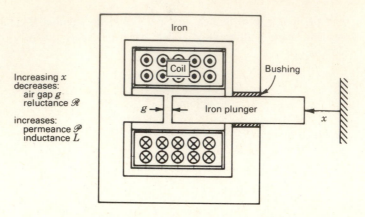

FIGURE 8.4 Magnetically operated plunger.

CASE I. COIL CURRENT OR MMF IS KNOWN

The current is assumed constant at the known value. The plunger is allowed to move a differential distance dx in the positive x direction. This motion will cause the air gap g to decrease, with a resultant decrease in reluctance and a differential increase in flux, and a corresponding differential increase in flux linkage with the coil, $d\lambda$. When there is no motion, the voltage applied to the coil is all used up as ir drop, and no energy is supplied to the magnetic field by the electric circuit. As the flux changes

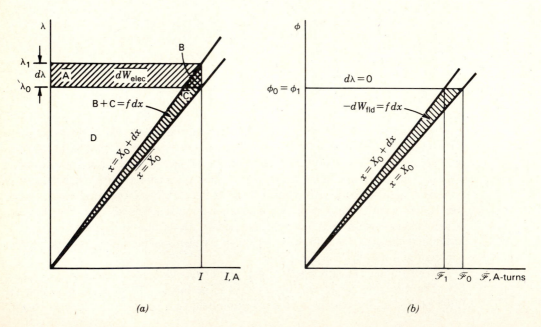

FIGURE 8.5 Differential energy changes in a magnetically linear system as the slope of the magnetization curve changes with displacement (*a*) for constant current (*b*) for constant flux.

during the motion, however, a voltage is induced in the coil, which allows the electric circuit to supply some energy to the field:

$$dW_{elec} = eI \, dt = \frac{d\lambda}{dt} I \, dt = I \, d\lambda \quad J \tag{8.19}$$

This energy input is shown as the horizontal shaded area in Figure 8.5a. In this figure, four energy-related areas are labeled A, B, C, and D. Thus the differential electrical energy input is given by

$$dW_{elec} = d\lambda \, I = (\lambda_1 - \lambda_0)I = A + B \quad J \tag{8.20}$$

Since core losses are being neglected, the law of conservation of energy indicates that this input energy must either be stored in the magnetic field or appear as mechanical output:

$$dW_{elec} = dW_\phi + f \, dx \tag{8.21}$$

Referring to Figure 8.5a, the decrease in gap length results in an increase in slope of the magnetization curve. The field energy before the movement was

$$W_0 = \frac{\lambda_0 I}{2} = C + D$$

while that after the movement is

$$W_1 = \frac{\lambda_1 I}{2} = A + D$$

The differential increase in field storage is

$$dW_\phi = W_1 - W_0 = \frac{(\lambda_1 - \lambda_0)I}{2} = \frac{dW_{elec}}{2}$$
$$= (A + D) - (C + D) = A - C, \, J \tag{8.22}$$

Substituting the area relationships into the conservation of differential energy equation, Equation 8.21,

$$A + B = A - C + f \, dx \tag{8.23}$$

Solving for the differential mechanical output,

$$f \, dx = B + C \tag{8.24}$$

Note that the mechanical output is the area bounded by the initial and final magnetization curves and the current locus.

Equation 8.22 in its first form shows that *in this case* exactly half of the electrical energy input is stored in the magnetic field and half is available for mechanical output. Equation 8.21 becomes

$$2 \, dW_\phi = dW_\phi + f \, dx$$

Solving for the force,

$$f = \frac{dW_\phi}{dx} \quad \text{N} \tag{8.25}$$

Thus in a linear system with a given current, the force is equal to the rate at which the field energy changes with displacement. Equation 8.25 is sign sensitive. That is, if the field energy decreases as x increases, the force is in the negative-x direction.

If the mechanical output is rotational, the torque is found by taking the derivative of the field energy with respect to the output angle of rotation, θ. Also since the field energy is a function of current as well as displacement, better bookkeeping results if the force or torque is expressed as a partial derivative, with an indication as to what quantity is assumed constant:

$$f = \left.\frac{\partial W_\phi}{\partial x}\right|_I \quad \text{N} \qquad \tau = \left.\frac{\partial W_\phi}{\partial \theta}\right|_I \quad \text{Nm} \tag{8.26}$$

If the field energy is written in terms of the inductance, an interesting and instructive result is obtained:

$$f = \frac{d}{dx}\left(\frac{1}{2}LI^2\right) = +\frac{I^2}{2}\frac{dL}{dx} \tag{8.27}$$

Note that in a system with a single electrical input (and no permanent magnets), the force developed is proportional to the square of the current and is in a direction to *increase* the inductance, that is, to increase the flux linkage per ampere. Since the square of the current is involved, the direction of the force is independent of the direction of the current.

■ Example 8.1

What is the direction of the electromagnetic force on the conductors in a coil side located in a slot in the rotor or stator core of a machine? Conductors in the slots in machines are largely shielded from the air gap flux field by the iron teeth. If a slight

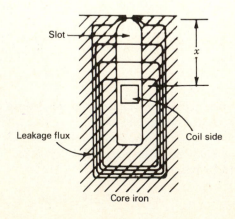

motion of a current-carrying coil within its slot is imagined, either in the direction of rotation or in the opposite direction, hardly any change in field energy results. Therefore there is essentially no force on the conductors in the direction of rotation. However the slot leakage flux due to the coil current crosses the slot above the conductor and returns through the core iron below the slot, as shown in the figure. By Ampère's law, the MMF across the slot above the coil side is simply $N_c i$, the ampere-conductors of the coil side, and there is no MMF across the slot below the coil. If the coil were to move upward toward the slot opening, the leakage flux linking the coil would be reduced; that is, dL/dx is negative. Therefore the electromagnetic force on a current-carrying coil side is such as to push it downward into the slot. Because this force is proportional to the square of the current, its direction is independent of the current direction.   ■

■ **Example 8.2**

A solenoid such as that of Figure 8.4 is provided with a laminated core and may be used with either ac or dc. Over a limited range of displacement x, the coil inductance is given by

$$L(x) = 0.30 + 270x^2 \text{ H}$$

where x is in meters. For $x = 1$ cm, (a) find the instantaneous, average and peak forces on the plunger when the coil current is 1.0 A, rms, at 60 Hz, (b) find the approximate voltage which must be applied to the coil to cause this current to flow, (c) find the force for a coil current of 1.0 A dc and compare with the average force found in part (a).

Solution

(a) From Equation 8.26,

$$f = 0.5i^2 \frac{dL}{dx}$$

$$i^2 = (\sqrt{2} \cdot 1.0 \sin \omega t)^2 = 2(0.5 - 0.5 \cos 2 \cdot 377t)$$

$$\frac{dL}{dx} = \frac{d}{dx}(0.30 + 270x^2) = 540x \quad \text{H/m} \qquad x = 0.01 \text{ m}$$

$$f = 0.5(1 - \cos 754t) \cdot 5.4 = 2.7 - 2.7 \cos 754t \quad \text{N} \qquad \text{Answer}$$

It is seen that the force varies from zero to 5.4 N peak (Answer) at twice the frequency of the current and has an average value of 2.7 N (Answer).

(b) The coil inductance at $x = 0.01$ m is

$$L(0.01) = 0.03 + 270 \cdot 0.01^2 = 0.057 \quad \text{H}$$

$$X_L = \omega L = 377 \cdot 0.057 = 21.5 \ \Omega$$

Neglecting coil resistance, the voltage to be applied for 1.0 A current is

$$V = IX_L = 1.0 \cdot 21.5 = 21.5 \text{ V} \qquad \text{Answer}$$

In a device of these dimensions, the resistance would not be a negligible quantity.

For a direct current of 1 A,

$$f = \frac{(1.0)^2}{2} (540 \cdot 0.01) = 2.7 \text{ N} \qquad \text{Answer}$$

Note that when force is proportional to the square of the current, a given value of rms ac current will produce an average force equal to that produced by the same value of dc current. This will not hold true if the iron of the device is driven into saturation. ∎

CASE II. FLUX OR FLUX LINKAGE IS KNOWN

If in a magnetically linear system the force is to be calculated for a known flux, a differential motion is assumed, with the flux constant at the known value. The effect is shown in Figure 8.5*b*. In this case, since $d\lambda$ is zero, no voltage is induced in the coil and there can be no electrical input. The mechanical output must then be at the expense of a decreased magnetic-field energy:

$$dW_{\text{elec}} = 0 = dW_\phi + f\,dx \tag{8.27}$$
$$f\,dx = -dW_\phi$$

In this case, the force is given by

$$f = -\left.\frac{\partial W_\phi}{\partial x}\right|_\Phi \text{ N} \tag{8.28}$$

Or for rotation,

$$\tau = -\left.\frac{\partial W_\phi}{\partial \theta}\right|_\Phi \text{ N-m} \tag{8.29}$$

RELUCTANCE FORCE OR TORQUE

There are essentially two kinds of force developed in electromechanical devices. The first is that due to the interaction of the fields produced by the currents in two windings which may move relative to each other. The second, usually called *reluctance force*, or *reluctance torque*, is dependent on the current in only one winding and is the result of variations in the reluctance of the air gap in the magnetic circuit carrying the flux which links that winding. Both of these phenomena are often active simultaneously.

For a given displacement, the field energy in a linear system is given by Equation 8.13:

$$W_\phi = \frac{\mathscr{F}\phi}{2} \text{ J} \tag{8.30}$$

Since the MMF may be written $\mathcal{F} = \phi\mathcal{R}(x)$, where $\mathcal{R}$ is the magnetic reluctance,

$$W_\phi = \frac{\phi^2 \mathcal{R}(x)}{2} \tag{8.31}$$

and

$$f = -\left.\frac{\partial W_\phi}{\partial x}\right|_\Phi = -\frac{\phi^2}{2}\frac{d\mathcal{R}}{dx}\ \text{N} \qquad \text{or} \qquad \tau = -\frac{\phi^2}{2}\frac{d\mathcal{R}}{d\theta}\ \text{N-m} \tag{8.32}$$

Example 8.3

A small air gap separates a piece of iron from the pole piece of a magnet. The area of the pole piece opposite the iron is A m^2. The flux density in the air gap is B T. The length of the gap is g m. Find the force by which the iron is attracted to the magnet.

Solution. The reluctance of the gap is given by

$$\mathcal{R} = \frac{g}{\mu_0 A}$$

Applying Equation 8.32 to find the force acting to increase g and taking into account that $\phi = BA$,

$$f = -\frac{B^2 A^2}{2}\frac{d}{dg}\left(\frac{g}{\mu_0 A}\right) = -\frac{B^2 A^2}{2}\frac{1}{\mu_0 A} = -\frac{B^2 A}{2\mu_0}\ \text{N}$$

The minus sign in this well-known expression indicates that the force is in the $-g$ direction, that is, in a direction to *decrease* the gap. ∎

8.4 LINEAR SYSTEMS WITH TWO ELECTRICAL INPUTS

Additional, generally useful insights into the behavior of electromagnetic devices may be gained by studying a doubly excited, magnetically linear system, that is, one with two electrical input terminal pairs. An example of such a system is depicted in Figure 8.6. Again the resistances of the two windings, r_1 and r_2, have been drawn outside of the windings with the assistance of Thévenin and are not considered part of the energy conversion analysis. Since the reluctance of the magnetic circuit will vary with rotor position, the self-inductance of the stator winding, L_{11}, the self-inductance of the rotor winding, L_{22}, and the mutual inductance L_{12} will all be functions of θ.

How can the induced voltages be calculated when the inductances are variables? The secret is: Write expressions for the flux linkages. The development leading to Equations 3.26 and 3.32 in Chapter 3 relates the flux linkages to the inductances and permeances of a doubly excited system, in that case a transformer. The difference between a transformer and a device like that of Figure 8.6 is that the permeances and mutual inductance of the latter are variables. Also as a matter of convenience, positive

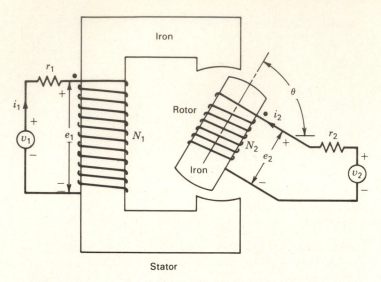

FIGURE 8.6 Example of a doubly excited electromagnetic system.

secondary current in a transformer was taken as coming out of the dotted terminal. For reasons of symmetry, i_2 in Figure 8.6 is taken as entering the dotted terminal, thus changing the sign on the mutual-inductance voltages relative to those in Chapter 3. With these things in mind, the following expressions for the flux linkages of the windings in Figure 8.6 are implied by Equations 3.26 and 3.32:

$$\lambda_1 = L_{11}(\theta)\, i_1 + L_{12}(\theta)\, i_2$$
$$\lambda_2 = L_{22}(\theta)\, i_2 + L_{12}(\theta)\, i_1 \tag{8.33}$$

The induced voltages are found by taking the total derivatives of these flux linkages with respect to time (Faraday's law):

$$e_1' = \frac{d\lambda_1}{dt} = L_{11}\frac{di_1}{dt} + i_1\frac{dL_{11}}{d\theta}\frac{d\theta}{dt} + L_{12}\frac{di_2}{dt} + i_2\frac{dL_{12}}{d\theta}\frac{d\theta}{dt} \tag{8.34}$$

$$e_2' = \frac{d\lambda_2}{dt} = L_{22}\frac{di_2}{dt} + i_2\frac{dL_{22}}{d\theta}\frac{d\theta}{dt} + L_{12}\frac{di_1}{dt} + i_1\frac{dL_{12}}{d\theta}\frac{d\theta}{dt} \tag{8.35}$$

Collecting terms and recognizing that $d\theta/dt$ is the angular velocity ω,

$$e_1' = L_{11}\frac{di_1}{dt} + L_{12}\frac{di_2}{dt} + \omega\left[i_1\frac{dL_{11}}{d\theta} + i_2\frac{dL_{12}}{d\theta}\right]$$

$$e_2' = L_{22}\frac{di_2}{dt} + L_{12}\frac{di_1}{dt} + \omega\left[i_2\frac{dL_{22}}{d\theta} + i_1\frac{dL_{12}}{d\theta}\right] \tag{8.36}$$

The first two terms in each of these expressions are the same as for a transformer, and their sum is called the *transformer voltage*. The remaining part, proportional to ω, is called the *speed voltage*. The quantities $dL_{11}/d\theta$, $dL_{22}/d\theta$, and $dL_{12}/d\theta$ are functions of θ that are fixed by the design of the device.

The electrical power input to the device is

$$P_{\text{elec}} = \frac{dW_{\text{elec}}}{dt} = e_1' i_1 + e_2' i_2$$

$$= \left[L_{11}\frac{di_1}{dt} + L_{12}\frac{di_2}{dt} + \left(i_1\frac{dL_{11}}{d\theta} + i_2\frac{dL_{12}}{d\theta} \right) \frac{d\theta}{dt} \right] i_1 \qquad (8.37)$$

$$+ \left[L_{22}\frac{di_2}{dt} + L_{12}\frac{di_1}{dt} + \left(i_2\frac{dL_{22}}{d\theta} + i_1\frac{dL_{12}}{d\theta} \right) \frac{d\theta}{dt} \right] i_2$$

Then the differential electrical-input energy is given by

$$dW_{\text{elec}} = P_{\text{elec}}\, dt$$

$$= L_{11}i_1\, di_1 + L_{12}i_1\, di_2 + i_1^2\, dL_{11} + i_1 i_2\, dL_{12}$$

$$+ L_{22}i_2\, di_2 + L_{12}i_2\, di_1 + i_2^2\, dL_{22} + i_1 i_2\, dL_{12} \qquad (8.38)$$

$$= L_{11}i_1\, di_1 + L_{22}i_2\, di_2 + i_1^2\, dL_{11} + i_2^2\, dL_{22}$$

$$+ 2i_1\, i_2\, dL_{12} + L_{12}\, (i_1\, di_2 + i_2\, di_1)$$

FIELD ENERGY

To find an expression for the energy stored in the magnetic field of the device, the same technique will be applied as in the case of the singly excited system. The rotor will be held at a fixed angle. As a result, there will be no mechanical output ($\tau\, d\theta = 0$) and all inductances will be constant; that is, the differentials of inductances will all be zero. Under these conditions,

$$dW_{\text{elec}} = dW_\phi + 0 \qquad (8.39)$$

and

$$dW_{\text{elec}} = L_{11}i_1\, di_1 + L_{22}i_2\, di_2 + L_{12}(i_1\, di_2 + i_2\, di_1) \qquad (8.40)$$

Note that $(i_1\, di_2 + i_2\, di_1) \equiv d(i_1 i_2)$. Then the field energy is given by

$$W_\phi = \int dW_{\text{elec}} = \frac{1}{2}i_1^2\, L_{11} + \frac{1}{2}i_2^2\, L_{22} + i_1\, i_2\, L_{12}, \quad J \qquad (8.41)$$

TORQUE

The torque exerted on the rotor is found by conservation of energy:

$$dW_{\text{elec}} = dW_\phi + \tau\, d\theta \qquad (8.42)$$

where dW_{elec} is expressed in Equation 8.38 and dW_ϕ is the total differential of Equation 8.41:

$$dW_\phi = d\left(\frac{1}{2}i_1^2 L_{11} + \frac{1}{2}i_2^2 L_{22} + i_1 i_2 L_{12}\right)$$

$$= i_1 L_{11}^* \, di_1 + \frac{1}{2}i_1^2 \, dL_{11} + i_2 L_{22}^* \, di_2 + \frac{1}{2}i_2^2 \, dL_{22} \qquad (8.43)$$

$$+ i_2 L_{12}^* \, di_1 + i_1 L_{12}^* \, di_2 + i_1 i_2 \, dL_{12}$$

When Equations 8.38 and 8.43 are substituted into Equation 8.42, the conservation-of-energy equation, it is found that the terms marked with an asterisk in the field energy Equation 8.43 cancel identical terms in the electrical input Equation 8.38. These terms each involve differential currents, and these components of the electrical input go into storage without affecting the mechanical output. (Such a phenomenon did not occur in the singly excited system.) The remaining terms in dW_ϕ are exactly half of the remaining terms of dW_{elec}. Then Equation 8.42 is reduced to

$$\tau \, d\theta = \frac{1}{2}i_1^2 \, dL_{11} + \frac{1}{2}i_2^2 \, dL_{22} + i_1 i_2 \, dL_{12}$$

or

$$\tau = \frac{1}{2}i_1^2 \frac{dL_{11}}{d\theta} + \frac{1}{2}i_2^2 \frac{dL_{22}}{d\theta} + i_1 i_2 \frac{dL_{12}}{d\theta} \text{ N-m} \qquad (8.44)$$

The first two terms are reluctance torques. The last is the torque due to interaction of the two current-carrying windings. Note that the directions of the two reluctance torques are such as to increase the self-inductances of the windings and are independent of current directions. The torque that results from the variation of L_{12} with θ is in a direction to increase the mutual inductance when the two currents are aiding and to decrease the mutual inductance when the currents are opposing. The currents are aiding when they are both flowing into dotted terminals or both out of dotted terminals. They are opposing when one current is entering a dotted terminal and the other is leaving its dotted terminal.

8.5 NONLINEAR SYSTEMS

Figure 8.7 shows magnetization curves for an electromagnetic system, such as the solenoid of Figure 8.4, when iron is operated in the saturated region. Curves are shown for air gaps corresponding to displacements $x = X_0$, the initial plunger position, and after an additional differential displacement $x = X_0 + dx$. These curves could just as well have been shown as functions of λ vs. i.

Equations 8.1 through 8.10 apply to both linear and nonlinear systems, as do Equations 8.19 through 8.24, with the obvious exception of Equation 8.22, which assumes straight-line magnetization curves. (**Equations 8.25 through 8.27 do NOT apply to a nonlinear system.**) Equation 8.20 may be written

$$dW_{\text{elec}} = NI \, d\phi = N(\phi_1 - \phi_0)I = (\lambda_1 - \lambda_0)I = \text{A} + \text{B} \quad \text{J} \qquad (8.45)$$

Again the magnetic-field energy before the change in x is

$$W_{\phi 0} = \text{area C} + \text{area D}$$

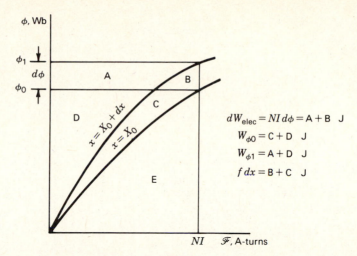

FIGURE 8.7 Effect of mechanical motion on the magnetization curve of a nonlinear electromagnetic system, showing energy relationships.

while after the differential displacement dx occurs, the field energy is

$$W_{\phi 1} = A + D$$

The change in field energy resulting from the displacement is thus

$$dW_\phi = W_{\phi 1} - W_{\phi 0} = A - C \quad J \tag{8.46}$$

and by Equations 8.21, 8.23, and 8.24 (conservation of energy), the differential mechanical output is

$$f \, dx = dW_{elec} - dW_\phi = B + C \quad J \tag{8.47}$$

As in the linear system, the differential mechanical output is the area between the two magnetization curves.

It should be noted that the area B is actually negligibly small, since its vertical dimension is $d\phi$ and its horizontal dimension is $d\mathscr{F}$.

Force or Torque in a Nonlinear System Having One Electrical Input

As in a linear system, the approach to be used in calculating force or torque will depend on whether the MMF (or current) is known, or the flux (or flux linkage) is known. These will be called Case I and Case II, respectively, as before.

CASE I. MMF (OR CURRENT) IS KNOWN

When the MMF is taken to be constant, the energy situation is that illustrated in Figure 8.8b. Nonlinear magnetization curves destroy the neat division of electrical energy into two equal parts, which was found in Equation 8.22. A simple expression for the force cannot be written in terms of the field energy. However it is interesting to see

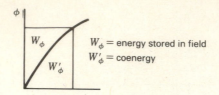

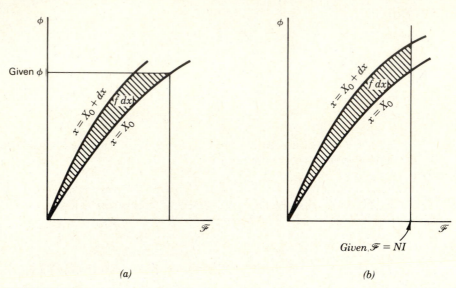

FIGURE 8.8 Differential energy relationships in a nonlinear electromagnetic system. (*a*) For fixed ϕ, $fdx = -dW_\phi$. (*b*) For fixed $\mathcal{F}$, $fdx = +dW_\phi'$.

what happens to the coenergy as a result of displacement. Again refer to Figure 8.7. The initial coenergy is

$$W_{\phi 0}' = \phi_0 NI - W_{\phi 0} = \text{area E} \tag{8.48}$$

After the displacement, the coenergy has become

$$W_{\phi 1}' = \phi_1 NI - W_{\phi 1} = E + B + C \tag{8.49}$$

Then the change in coenergy resulting from the differential motion dx is given by

$$dW_\phi' = W_{\phi 1}' - W_{\phi 0}' = (E + B + C) - E = B + C = f \, dx$$

or

$$f \, dx = dW_\phi' \tag{8.50}$$

Then for Case I,

$$f = \left.\frac{\partial W_\phi'}{\partial x}\right|_I \text{N} \quad \text{or} \quad \tau = \left.\frac{\partial W_\phi'}{\partial \theta}\right|_I \text{Nm} \tag{8.51}$$

CASE II. FLUX (OR FLUX LINKAGE) IS KNOWN

If the flux or flux linkage is known or specified, it is taken to be constant. Then $d\lambda = 0$, no voltage can be induced in the coil, and there can be no electrical input to the system as a result of the mechanical displacement. As illustrated in Figure 8.8a, the mechanical output energy results in a depletion of the energy stored in the magnetic field. There is little difference between this situation and that of Case II for the linear system:

$$f \, dx = -dW_\phi \tag{8.52}$$

and as in the linear system,

$$f = -\left.\frac{\partial W_\phi}{\partial x}\right|_\Phi \quad \text{N} \quad \text{or} \quad \tau = -\left.\frac{\partial W_\phi}{\partial \theta}\right|_\Phi \quad \text{N-m} \tag{8.53}$$

FOR FURTHER STUDY

FORCES/TORQUES IN A GENERAL CONSERVATIVE ELECTROMAGNETIC SYSTEM

To obtain a somewhat deeper insight and a more solid basis for the application of the principle of virtual work, a general, nonlinear but conservative system will be examined. Such a system is shown schematically in Figure 8.9. "Conservative" means that there are no energy losses within the system. Although nonlinear, the system has no hysteresis, and i^2r and eddy current losses are excluded. As indicated, the system has n electrical input terminal pairs and m mechanical outputs. An

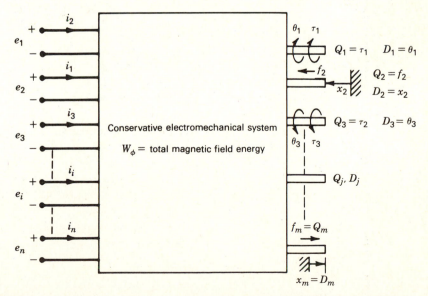

FIGURE 8.9 General conservative electromagnetic system having n electrical input terminal pairs and m mechanical outputs.

example of a system with two mechanical outputs is a motor having equal rotor and stator stack lengths, with the rotor displaced axially from the stator. In addition to the normal torque there is an axial force developed in a direction to align the rotor and stator cores, hence two outputs.

Notation and Definitions

Mechanical outputs may be rotational or involve linear motion or even helical or other kinds of motion. For generality, generalized forces Q_j and generalized displacements D_j will be used as mechanical output quantities, such that the power output at the jth mechanical terminal is

$$p_j = Q_j \frac{dD_j}{dt} \quad \text{W} \qquad \text{and} \qquad dW_j = p_j \, dt = Q_j \, dD_j \quad \text{J} \qquad (8.54)$$

For example, if the fifth mechanical terminal is a rotating shaft, then $Q_5 = \tau_5$, the torque delivered by that shaft, and $D_5 = \theta_5$, the angle of rotation of that shaft. Note that it is assumed that Q_j is in the direction of positive D_j. The use of these generalized quantities is valid if it can be shown that the output forces can be determined independently.

The differential mechanical output from the entire system is then

$$dW_{\text{mech}} = \sum_{j=1}^{m} Q_j \, dD_j \qquad (8.55)$$

A fixed or chosen set of displacements will be represented by a vector $[D_j]_k$ (k implying constant values) such that

$$[D_j]_k = \begin{bmatrix} D_{1k} \\ D_{2k} \\ \vdots \\ D_{jk} \\ \vdots \\ D_{mk} \end{bmatrix}$$

where each D_{jk} is a fixed value of that displacement. The power input to the ith electrical terminal pair is

$$p_i = e_i i_i = \frac{d\lambda_i}{dt} i_i \quad \text{W} \qquad \text{and} \qquad dW_i = e_i i_i \, dt = i_i \, d\lambda_i \quad \text{J} \qquad (8.56)$$

where λ_i is the flux linkage of the system's magnetic field with that part of the internal winding connected between the ith pair of terminals. Then the differential electrical input to the system is

$$dW_{\text{elec}} = \sum_{i=1}^{n} i_i \, d\lambda_i \quad \text{J} \qquad (8.57)$$

Finally, the expression

$$W = W(x, y, z)$$

simply means that the quantity W is a function of the variables x, y, and z.

Derivation of Force Relationships

In a conservative system such as the one being investigated, a differential mechanical output may cause a corresponding differential quantity of electrical energy to be drawn from the sources connected to the electrical terminals. If the amount of this electrical energy differs from that of the mechanical output energy, the difference will either be stored as magnetic-field energy or will result in a reduction in the stored energy:

$$dW_\phi = dW_{elec} - dW_{mech} \tag{8.58}$$

In this expression, dW_ϕ is the differential increase in energy stored in the magnetic field. Substituting from Equations 8.55 and 8.57,

$$dW_\phi = \sum_{i=1}^{n} i_i \, d\lambda_i - \sum_{j=1}^{m} Q_j \, dD_j \tag{8.59}$$

Another equation is needed to solve for the forces Q_j. This will require mathematical blue smoke and mirrors. Obviously,

$$W_\phi = W_\phi(\lambda_1, \lambda_2, \ldots, \lambda_i, \ldots, \lambda_n, D_1, \ldots, D_j, \ldots, D_m) \tag{8.60}$$

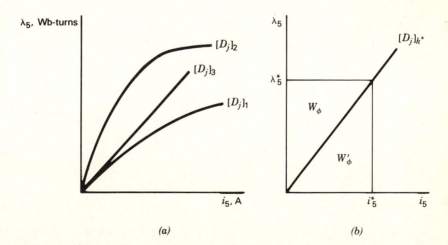

FIGURE 8.10 Illustration of possible conditions at a typical electrical terminal pair ($i = 5$). (a) Effect of displacement vector on the magnetization curve, showing that for given displacements, the flux linkage is dependent on the current. (b) Relationship of energy to coenergy at a typical terminal pair in a linear system, for a given displacement vector $[D_j]_k^*$ and a given current i_5^*.

or

$$W_\phi = W_\phi(i_1, i_2, \ldots, i_i, \ldots, i_n, D_1, \ldots, D_j, \ldots, D_m) \qquad (8.61)$$

(Why is W_ϕ not a function of the λ_i's, *and* the i_i's and the D_j's? The reason is that the λ_i's and the i_i's are not independent. For a given set of displacements $[D_j]_k$, there is a fixed magnetization curve for each electrical terminal pair. This fact is illustrated by Figure 8.10a for three hypothetical displacement vectors, taking the flux linkage and current at the fifth terminal pair ($i = 5$) as an example.)

Equation 8.60 will be used as a basis for writing an expression for the differential increase in field energy. This is equivalent to specifying that the forces will be determined for a given set of λ's, $[\lambda_i]_k$:

$$dW_\phi = \frac{\partial W_\phi}{\partial \lambda_1} d\lambda_1 + \frac{\partial W_\phi}{\partial \lambda_2} d\lambda_2 + \cdots + \frac{\partial W_\phi}{\partial D_1} dD_1 + \cdots + \frac{\partial W_\phi}{\partial D_m} dD_m$$

$$= \sum_{i=1}^{n} \frac{\partial W_\phi}{\partial \lambda_i} d\lambda_i + \sum_{j=1}^{m} \frac{\partial W_\phi}{dD_j} dD_j \qquad (8.62)$$

This is the second equation needed to find the forces. Subtracting Equation 8.62 from Equation 8.59,

$$0 = \sum_{n=1}^{n} \left(i_i - \frac{\partial W_\phi}{\partial \lambda_i} \right) d\lambda_i - \sum_{m=1}^{m} \left(Q_j + \frac{\partial W_\phi}{\partial D_j} \right) dD_j \qquad (8.63)$$

Now $d\lambda_i$ and dD_j are independent, since i_i can have any value independent of $[D_j]$. Then the two summations in Equation 8.44 are independent of each other; and since their difference is zero, they must both be zero. For the present purpose, take the second summation as being equal to zero:

$$\sum_{j=1}^{m} \left(Q_j + \frac{\partial W_\phi}{\partial D_j} \right) dD_j = 0 \qquad (8.64)$$

Again since all of the displacements (D_j's) are independent of each other, and the sum of all the m terms is zero, each term must be zero:

$$\left(Q_j + \frac{\partial W_j}{\partial D_j} \right) dD_j = 0$$

Solving for the force,

$$Q_j = -\frac{\partial W_\phi}{\partial D_j} \bigg|_{[\lambda_i]_k} \qquad (8.65)$$

This remarkable result shows that if the flux linkages are known or are known to be constant, the force at any of the mechanical terminals is equal to the negative of the rate of change of field energy that results from a displacement at that terminal, alone. The fact that the force expressions are decoupled from each other validates the use of generalized forces and displacements in the analysis.

It is often more convenient to find the forces for a given set of currents rather than a given set of flux linkages. It might be supposed that, for this purpose, the expression for the differential change in field energy should be derived from Equation 8.61. The problem with that approach is that the resulting expression does not contain the common factor $d\lambda_i$ found in Equation 8.59. A more fruitful result is obtained by working with the coenergy. The coenergy introduced into the system through the ith terminal pair is given by

$$W'_{\phi i} = \lambda_i i_i - W_{\phi i} \tag{8.66}$$

and the system coenergy is defined as

$$W'_{\phi} = \sum_{i=1}^{n} \lambda_i i_i - W_{\phi} \tag{8.67}$$

where W_{ϕ}, as before, is the system stored energy, given by $\sum_i W_{\phi i}$. The differential coenergy may be written

$$dW'_{\phi} = \sum_{i=1}^{n} (\lambda_i \, di_i + i_i \, d\lambda_i) - dW_{\phi} \tag{8.68}$$

Substituting for dW_{ϕ} from Equation 8.59 and splitting the summation of Equation 8.68 into two parts,

$$dW'_{\phi} = \sum_{i=1}^{n} \lambda_i \, di_i + \sum_{i=1}^{n} i_i \, d\lambda_i - \sum_{i=1}^{n} i_i \, d\lambda_i + \sum_{j=1}^{m} Q_j \, dD_j$$

or

$$dW'_{\phi} = \sum_{i=1}^{n} \lambda_i di_i + \sum_{j=1}^{m} Q_j \, dD_j \tag{8.69}$$

Another expression for the differential coenergy may be derived from the fact that

$$W'_{\phi} = W'_{\phi}(i_1, i_2, \ldots, i_i, \ldots, i_n, D_1, \ldots, D_j, \ldots, D_m) \tag{8.70}$$

Then

$$dW'_{\phi} = \sum_{i=1}^{n} \frac{\partial W'_{\phi}}{\partial i_i} \, di_i + \sum_{j=1}^{m} \frac{\partial W'_{\phi}}{\partial D_j} \, dD_j \tag{8.71}$$

Subtracting this equation from Equation 8.69,

$$0 = \sum_{i=1}^{n} \left(\lambda_i - \frac{\partial W'_{\phi}}{\partial i_i} \right) di_i + \sum_{j=1}^{m} \left(Q_j - \frac{\partial W'_{\phi}}{\partial D_j} \right) dD_j \tag{8.72}$$

Since di_i and dD_j are independent and the sum of the two summations is zero, by the previous argument each summation must equal zero, so

$$\sum_{j=1}^{m} \left(Q_j - \frac{\partial W'_\phi}{\partial D_j} \right) dD_j = 0 \qquad (8.73)$$

Again since the displacements are independent of each other, each term in the summation must be zero. Therefore

$$Q_j - \frac{\partial W'_\phi}{\partial D_j} = 0$$

and

$$Q_j = \frac{\partial W'_\phi}{\partial D_j} \bigg|_{[i_i]_k} \qquad (8.74)$$

This shows that when currents are specified or known, the force, torque, or what have you, being developed at any mechanical terminal is equal to the rate of change of coenergy with respect to displacement at that particular mechanical terminal.

Special Relationships in a Magnetically Linear System

Actually, magnetic-field energy in the cores of devices is rarely taken into account in calculating the output force or torque. It is usually assumed that all of the force is developed in the air gap. Now the air gap is magnetically linear, having a constant permeability of μ_0. As a result, the assumption of linearity in force calculations is an assumption useful in many practical applications. If one refers to Figure 8.10b, which shows the relationship between energy and coenergy at a typical electrical terminal pair, it is evident that for a linear system the field energy that has been delivered through those terminals is

$$W_{\phi i} = \frac{\lambda_i i_i}{2} \qquad (8.75)$$

In the figure, $i = 5$. By conservation of energy, the total system energy is the sum of the individual contributions:

$$W_\phi = \sum_{i=1}^{n} W_{\phi i} = \frac{1}{2} \sum_{i=1}^{n} \lambda_i i_i \qquad (8.76)$$

Coenergy was defined in Equation 8.67 as $\sum_i i_i \lambda_i - W_\phi$. Then for a *linear* system,

$$W'_\phi = \sum_{i=1}^{n} i_i \lambda_i - \frac{1}{2} \sum_{i=1}^{n} \lambda_i i_i = W_\phi \qquad (8.77)$$

Thus for a linear system the energy and coenergy may be used interchangeably in Equation 8.74, and the force at the jth mechanical output terminal, for given currents, is

$$Q_j = \frac{\partial W_\phi}{\partial D_j} \bigg|_{[i_i]_k} \qquad (8.78)$$

> Equation 8.78 is probably the most useful of all of the force expressions developed in this section.

8.6 TORQUE IN MACHINES WITH CYLINDRICAL AIR GAPS

The expressions for torque that have been developed so far may be used to find the torque developed by machines in terms of their dimensions and winding specifications. Consider a machine with a uniform cylindrical air gap, having sinusoidally distributed rotor and stator MMFs. The dimensions of such an air gap are illustrated in Figure 2.28. The field distributions are shown in Figure 2.24. The sinusoidal stator field has an amplitude of F_1 ampere-turns, and the amplitude of the rotor field is F_2 ampere-turns. The peaks of the two fields are displaced from each other by an angle of β electrical degrees. When a machine is operating in the steady state, β is constant. The point-by-point sum of these two fields is the MMF across the air gap at each point. A plot of this "resultant" field is also a sinusoid having an amplitude of R ampere-turns. Since the rotor and stator MMFs are sinusoidal functions of a space angle θ, measured in electrical degrees, they may be resolved as space phasors or two-dimensional vectors. This resolution is shown for a cylindrical-rotor synchronous machine in Figures 2.25 and 2.26 and for an induction machine in Figures 4.8 and 4.12. A similar construction will be found in Figure 8.11.

Applying the law of cosines in Figure 8.11, an expression is found for the square of the peak air gap MMF R:

$$R^2 = F_1^2 + F_2^2 - 2F_1F_2 \cos \alpha \tag{8.79}$$

where $\alpha = \pi - \beta$. Then in terms of the displacement angle β between the stator and rotor fields,

$$R^2 = F_1^2 + F_2^2 + 2F_1F_2 \cos \beta \tag{8.80}$$

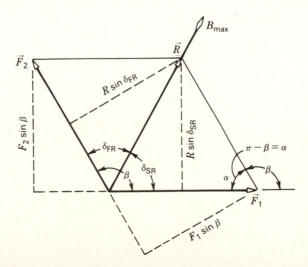

FIGURE 8.11 MMF relationships in a machine with a cylindrical air gap.

MAGNETIC-FIELD ENERGY IN THE AIR GAP

Let the space angle θ, in electrical degrees, be measured from a point of maximum, outward-directed, air gap MMF, that is, where $R(\theta) = +R$, as shown in Figure 8.12. Then

$$R(\theta) = R \cos \theta \tag{8.81}$$

By Equation 8.15, the field energy density in the air gap as a function of θ is given by

$$w_{\text{fld}}(\theta) = \frac{\mu_0 H^2(\theta)}{2} \tag{8.82}$$

The radial length of the air gap is g m. Then the variation of field intensity around the air gap is

$$H(\theta) = \frac{R \cos \theta}{g} \text{ A-turns/m} \tag{8.83}$$

Substituting into Equation 8.82,

$$w_{\text{fld}}(\theta) = \frac{\mu_0 R^2 \cos^2 \theta}{2g^2} = \frac{\mu_0 R^2}{2g^2}\left(\frac{1}{2} + \frac{1}{2} \cos 2\theta\right) \tag{8.84}$$

The *average* energy density in the air gap, by inspection of Equation 8.84, is

$$\overline{w}_{\text{fld}} = \frac{\mu_0 R^2}{4g^2} \tag{8.85}$$

Then the total magnetic-field energy in the air gap is

$$W_\phi = (\text{gap volume})(\overline{w}_{\text{fld}}) = (\pi D L g)\frac{\mu_0 R^2}{4g^2} \tag{8.86}$$

In this expression, D is the mean diameter of the air gap in meters and L is its axial length.

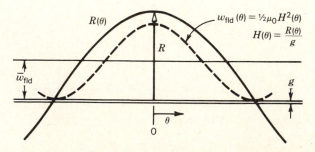

FIGURE 8.12 MMF across a uniform air gap and magnetic-field energy density in the air gap as functions of angular distance along the gap in electrical degrees.

Finding the Torque

Substituting the expression for R^2 found in Equation 8.80 into Equation 8.86 and at the same time expressing β in mechanical radians,

$$W_\phi = \frac{\pi\mu_0}{4}\frac{DL}{g}\left(F_1^2 + F_2^2 + 2F_1F_2 \cos\frac{p}{2}\beta_{\text{mech}}\right) \qquad (8.87)$$

There are several legitimate concepts as to how torque is produced. First, it will be assumed that torque results from the attraction of stator MMF poles to rotor MMF poles. Then by Equation 8.78,

$$\tau = \frac{\partial W_\phi}{\partial \beta_{\text{mech}}} = \frac{\pi\mu_0}{4}\frac{DL}{g}\left(-2F_1F_2\frac{p}{2}\sin\frac{p}{2}\beta_{\text{mech}}\right)\text{ N-m}$$

or

$$\tau = -\frac{\pi\mu_0}{4}\frac{DL}{g}pF_1F_2 \sin\beta \quad \text{N-m} \qquad (8.88)$$

where β is in electrical radians.

Alternatively, the torque may be considered to result from the attraction of the rotor poles to those of the resultant field. In Figure 8.11 it is seen that

$$F_1 \sin\beta = R \sin\delta_{\text{FR}} \qquad (8.89)$$

With this substitution, the expression for the torque becomes

$$\tau = -\frac{\pi\mu_0}{4}\frac{DL}{g}pF_2R \sin\delta_{\text{FR}} \quad \text{N-m} \qquad (8.90)$$

Usually, however, the torque is thought to result from the interaction of the air gap flux and the rotor MMF F_2. When Equation 2.25 is written in terms of the diameter D, the air gap flux per pole is given by

$$\phi = \frac{2DLB_{\text{max}}}{p} \quad \text{Wb} \qquad (8.91)$$

Now

$$B_{\text{max}} = \mu_0 H_{\text{max}} = \mu_0 \frac{R}{g} \quad \text{T} \qquad (8.92)$$

Substitution of Equation 8.92 into Equation 8.91 provides an expression for the flux per pole in terms of the resultant of the stator and rotor MMFs, R:

$$\phi = \frac{2DL\mu_0}{pg}R \qquad (8.93)$$

Since the effective permeance per pole was defined by the relationship $\phi = \mathscr{P}_p R$, a corollary to this development is an expression for $\mathscr{P}_p$ in terms of machine dimensions:

$$\mathscr{P}_p = \frac{2DL\mu_0}{pg} \tag{8.94}$$

This expression does not take iron reluctance into account, however.

Solving Equation 8.93 for R and substituting into Equation 8.90 gives a very useful expression for the torque in machines having cylindrical air gaps:

$$\tau = -\frac{\pi}{8}p^2 F_2 \phi \, \sin \delta_{FR} \tag{8.95}$$

NOTE: **The only significance of the minus sign in the three torque expressions derived in this section is that the torque is in a direction to reduce the displacement between the two magnetic quantities involved in each case.** In Equation 8.95, for example, the torque is in a direction to reduce δ_{FR}.

Torque in Terms of Design Parameters

Figure 8.11 demonstrates that $F_2 \sin \beta = R \sin \delta_{SR}$, where δ_{SR} is the displacement of the stator MMF field F_1 from the resultant air gap field R. If this substitution is made in Equation 8.88 and, as in Equations 8.91 through 8.95, an expression is developed for the torque in terms of F_1 and the flux per pole, ϕ, the result is useful from the point of view of the designer:

$$\tau = \frac{\pi}{8}p^2 \phi F_1 \sin \delta_{SR} \tag{8.96}$$

This equation for the torque is of interest because it is in terms of the flux, which is limited by the maximum flux density permitted by iron saturation, and also in terms of the armature winding MMF, which is related to the number of poles and the number of ampere conductors in the winding.

The *electrical loading* (EL) of a machine is the total number of ampere conductors in its armature winding. In this discussion, a three-phase winding is assumed. If N_ϕ is the number of series turns per phase, then the number of series conductors per phase is $2N_\phi$ and the total number of ampere conductors in a three-phase armature is

$$(EL) = 6N_\phi |I_1| \tag{8.97}$$

Now the peak armature MMF is given by

$$F_1 = N_{f1}I_1 = \frac{3}{2}\frac{4\sqrt{2}}{\pi}\frac{N_\phi}{p}k_w|I_1|$$

$$= \frac{6\sqrt{2}}{\pi p}\frac{(EL)}{6}k_w = \frac{\sqrt{2}}{\pi p}k_w(EL) \quad \text{A-turns} \tag{8.98}$$

The "total flux" in the machine is sometimes called the *magnetic loading* (ML).

$$(ML) = p\phi \quad Wb \tag{8.99}$$

where ϕ is the flux per pole.

Substituting from Equations 8.98 and 8.99 into the torque expression,

$$\tau = \frac{\sqrt{2}}{8} (ML)(EL)k_w \sin \delta_{SR} \tag{8.100}$$

The torque may be expressed in terms of two other meaningful quantities, the *specific electrical loading* (SEL) and the *specific magnetic loading* (SML). Definitions of these terms vary, but here,

$$(SEL) = \text{ampere-conductors per meter} = \frac{(EL)}{\pi D} \tag{8.101}$$

$$(SML) = \text{rms air gap flux density}$$

Assuming the air gap flux density to be sinusoidally distributed,

$$(SML) = \frac{B_{max}}{\sqrt{2}} \tag{8.102}$$

In Chapter 2, Equation 2.25, the flux per pole was found in terms of B_{max}:

$$\phi = \frac{4LrB_{max}}{p} = \frac{2DLB_{max}}{p}$$

Solving for B_{max},

$$B_{max} = \frac{p\phi}{2DL} = \frac{(ML)}{2DL} \tag{8.103}$$

Then by Equation 8.102,

$$(SML) = \frac{(ML)}{2\sqrt{2}DL} \quad \text{or} \quad (ML) = 2\sqrt{2}DL(SML) \text{ T} \tag{8.104}$$

The torque of a cylindrical air gap machine may now be written in terms of the specific electrical loading and the specific magnetic loading:

$$\tau = \frac{\sqrt{2}}{8} 2\sqrt{2}DL(SML)\pi D(SEL)k_w \sin \delta_{SR}$$

$$= \frac{\pi}{2} D^2L(SML)(SEL)k_w \sin \delta_{SR} \text{ N-m} \tag{8.105}$$

Now (SEL) is limited by temperature rise, and (SML) is limited by saturation. These quantities can be chosen by the designer over a rather narrow range. Thus it is seen that the torque is proportional to the length of the core stack and, for a given length, is proportional to the square of the diameter.

8.7 TORQUE OF CYLINDRICAL-ROTOR SYNCHRONOUS MACHINES

The torque developed in any machine having a uniform cylindrical air gap and sinusoidally distributed magnetic fields was found to be, by Equation 8.95,

$$\tau = \frac{\pi}{8} p^2 F_2 \phi \sin \delta_{FR} \qquad (8.106)$$

The minus sign has been omitted because it tells nothing about the direction of shaft torque. The sign only indicates that the torque is in a direction to decrease δ_{FR}. In this expression, p is the number of poles, F_2 is the peak rotor pole MMF, and ϕ is the air gap flux per pole due to the resultant of the rotor and stator MMFs.

The first step in applying Equation 8.95 specifically to a cylindrical-rotor synchronous machine will be to relate the flux per pole to the voltage induced in each armature phase winding, E_ϕ. From Equation 2.46,

$$\phi = \frac{|E_\phi|}{N_{E1}\omega_e} \qquad (8.107)$$

The voltage E_f is the model of the rotor MMF, F_2 in Chapter 2, and by Equation 2.63,

$$E_f = -jmF_2$$

where

$$m = \omega_e N_{E1} \, \mathscr{P}_p \qquad (8.108)$$

Then

$$F_2 = \frac{|E_f|}{m} \qquad (8.109)$$

Also by Equations 2.65, 2.17, and 2.18,

$$m = \frac{|I_1|X_m}{F_1} = \frac{|I_1|X_m}{N_{I1}|I_1|} = \frac{X_m}{N_{I1}} \qquad (8.110)$$

where

$$N_{I1} = \frac{3}{2} \frac{4\sqrt{2}}{\pi} \frac{N_\phi k_w}{p} \qquad (8.111)$$

and N_ϕ is the number of coil turns in series in one phase winding (N_ϕ/p is the number of turns per pole per phase and is equal to the number of turns per phase group) and k_w is the product of the pitch and distribution factors. Substituting from Equations 8.107 and 8.109 into Equation 8.106 yields

$$\tau = \frac{\pi}{8} p^2 \frac{|E_\phi|}{\omega_e N_{E1}} \frac{|E_f|}{m} \sin \delta_{FR} \qquad (8.112)$$

Now

$$N_{E1} = \frac{N_\phi k_w}{\sqrt{2}} \tag{8.113}$$

and comparing Equations 8.113 and 8.111, the relationship between the voltage and current turns per phase is found to be

$$N_{E1} = \frac{\pi p}{12} N_{I1} \tag{8.114}$$

Making this substitution into Equation 8.75,

$$\tau = \frac{\pi}{8} p^2 \frac{|E_\phi||E_f|}{\omega_e(\pi p/12) \, mN_{I1}} \sin \delta_{FR} \tag{8.115}$$

The quantity mN_{I1} is recognized as the magnetizing reactance X_m. Making appropriate cancellations and recalling that the synchronous speed $\omega_s = 2\omega_e/p$ yields the following gratifying result:

$$\tau = \frac{3}{\omega_s} \frac{|E_\phi||E_f|}{X_m} \sin \delta_{FR} \tag{8.116}$$

This same expression for the torque of a synchronous machine was obtained as Equation 2.77 in Chapter 2!

8.8 TORQUE RELATIONS IN THREE-PHASE INDUCTION MOTORS

Equation 8.95 will now be applied to a three-phase induction motor. Referring to Chapter 4, Figure 4.14, it will be observed that δ_{FR}, the displacement between the rotor MMF vector F_2 and the resultant R is $90° + \theta_2$, where θ_2 is the power factor angle of the rotor circuit. Then

$$\sin \delta_{FR} = \sin(90° + \theta_2) = \cos \theta_2 \tag{8.117}$$

Now $\cos \theta_2$ is the power factor of the rotor circuits and may be related to the impedances of one rotor phase:

$$\cos \theta_2 = \frac{R_R}{\sqrt{R_R^2 + (sX_{RB})^2}} \tag{8.118}$$

In this expression, R_R is the resistance of one rotor phase and X_{RB} is the leakage reactance of one rotor phase at blocked rotor ($s = 1; f_2 = f_1$). See Figure 4.13. Recall that a_E is the stator/rotor voltage-turns ratio and that a_I is the corresponding current-turns ratio. If the numerator and denominator of Equation 118 are both multiplied by the product $a_E \cdot a_I$, the impedances are referred to the stator:

$$\cos \theta_2 = \frac{r_2}{\sqrt{r_2^2 + (sx_2)^2}}$$

$$= \frac{r_2/s}{\sqrt{(r_2/s)^2 + x_2^2}} \tag{8.119}$$

The amplitude of the rotor MMF, F_2, may be expressed in terms of the effective current turns per pole of the rotor, N_{I2}, Equation 4.32:

$$F_2 = N_{I2}|I_R| \tag{8.120}$$

By Equation 4.15,

$$|E_1| = \omega_e N_{E1} \phi$$

or

$$\phi = \frac{|E_1|}{\omega_e N_{E1}} \tag{8.121}$$

Substituting Equations 8.119, 8.120, and 8.121 into Equation 8.117, the torque of an induction machine is found to be

$$\tau = \frac{\pi p^2}{8} \frac{|E_1|N_{I2}}{\omega_e N_{E1}} |I_R| \frac{r_2/s}{\sqrt{(r_2/s)^2 + x_2^2}} \tag{8.122}$$

Note that

$$N_{E1} = \frac{1}{\sqrt{2}} N_\phi k_w \qquad \text{and} \qquad N_{I1} = \frac{3}{2} \frac{4}{\pi} \sqrt{2} \frac{N_\phi}{p} k_w$$

Then

$$N_{E1} = \frac{\pi p}{12} N_{I1}$$

Making this substitution into Equation 8.122,

$$\tau = 3 \frac{p}{2\omega_e} \frac{N_{I1}}{N_{I2}} I_R \frac{|E_1|}{\sqrt{(r_2/s)^2 + x_2^2}} \frac{r_2}{s} \tag{8.123}$$

Recall that

$$\frac{p}{2\omega_e} = \frac{1}{\omega_s} \qquad \text{and} \qquad |I_R| \frac{N_{I2}}{N_{I1}} = |I_2| = \frac{|E_1|}{\sqrt{(r_2/s)^2 + x^2}}$$

where $|I_2|$ is the rotor current referred to one phase of the stator winding. Making these substitutions results in the expression for the torque derived in Chapter 4, Equation 4.52:

$$\tau = \frac{3}{\omega_s} |I_2|^2 \frac{r_2}{s} = \frac{P_g}{\omega_s} \quad \text{N-m} \tag{8.124}$$

8.9 TORQUE IN DC MACHINES

Direct-current machine armature windings are uniformly distributed in slots in the surface of the laminated-iron rotor. The stator has salient magnetic poles mounted on an iron yoke and projecting inward toward the rotor. The curved pole faces are separated from the rotor surface by a small air gap, and the angle subtended by each pole face is such that they cover approximately 60 percent of the rotor surface. As seen from the rotor, the magnetic polarity of these poles is alternately north, south, north, south. The coils of the armature winding are connected to a ring of insulated copper segments called a *commutator,* which is mounted on the rotor shaft. Current is conducted to the armature through *brushes*, which are blocks of carbon compound that make sliding electrical contact with the bare outer surface of the commutator. The action of the commutator is such that, regardless of armature rotation, the armature currents under all the north stator poles are in one direction, while the currents under the south poles are in the opposite direction. Thus the interaction of the rotor and stator magnetic fields in the air gap under each of the poles produces torque in the same direction.

The MMFs applied to the air gap of a dc machine are plotted in Figure 5.25. The armature MMF as a function of angular distance around the air gap may be approximated by a triangular wave having an amplitude of

$$F_a = \frac{ZI_a}{2pa} \quad \text{A-turns} \tag{8.125}$$

where Z is the total number of conductors in the armature winding, I_a is the dc armature current in amperes, p is the number of poles, and a is the number of current paths through the armature winding. The current in each conductor is I_a/a. The position of this rotor MMF wave relative to the stator poles depends on the position of the stationary brushes. Normally, the brushes are positioned so that the peaks of the rotor MMF (which correspond to the centers of magnetic poles set up on the surface of the rotor by the armature currents) are exactly halfway between the stator poles. Then the rotor MMF goes through zero under the center of each stator pole. As discussed in Section 5.12, the brushes are sometimes shifted from this position to reduce sparking at the brush contacts. Brush shift moves the rotor MMF wave relative to the stator poles.

The amplitude of that component of the air gap MMF due to the stator poles, often called the *field poles,* is F_f ampere-turns. It will be assumed to be constant over the pole arc α, resulting in a rectangular wave, as shown in Figure 8.13.

Angular distance around the air gap, in mechanical radians, will be measured from the center of one of the south stator poles, as shown in Figure 8.13. If there is no brush shift, the rotor MMF will be zero at that point.

Applying Ampère's law in circuital form as indicated in Figure 5.25, the slope of the triangular armature MMF wave is equal to the ampere-conductors per radian of angular distance along the rotor surface. Since the peak occurs at one half of one pole pitch from the center of the pole, that is, $\theta = 2\pi/2p$ radians, the slope of the armature MMF at $\theta = 0$ is

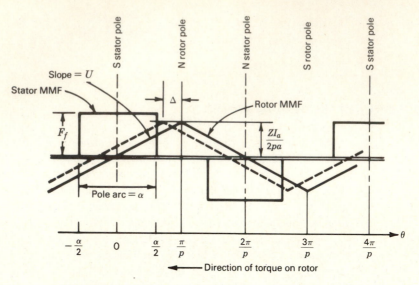

FIGURE 8.13 MMF fields applied to the air gap of a dc machine.

$$U = \frac{dF_2}{d\theta} = \frac{1}{\pi/p} \frac{ZI_a}{2pa} = \frac{ZI_a}{2\pi a} \quad \text{A/radian} \tag{8.126}$$

Then over the range $(-\pi/p) \leq \theta \leq (+\pi/p)$ the expression for the rotor MMF is

$$F_2(\theta) = U\theta = \frac{ZI_a}{2\pi a} \theta \quad \text{A-turns} \tag{8.127}$$

If α is the pole arc in radians, then over the range $(-\alpha/2) \leq \theta \leq (+\alpha/2)$ the total MMF applied to the air gap between the pole face and the rotor is the sum of the pole MMF and that of the armature:

$$F_{\text{gap}} = F_f + U\theta \tag{8.128}$$

Then the field intensity in the air gap between the pole face and the rotor surface is

$$H(\theta) = \frac{F_{\text{gap}}}{g} = \frac{F_f}{g} + \frac{U}{g} \theta \quad \text{A-turns/m} \tag{8.129}$$

where g is the radial distance between the rotor and pole surfaces, namely, the air gap length, in meters.

To apply the principle of virtual work, assume that the rotor field is permitted to rotate a small angle Δ in a direction to align the north rotor pole with the south stator pole. (If the brushes were allowed to rotate with the commutator, the rotor magnetic poles would rotate with the rotor one half pole pitch so that the rotor and stator poles would be aligned, at which point the developed torque would be zero.)[1] Let that

---

[1]A similar shift in the rotor field pattern would result from shifting the brushes Δ mechanical degrees in the direction of developed torque. However to correct sparking at the brushes, the brushes should be shifted in a direction opposite to developed torque, that is, in a direction opposite to that shown in Figure 8.13.

component of the air gap field intensity contributed by the stator poles be H_f. Then with the rotor field shifted Δ mechanical degrees,

$$H_{\text{gap}} = H_f + \frac{U}{g}(\theta + \Delta) \tag{8.130}$$

$$= \left(H_f + \frac{U}{g}\Delta\right) + \frac{U}{g}\theta$$

By Equation 8.15, the magnetic-field energy density in the gap is

$$w_{\text{fld}} = \frac{\mu_0 H_{\text{gap}}^2}{2} = \frac{\mu_0}{2}\left[\left(H_f + \frac{U}{g}\Delta\right)^2 + 2\left(H_f + \frac{U}{g}\Delta\right)\frac{U}{g}\theta + \frac{U^2}{g^2}\theta^2\right] \tag{8.131}$$

To find the air gap field energy per pole, this expression will be integrated over the air gap volume under one pole. If L is the axial length of the poles and r is the mean radius of the air gap, then

$$dv = gLr\,d\theta$$

and the field energy per pole is

$$W_{\text{fld}} = \int_{-\alpha/2}^{+\alpha/2} w_{\text{fld}}gLr\,d\theta$$

$$= \frac{gLr\mu_0}{2}\int_{-\alpha/2}^{+\alpha/2}\left[\left(H_f + \frac{U}{g}\Delta\right)^2 + 2\left(H_f + \frac{U}{g}\Delta\right)\frac{U}{g}\theta + \frac{U^2}{g^2}\theta^2\right]d\theta$$

$$= \frac{gLr\mu_0}{2}\left[\left(H_f + \frac{U}{g}\Delta\right)^2\alpha + \frac{U^2}{g^2}\frac{\alpha^3}{12}\right] \tag{8.132}$$

$$= \frac{gLr\mu_0}{2}\left[H_f^2\alpha + \frac{U^2}{g^2}\frac{\alpha^3}{12} + \left(2\frac{H_f U}{g}\alpha\Delta + \frac{U^2}{g^2}\alpha\Delta^2\right)\right]$$

This is a linear system because the energy being considered is all in the air gap. Since the currents are the fixed quantities, the torque is found by applying Equation 8.26 or better Equation 8.78:

$$\tau = +\left.\frac{\partial W_\phi}{\partial \Delta}\right|_{[i_i]_k} = \frac{gLr\mu_0}{2}\left[\frac{2H_f U\alpha}{g} + \frac{2U^2\alpha}{g^2}\Delta\right]$$

$$= Lar\cdot\mu_0 H_f\cdot U + Lar\mu_0\frac{U^2}{g}\Delta \tag{8.133}$$

Now αr is the pole arc, in meters, so $L\alpha r$ is the area of the pole face, A_{pole}, in square meters. Also $\mu_0 H_f$ is the flux density at the pole face. So the flux per pole, ϕ, in webers, is given by

$$\phi = L\alpha r\mu_0 H_f \quad \text{Wb} \tag{8.134}$$

Then by substituting for the slope U from Equation 8.126,

$$\tau = \phi \frac{ZI_a}{2\pi a} + \frac{\mu_0 A_{\text{pole}}}{g} \frac{Z^2 I_a^2}{4\pi^2 a^2} \Delta \quad \text{N-m} \qquad (8.135)$$

The torque found in Equation 8.135 was calculated from the air gap energy under only one pole. To obtain the full torque of the machine, the terms of the equation must be multiplied by the number of poles, p:

$$\tau = p\phi \frac{ZI_a}{2\pi a} + \frac{p\mu_0 A_{\text{pole}}}{g} \frac{Z^2 I_a^2}{4\pi^2 a^2} \Delta \quad \text{N-m} \qquad (8.136)$$

As Δ approaches zero, only the first term of Equation 8.136 is of significance, and the resulting expression for the torque is the same as that derived in Chapter 5 (Equation 5.26, where K_a is defined by Equation 5.12):

$$\tau = \frac{Zp}{2\pi a} \phi I_a = K_a \phi I_a \quad \text{N-m} \qquad (8.137)$$

The second term in Equation 8.136, however, is a bonus, because it tells what happens to the torque of a dc machine when the brushes are shifted. To reduce sparking, the brushes are shifted in a direction opposite to the direction of the developed torque, that is, in the negative Δ direction. Refer to Figure 5.32. The angle Δ may be replaced with $-\beta$ in the second term of Equation 8.136 to determine the effect of brush shift on the torque. Both β and Δ are in mechanical degrees. From Figure 5.32, the fraction of total armature conductors tending to demagnetize each pair of poles is $4\beta/2\pi = 2\beta/\pi$. The number of demagnetizing ampere-conductors per pole is then $\beta ZI_a/\pi a$, since the current in each conductor is I_a/a amperes. The demagnetizing *ampere-turns* per pole is then $\beta ZI_a/2\pi a$, and the change in MMF applied to the air gap under one pole is

$$\Delta \mathcal{F} = \frac{-\beta ZI_a}{2\pi a} \quad \text{A-turns} \qquad (8.138)$$

The corresponding change in flux per pole is given by $\Delta \mathcal{F}$ times the air gap permeance per pole:

$$\Delta \phi = -\frac{\beta ZI_a}{2\pi a} \frac{\mu_0 A_{\text{pole}}}{g} \quad \text{Wb} \qquad (8.139)$$

and the change in torque would be, from Equation 5.26,

$$\Delta \tau = K_a (\Delta \phi) I_a = \frac{Zp}{2\pi a} \frac{-\beta ZI_a \mu_0 A_{\text{pole}}}{2\pi a g} I_a \qquad (8.140)$$

or

$$\Delta \tau = -\left(\frac{p\mu_0 A_{\text{pole}} Z^2 I_a^2}{4\pi^2 a^2 g} \right) \beta \quad \text{N-m} \qquad (8.141)$$

Equation 8.141 shows that the change in torque that results from brush shift is given by the second term of Equation 8.136, with Δ made equal to $-\beta$.

Torque in Terms of Electrical and Magnetic Loadings

The expression for dc machine torque without brush shift is found in Equation 8.137. In a dc machine, the electrical loading (EL) is simply the total number of ampere-conductors in the armature winding:

$$\text{(EL)} = Z\frac{I_a}{a} \tag{8.142}$$

The magnetic loading (ML) is the "total flux":

$$\text{(ML)} = p\phi \tag{8.143}$$

The product of the two loadings yields

$$\text{(EL)(ML)} = \frac{Zp\phi I_a}{a} \tag{8.144}$$

Comparing Equations 8.144 and 8.137, the developed torque may be expressed in terms of the loadings:

$$\tau = \frac{1}{2\pi}\text{(EL)(ML)} \quad \text{N-m} \tag{8.145}$$

The specific electrical loading is the number of ampere-conductors per meter around the circumference of the rotor:

$$\text{(SEL)} = \frac{\text{(EL)}}{\text{rotor circumference}} = \frac{ZI_a}{\pi Da} \tag{8.146}$$

The specific magnetic loading will be defined as the *average* flux density in the air gap under a pole. (This differs from the definition used for machines having sinusoidally distributed air gap flux densities, in which case the rms flux density is used. See Equation 8.101.) If the arc subtended by one pole face is α mechanical degrees,

$$\text{(SML)} = \frac{\text{flux per pole}}{\text{pole face area}} = \frac{\phi}{(D/2)\alpha L} = \frac{2\phi}{\alpha DL} \tag{8.147}$$

In terms of these specific loadings, the torque of a dc machine is

$$\tau = D^2L\frac{\alpha p}{4}\text{(SEL)(SML)} \tag{8.148}$$

As in the case of cylindrical rotor machines, torque of a dc machine is proportional to volume, or proportional to the square of the diameter for constant core length, if the loadings are held constant. Tooth saturation would limit (SML) to about 1 T in wound-pole machines. Magnet design places a much lower limit on (SML) in PM machines. Specific electrical loading (SEL) may be limited to avoid demagnetization in PM dc machines; while in wound-pole machines, thermal considerations impose the limit. For a 100-kW machine the (SEL) would be in the order of 33,500 A/m. For a 1000-kW machine it would be about one third more.

A fleeting glance at Equation 8.148 might leave one with the impression that the torque of dc machines is proportional to the number of poles. That is not the case. As the number of poles increases, the arc subtended by each pole, α, must decrease. As a result, the quantity αp is practically constant and has a value of about 1.2π. As in the case of the ac machines, torque capability is proportional to the volume of the machine, and for a fixed volume the power rating varies directly with rated speed.

PROBLEMS

8.1. An air gap is cut into the magnetic circuit of an audio transformer to improve linearity. The air gap is 1.00 mm long and the cross section of the magnetic circuit at that point is 2.00 cm × 2.00 cm. At rated sinusoidal voltage, 1000 Hz, the flux density in the air gap also varies sinusoidally with time and has a peak value of 0.45 T. Find the peak and average forces acting on the core to close the air gap. State answers in both newtons and pounds.

8.2. In an inexpensive relay, the contacts are operated by a hinged piece of iron which is attracted by a magnetic field to a cylindrical iron core that protrudes from an actuating coil. The force of attraction moves the iron piece and closes the contacts. The air gap between the movable part and the end of the core is 0.50 mm in the closed position. The diameter of the core at the air gap is 8.0 mm. If at rated coil current the flux density at the end of the core facing the air gap is 0.10 T, find the force of attraction on the movable part.

8.3. A crossing gate for a model railroad is operated by a laminated iron plunger suspended in a coil of wire. The plunger is free to move over a distance $x = 0$ to $x = 3.0$ cm. As the plunger enters the coil, the coil inductance varies as follows:

x, cm:	0	1.0	2.0	3.0	4.0
L, mH:	1.0	3.0	7.2	13.0	20.3

When $x = 0$, $(dL/dx) = 0.133$ H/m. The coil has a resistance of 3.5 Ω. Find the force exerted on the plunger when $x = 0$ and $x = 3$ cm under the following conditions: (a) Coil current is 2.0 A, dc. (b) Coil current is 2.0 A, rms ac. (c) Coil voltage is 12.0 V at 60 Hz. In the two ac cases the *average* force is required.

8.4. The accompanying figure shows two square U-shaped iron cores separated by equal air gaps g m long. The cross-sectional area of each core is A m². The gaps are so small that fringing of the magnetic flux may be neglected; however they are not so small that the reluctance of the iron need be considered in calculating the permeance of the magnetic circuit. Each core is wound with N turns of wire, and the two coils are connected in series, opposing. When current flows in the coils, the forces due to the variation of the two self-inductances with g tend to close the gaps, while that due to the variation of mutual inductance with g tends to open the gaps. (a) Write expressions for $L_{11}(g)$, $L_{22}(g)$, and $L_{12}(g)$. (b) Is there a value of g at which the positive and negative forces cancel, resulting in zero net force? Show equations to prove your answer and explain in words.

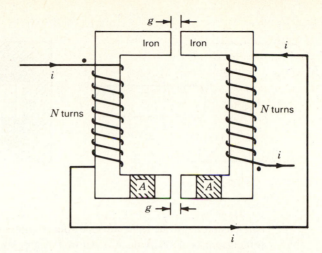

8.5. In a device like that of Figure 8.4, data for magnetization curves may be obtained by inserting a Hall effect probe in the air gap to measure flux density and measuring the coil current as it is varied. If the air gap is short enough, fringing may be neglected and the flux taken to be equal to $B \cdot A$. The number of coil turns is usually known, so $\mathscr{F} = N \cdot i$. Then the flux ϕ may be plotted as a function of $\mathscr{F}$. Magnetization curves for such a device were determined for two gap lengths, $g = 4.50$ mm and $g = 5.00$ mm. Now magnetization curves exhibiting saturation may be approximated by an empirical formula called Froelich's equation:

$$\phi = \frac{a\mathscr{F}}{b + \mathscr{F}}$$

Two points are selected on the curve to be approximated, and simultaneous equations are set up to solve for a and b. This was done for the two measured curves, with the following results:

g, mm	a	b
4.50	0.756	5250
5.00	0.850	7200

(a) Plot the two curves over the range $0 \leqslant \mathscr{F} \leqslant 3000$ A-turns. (b) For $\mathscr{F} = 3000$ A-turns, find the coenergy for each gap length. (c) Find the force on the plunger at $g = 4.75$ mm by $\Delta W_{\phi}'/\Delta g$. (d) What is the direction of this force, based on the sign of your answer?

8.6. A six-pole three-phase, 60-Hz induction motor is operating at full load. The resultant air gap flux density is sinusoidally distributed, and the flux per pole of this field is 0.01 Wb. The rotor MMF field has an amplitude of 500 A-turns. It is also sinusoidally distributed, rotates at the same speed as the resultant field, and lags behind it by 35 mechanical degrees. What torque is being developed by this motor?

8.7. Refer to the accompanying sketches of a solenoid. The coil has 500 turns. The cross-sectional area of the air gap is 2.00 cm². Neglect flux fringing,

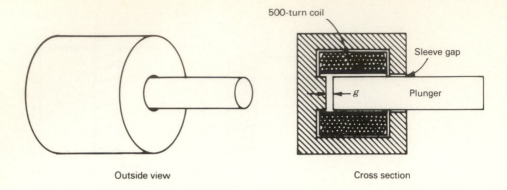

500-turn coil

Sleeve gap

g Plunger

Outside view   Cross section

flux leakage, and the reluctances of the iron and of the sleeve gap. Let the coil current be 5.00 A dc. (a) Allow the gap length g to vary from 1 mm to 1 cm, and plot $\phi(g)$, $\lambda(g)$, coil $L(g)$, $W_\phi(g)$ in the gap, and force on the plunger. (b) Write an expression for $dL(g)/dg$. Show by means of algebra and definitions that $(i^2/2)(dL(g)/dg)$ is the same as $B^2A/2\mu_0$.

8.8. A four-pole dc motor is simplex-wave-wound (i.e., $a = 2$). There are 45 slots in the armature core and 8 conductors per slot. (Neglect the fact that some coils are undergoing commutation and the possibility of dummy coils.) Armature winding resistance between terminals is $0.100\ \Omega$. When connected to a 125-V dc source and operating under load, the armature current is 50.0 A and the speed is 1000 rev/min, with the field current adjusted so that the MMF applied to the air gap by each field pole is 2000 A-turns. (a) Find the reluctance of the air gap under one pole and the flux per pole. (b) Find the peak armature MMF F_a. (c) Find the developed torque and developed horsepower, assuming the brushes to be on the neutral axis.

8.9. Repeat Problem 8.8 with the brushes shifted 10 mechanical degrees in a direction to reduce sparking. Assume the same terminal voltage and armature current. What is the speed with the brushes shifted?

8.10. A three-phase, Y-connected, four-pole induction motor has an air gap diameter of 20 cm, a stack length of 15 cm, and an air gap of 1.0 mm. The stator winding has an N_{f1} of 400 turns. Under certain operating conditions, the magnetizing current is 2.00 A rms, the rotor MMF is 1800 A-turns per pole, and the angle between the resultant and rotor MMFs is 110 electrical degrees. (a) Find the torque being developed. (b) Find I_1, the rms stator phase current in rms amperes.

8.11. A coincidence in dc machine theory:

The armature MMF of a dc machine as a function of angular distance along the rotor surface is triangular and has an amplitude of F_a A-turns. (a) Using Fourier analysis, show that the amplitude of the fundamental component of this wave is $8F_a/\pi^2$. Now

$$F_a = \frac{ZI_a}{2pa}$$

In Equation 8.95, the angle δ_{FR} is the angle between the center of one pole of air gap flux ϕ and the peak rotor MMF, F_2. In a dc machine without brush shift, this angle is maintained at 90° electrical by the commutator. However Equation 8.95 does not apply to dc machines because it is derived on the assumptions that the air gap is of uniform radial length and that the MMFs applied to the air gap are sinusoidally distributed. (b) Neglecting the sign in Equation 8.95, substitute the amplitude of the fundamental component of dc machine rotor MMF for F_2 and 90° for δ_{FR} in this equation, to show that by strange coincidence the correct torque expression results (Equation 8.137).

8.12. Figure 5.27 shows the relationship between the pole arc of a dc machine α and the pole pitch $\rho = 2\pi/p$. If the ratio α/ρ is taken to be 0.6, then Equation 8.148 reduces to

$$\tau = 0.9425{\cdot}D^2L(\text{SEL})(\text{SML})$$

If the machine having this α/ρ ratio the average flux density under a pole is 0.80 T and $D^2L = 0.00354$ m³, (a) find the (SEL) required to develop 15 hp at 1800 rev/min. (b) If $D = L$, find the rotor diameter in inches.

8.13. In an induction motor, the angle between F_1 and R (Figure 4.14) is δ_{SR} and is equal to $90° - \theta_1'$, where θ_1' is the angle between the stator phase current and the induced phase voltage E_1. In Figure 4.12, it is seen that θ_1' does not differ greatly from the power factor angle $\theta_{\phi m}$. So

$$\sin \delta_{SR} = \sin(90° - \theta_1') = \cos \theta_1' \approx \text{power factor}$$

A certain induction motor has D^2L of $1.647{\cdot}10^{-3}$ m³. (a) If $L = 1.57D$, what are D and L? The stator winding of this machine is designed to have a k_w of 0.923. (b) For a specific electrical loading of 16,500 A/m, approximately what must be the rms flux density in the air gap to produce a torque of 20 N-m at a power factor of 0.9?

APPENDIXES

APPENDIXES

A

A REVIEW OF MAGNETIC CIRCUITS

A.1 MAGNETIC CONCEPTS

When an electron moves, it produces a magnetic field. Magnetic fields cannot be seen, felt, heard, smelled, or tasted. We know they exist by their effects on objects around us. A compass needle points north, for example. We can *describe* a magnetic field, however, in terms of a consistent set of concepts. Expressing these concepts mathematically allows us to assign values to magnetic quantities and thus to design useful devices that operate magnetically.

One such concept is magnetic polarity. A magnetic-field source such as current-carrying coil or a piece of magnetized iron will tend to align itself within the earth's magnetic field. That is, one end of the source will tend to point north and is thus designated as the "north" pole of the field source. Closely related to the concept of polarity is the concept of field *flow*. This concept visualizes the field as circulating through the sources: coming out of the north pole, following some sort of path, and reentering the south pole. A corollary to these concepts is that the field has *direction* at any point in space.

Figure A.1 shows a current-carrying coil embedded in a sheet of white cardboard on which iron particles have been sprinkled. These particles have become magnetized by a field produced by the electrons circulating in the coil and have aligned themselves with the field. The pattern thus formed shows the shape of the field surrounding the coil, and thus the direction of the field at various points in the plane of the cardboard. To completely describe the field direction in this plane, it would be necessary to know the magnetic polarity of the coil. It has been found that if the right hand is wrapped around the coil with the fingers pointing in the direction of conventional current flow (opposite to the direction of electron flow), the thumb will point to the *north* end of the coil (that is, the end of the coil that would point north if the coil were free to turn, and the end from which it is assumed that the field flows outward).

The experiments of the type shown in Figure A.1 have led to the concept of magnetic *lines of force*. However the lines visible in the figures are the result of the fact that iron is a good magnetic conductor, and for that reason the field tends to concentrate along the paths developed by the iron particles as they align themselves with the field.

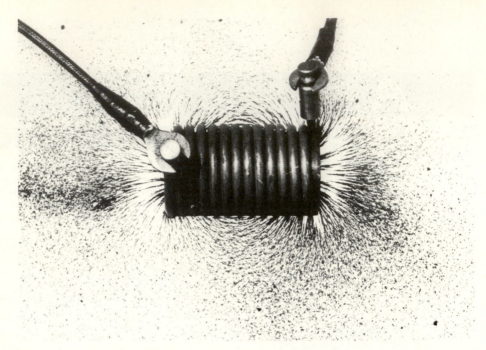

FIGURE A.1 Magnetic field of a current-carrying coil.

Lines of force actually do not exist, but the concept is sometimes useful in describing the properties of magnetic fields.

The total flow of a field is called its *flux*. For example, the total light emitted from a light source is called the *luminous flux* of the source and is expressed in lumens. Light fields radiate outward; however magnetic fields always circulate. For the coil of Figure A.1, the path of circulation is easily traced. The older unit of magnetic flux was the *line,* or *maxwell*. Thus the total flux of a magnetic pole can be expressed as so many lines, or maxwells. It is perfectly possible for the flux emanating from a magnetic pole to be less than one maxwell, and here the concept of "lines" breaks down, because a fraction of a line cannot be visualized as describing the flux pattern of a weak magnetic pole. The S.I. units for flux are webers (Wb):

$$1 \text{ weber (Wb)} = 10^8 \text{ maxwells} = 10^8 \text{ lines}$$

The symbol for magnetic flux is ϕ.

The strength of a magnetic field at any point in space is expressed in terms of *flux density* (i.e., the flux per unit area). The symbol for flux density is B. The units of flux density are teslas (T):

$$1 \text{ tesla (T)} = 1 \text{ Wb/m}^2$$

Other units of B now in use are lines/in.^{2}, kilolines/in.^{2}, and gausses (G).

$$1 \text{ gauss (G)} = 1 \text{ maxwell/cm}^2$$

Note that

$$1 \text{ T} = 10 \text{ kG} = 10^4 \text{ G}$$

A.2 MAGNETIC CIRCUITS

The field of a coil or other magnetic source can be transferred to some more useful location by means of magnetic conductors. Materials available for magnetic conductors are not nearly as good as those available for electrical conductors. Hence there is considerable leakage of flux involved in magnetic circuits, whereas current leakage is usually negligible in electric circuits. Figure A.2 is a photograph of a magnetic circuit. Figure A.3 is a schematic diagram of a similar circuit and its electrical analog. Note that the magnetic analog of resistance is *reluctance,* given the symbol $\mathcal{R}$, and the analog of electromotive force is *magnetomotive force* (MMF), given the symbol $\mathcal{F}$. The magnetic analog of current is the flux ϕ. Thus magnetic Ohm's law may be written as

$$\phi = \frac{\mathcal{F}}{\mathcal{R}} \quad \text{Wb} \tag{A.1}$$

The units of ϕ are webers, and those of $\mathcal{F}$ are ampere-turns, or simply amperes (A), since turns have no dimensions. However when the MMF is to be determined for a coil,

$$\mathcal{F} = NI \quad \text{A} \tag{A.2}$$

where N is the number of turns and I is the current in each turn. No particular unit name has been assigned to reluctance in the S.I. system. The magnetic analog of electrical conductance is *permeance* $\mathcal{P}$.

FIGURE A.2 Magnetic field of a simple magnetic circuit.

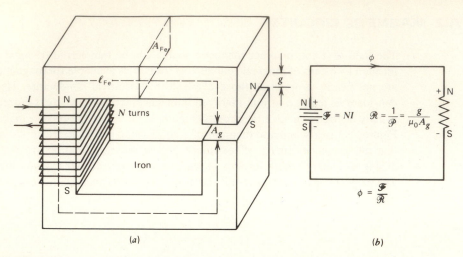

FIGURE A.3 A magnetic circuit and its electrical analog. (*a*) A simple magnetic circuit. (*b*) Electric analog.

The resistance of a wire of length ℓ and cross-sectional area A is given by $R = \ell/\sigma A$, where σ is the *conductivity* of the material in mhos/m. Similarly for a magnetic conductor,

$$\mathcal{R} = \frac{\ell}{\mu A} \tag{A.3}$$

where ℓ and A are in meters and square meters, respectively, and μ is the *permeability* of the material in henrys per meter (H/m). The permeabilities of air and free space are practically identical. This permeability is given the symbol μ_0:

$$\mu_0 = 4\pi \cdot 10^{-7} \text{ H/m}$$

If it is assumed that the reluctance of the magnetic conductors in Figure A.3 is negligible, then the MMF across the air gap is $\mathcal{F} = NI$ A, or A-turns. The reluctance of the gap is

$$\mathcal{R} = \frac{g}{\mu_0 A_g} \tag{A.4}$$

and the flux in the gap is

$$\phi_{\text{gap}} = \frac{\mathcal{F}\mu_0 A_g}{g} \tag{A.5}$$

Note in Figure A.2 that the flux near the edges of the gap does not go straight across, but "fringes" around the gap. If g is short, the fringing may be neglected. Then the flux density in the gap is given very nearly by

$$B_{\text{gap}} = \frac{\phi_{\text{gap}}}{A_g} = \mu_0 \frac{\mathcal{F}}{g} \quad \text{Wb/m}^2, \text{ or T}$$

The quantity $\mathcal{F}/g$ is called the *field intensity H*. In a uniform air gap,

$$H = \frac{\mathcal{F}}{g} \text{ A/m} \tag{A.6}$$

Equation A.6 does not apply to all situations. In general, however,

$$B = \mu H \tag{A.7}$$

Another general expression, which is a magnetic analog of Kirchhoff's voltage law, relates $\mathcal{F}$ to H:

$$\mathcal{F} = \oint H \cdot d\ell \tag{A.8}$$

A.3 TWO TYPES OF MAGNETIC-CIRCUIT PROBLEMS

There are two basic classes of problems in nonlinear magnetic circuits. These classes may be characterized as follows:

Class I: Given ϕ, find $\mathcal{F}$ (straightforward).
Class II. Given $\mathcal{F}$, find ϕ (requires graphical or iterative approach).

Examples of Class I Problems

■ **Example A.1**

Refer to Figure A.3. Suppose that $A_{\text{Fe}} = A_g = 0.01 \text{ m}^2$, $g = 0.005 \text{ m}$, and $N = 1000$ turns. What current would be required to produce a flux of 0.01 Wb in the air gap, neglecting iron reluctance?

Solution

1. $\mathcal{R} = \dfrac{g}{\mu_0 A_g} = \dfrac{0.005}{4\pi \cdot 10^{-7} \cdot 0.01} = 3.98 \cdot 10^5$

2. $\mathcal{F} = NI = 1000I$

3. $\phi = \dfrac{\mathcal{F}}{\mathcal{R}} = \dfrac{1000I}{3.98 \cdot 10^5} = 0.01 \text{ Wb}$

$I = \dfrac{3.98 \cdot 10^3}{1000} = 3.98 \text{ A}$

$\mathcal{F} = NI = 3980 \text{ A}$ ■

■ **Example A.2**

For the same circuit, what current is required to produce a flux density in the air gap of 1.5 T?

Solution. $\phi = BA = 1.5 \cdot 0.01 = 0.015$ Wb, or 1.5 times the flux of Example A.1, which implies $I = 1.5 \cdot 3.98 = 5.97$ A;

or

1. $H = \dfrac{B}{\mu_0} = \dfrac{1.5}{4\pi \cdot 10^{-7}} = 1.194 \cdot 10^6$ A/m

2. $\mathscr{F} = Hg = 1.194 \cdot 10^6 \cdot 5 \cdot 10^{-3} = 5.97 \cdot 10^3$ A

3. $\mathscr{F} = NI = 10^3 I = 5.97 \cdot 10^3$ A

$I = 5.97$ A   ■

■ Example A.3

Repeat A.1, taking the MMF drop in the iron into account. The total iron cross section is $A_{\text{Fe}} = 0.01$ m^2 and the length of the iron flux path is $\ell_{\text{Fe}} = 0.25$ m. (a) Assume the iron to be a soft steel casting. (b) Assume the iron to be a stack of 29-gage, M-19 steel laminations with a stacking factor of 0.85. Curve sheets for the iron are shown in Figure A.4.

Solution

(a) Since the iron area is equal to the air gap area and fringing is to be neglected,

1. $B_{\text{Fe}} = B_g = \dfrac{\phi}{A} = \dfrac{0.01}{0.01} = 1.0$ T

2. From the curve for soft steel castings, $H_{\text{Fe}} = 900$ A/m

3. $\mathscr{F}_{\text{Fe}} = H_{\text{Fe}}\ell_{\text{Fe}} = 900 \cdot 0.25 = 225$ A

4. $\mathscr{F} = \mathscr{F}_{\text{Fe}} + \mathscr{F}_g = 225 + 3.98 \cdot 10^3 = 4205$ A

5. $I = \dfrac{4205}{N} = 4.20$ A

(b) When a laminated core is used, a *stacking factor* takes into account the insulating space between laminations. The actual iron area is equal to the stacking factor times the gross area:

$$A_{\text{Fe}} = (\text{stacking factor}) \times (\text{gross area}) \tag{A.9}$$

1. $A_{\text{Fe}} = 0.85 \cdot 0.01 = 0.0085$ m^2

2. $B_{\text{Fe}} = \dfrac{\phi}{A_{\text{Fe}}} = \dfrac{0.01}{0.0085} = 1.18$ T

3. From the curve for M-19 sheets, $H_{\text{Fe}} \cong 260$ A/m

4. $\mathscr{F}_{\text{Fe}} = H_{\text{Fe}}\ell_{\text{Fe}} = 260 \cdot 0.25 \cong 65$ A

5. $I = \dfrac{\mathscr{F}}{N} = \dfrac{1}{1000}(\mathscr{F}_{\text{Fe}} + \mathscr{F}_g) = (65 + 3980) \cdot 10^{-3} = 4.05$ A   ■

NOTE: *For a flux density of 1.0 T in the air gap, an increase in the current of only 6 percent for soft steel and 1.8 percent for M-19 is required to overcome iron reluc-*

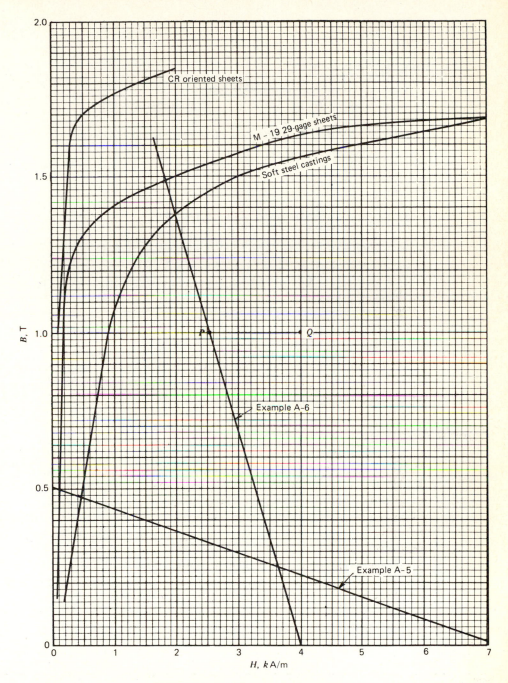

FIGURE A.4 B–H curves for three materials.

tance. Thus in this case, neglecting the iron is a fairly good approximation for low to medium flux densities.

■ Example A.4

Repeat Example A.2, taking into account the iron, assuming (a) a soft steel casting, and (b) M-19 sheets with a stacking factor of 0.90.

Solution

(a)

1. $B_{Fe} = B_g = 1.5$ T
2. From the curve, $H_{Fe} = 3000$ A/m
3. $\mathcal{F} = \mathcal{F}_g + H_{Fe}\ell_{Fe} = 5970 + 3000 \cdot 0.25 = 6720$ A
4. $I = \dfrac{\mathcal{F}}{N} = 6.72$ A, an increase of 13 percent above the value obtained with iron neglected

(b)

1. $B_{Fe} = \dfrac{B_{gross}}{\text{stacking factor}} = \dfrac{1.5}{0.90} = 1.67$ T
2. From the curve, H_{Fe} is very indefinite, say, 5500 A/m
3. $\mathcal{F}_{Fe} = H_{Fe}\ell_{Fe} = 5500 \cdot 0.25 = 1375$ A
4. $I = 1.375 + 5.97 = 7.34$ A, an increase of 23 percent over that required for the air gap alone ∎

NOTE: *Modern machines and transformers operate at flux densities in the iron of 1.3–1.5 T. Saturation makes it difficult to predict the performance of the iron, because different batches may have slightly different saturation characteristics. Note how indefinite the value of* H_{Fe} *is for high flux densities.*

Class II Problems

These are fundamentally different problems because of the nonlinearity of the iron. The previous examples were of the general form *Given* ϕ, *find* $\mathcal{F}$. Class II problems are of the form *Given* $\mathcal{F}$, *find* ϕ. There are at least three approaches to this type of problem. (1) *Cut and try,* that is, choose a value of ϕ and calculate $\mathcal{F}$. Compare with *NI,* choose a revised value of ϕ, and calculate a new value of $\mathcal{F}$. Repeat until $\mathcal{F}$ is a reasonably close match to *NI*. (2) *Magnetization curve;* choose values of ϕ and find the corresponding values of $\mathcal{F}$. Plot ϕ versus $\mathcal{F}$. This is the *magnetization curve* of the device. Using this curve and the given value of *I*, find ϕ corresponding to $\mathcal{F} = NI$. (3) *Graphical method;* this method is demonstrated below.

GRAPHICAL METHOD

■ Example A.5

For the magnetic device of the previous examples, how much flux would be produced in the air gap for a coil current of 1.8 A? The iron is M-19, 29-gage steel laminations, stacking factor = 0.9.

Solution

1. $B_{Fe} = \dfrac{\phi}{A_{Fe}}$, where ϕ is the desired flux

2. $H_{Fe} = \dfrac{\mathscr{F}_{Fe}}{\ell_{Fe}} = \dfrac{NI - \mathscr{F}_g}{\ell_{Fe}} = \dfrac{\mathscr{F}_{coil}}{\ell_{Fe}} - \dfrac{\phi_g \mathscr{R}_g}{\ell_{Fe}}$

3. From (1): $H_{Fe} = \dfrac{\mathscr{F}_{coil}}{\ell_{Fe}} - \dfrac{B_{Fe} A_{Fe} \mathscr{R}_g}{\ell_{Fe}}$

$$H_{Fe} = \frac{NI}{\ell_{Fe}} - B_{Fe} \frac{1}{\mu_0} \frac{A_{Fe}}{A_g} \frac{g}{\ell_{Fe}} \qquad (A.10)$$

Note that Equation A.10 is that of a straight line with an intercept $\mathscr{F}_{coil}/\ell_{Fe}$ and a slope $-A_{Fe}g/\mu_0 A_g \ell_{Fe}$. This is the air gap characteristic in terms of B_{Fe} and H_{Fe}. For this example,

$$\frac{NI}{\ell_{Fe}} = \frac{1000 \cdot 1.8}{0.25} = 7200 \text{ A/m}$$

The gross iron area is equal to the gap area. The stacking factor makes $A_{Fe}/A_g = 0.9$.

$$\frac{A_{Fe}g}{\mu_0 A_g \ell_{Fe}} = 0.9 \cdot \frac{0.005}{4\pi \cdot 10^{-7} \cdot 0.25} = 1.432 \cdot 10^4$$

and Equation A.10 becomes

$$H_{Fe} = 7200 - 1.432 \cdot 10^4 B_{Fe}$$

To find the intercept on the B axis, let $H_{Fe} = 0$:

$$B_{Fe} = \frac{7200}{1.432 \cdot 10^4} = 0.503 \text{ T}$$

This line is plotted on Figure A.4. The intersection of the line with the M-19 curve is the simultaneous solution of the nonlinear equation for the iron and the linear equation for the air gap. The intercept shows that for 1.8 coil amperes, B_{Fe} is 0.50 T. Then

$$\phi = B_{Fe} A_{Fe} = 0.50 \cdot (0.9 \cdot 0.01) = 4.5 \cdot 10^{-3} \text{ Wb} \qquad \blacksquare$$

ALTERNATE METHOD FOR CLASS II PROBLEMS

The B-axis intercept will often be off the paper. In this case, the air gap characteristic may be drawn as indicated in Figure A.5.

1. Locate the H-axis intercept, as before, at

$$H \text{ intercept} = \frac{\mathscr{F}_{coil}}{\ell_{Fe}} \equiv \frac{NI}{\ell_{Fe}} \qquad (A.11)$$

2. By Equation A.10, the slope of the air gap line is

$$\frac{\Delta H}{\Delta B} = \frac{-1}{\mu_0} \left(\frac{A_{Fe}}{A_g} \right) \left(\frac{g}{\ell_{Fe}} \right) \qquad (A.12)$$

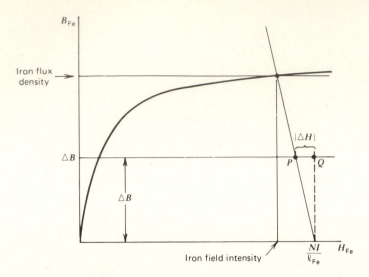

FIGURE A.5 A method for locating the air gap characteristic of the B–H characteristic of the iron.

Then

$$\Delta H = -\Delta B \left[\frac{1}{\mu_0} \left(\frac{A_{\mathrm{Fe}}}{A_g} \right) \left(\frac{g}{\ell_{\mathrm{Fe}}} \right) \right] \qquad (A.13)$$

Choose some convenient value of ΔB (often $\Delta B = 1.0$ T). Draw a line parallel to the H axis at $B = \Delta B$, extending over the air gap line intercept. Draw a line up from the $\mathcal{F}_{\mathrm{coil}}/\ell_{\mathrm{Fe}}$ intercept to locate the point Q in Figure A.5.

3. Calculate the coefficient of ΔB in Equation A.13, and multiply by the chosen value of ΔB to get the corresponding value of ΔH. Since ΔH is negative, it will be marked off to the left of Q to locate a second point on the air gap line, F. This point and the H_{Fe} intercept determine the air gap line.

4. The intersection of the air gap line with the B–H curve for the iron determine the operating levels of B and H in the iron of the magnetic circuit.

■ **Example A.6**

In the magnetic circuit of Example A.5, what will the flux in the air gap be if the air gap length is reduced to 0.0005 m and the coil current is 1.00 A?

Solution

1. $\mathcal{F}_{\mathrm{coil}} = NI = 1000$ A

   $\dfrac{\mathcal{F}_{\mathrm{coil}}}{\ell_{\mathrm{Fe}}} = \dfrac{1000 \text{ A}}{0.25 \text{ m}} = 4000$ A/m

2. Let $\Delta B = 1.0$ T

   Q is at $(H, B) = (4000, 1)$

3. $\left| \dfrac{\Delta H}{\Delta B} \right| = \dfrac{1}{\mu_0} \left(\dfrac{A_{\mathrm{Fe}}}{A_g} \right) \left(\dfrac{g}{\ell_{\mathrm{Fe}}} \right) = \dfrac{1}{4\pi \cdot 10^{-7}} (0.9) \left(\dfrac{0.0005}{0.25} \right) = 1.43 \cdot 10^3.$

   Compare with $1.432 \cdot 10^4$ found previously for $g = 0.005$ m

4. $|\Delta H| = \Delta B \left| \dfrac{\Delta H}{\Delta B} \right| = 1 \cdot 1.43 \cdot 10^3$ A/m

P is at $(H, B) = ([4000 - 1433], 1) = (2570, 1)$

This point is plotted in Figure A.4. The air gap line shows the flux density in the iron to be 1.49 T for M-19 sheets:

$$\phi = B_{Fe}A_{Fe} = 1.49 \times 0.009 = 0.0134 \text{ Wb}$$ ∎

A.4 THE MAGNETIZATION CURVE

The magnetization curve is a plot of useful flux as a function of coil MMR or coil current for a specific device. The useful flux is usually the air gap flux. Data for the curve are obtained by working the Class I problem several times for values of ϕ chosen within the range of interest. It is well to organize the calculations by means of a data table. Using the data of Example A.6: $A_g = 0.01$ m^2, $A_{Fe} = 0.009$ m^2, $g = 0.0005$ m, $\ell_{Fe} = 0.25$ m, $\mathcal{R}_g = 0.0005/\mu_0 \cdot 0.01 = 3.98 \cdot 10^4$, $N = 1000$:

ϕ_g Wb	$\mathcal{F}_g = \phi_g \mathcal{R}_g$ A	$B_{Fe} = \dfrac{\phi_g}{0.009}$ T	H_{Fe} (B–H curve) A/m	$\mathcal{F}_{coil} =$ $H_{Fe} \cdot 0.25 + \mathcal{F}_g$ A	$I = \mathcal{F}_{coil}/N$ A
0.002	79.6	0.222	80	100	0.100
0.005	199	0.555	110	227	0.227
0.010	398	1.11	190	444	0.444
0.013	517	1.44	1300	842	0.842
0.014	557	1.56	2800	1275	1.257

The resulting magnetization curve is shown in Figure A.6. Note that the flux for 1 A coil current is 0.0133 Wb, as in Example A.6.

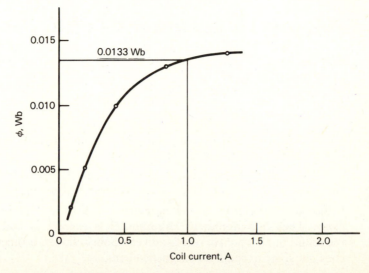

FIGURE A.6 Magnetization curve of Example A.6.

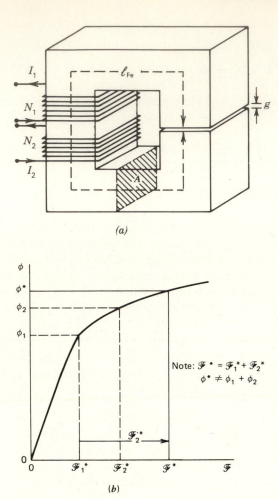

FIGURE A.7 (*a*) A simple magnetic circuit with two coils. (*b*) A typical magnetization curve for the circuit of part (*a*).

A.5 SUPERPOSITION IN NONLINEAR MAGNETIC CIRCUITS

Consider a magnetic circuit with two magnetizing coils, such as Figure A.7. The total MMF applied to the circuit is

$$\mathscr{F} = \mathscr{F}_1 + \mathscr{F}_2 = N_1 I_1 + N_2 I_2$$

Given a value of ϕ, $\mathscr{F}$ may be calculated by the method described in Section A.3 for Class I problems.

Let $\mathscr{F}_1^*$ and $\mathscr{F}_2^*$ be particular values of $\mathscr{F}_1$ and $\mathscr{F}_2$. Then $\mathscr{F}^* = \mathscr{F}_1^* + \mathscr{F}_2^*$ and the correct value of ϕ is ϕ^*. Note, however, that ϕ^* is *not* equal to $\phi_1 + \phi_2$, where ϕ_1 is the flux produced by $\mathscr{F}_1^*$ acting alone and ϕ_2 is the flux produced by $\mathscr{F}_2^*$ acting alone. **Thus in a nonlinear situation, it is correct to add MMFs to obtain a resultant, but it is not correct to add component fluxes.** In other words, superposition does not apply to nonlinear magnetic circuits.

■ Example A.7

In Figure A.7, let $A = 0.005$ m^2, $g = 0.001$ m, $\ell_{Fe} = 0.5$ m, $N_1 = 500$ turns, $N_2 = 300$ turns. The core iron is a soft steel casting.

(a) Find $\mathscr{F}$ for a flux of 0.006 Wb in the air gap, neglecting leakage and fringing.
(b) If the two coils are in series, aiding, what must $I_1 = I_2$ be?
(c) If the two coils are in series, opposing, what must $I_1 = I_2$ be?
(d) If $I_1 = 3.0$ A, what must I_2 be?
(e) If only N_1 is excited, what must I_1 be?
(f) If only N_2 is excited, what must I_2 be?

Solution

(a)

1. $B_g = \dfrac{\phi}{A_g} = \dfrac{0.006}{0.005} = 1.2$ T
2. $B_{Fe} = B_g = 1.2$ T
3. $H_g = \dfrac{B_g}{\mu_0} = \dfrac{1.2}{4\pi \cdot 10^{-7}} = 9.55 \cdot 10^5$ A/m
4. $\mathscr{F}_g = g \cdot H_g = 0.001 \cdot 9.55 \cdot 10^5 = 955$ A
5. From Figure A.4 and $B = 1.2$ T, $H_{Fe} = 1250$ A/m
6. $\mathscr{F}_{Fe} = 1250(0.5) = 625$ A
   $\mathscr{F} = \mathscr{F}_g + \mathscr{F}_{Fe} = 955 + 625 = 1580$ A

(b) $\mathscr{F} = N_1 I_1 + N_2 I_2 = I(N_1 + N_2)$

$$1580 = I(800)$$

$$I_1 = I_2 = I = 1.98 \text{ A}$$

(c) $\mathscr{F} = N_1 I - N_2 I = I(500 - 300)$

$$200 I = 1580 \text{ A}$$

$$I_1 = I_2 = I = 7.90 \text{ A}$$

(d) $\mathscr{F} = 500 \cdot 3.0 + 300 I_2 = 1580$ A

$$300 I_2 = 80 \text{ A}$$

$$I_2 = 0.27 \text{ A}$$

(e) $\mathscr{F} = 1580 = N_1 I_1 = 500 I_1$

$$I_1 = 3.16 \text{ A}$$

(f) $1580 = N_2 I_2 = 300 I_2$

$$I_2 = 5.27 \text{ A}$$ ■

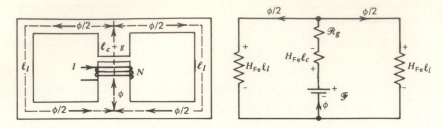

FIGURE A.8 A symmetrical, three-limbed, magnetic circuit and its electric analog.

A.6 PARALLEL MAGNETIC CIRCUITS

Previous sections and examples have considered series magnetic circuits, that is, circuits where the same flux passes through all branches of the circuit. Obviously, magnetic devices may be constructed with geometries such that the total flux divides between parallel paths. The symmetrical, three-limbed circuit of Figure A.8 illustrates the point. The total flux ϕ flows through the common leg. Since both outside branches have the same length and cross-sectional area, symmetry dictates that $\phi/2$ webers flow through each. Combining the two parallel legs (like parallel resistors in the electrical analog) simplifies the circuit. Thus the common leg and air gap may be considered to be in series with a branch of area $2A_l$ and length ℓ_l. Or since the same current penetrates the planes of both loops, Ampère's Law requires that the sum of the MMFs around either loop must equal the coil MMF.

In general, the parallel legs will have a different flux density than the common leg. Because of saturation, the permeabilities will also be different. Solutions to Class I problems where ϕ is known are still straightforward (see Example A.8). Class II problems, however, must be solved through iteration or construction of the magnetization curve. The graphical approach is no longer valid since the various branches are operating at different points on the saturation curve. The one exception to this is the special case when $A_l = A_c/2$. Then the flux density is equal to all parts of the magnetic circuit.

■ **Example A.8**

For the magnetic circuit of Figure A.8, $A_c = 10^{-3}$ m², $\ell_c = 6$ cm, $g = 2$ mm, $A_l = 6 \cdot 100^{-4}$ m², $\ell_l = 20$ cm, and N = 500 turns. If $\phi = 1.6$ mWb, find I.

Solution

1. $B_c = \dfrac{1.6 \cdot 10^{-3}}{10^{-3}} = 1.6$ T

   $H_c = 4950$ A/m (from Figure A.4)

2. $B_l = \dfrac{8 \cdot 10^{-4}}{6 \cdot 10^{-4}} = 1.33$ T

   $H_l = 1700$ A/m (from Figure A.4)

3. $\mathscr{F} = 500I = \phi\mathscr{R}_g + H_c\ell_c + H_l\ell_l$

   $= 1.6 \cdot 10^{-3}\dfrac{2 \cdot 10^{-3}}{\mu_0 10^{-3}} + 4950 \cdot 0.06 + 1700 \cdot 0.2$

$$= 2546 + 297 + 340 = 3183 \text{ A-turns}$$

$$I = 6.37 \text{ A} \qquad \blacksquare$$

If the parallel paths were not symmetrical, the problem would be much more difficult. Flux would divide unequally, and varying degrees of saturation would exist in the two branches. Even a Class I problem would require a tedious trial-and-error iterative approach. The procedure is:

1. Estimate (make a reasonable guess) the flux for each parallel path, realizing that they must sum to ϕ.
2. Calculate the corresponding MMFs (H_{Fe}) for each path.
3. Adjust the flux estimates to decrease any differences in the calculated MMFs, and repeat (2) until all parallel MMFs are equal.

PROBLEMS

A.1. The problem is to find the coil ampere-turns required to produce a given magnetic flux in a magnetic circuit such as that shown in the accompanying figure. The cross-sectional area of the iron is equal to that of the air gap, A. The air gap length is g. Let ℓ_{Fe} be the mean length of flux path through the steel part of the circuit. Solve this problem for the following conditions: desired flux $= 1.80 \cdot 10^{-3}$ Wb, $A = 2.00$ in.$^2 = 1.29 \cdot 10^{-3}$ m^2, $g = 3.00 \cdot 10^{-3}$ m, $\ell_{Fe} = 10.0$ in. $= 0.254$ m.

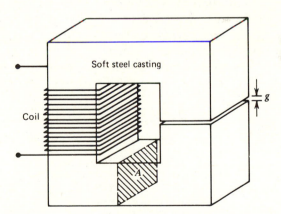

A.2. Plot a magnetization curve for the magnetic circuit of Problem A.1, assuming the coil to have 1000 turns. Let the maximum flux density be 1.5 T.

A.3. What is the reluctance of the air gap in Problem A.1? Neglecting the iron reluctance, what flux would flow in the circuit if the MMF of the coil is 1500 A-turns?

A.4. If the coil of Problem A.1 has 1000 turns and carries a current of 1.500 A, find the flux in the circuit by the graphical method.

A.5. Repeat Problems A.1 through A.4 for the same casting, except with a 1-mm air gap.

A.6. The force between two parallel magnetic pole faces of equal area is given by

$$f = \frac{B^2 A}{2\,\mu_0}\ \text{N}$$

where B is in teslas and A is in square meters.

A dc solenoid magnet for operating a gas valve has a plunger that moves $\frac{1}{2}$in. When the valve is closed, the air gap g is 0.5 in.; when open, g is 0.01 in. The cross-sectional area of the air gap is 1.0 in^2.

(a) What flux density is required in the air gap to produce a pull of 10 lb when the gap length is 0.5 in.? What is the total flux in webers?

(b) The flux path in the iron parts of the solenoid is made of 29-gage, M-19 laminations. The core is designed so that the iron flux density is the same as that of the gap. The length of the flux path in iron varies, and is given by $\ell_{\text{Fe}} = (9.00 - g)$ in. Find the MMF required of the solenoid coil to produce the 10-lb pull at $g = 0.5$ in. (1 pound = 4.445 N).

(c) Plot the magnetization curves of the device in the closed and in the open positions. (Plot ϕ as a function of $\mathcal{F}$.)

(d) For the same coil MMF as that required to produce 10 lb at $g = 0.5$ in., find the force when the value is in the open position: $g = 0.01$ in. (*Hint:* The iron will be very highly saturated. Estimate the saturation flux density and use that value to calculate the force.)

A.7. The center leg of an iron core made from a 3-cm stack of this lamination ($S.F. = 0.90$) is wrapped with a single coil. (a) What coil MMF is required to produce $B_g = 1.1$ T? (b) If the mean length of a turn is 20 cm, specify the wire size and number of turns to approximate the MMF found in part (a). (*Hint:* Applying Ohm's law to the coil circuit, in combination with the MMF requirement, permits calculation of the wire's resistance per meter. Choose N so that I is within the ampacity of the wire size thus determined.)

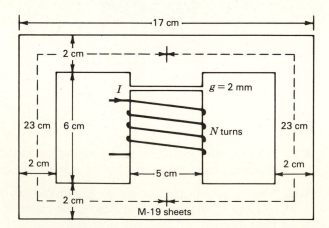

A.8. An MMF of 2500 A is applied to the laminated core of Problem A.7. Find ϕ.

A.9. If the width of the center leg of the laminated core of Problem A.7 is reduced to 4 cm (making the flux density approximately uniform in the iron) and the coil MMF is 2500 A, find ϕ.

B

BALANCED THREE-PHASE CIRCUITS

B.1 WHY STUDY BALANCED THREE-PHASE CIRCUITS?

Nearly all of the electrical power generated in the world is generated as three-phase ac. A three-phase circuit is just a particular configuration of electrical circuit elements and may be solved by straightforward application of electric circuit theory. Such an approach is appropriate when a three-phase circuit is unbalanced. However power equipment works better when system voltages are balanced, and every effort is made to maintain balanced operation of three-phase power circuits. Special techniques are available that make the solution of balanced three-phase circuits extremely simple. These techniques are easily learned, so it would be very inefficient to use general circuit theory to solve the problems of balanced three-phase circuits.

B.2 SINGLE-PHASE POWER

Consider a simple single-phase circuit, consisting of a voltage source and a load (Figure B.1). The voltage V and the current I are phasor quantities:

$$V = |V| \, \underline{/\alpha}$$
$$I = |I| \, \underline{/\alpha - \theta} \tag{B.1}$$

and the impedance is complex:

$$Z = |Z| \, \underline{/\theta} = |Z|(\cos \theta + j \sin \theta) = R + jX \tag{B.2}$$

Applying Ohm's law, it is seen that a positive impedance angle results in a current that *lags* the voltage:

$$I = \frac{V}{Z} = \frac{|V| \, \underline{/\alpha}}{|Z| \, \underline{/\theta}} = |I| \, \underline{/\alpha - \theta} \tag{B.3}$$

The angle of lag equals the impedance angle.

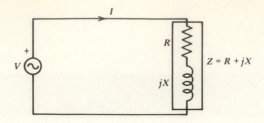

FIGURE B.1 A single-phase ac circuit.

The *apparent power* $|S|$ is the product of the *magnitudes* of the voltage and current:

$$|S| = |V|\,|I| \quad \text{VA} \tag{B.4}$$

The power P is the apparent power multiplied by the *power factor*. Equation B.5 defines the power factor (p.f.):

$$P = |S| \times (\text{p.f.}) \quad \text{W} \tag{B.5}$$

The *apparent power* is the magnitude of the *complex power* S. The complex power is defined by

$$S \triangleq VI^* \tag{B.6}$$

where I^* is the complex conjugate of the I phasor. (This quantity could just as well be defined as V^*I from a theoretical point of view, but Equation B.6 is the usually accepted definition.)

Since

$$I^* = |I|\,\underline{/-(\alpha - \theta)}$$

then

$$S = (|V|\,\underline{/\alpha}) \cdot (|I|\,\underline{/-\alpha + \theta})$$

$$= |V|\,|I|\,\underline{/\theta}$$

$$= |S|\,\underline{/\theta} = |S|\cos\theta + j|S|\sin\theta \tag{B.7}$$

The real part of S is the power P, in watts, and the imaginary part is called the *reactive volt-amperes* and is given the symbol Q. The units of Q are VARs:

$$S = P + jQ \quad \text{VA}$$

$$P = |S|\cos\theta \quad \text{W} \tag{B.8}$$

$$Q = |S|\sin\theta \quad \text{VAR}$$

Comparing Equations B.5 and B.8:

$$\text{p.f.} \triangleq \cos\theta \tag{B.9}$$

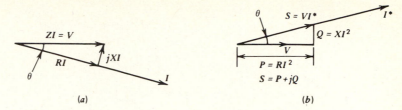

FIGURE B.2 Complex power. (*a*) Voltage diagram. (*b*) Volt-ampere diagram.

The sine of θ is sometimes called the *reactive factor* (**rf**),

$$Q = |S| \cdot (\text{r.f.})$$
$$\text{r.f.} \triangleq \sin \theta$$

$$(B.10)$$

It is important to note that a positive impedance angle results in *positive Q* and *lagging* current. *Thus lagging currents results in positive* Q. Figure B.2 illustrates this convention.

B.3 DESCRIPTION OF BALANCED, THREE-PHASE CIRCUITS

A three-phase generator has three sets of coils, so placed in the machine that the three equal voltages produced are 120° out of phase with each other. These coils sets may be represented by three voltage sources, and may be connected in one of two ways, Y or Δ, as shown in Figure B.3. Sometimes Y is spelled out "wye," and Δ, "delta," but it is obvious what is meant. The angle α is chosen for convenience in solving a problem. When a given phase voltage is taken to be at an angle zero, it is said to be the *reference voltage*.

B.4 DEFINITION OF A PHASE

A phase is one of the three branch circuits making up a three-phase circuit (refer to Figure B.3). In a Y connection, a phase consists of those circuit elements connected between one line and neutral. In a Δ circuit, a phase consists of those circuit elements connected between two lines. In a balanced system, all three phase voltages are equal in magnitude but differ from each other in phase by 120°. The *magnitudes* of the phase voltages and currents will be given the symbols V_ϕ and I_ϕ.

B.5 RELATIONSHIPS BETWEEN LINE AND PHASE VOLTAGES

When the voltage of a three-phase system is given, it is given as the *magnitude* of the *line-to-line voltage* V_L. The three line voltages in a balanced, three-phase system are equal in magnitude and are also 120° out of phase with each other:

$$|V_{ab}| = |V_{bc}| = |V_{ca}| \triangleq V_L$$

$$(B.11)$$

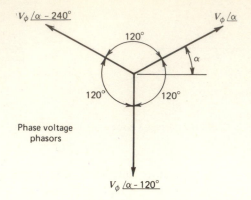

Phase voltage
phasors

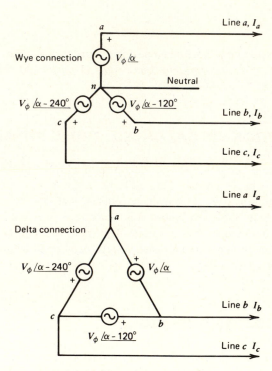

FIGURE B.3 Three-phase voltage sources.

if

$$V_{ab} = V_L \underline{/\alpha}$$

then

$$V_{bc} = V_L \underline{/\alpha - 120°}$$

and

$$V_{ca} = V_L \underline{/\alpha - 240°} = V_L \underline{/\alpha + 120°}$$

In a Δ connection, it is obvious that V_{ab} is both the line and phase voltage from a to b. Then for a Δ,

$$V_{\phi\Delta} = V_L \tag{B.12}$$

In a Y connection, we see that

$$V_{ab} = V_{an} + V_{nb} = V_{an} - V_{bn}$$
$$V_{bc} = V_{bn} + V_{nc} = V_{bn} - V_{cn} \tag{B.13}$$
$$V_{ca} = V_{cn} + V_{na} = V_{cn} - V_{an}$$

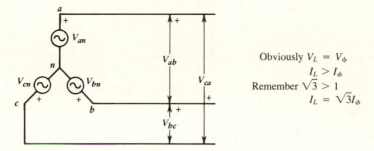

Obviously $V_L = V_\phi$
$$I_L > I_\phi$$
Remember $\sqrt{3} > 1$
$$I_L = \sqrt{3}I_\phi$$

Figure B.4 shows the sinusoidal phase voltages and illustrates the instantaneous addition, $v_{ab} = v_{an} - v_{bn}$. The phasor additions are shown in Figure B.5. Note that the line voltages are all equal, that they are displaced from each other by 120°, and that the set of line voltage phasors is displaced from the phase voltage set by 30°. Then for a Y connection,

$$V_{\phi Y} = \frac{V_L}{\sqrt{3}} \tag{B.14}$$

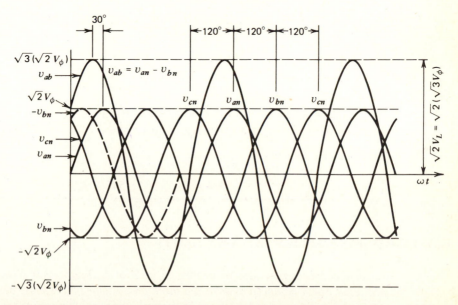

FIGURE B.4 Instantaneous phase voltages in a Y-connected generator, showing $v_{ab} = v_{an} - v_{bn}$, where the lowercase variables indicate sinusoidal waveforms corresponding to their respective phasors.

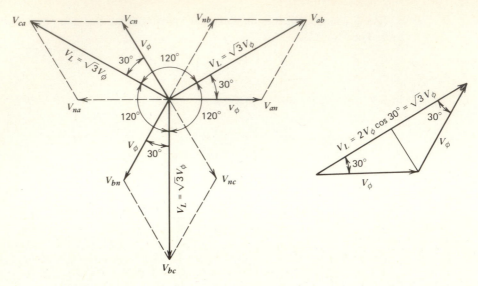

FIGURE B.5 Relationship between line and phase voltages in a Y connection.

■ Example B.1

A Y-connected generator is to be designed to supply a 20,000-V, three-phase line. What must be the terminal voltage of each phase winding?

Solution. 20,000-V, three-phase implies that the *line* voltage is 20,000 V rms:

$$V_L = 20,000 \text{ V}$$

Then by Equation B.14,

$$V_{\phi Y} = \frac{20,000}{\sqrt{3}} = 11,547 \text{ V} \qquad ■$$

■ Example B.2

If the generator designed to meet the needs of Example B.1 were reconnected in Δ, what would be its output line voltage when its phase voltage is 11,547 V?

Solution. By Equation B.12,

$$V_L = V_{\phi\Delta} = 11,547 \text{ V} \qquad ■$$

B.6 LOAD IMPEDANCE CONNECTIONS

Load impedances may also be connected in Y or Δ. Figure B.6 shows typical connections. *For a balanced system, the three phase impedances must be equal to each other.* Since these impedances are in general complex, this means that not only the magnitudes must be equal, but also the impedance angles must all be the same.

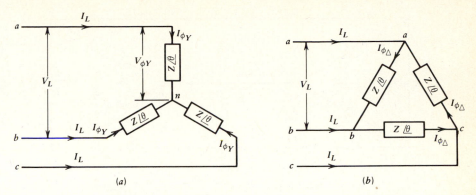

FIGURE B.6 Three-phase loads. (*a*) Y connection. (*b*) Δ connection.

Large three-phase motors are designed to act as balanced loads, so industrial loads are inherently well balanced. Residential loads are usually single-phase, however. Three-phase power is delivered to substations in residential areas. An attempt is made to have the same number of houses connected in parallel to form each phase load, and thus maintain balanced conditions.

B.7 SOLVING A Δ LOAD

Given the phase impedance and the line voltage, it is important to know the line current I_L, the phase currents I_ϕ, the power P absorbed by the load, the power factor, the total volt amperes S, and the total reactive volt-amperes Q. It is extremely easy to find these quantities in a balanced, three-phase system.

For a Δ load, first draw a sketch similar to Figure B.6*b*. It is obvious that the phase voltage is equal to the line voltage. The magnitude of the phase current is thus

$$I_{\phi\Delta} = \frac{V_L}{|Z|} \tag{B.15}$$

From the sketch you have drawn, it would seem reasonable that the incoming line current should be greater than $I_{\phi\Delta}$. Figure B.7 shows the phasor relationships between phase and line currents in a Δ load. The *magnitude* relationship is

$$I_L = \sqrt{3}I_{\phi\Delta} \tag{B.16}$$

Since $\sqrt{3} = 1.732$ is greater than 1.0, we see that the line current *is* greater than the phase current in a Δ.

By ordinary circuit theory, the power in one phase of a Δ is given by

$$P_\phi = V_\phi I_\phi \cos\theta = V_L \frac{I_L}{\sqrt{3}} \cos\theta \quad \text{W} \tag{B.17}$$

where θ is the angle of the phase impedance and

$$\text{Power factor} \equiv \text{p.f.} = \cos\theta \tag{B.18}$$

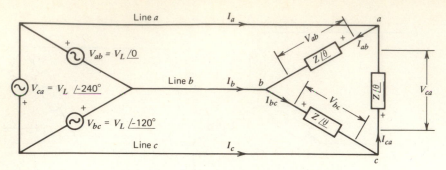

$$\text{Load phase } a\text{-}b: I_{ab} = \frac{V_{ab}}{Z_{ab}} = \frac{V_L\,\underline{|0°}}{Z\,\underline{|\theta}} = I_\phi\underline{|-\theta}$$

$$\text{where } I_\phi \overset{\Delta}{=} \frac{V_L}{|Z|}$$

$$\text{Similarly, } I_{bc} = \frac{V_{bc}}{Z\,\underline{|\theta}} = \frac{V_L\,\underline{|-120°}}{Z\,\underline{|\theta}} = I_\phi\,\underline{|-120° - \theta}$$

$$I_{ca} = I_\phi\,\underline{|-240° - \theta} = I_\phi\,\underline{|+120° - \theta}$$

$$\text{KCL at } a: I_a = I_{ab} - I_{ca} = I_\phi\,\underline{|-\theta} + I_\phi\,\underline{|-60° - \theta} = \sqrt{3}I_\phi\,\underline{|-30° - \theta}$$

$$\text{KCL at } b: I_b = I_{bc} - I_{ab} = I_\phi\,\underline{|-120° - \theta} + I_\phi\,\underline{|180° - \theta} = \sqrt{3}I_\phi\,\underline{|-150° - \theta}$$

$$\text{KCL at } c: I_c = I_{ca} - I_{bc} = I_\phi\,\underline{|120° - \theta} + I_\phi\,\underline{|-60° - \theta} = \sqrt{3}I_\phi\,\underline{|90° - \theta}$$

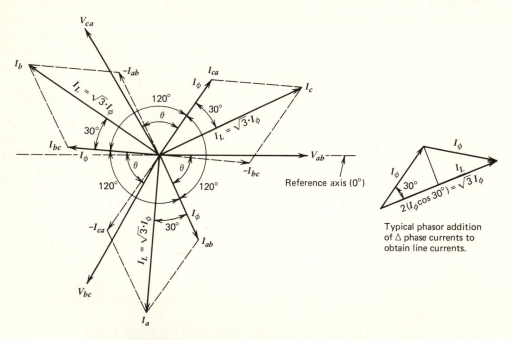

FIGURE B.7 Phasor relationships for a balanced Δ load.

The volt-amperes per phase is given by

$$|S_\phi| = V_\phi I_\phi = V_L \frac{I_L}{\sqrt{3}} \quad \text{VA} \tag{B.19}$$

and the reactive volt-amperes are given by

$$Q_\phi = V_\phi I_\phi \sin\theta = V_L \frac{I_L}{\sqrt{3}} \sin\theta \quad \text{VAR} \tag{B.20}$$

The total complex power quantities for the entire delta load are three times the values given by Equations B.17, B.19, and B.20:

$$P = 3P_\phi = 3 V_\phi I_\phi \cos\theta = \sqrt{3} V_L I_L \cos\theta \quad \text{W} \tag{B.21}$$

$$Q = 3Q_\phi = 3 V_\phi I_\phi \sin\theta = \sqrt{3} V_L I_L \sin\theta \quad \text{VAR} \tag{B.22}$$

$$|S| = \sqrt{P^2 + Q^2} = 3 V_\phi I_\phi = \sqrt{3} V_L I_L \; \text{VA} \tag{B.23}$$

Note that

$$P = |S| \cos\theta \qquad Q = |S| \sin\theta$$

Summary, for Δ load; given V_L and $Z_L \underline{|\theta}$:

$$I_\phi = \frac{V_L}{|Z_L|} \quad \text{A} \tag{B.15}$$

$$I_L = \sqrt{3} \, I_\phi \; \text{A} \tag{B.16}$$

$$|S| = \sqrt{3} \, V_L I_L \; \text{VA} \tag{B.23}$$

$$P = |S| \cos\theta \; \text{W} \tag{B.24}$$

$$Q = |S| \sin\theta \; \text{VAR} \tag{B.25}$$

■ **Example B.3**

Three impedances of $3 + j4 \; \Omega$ are connected in Δ to a 220-V, three-phase line. Find the line current, the power factor, and the complex power.

Solution

1. Draw a sketch:

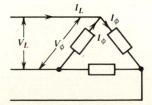

Obviously $V_L = V_\phi$
$I_L > I_\phi$
Remember $\sqrt{3} > 1$
$I_L = \sqrt{3} I_\phi$

2. $Z_L = 3 + j4 = 5\underline{|53.1°}$

3. $I_\phi = \dfrac{V_\phi}{|Z_\phi|}; \dfrac{220}{5} = 44\text{A}$

4. $I_L = \sqrt{3} \cdot 44 = 76.2\text{A}$

5. p.f. $= \cos\theta = \cos 53.1° = 0.600$
   $\sin\theta = 0.800$

        6. $|S| = \sqrt{3}\, V_L I_L = \sqrt{3} \cdot 220 \cdot 76.2 = 29{,}000 \text{ VA}$
        $S = |S|(\cos\theta + j\sin\theta) = 17{,}400 + j23{,}200 \text{ VA}$

Wasn't that easy?   ∎

B.8 SOLVING A Y LOAD

Consider a Y-connected set of equal impedances connected to a Y-connected generator by three lines and a neutral, as in Figure B.8. The neutral is shown as a broken line because, as will soon be apparent, a neutral connection is unnecessary when the load is balanced.

With the neutral wire connected, it is obvious that the current I_{an} is

$$I_{an} \equiv I_a = \frac{V_{an}}{Z\underline{|\theta}} = \frac{V_L\underline{|0°}}{\sqrt{3}|Z|}\underline{|-\theta}$$

Similarly,

$$I_{bn} \equiv I_b = \frac{V_L}{\sqrt{3}|Z|}\underline{|-120° - \theta} \qquad I_{cn} \equiv I_c = \frac{V_L}{\sqrt{3}|Z|}\underline{|-240° - \theta} \quad \text{(B.26)}$$

It is also obvious that I_a and I_{an} are the same, identical current. The same can be said about $I_b \equiv I_{bn}$ and $I_c \equiv I_{cn}$. Since I_L is defined as the magnitude of the line current, from Equations B.11 and B.26,

$$I_L = |I_a| = |I_b| = |I_c| = \frac{V_L}{\sqrt{3}|Z|} \qquad \text{(B.27)}$$

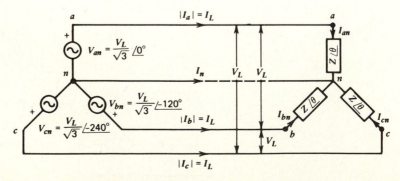

FIGURE B.8 Y-connected generator with Y-connected load.

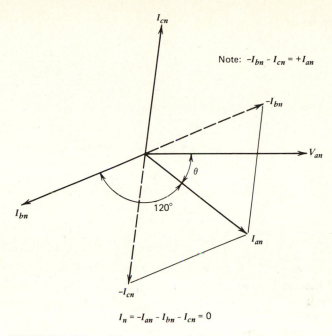

FIGURE B.9 Balanced line currents result in zero neutral current.

Or, from Equation B.14,

$$I_L = \frac{V_{\phi Y}}{|Z|} \equiv I_{\phi Y} \tag{B.28}$$

The neutral current I_n is given by Kirchhoff's current law at the load neutral:

$$I_n = -I_{an} - I_{bn} - I_{cn} \equiv -I_a - I_b - I_c \tag{B.29}$$

But, as Figure B.9 shows, the three line currents are equal in magnitude and 120° out of phase with each other. The sum of such a set of three phasors is always zero. *Thus $I_n = 0$; hence the neutral connection may be removed without changing the solution of the system.*

The procedure for solving a Y load, then, is as follows: Draw a sketch of the circuit similar to Figure B.6a. The sketch makes it obvious that

$$I_L = I_{\phi Y} \tag{B.30}$$

and

$$V_{\phi Y} < V_L$$

Remembering $\sqrt{3} > 1$,

$$V_{\phi Y} = \frac{V_L}{\sqrt{3}} \tag{B.14}$$

Then

$$I_L = I_{\phi Y} = \frac{V_{\phi Y}}{|Z|} = \frac{V_L}{\sqrt{3}|Z|}$$

The complex power components, per phase, are

$$P_\phi = V_{\phi Y} I_{\phi Y} \cos\theta = \frac{V_L}{\sqrt{3}} I_L \cos\theta \quad \text{W} \tag{B.31}$$

$$|S_\phi| = V_{\phi Y} I_{\phi Y} = \frac{V_L I_L}{\sqrt{3}} \quad \text{VA} \tag{B.32}$$

$$Q_\phi = V_{\phi Y} I_{\phi Y} \sin\theta = \frac{V_L I_L}{\sqrt{3}} \sin\theta \quad \text{VAR} \tag{B.33}$$

For all three phases,

$$P = 3P_\phi = \sqrt{3}\, V_L I_L \cos\theta \quad \text{W} \tag{B.34}$$

$$|S| = \sqrt{3}\, V_L I_L \quad \text{VA} \tag{B.35}$$

$$Q = \sqrt{3}\, V_L I_L \sin\theta \quad \text{VAR} \tag{B.36}$$

Note that these are identically the same expressions for $|S|$, P, and Q obtained for the Δ load! This is a reasonable result, since the power flowing in a line should be determined by the line voltages and currents, regardless of whether a Y or Δ load is being supplied.

Note also that the power factor is the cosine of the angle between *phase* voltage and *phase* current, and not the angle between line voltage and line current. Again, this is determined by the load impedance angle and not by whether the impedances are connected in Y or Δ.

Summary for a Y load, given V_L and $Z \underline{|\theta}$:

$$V_{\phi Y} = \frac{V_L}{\sqrt{3}} \quad \text{V} \tag{B.14}$$

$$I_{\phi Y} = I_L = \frac{V_{\phi Y}}{|Z|} \quad \text{A} \tag{B.30}$$

p.f. $= \cos\theta$

$$P = \sqrt{3} V_L I_L \cos\theta \quad \text{W} \tag{B.34}$$

$$|S| = \sqrt{P^2 + Q^2} = \sqrt{3}\, V_L I_L \quad \text{VA} \tag{B.35}$$

$$Q = \sqrt{3} V_L I_L \sin\theta \quad \text{VAR} \tag{B.36}$$

■ **Example B.4**

Three impedances, each $4 - j3 \ \Omega$, are connected in Y to a 208-V, three-phase line. Find the line current, power factor, and the complex power.

Solution

1. Draw a sketch:

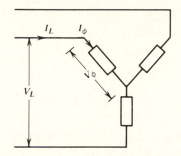

Obviously $I_L = I_\phi$

$V_\phi < V_L$

Remember $\sqrt{3} > 1$

$$V_\phi = \frac{V_L}{\sqrt{3}}$$

2. $Z_L = 4 - j\,3 = 5\,\underline{|-36.9°}$

3. $V_\phi = \dfrac{208}{\sqrt{3}} = 120$ V

4. $I_L = I_\phi = \dfrac{V_\phi}{|Z|} = \dfrac{120}{5} = 24$ A

5. Power factor $= \cos(-36.9°) = 0.800$, leading

6. $|S| = \sqrt{3}V_L I_L = \sqrt{3} \cdot 208 \cdot 24 = 8.65$ kVA

   $P = |S|\cos(-36.9°) = 6.92$ kW

   $Q = |S|\sin(-36.9°) = -5.19$ kVAR

   $S = P + jQ = 6.92 - j5.19$ kVA

B.9 THREE-PHASE PROBLEMS REQUIRING PHASOR NOTATION

Certain types of problems in three-phase circuits make it necessary to write the line and phase voltages and currents in phasor notation. Examples are circuits involving a single-phase load in combination with a three-phase load, circuits for measuring three-phase power with two wattmeters, and transformer banks with only two transformers.

Setting Up the Voltage Phase Diagrams

1. Draw the circuit and label the voltages and currents consistently and completely. Much confusion can be eliminated if this is done carefully. Double subscript notation is recommended for all quantities except line currents:

   I_a is the current in line a. (Show assumed positive flow direction on the diagram.)

   I_{ab} is the current flowing from point a to point b.

   V_{ab} is the voltage of a with respect to b, meaning that a is assumed to be more positive than b.

   Choose a consistent set of subscripts. The subscripts should follow in sequence:

   Correct: 1. *ab, bc, ca*   Invalid: *ab, ba, ca*

   2. *ac, cb, ba*   *ac, cb, ca*

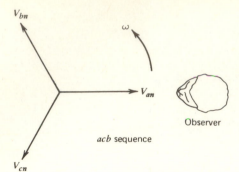

FIGURE B.10 Illustrating **acb** phase sequence.

To be sure, set up the letters in a triangle and go around in either the clockwise or counterclockwise direction:

$$\begin{matrix} a \\ c \quad b \end{matrix} \; ab,bc,ca \qquad \begin{matrix} a \\ c \quad b \end{matrix} \; ac,cb,ba$$

2. Choose one voltage as a reference phasor, say, a line-to-line voltage, such as V_{ab} or a line-to-neutral voltage, such as V_{an}, and assign $\underline{0^\circ}$ to that phasor.
3. Determine the phase sequence. If the sequence is optional, choose one. *Phase sequence* is the time sequence of the voltage waves. Only two sequences are possible: *abc* or *acb*. The phasors representing the voltage waves may be considered to be rotating past an observer in the order of the phase sequence, as in Figure B.10.

   When line-to-line voltages are being considered, both the first and second subscripts will have the same sequence, as in Figure B.11.
4. Set up the voltage phasor diagram. There are some techniques to simplify this procedure.

Line-to-Neutral Voltage As Reference

1. Draw line-to-neutral voltages in proper sequence. Sketch in neutral-to-line voltages as the negatives of the line-to-neutral voltages.

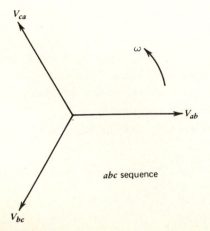

FIGURE B.11 Line-to-line voltages in *abc* sequence.

2. Sketch in lines connecting tips of neutral line phasors to determine line-to-line phasors.
3. Transfer the line-to-line and line-to neutral voltage phasors to a diagram in which they radiate from a common origin, and write expressions for them in polar form.

■ **Example**

$V_{bn} = V_{Ln}\underline{|0°}$; *acb* sequence. See Figure B.12.

NOTE: *Line-to-line and line-to-neutral voltages have the same phase sequence.* ■

Line-to-Line Voltage as a Reference

1. Draw the three line-to-line voltages in the proper sequence, with the reference voltage at angle zero.
2. Sketch the same phasors in delta form and locate the neutral in the center of the delta. Sketch in the line-to-neutral voltages.
3. Transfer the line-to-neutral voltages to your phasor diagram. See Figure B.13 as an example in which V_{ab} is the reference phasor and the sequence is *abc*.
4. Sketch in the line and phase currents.

In a Y, the line and phase currents are identical, and lag the line-to-neutral voltages by the impedance angle.

In a delta, the phase currents lag the phase voltages having the same subscripts by the impedance angle. The line currents are $\sqrt{3}$ times as large as the delta phase currents and are displaced from the phase currents by $\pm 30°$, depending on the phase sequence. (See Figure B.7.) The student may readily verify the following principle:

PRINCIPLE: REGARDLESS OF WHETHER THE CIRCUIT IS Y OR Δ, THE LINE CURRENTS LAG THE *LINE-TO-NEUTRAL* VOLTAGES BY THE PHASE IMPEDANCE ANGLE.

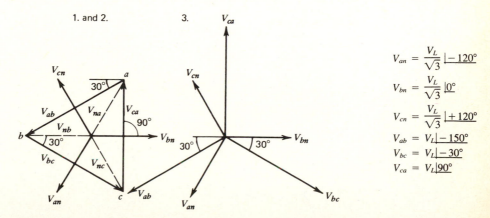

$$V_{an} = \frac{V_L}{\sqrt{3}}\underline{|-120°}$$

$$V_{bn} = \frac{V_L}{\sqrt{3}}\underline{|0°}$$

$$V_{cn} = \frac{V_L}{\sqrt{3}}\underline{|+120°}$$

$$V_{ab} = V_L\underline{|-150°}$$

$$V_{bc} = V_L\underline{|-30°}$$

$$V_{ca} = V_L\underline{|90°}$$

FIGURE B.12 Method of establishing system voltage phasors, with one line-to-neutral voltage as a reference.

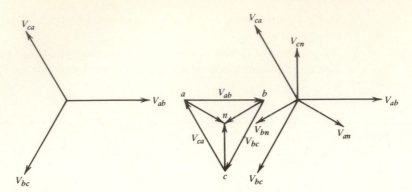

FIGURE B.13 System voltages with a line-to-line voltage as a reference. Phase sequence *abc* illustrated.

The line currents are displaced from each other by 120° and have the same phase sequence as the voltages. The delta phase currents are displaced from each other by 120° and have the same phase sequence as the voltages.

■ Example B.5 Delta Load

A 208-V, three-phase line is feeding a Δ-connected motor. The current in each phase of the motor winding is 5 A at 0.8 power factor, lagging (i.e., θ = 36.9°). If V_{an} is taken as a reference and the phase sequence is *acb*, write all voltages and currents as phasors.

Solution

1. Circuit:

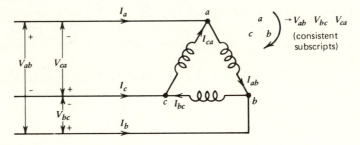

2. Current magnitudes:

$$I_{\phi\Delta} = |I_{ab}| = |I_{bc}| = |I_{ca}| = 5 \text{ A}$$

$$I_L = \sqrt{3}I_{\phi\Delta} = |I_a| = |I_b| = |I_c| = 8.66\text{A}$$

3. Line-to-neutral voltages:

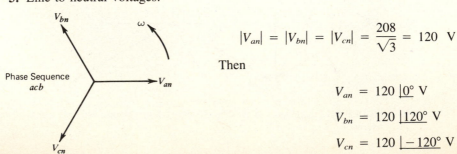

$$|V_{an}| = |V_{bn}| = |V_{cn}| = \frac{208}{\sqrt{3}} = 120 \text{ V}$$

Then

$$V_{an} = 120 \underline{|0°} \text{ V}$$

$$V_{bn} = 120 \underline{|120°} \text{ V}$$

$$V_{cn} = 120 \underline{|-120°} \text{ V}$$

4. Line-to-line voltages:

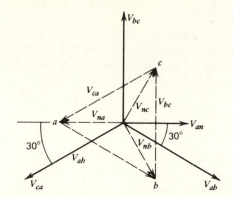

$$V_{ab} = 208 \underline{|-30°}$$

$$V_{bc} = 208 \underline{|90°}$$

$$V_{ca} = 208 \underline{|-150°}$$

5. Phase currents:

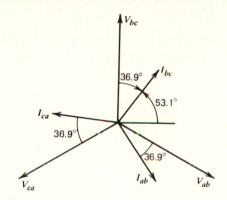

$$I_{ab} = 5 \underline{|-66.9°}$$

$$I_{bc} = 5 \underline{|53.1°}$$

$$I_{ca} = 5 \underline{|173.1°}$$

6. Line currents:

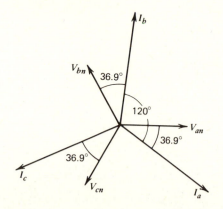

$$I_a = 8.66 \underline{|-36.9°}$$

$$I_b = 8.66 \underline{|83.1°}$$

$$I_c = 8.66 \underline{|-156.9°}$$

Line currents always lag line-to-neutral voltages by the load impedance angle θ. When θ is negative, they lead. ■

■ **Example B.6 Loads in Parallel**

Find the total line current for three balanced, three-phase loads connected in parallel to a 208-V, three-phase line:

Load 1: Three 20.8-Ω resistors in Δ

Load 2: Three impedances, each $12 \underline{|-53.1°}\ \Omega$ in Y

Load 3: The motor of Example B.5

Solution. Since the loads are all balanced, the line currents all have the same magnitude. Then only one line current need be found. Let that line current be I_a, and let V_{an} be taken as a reference. Remember that the line current for each load will lag the line-to-neutral voltage by the phase impedance angle.

$$\text{Load 1: } |I_{\phi\Delta 1}| = \frac{V_{\phi\Delta}}{|Z_{\phi\Delta}|} = \frac{208}{20.8} = 10 \text{ A}$$

$$I_{L1} = \sqrt{3} \cdot 10 = 17.32 \text{ A}$$

Angle of I_{a1} with $V_{an} = 0°$

$$\text{Load 2: } I_{L2} = I_{\phi Y2} = \frac{V_{\phi Y2}}{|Z_{\phi Y}|} = \frac{208/\sqrt{3}}{12} = \frac{120}{12} = 10 \text{ A}$$

I_{a2} lags V_{an} by $-53.1°$ or *leads* V_{an} by $+ 53.1°$

Load 3: $I_{L3} = 8.66$ A from Example B.5; I_{a3} lags V_{an} by $36.9°$

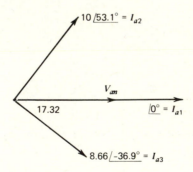

Total line current in a line a equals $I_{a1} + I_{a2} + I_{a3}$

$$I_{a1} = 17.32 + j0 \text{ A}$$

$$I_{a2} = 6 \quad + j8 \text{ A}$$

$$I_{a3} = 6.93 - j5.20 \text{ A}$$

$$I_a = 30.25 + j2.80 \text{ A} = 30.38 \underline{|5.29°}$$

Total line current $= |I_a| = 30.38$ A ∎

■ Example B.7 Combined Single-Phase and Three-Phase Loads

A single-phase motor drawing 6 A at 0.707 power factor, lagging, is connected across lines a and b of a three-phase line connected to a three-phase motor drawing 10 A at 0.8 power factor, lagging. Find the current in each line. The currents in lines a and b will depend on the phase sequence. Assume the sequence to be abc.

Solution

1. Draw the circuit:

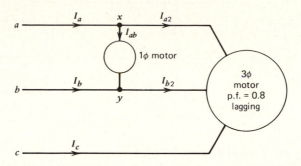

2. Set up the voltages:

$$V_{an}, V_{bn}, V_{cn}, V_{ab}:$$

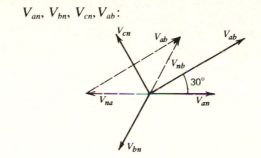

3. The single-phase motor current I_{ab} will lag V_{ab} by $\cos^{-1} 0.707$ or $45°$. Then
$I_{ab} = 6\underline{|-15°} = 5.80 - j1.55$ A.

4. The three-phase motor currents will lag the line-to-neutral voltages by $\cos^{-1} 0.8$ or $36.9°$. Then

$$I_c = 10 \underline{|120° - 36.9°} = 10 \underline{|83.1°} \text{ A}$$

$$I_{a2} = 10 \underline{|-36.9°} = 8 - j6 \text{ A}$$

$$I_{b2} = 10 \underline{|-120° - 36.9°} = 10 \underline{|-156.9°} = -9.20 - j3.92 \text{ A}$$

5. Kirchhoff's current law at point x:

$$I_a = I_{ab} + I_{a2} = 13.80 - j7.55 = 15.73 \underline{|-28.7°} \text{ A}$$

$$|I_a| = 15.73 \text{ A}$$

KCL at point y:

$$I_b = I_{b2} - I_{ab} = -15.00 - j2.37 = 15.19 \underline{|-171.0°}$$

$$|I_b| = 15.19 \text{ A}$$

The student may wish to demonstrate that, with acb sequence, $|I_a| = 15.19$ A and $|I_b| = 15.73$ A. ∎

■ **Example B.8 Measuring Three-Phase Power with Two Wattmeters**

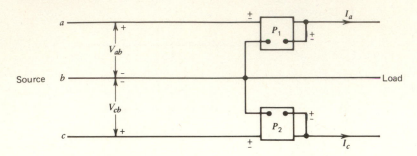

Proper Connection of Wattmeters

Connect ± current terminal of each meter toward the *source*.

Connect ± potential terminal to the line containing current coil.

If one wattmeter reads backward, reverse its *current coil* and subtract its reading from the reading of the other wattmeter.

The reading of wattmeter P_1 is determined by V_{ab} and current I_a. The reading of P_2 is determined by V_{cb} and current I_c. While the sum of the two readings depends only on the total power of the load, the individual readings depend on the phase sequence. Let V_{ab} be the reference phasor. A consistent set of subscripts is *ab, bc, ca*.

Wattmeter Readings

abc Sequence:

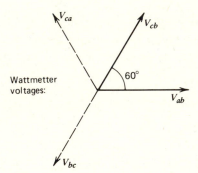

To find the currents for a load power-factor angle θ, first find the line-to-neutral voltages:

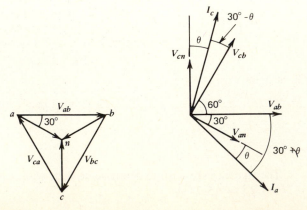

$$P_1 = |V_{ab}| |I_a| \cos \underset{I_a}{\overset{V_{ab}}{\rfloor}} = V_L I_L \cos(30° + \theta)$$

$$P_2 = |V_{cb}| |I_c| \cos \underset{I_c}{\overset{V_{cb}}{\rfloor}} = V_L I_L \cos(30° - \theta)$$

acb Sequence:

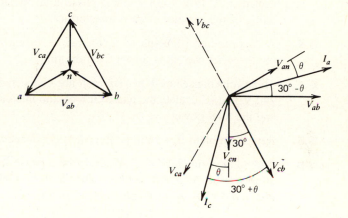

$$P_1 = |V_{ab}| |I_a| \cos \underset{I_a}{\overset{V_{ab}}{\rfloor}} = V_L I_L \cos(30° - \theta)$$

$$P_2 = |V_{cb}| |I_c| \cos \underset{I_c}{\overset{V_{cb}}{\rfloor}} = V_l I_L \cos(30° + \theta)$$

Total P and Q

abc Sequence:

$$P_1 + P_2 = V_L I_L [\cos(30° + \theta) + \cos(30° - \theta)] = \sqrt{3} V_L I_L \cos \theta$$
$$= \text{Total power}$$

$$\sqrt{3}(P_2 - P_1) = \sqrt{3} V_L I_L [\cos(30° - \theta) - \cos(30° + \theta)] = \sqrt{3} V_L I_L \sin \theta$$
$$= \text{Total reactive power}$$

acb Sequence:

$$P_1 + P_2 = \text{Total power}$$
$$\sqrt{3}(P_1 - P_2) = \text{Total reactive power}$$

PROBLEMS

B.1. (a) Given an impedance $5 \underline{|30°} \; \Omega$. Compute the equivalent series resistance and reactance. Is the reactance inductive or capacitive?

(b) This impedance is connected across a 120-V (rms), 60-Hz, single-phase line. Taking the line voltage as a reference, draw a phasor diagram of the following rms quantities: line voltage, current, IR drop, IX drop.

(c) Carefully plot the voltage, current, and power as functions of ωt over $1\frac{1}{2}$ cycles, and indicate maxima, minima, and average values of all three functions, giving numerical values.

B.2. The primaries of three transformers having identical ratings are to be connected in Y to a 138-kV, three-phase line. What should be the voltage rating of the primary of each transformer?

B.3. A Δ-connected, three-phase motor has a current of 5 A flowing in each phase. What current is drawn from the line?

B.4. Three impedances of $4 + j3\Omega$ each are connected in Δ to a three-phase, 240-V line. Calculate the phase current, the line current, the power factor, and the total power.

B.5. Repeat Problem B.4 with the impedances connected in Y.

B.6. The voltage supplied to a balanced, three-phase load is 2300 V. The line current is 60 A, and the power factor is 0.90. Calculate the input power, kVA, and kVARs for lagging power factor.

B.7. With V_{ab} as reference and phase sequence ab, bc, ca, draw phasor diagrams for Problems B.4 and B.5, showing all line and phase voltages and currents.

B.8. Three equal load impedances, $Z = 3 + j4 \ \Omega$, are to be connected to a 173.2-V, three-phase line. Calculate the magnitudes of the line voltage, phase voltage, line current, phase current, power per phase, and total power, if the loads are connected (a) in Y, (b) in Δ.

B.9. Draw a complete phasor diagram for Problem B.8(b). Assume the phase sequence to be V_{ab}, V_{ca}, V_{bc}.

B.10. Three hundred megavolt-amperes at 0.800 power factor, lagging, are being transmitted over a 345-kV, three-phase transmission line. (a) If the receiving end voltage is 345 kV, what is the complex phase load impedance if the load is assumed to be connected in Δ? (b) In Y? (c) Calculate the line current and phase current in each phase. (d) What are the power per phase and the total power?

B.11. Two wattmeters are properly connected to read total power being absorbed by a balanced load. The phase sequence is abc. The current coils of the wattmeters are connected in lines a and b. Let the readings of two meters be P_a and P_b, respectively. The line voltage is 2400 V and the load is 30 kVA.

(a) Show a wiring diagram.

(b) Calculate P_a and P_b if the power factor is 1.0.

(c) Calculate P_a and P_b if the power factor is 0.2, lagging.

(d) Calculate P_a and P_b if the power factor is 0.5, leading.
Sketch phasor diagrams for each case.

B.12. (a) Repeat parts (b), (c), and (d) of Problem B.11 for *acb* phase sequence.

(b) Make a statement regarding the effect of phase sequence on the meter readings.

B.13. As a phasor diagram exercise, what would P_a be in Problem B.11(c) if the wattmeter's potential lead were moved from line *c* to line *b?*

B.14. From the diagram below,

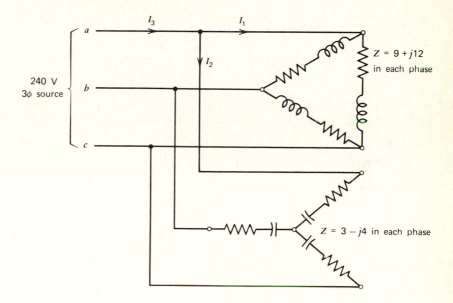

(a) Find the magnitudes of I_1 and I_2.

(b) Using V_{an} as a reference, draw a phasor diagram showing V_{an}, I_1, I_2, and I_3.

(c) Express the above quantities as phasors in both polar and rectangular forms.

(d) Find the magnitude of I_3.

C

SALIENT-POLE THEORY OF SYNCHRONOUS MACHINES

C.1 DERIVATION OF THE *d–q* MODEL

Cylindrical-rotor theory does a fairly satisfactory job of predicting excitation requirements for salient-pole machines. However it is not always satisfactory in determining δ and in solving system stability problems.

Since cylindrical-rotor theory needs only slight adjustment to be accurate for salient-pole machines, Blondel suggested that the armature MMF be resolved into two components, one along the direct axis and the other along the quadrature axis of the rotor. If these two components are called F_{1d} and F_{1q}, then the total *d*-axis MMF is $F_{1d} + F_2$ and the total *q*-axis MMF is F_{1q}.

Two fluxes are visualized as existing in the machine, determined by the *d*- and *q*-axis MMFs and the effective permeances along the *d* and *q* axes:

$$\phi_d = F_d \mathscr{P}_d = (F_{1d} + F_2)\mathscr{P}_d$$

$$\phi_q = F_{1q} \mathscr{P}_q$$

Superposition of fluxes in a *linear* medium is allowable from a mathematical point of view. However when saturation is taken into account, this method is not as theoretically respectable as cylindrical-rotor theory. The two axes are considered to be decoupled since they are electrically at right angles to each other (orthogonal). The Blondel *two-reaction theory* has been greatly elaborated by Doherty and Nickle. Its linear adaptation to transient problems has been the subject of many books, papers, and theses.

The salient-pole phasor diagram will be derived on a basis similar to that of the cylindrical-rotor machine. Figure C.1 shows a schematic representation of the machine at the instant the *a*-phase current is at its peak. At this instant, the stator MMF $\vec{F}_1$ is aligned with the axis of phase *a*. The *d* axis is instantaneously at some angle β with respect to the *a* axis. Generator operation is illustrated. Then the direct and quadrature axis components of $\vec{F}_1$ are

$$F_{1d} = F_1 \cos \beta \tag{C.1}$$

$$F_{1q} = F_1 \sin \beta \tag{C.2}$$

The total d-axis MMF R_d is the algebraic sum of F_2 and F_{1d}, or, in vector form,

$$\vec{R}_d = \vec{F}_2 + \vec{F}_{1d} \tag{C.3}$$

The d-axis flux is

$$\phi_d = \mathcal{P}_d R_d = \mathcal{P}_d(F_2 + F_{1d}) \tag{C.4}$$

where $\mathcal{P}_d$ is the effective d-axis magnetic-circuit permeance. The q-axis flux is

$$\phi_q = \mathcal{P}_q F_{1q} \tag{C.5}$$

where $\mathcal{P}_q$ is the effective permeance of the q-axis magnetic circuit, obviously considerably less than $\mathcal{P}_d$ as a result of the large air gap along the q axis.

As in the case of cylindrical-rotor theory, and for the same reasons, it is assumed that voltages lag the fluxes that produce them by 90°. Voltages along the d axis are assigned a subscript d, and those along the q axis are given the subscript q. Then by superposition, let the voltage induced in each phase be the sum of the voltages induced by the two fluxes ϕ_d and ϕ_q:

$$E_\phi = E_{ad} + E_q \tag{C.6}$$

where

$$|E_{ad}| = \sqrt{2}\pi N_\phi k_w f \phi_q = \omega_e N_{E1} \phi_q \tag{C.7}$$

and

$$|E_q| = \sqrt{2}\pi N_\phi k_w f \phi_d = \omega_e N_{E1} \phi_d \tag{C.8}$$

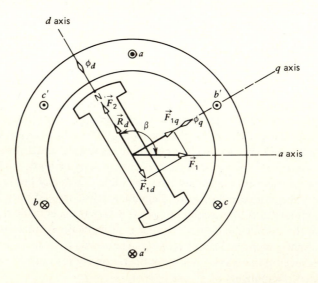

FIGURE C.1 Salient-pole synchronous machine at the instant current in phase a is maximum.

Combining Equations C.5 and C.7 and establishing the phasor relationship between $\vec{F}_{1q}$ and E_d,

$$E_{ad} = -jm_q\vec{F}_{1q} \tag{C.9}$$

where

$$m_q \overset{\Delta}{=} \sqrt{2}\pi N_\phi k_w f \mathcal{P}_q = \omega_e N_{E1}\mathcal{P}_q \tag{C.10}$$

Similarly

$$E_q = -jm_d\vec{R}_d$$

where

$$m_d \overset{\Delta}{=} \sqrt{2}\pi N_\phi k_w f \mathcal{P}_d = \omega_e N_{E1}\mathcal{P}_d \tag{C.11}$$

Then

$$E_q = -jm_d(\vec{F}_2 + \vec{F}_{1d}) = E_f + E_{aq} \tag{C.12}$$

where

$$E_f \overset{\Delta}{=} -jm_d\vec{F}_2 \tag{C.13}$$

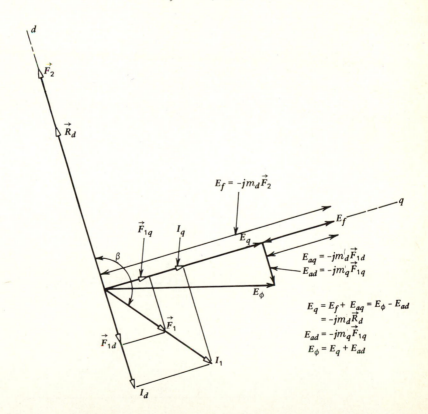

FIGURE C.2 Salient pole phasor/vector diagram.

and

$$E_{aq} \overset{\Delta}{=} -jm_d\vec{F}_{1d} \tag{C.14}$$

The phasor/vector diagram resulting is shown in Figure C.2, where the expression for E_ϕ has been rearranged:

$$E_\phi = E_q + E_{ad}$$
$$= E_f + E_{aq} + E_{ad} \tag{C.15}$$

A circuit model may now be drawn (Figure C.3).

Blondel noted that since $\vec{F}_1$ is proportion to I_1 and in phase with I_1, I_1 could also be resolved into two components: I_d, lying along the d axis so that F_{1d} is proportional to I_d; and I_q, lying along the q axis, with F_{1q} proportional to I_{1q}. Then

$$|I_d| = |I_1| \cos \beta$$
$$|I_q| = |I_1| \sin \beta \tag{C.16}$$

The concept of magnetizing reactance will now be employed as was done in cylindrical-rotor theory. Note that $-E_{ad}$ is perpendicular to I_q and, by Equation C.9, is proportional to I_q, neglecting saturation. Then let

$$-E_{ad} = +jkm_qF_{1q} \overset{\Delta}{=} jX_{mq}I_q \tag{C.17}$$

Similarly, let

$$-E_{aq} = +jm_dF_{1d} \overset{\Delta}{=} jX_{md}I_d. \tag{C.18}$$

The phasor diagram of the model of Figure C.3, including the information of Equations C.17 and C.18, is shown in Figure C.4, with r_1 neglected.

Attention is called to the two similar shaded triangles. Note that the sides of the smaller one are the sides of the larger, current triangle multiplied by x_1, the leakage reactance. Note also that jI_dx_1 is in phase with jI_dX_{md} and that jI_qx_1 is in phase with jI_qX_{mq}. Then define

$$x_d \overset{\Delta}{=} x_1 + X_{md}$$
$$x_q \overset{\Delta}{=} x_1 + X_{mq} \tag{C.19}$$

where x_d is the *direct-axis synchronous reactance* and x_g is the *quadrature-axis synchronous reactance*.

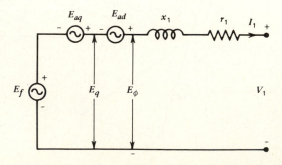

FIGURE C.3 Preliminary circuit model of a salient-pole machine.

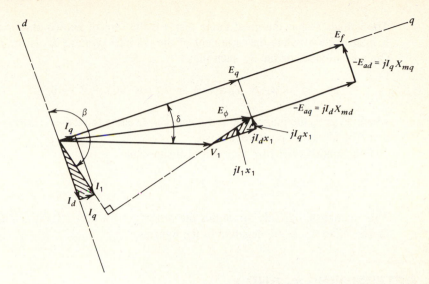

FIGURE C.4 Resolution of armature current and leakage reactance drop into d and q components.

The model and phasor diagram may now be redrawn to take into account these new reactances, as in Figure C.5.

The angle β must be known in order to break I_1 down into its components I_d and I_q; and these appear to be essential to finding E_f. The problem is that

$$\beta = 90° + \delta - \theta_\phi \qquad (lagging\ \theta_\phi\ assumed\ negative) \qquad (C.20)$$

But δ is not known, initially. The solution to this dilemma lies in the triangle shown in dashed lines, which is similar to the shaded current triangle. Note that this auxiliary triangle is the current triangle with all sides multiplied by jx_q. This permits the definition of a voltage E':

$$E' = V_1 + jI_1 x_q \qquad (C.21)$$

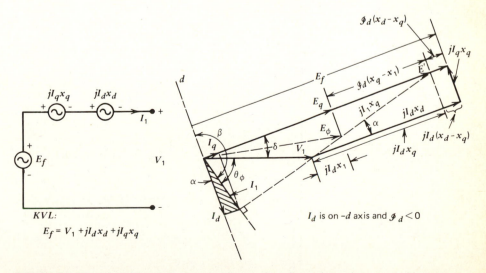

FIGURE C.5 Blondel model and phasor/vector diagram for salient-pole synchronous machines (stator winding resistance neglected).

The voltage E' has the great virtue that it is the sum of known phasor quantities and it is at angle δ! Thus the argument of the complex voltage E' is the needed angle. Once δ is found, define $\mathscr{I}_d$ such that

$$\mathscr{I}_d = |I_1| \cos \beta = |I_1| \cos (90° + \delta - \theta_\phi) \tag{C.22}$$

where $\mathscr{I}_d$ gives the magnitude and polarity of I_d along the d axis. (Positive $\mathscr{I}_d$ indicates that I_d is along the positive d axis.)

Examination of Figure C.5 will also show that

$$|E_f| = |E'| - \mathscr{I}_d(x_d - x_q) \tag{C.23}$$

This relationship greatly simplifies the procedure for determining excitation requirements. Note that $\mathscr{I}_d$ is negative in the figure.

C.2. DETERMINING x_d AND x_q

The reactance x_d is the direct-axis synchronous reactance of cylindrical-rotor theory. That this is so is evident when open-circuit and short-circuit conditions are considered for a salient-pole machine. Examining the circuit model of Figure C.5 for open-circuit conditions, it is seen that when $I_1 = 0$, $I_d = 0$ and $I_q = 0$, and as a result, $V_{1oc} = E_f$, as in the cylindrical-rotor case.

The salient-pole phasor diagram on short circuit is as shown in Figure C.6. It is seen that $\vec{F}_1$ is along the d axis, and there is no MMF along the q axis. Thus the ratio of open-circuit voltage to short-circuit current gives the *direct*-axis synchronous reactance, as in the cylindrical-rotor case.

Correcting x_d for saturation requires a knowledge of the degree of saturation along the d axis. The two MMFs acting along the axis are $\vec{F}_2$ and $\vec{F}_{1d}$, having a resultant

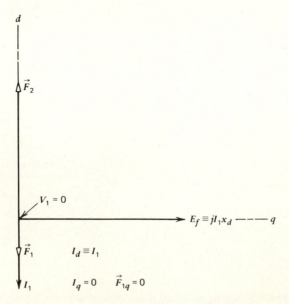

FIGURE C.6 Phasor/vector diagram of a short-circuited, salient-pole machine.

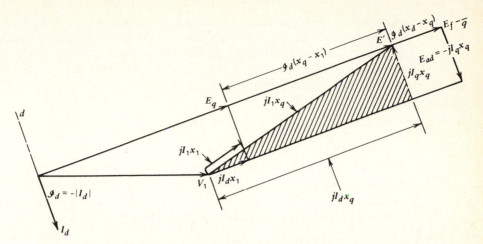

FIGURE C.7 Showing that the q-axis induced voltage E_q is given by $|E'| + \mathcal{I}_d(x_q - x_1)$.

$\vec{R}_d$. Recall that these MMFs are modeled by E_f and E_{aq}, respectively, having a resultant E_q. Then locating I_{fs} on the OCC corresponding to $\sqrt{3}|E_q|$ will permit a value to be assigned to m_d' as in Figure 2.66:

$$m_d' = \frac{|E_q|}{I_{fs}} \tag{C.24}$$

Locating I_{fu} on the air gap line, corresponding to $\sqrt{3}|E_q|$, allows the direct-axis saturation factor to be determined:

$$k_{sd} = \frac{I_{fs}}{I_{fu}} \tag{C.25}$$

To determine $|E_q|$, compare Figure C.7 with Figures C.2, C.3, and C.4. Then note that the shaded triangle of Figure C.7 is the same as the dashed-lines triangle of Figure C.5. Here $\mathcal{I}_d$ is negative, and it will be seen that

$$|E_q| = |E'| + \mathcal{I}_d X_{mq} \equiv |E'| + \mathcal{I}_d(x_q - x_1) \tag{C.26}$$

The Potier method may be used to find x_1.

The Slip Test

It is usually assumed that the q-axis magnetic circuit remains unsaturated, due to the large air gap involved. The ratio

$$\frac{x_{qu}}{x_{du}} \cong \frac{x_q}{x_{du}} \tag{C.27}$$

may be obtained from a so-called *slip test*. In this test, reduced three-phase voltages are applied to the terminals of the machine. The shaft is mechanically driven at a speed very slightly different from synchronous speed, so that the rotor poles are slipping through the stator field. The field circuit is open. The currents of the three phases are measured either by ammeters or by means of an oscilloscope. When the pole d axes are aligned with the poles of the rotating stator field, the phase impedance is x_d, neglecting r_1. When the q axes are aligned with the poles of the stator field, the terminal impedance of each phase is x_q. Since $x_q < x_d$, maximum current corresponds to x_q and minimum current corresponds to x_d. Then

$$\frac{x_{qu}}{x_{du}} = \frac{I_{min}}{I_{max}} \qquad (C.28)$$

The low voltage applied assures that the machine is unsaturated during the test. Approximate values for x_q and x_d can be obtained by dividing the applied phase voltage by I_{max} and I_{min}, respectively; however the accuracy is poor. It is much better to calculate x_{du} from the air gap line and SCC. Then let

$$x_q = x_{du}\left(\frac{x_{qu}}{x_{du}}\right) \qquad (C.29)$$

C.3 PROCEDURE FOR FINDING I_f

Given: The desired line-to-line voltage, kVA load and power factor, the leakage reactance x_1, OCC, and SCC curves and the ratio x_q/x_{du}.

1. From the air gap line and SCC, find x_{du}.
2. Using the slip-test ratio, find x_q.
3. Find $E' = V_1 + jI_1x_q = |E'|\ \underline{|\delta}$.
4. Find $\mathcal{I}_d = |I_1| \cos(90° + \delta - \theta_\phi) \equiv |I_1| \cos \beta$,
   where $\theta_\phi = \cos^{-1}$ (power factor). Note that lagging θ_ϕ is considered negative.
5. Find $|E_q| = |E'| + \mathcal{I}_d(x_q - x_1)$. The Potier method may be used to find x_1.
6. Locate $\sqrt{3}|E_q|$ on the OCC and air gap line and find I_{fs} and I_{fu}. Then
   $m'_d = |E_q|/I_{fs}$ and $k_{sd} = I_{fs}/I_{fu}$. (See Figure 2.66.)
7. Find $x_d = x_1 + (x_{du} - x_1)/k_{sd}$.
8. Find $|E_f| = |E'| - \mathcal{I}_d(x_d - x_q)$.
9. $I_f = |E_f|/m'_d$.

An example of this calculation is included in Example C.1 at the end of this appendix.

C.4 POWER-δ CHARACTERISTICS OF A SALIENT-POLE SYNCHRONOUS MACHINE

Figure C.8 is a restatement of the d and q components of the armature current I_1. The power output of all three phases is given by

$$P = 3|V_1||I_1| \cos \theta_\phi \qquad (C.30)$$

Figure C.8 demonstrates that

$$|I_1| \cos \theta_\phi = |I_q| \cos \delta + |I_d| \sin \delta$$

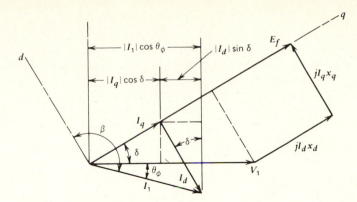

FIGURE C.8 Current components.

Then

$$P = 3|V_1||I_q| \cos \delta + 3|V_1||I_d| \sin \delta \qquad \text{(C.31)}$$

since

$$|I_q|x_q = |V_1| \sin \delta \quad \text{and} \quad |I_d|x_d = |E_f| - |V_1| \cos \delta$$

$$|I_q| = \frac{|V_1| \sin \delta}{x_q} \qquad \text{and} \qquad I_d = \frac{|E_f| - |V_1| \cos \delta}{x_d} \qquad \text{(C.32)}$$

Applying $\cos \alpha \sin \alpha = (\sin 2\alpha)/2$, the equation for power may be written

$$P = \frac{3}{2} \frac{|V_1|^2}{x_q} \sin 2\delta + \frac{3|V_1||E_f|}{x_d} \sin \delta - \frac{3}{2} \frac{|V_1|^2}{x_d} \sin 2\delta$$

$$= \frac{3|V_1||E_f|}{x_d} \sin \delta + \frac{3}{2} |V_1|^2 \left(\frac{1}{x_q} - \frac{1}{x_d} \right) \sin 2\delta \qquad \text{(C.33)}$$

The first term in this expression for the power is the same as that obtained for the cylindrical-rotor machine.

The second term of Equation C.33 depends on the *saliency,* defined by the quantity $[(1/x_q) - (1/x_d)]$ and is of interest for more than one reason. Note that it disappears when $x_d = x_q$ (i.e., for a cylindrical rotor). Also, this term exists even when there is no field current ($E_f = 0$). The torque that this term expresses,

$$\tau_R = \frac{1}{\omega} \left[\frac{3}{2} |V_1|^2 \left(\frac{1}{x_q} - \frac{1}{x_d} \right) \right] \sin 2\delta \qquad \text{(C.34)}$$

is due to the magnetic attraction between the protruding rotor poles and the rotating MMF field produced by stator currents ($\vec{F}_1$). With zero-field current, $\vec{F}_2$ is zero, the resultant $\vec{R}$ is identical with $\vec{F}_1$, and the air gap flux is centered on this rotating MMF. When the air gap flux is aligned with the pole axis ($\delta = 0$), the torque τ_R is zero. If an attempt is made to cause the machine to act as a motor or generator with no field current (V_1 supplied by the bus to which the machine is connected), the poles shift

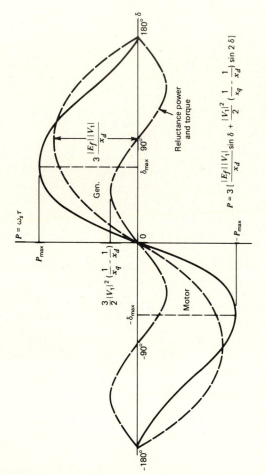

FIGURE C.9 Power as a function of power angle δ for a salient-pole synchronous machine.

relative to the stator field, thus increasing the reluctance of the flux path as the torque increases. For this reason the torque due to the second term of Equation C.33 is called *reluctance torque*. Its maximum value occurs for $\delta = 45°$, and is about 20 percent of the maximum torque available with the rotor windings excited. Reluctance torque is typically proportional to the square of the voltage or current and to the sine of twice the displacement angle.

Under steady-state conditions, a synchronous machine runs at constant speed, $\omega_s = 4\pi f/p$. The developed torque, in newton-meters, is given by the developed power, in watts, divided by ω_s. Thus the torque and power are proportional. Figure C.9 is a typical plot of torque and power as functions of δ for a salient-pole machine.

Note that the machine is "stiffer" than a cylindrical rotor machine. That is, its torque increases more rapidly for small values of δ. It is also evident from the figure that maximum power occurs at a value of δ less than 90°. The theoretical angular stability limit δ_{max} is thus less. Its value depends on the relative magnitudes of V_1 and E_f, as well as on the saliency.

C.5 THE SALIENT-POLE MACHINE AS A MOTOR

Very nearly all synchronous motors have salient poles. When the power flow is negative relative to generator operation, the machine is operating as a motor. Equation C.33 and Figure C.9 show that motor operation requires a negative δ.

It is usually more convenient to consider motor current in flowing *into* the assumed

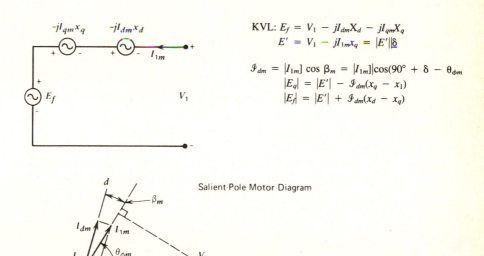

KVL: $E_f = V_1 - jI_{dm}X_d - jI_{qm}X_q$

$E' = V_1 - jI_{1m}x_q = |E'|\underline{\delta}$

$\mathcal{I}_{dm} = |I_{1m}]\cos\beta_m = |I_{1m}]\cos(90° + \delta - \theta_{\phi m}$

$|E_q| = |E'| - \mathcal{I}_{dm}(x_q - x_1)$

$|E_f| = |E'| + \mathcal{I}_{dm}(x_d - x_q)$

FIGURE C.10 Salient-pole model and phasor diagram, motor mode, leading power factor.

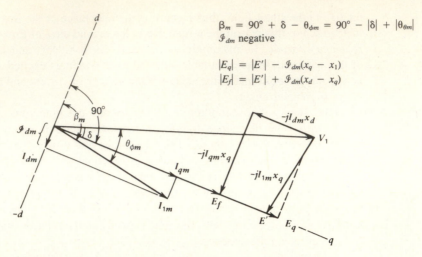

$$\beta_m = 90° + \delta - \theta_{\phi m} = 90° - |\delta| + |\theta_{\theta m}|$$
$$\mathscr{I}_{dm} \text{ negative}$$

$$|E_q| = |E'| - \mathscr{I}_{dm}(x_q - x_1)$$
$$|E_f| = |E'| + \mathscr{I}_{dm}(x_d - x_q)$$

FIGURE C.11 Motor diagram, phase current lagging the q axis.

positive terminal of the model, as indicated in Figure C.10. Motor current will be defined as the negative of generator current, with the following results:

$$I_{1m} \stackrel{\Delta}{=} -I_1$$

then

$$I_{dm} = -I_d \qquad \mathscr{I}_{dm} = -\mathscr{I}_d$$

and

$$I_{qm} = -I_q$$

Figure C.11 shows the phasor diagram for the case when the current is sufficiently lagging that a negative $\mathscr{I}_{dm}$ results.

The equations accompanying Figures C.10 and C.11 show how to calculate δ and the magnitudes of E_q and E_f. The procedure for finding the saturation factor m'_d and I_f are identical with those of the generator mode, steps 5 through 9 in Section C.3, with the exception of changes in sign in steps 5 and 8, to accommodate the definition of $\mathscr{I}_{dm}$ as the negative $\mathscr{I}_d$.

■ Example C.1

Figure C.12 provides characteristic curves for a 1000-kVA, 2300-V, 60-Hz salient-pole synchronous machine. By slip test, the ratio of x_q/x_{du} has been found to be 0.52.

(a) Find the unsaturated d-axis synchronous reactance and the quadrature-axis synchronous reactance.
(b) Find the leakage reactance per phase by the Potier method.
(c) Find the field current required to produce 2300 V at the terminals of the machine in the generator mode if the load is 1000 kVA at unity power factor.

(d) For motor operation, find the field current required for a power factor of 0.800, leading, when the machine is drawing 800 kW from a 2400-V three-phase line.

Solution

(a) Corresponding to $I_f = 20$ A, we find 2875 V on the air gap line and 314 A on the short-circuit characteristic. Then

$$x_{du} = \frac{V_{oc}/\sqrt{3}}{I_{sc}} = \frac{2875}{\sqrt{3} \cdot 314} = 5.29 \ \Omega$$

$$x_q = \left(\frac{x_q}{x_{du}}\right) \cdot x_{du} = 0.52 \cdot 5.29 = 2.75 \ \Omega$$

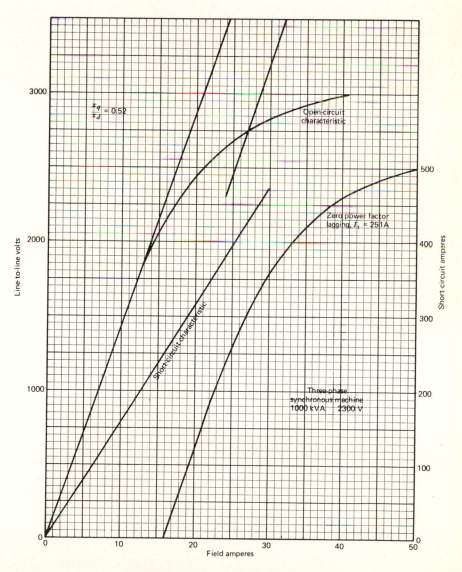

FIGURE C.12 Curves for Example C.1.

(b) A horizontal line at rated voltage intersects the zero-power-factor characteristic at $I_f = 40$ A. The I_f intercept of the zero-power-factor characteristic is 16 A. The left-hand apex of the auxiliary triangle is thus at $I_f = 40 - 16 = 24$ A; $V_{oc} = 2300$ V, or $(I_f, V_{oc}) = (24, 2300)$. A line drawn through this point parallel to the air gap line intersects the OCC at 2750 V. The altitude of the Potier triangle is then

$$\sqrt{3}I_1 x_2 = 2750 - 2300 = 450 \text{ V}$$

The zero-power-factor data were taken at $I_1 = 251$ A, so

$$x_1 = \frac{450}{\sqrt{3} \cdot 250} = 1.04 \ \Omega$$

(c) For 1000 kVA at 2300 V, unity power factor,

$$I_1 = \frac{10^6}{\sqrt{3} \cdot 2300} \underline{|0°} = 251\underline{|0} \text{ A}$$

Neglecting r_1,

$$E' = V_1 + jI_1 x_q = \frac{2300}{\sqrt{3}} \underline{|0°} + 251 \cdot 2.75\underline{|90°}$$

$$= 1328 + j690 = 1497\underline{|27.5°} \text{ V}$$

Note that $\delta = +27.5°$. Then $\beta = 90° + \delta - \theta_\phi = 90° + 27.5° - 0° = 117.5°$, and

$$\mathcal{I}_d = |I_1| \cos \beta = 251 \cdot (-0.462) = -116\text{A}$$
$$|E_q| = |E'| + \mathcal{I}_d X_{mq} = 1497 - 116(2.75 - 1.04)$$

$$= 1299 \text{ V}$$

Turning to Figure C.12, we find $I_{fu} = 15.7$ and $I_{fs} = 17.2$ A on the air gap line and OCC, respectively. The direct-axis saturation factor is

$$k_{sd} = \frac{I_{fs}}{I_{fu}} = \frac{17.2}{15.7} = 1.10$$

and saturated d-axis synchronous reactance is:

$$x_d = \frac{x_{du} - x_1}{k_{sd}} + x_1 = \frac{5.29 - 1.04}{1.10} + 1.04 = 4.90 \ \Omega$$

The voltage model of F_2 is then

$$|E_f| = |E'| - \mathcal{I}_d(x_d - x_q) = 1497 + 116(4.90 - 2.75)$$

$$= 1746 \text{ V}$$

Now

$$m'_d = \frac{|E_q|}{I_{fs}} = \frac{1299}{17.2} = 75.5 \ \Omega$$

$$I_f = \frac{|E_f|}{m'_d} = 23.1 \text{ A}$$

The phasor diagram for this condition is as follows:

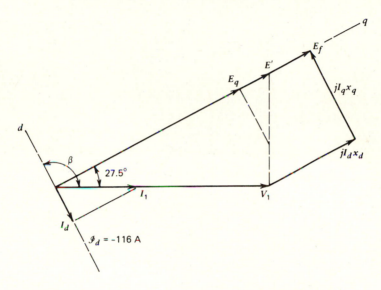

(d) For the motor case,

$$|S| = \frac{P}{\text{p.f.}} = \frac{800{,}000}{0.8} = 10^6 \quad \text{VA}$$

$$|I_{1m}| = \frac{|S|}{\sqrt{3}V_L} = \frac{10^6}{\sqrt{3} \cdot 2400} = 241\text{A}$$

$$I_{1m} = 241 \ \underline{|+ \cos^{-1}0.8} = 241\underline{|36.87°}$$

$$E' = V_1 - jI_{1m}x_q = \frac{2400}{\sqrt{3}} \ \underline{|0°} - 241 \cdot 2.75 \ \underline{|36.87 + 90°}$$

$$= 1386 - 663 \ \underline{|126.87°} = 1386 + 398 - j530$$

$$= 1861 \ \underline{|-16.56°} \qquad \delta = -16.56°$$

$$\beta = (90° + \delta - \theta_{\phi m}) = 90° - 16.56° - 36.87° = 36.57°$$

$$\mathcal{I}_{dm} = |I_{1m}| \cos \beta = 241 \cos 36.57° = +193.6 \text{ A}$$

$$|E_q| = |E'| - \mathcal{I}_{dm}(x_q - x_1) = 1861 - 193.6(2.75 - 1.04)$$

$$= 1530 \text{ V}$$

$$\sqrt{3}|E_q| = 2650\text{V}$$

Corresponding to this voltage, $I_{fu} = 18.6$ A and $I_{fs} = 24.2$ A, and

$$k_{sd} = \frac{24.2}{18.6} = 1.30 \qquad m_d' = \frac{1530}{24.2} = 63.2 \ \Omega$$

Then

$$x_d = \frac{5.29 - 1.04}{1.30} + 1.04 = 4.31 \ \Omega$$

$$|E_f| = |E'| + \mathcal{I}_{dm}(x_d - x_q) = 1861 + 193.6(4.31 - 2.75)$$
$$= 2163 \text{ V}$$

Finally, $I_f = \dfrac{|E_f|}{m_d'} = \dfrac{2163}{63.2} = 34.2$A

The phasor diagram for this condition is

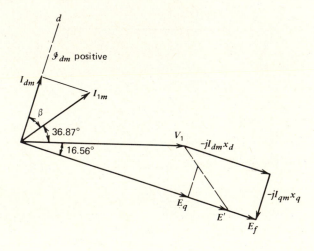

PROBLEMS

C.1. Use the machine data of Figure C.12, except that

$$\frac{x_q}{x_{du}} = 0.56$$

Using salient-pole theory, calculate the field current required under each of the following conditions, assuming rated terminal voltage in each case:

(a) Generator, full load, 0.800 power factor, *lagging*.

(b) Motor, full load, 0.800 power factor, *leading*. Compare results with (a).

(c) Generator, half rated current, 0.800 power factor, leading.

(d) Synchronous condenser, 900 kVAR.

C.2. Refer to the accompanying characteristic curves for a 400-kVA, 240-V, 60-Hz salient-pole synchronous machine. The q-axis synchronous reactance of this machine is 0.201 Ω. Neglect r_1 in solving this problem.

(a) Find the short-circuit ratio.

(b) Find the field current necessary to deliver rated load as a generator at 0.8 power factor, lagging, at rated voltage.

(c) Find the field current necessary to operate this machine as a synchronous condenser at half rated current, rated voltage.

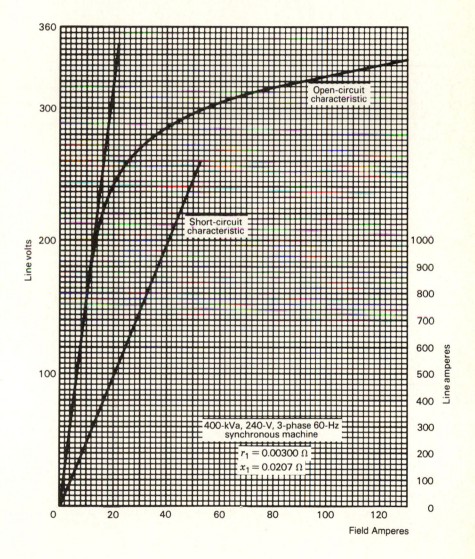

D
A SHORT BIBLIOGRAPHY

CHAPTER 2

Novotny, D. D. and King, T. L., "Equivalent circuit representation of current inverter driven synchronous machines," IEEE *Transactions on Power Apparatus and Systems,* Vol. PAS-100, June, 1981, pp. 2920–2926.

Liwschitz-Garik, M. and Whipple, C., *Alternating-Current Machines,* 2nd ed. (Princeton, N.J.: D. Van Nostrand, 1961).

Fitzgerald, A. E. and Kingsley, C. *Electric Machinery,* 2nd ed. (New York: McGraw-Hill, 1961), Chapter 5 and 9, Appendix C.

Kingsley, C., "Saturated Synchronous Reactance," AIEE *Transactions,* Vol. 54, No. 3 (March 1955), pp. 300–305.

Potier, A., "Sur la Réaction d'Induit de Alternateurs," *Revue d'Électricité,* July 28, 1900, pp. 133–141.

CHAPTER 3

MIT Staff, *Magnetic Circuits and Transformers* (New York: John Wiley, 1943).

Feinberg, R., *Modern Power Transformer Practice* (New York: John Wiley, 1979).

Stigant, S. A. and Franklin, A. C., *J & P Transformer Book,* 10th ed. (London: Newnes-Butterworths, 1973).

Westinghouse Staff, *Electrical Transmission and Distribution Reference Book* (East Pittsburgh, Pa.: Westinghouse Electric Corporation, 1964).

CHAPTER 4

National Electrical Manufacturers Association, *Publication No. MG1-1987, Motors and Generators,* (New York: NEMA, 1987).

IEEE, Std. 112-1984, *IEEE Standard Test Procedure for Polyphase Induction Motors and Generators,* (New York: IEEE, Inc., 1984).

Alger, P., *Induction Machines,* 2nd ed. (New York: Gordon and Breach, 1970).

CHAPTER 5

Millermaster, R., *Harwood's Control of Electric Motors*, 4th ed. (New York: Wiley-Interscience, 1970).

Siskind, C. *Direct Current Machinery* (New York: McGraw-Hill, 1952).

CHAPTER 6

Veinott, C., *Theory and Design of Small Induction Motors* (New York: McGraw-Hill, 1959).

Fitzgerald, A. E. and Kingsley, C., *ibid.*, Chapter 11.

CHAPTER 7

Slemon, G. R., Jackson, R. D., and Rohman, M. A., "Performance Predictions for Large Hysteresis Motors," IEEE *Transactions*, Vol. PAS-96, No. 6 (Nov./Dec. 1977).

Wernick, E. H., Ed., *Electric Motor Handbook* (London: McGraw-Hill (U.K.), 1978).

Yamamura, S., *Theory of Linear Induction Motors*, 2nd ed. (New York: Halstead Press Division, John Wiley, 1979).

Nassar, S. A. and Boldea, I., *Linear Electric Motors* (Englewood Cliffs, NJ: Prentice-Hall, 1987).

Harris, M. R., Finch, J. W., Mallick, J. A., and Miller, T. J. E., "A Review of the Integral Horsepower Switched Reluctance Drive," Procedings of the IEEE IAS 1985, Annual Meeting, pp. 783–789.

Small Motor, Gear Motor and Control Handbook, 4th ed. (Chicago: Bodine Electric Company, 1978).

Puchstein, A. F., Lloyd, T. C., and Conrad, A. G., *Alternating Current Machines*, 3rd ed. (New York: John Wiley, 1954).

CHAPTER 8

Levi, E., *Polyphase Motors* (New York: Wiley-Interscience, 1984).

Fitzgerald, A. E. and Kingsley, C., *ibid.*, Chapters 1 and 2.

White, D. C. and Woodson, H. H., *Electromechanical Energy Conversion* (New York: John Wiley, 1959).

APPENDIX C

Blondel, A., *Synchronous Motors and Converters*; *Theory and Methods of Calculation and Testing,* transl. by C. O. Mailloux (New York: McGraw-Hill, 1913).

GLOSSARY OF SYMBOLS

The abbreviations in parentheses indicate the applications of the symbols, as follows: D—d-c machines; I—induction machines; S—synchronous machines; T—transformers; sub—symbol used as a subscript; 1ϕ—single-phase machines; 1ϕ Alt.—single phase alternators. The numbers following the parentheses indicate the chapters in which the symbol occurs.

A		Area of a magnetic circuit: A_{Fe} in iron; A_g of an air gap
ACL	(D) 5	Armature copper loss
a	(sub)	Auxiliary winding, of phase a, approximate
a	(D) 5	Number of current paths through the armature winding
a	(T) 3	Ratio of primary to secondary turns
a	(1ϕ) 6	Effective turns ratio of auxiliary winding to main winding
a	(I) 4	Asynchronous generator: Ratio of operating frequency to rated frequency
a_v	(T) 2	Ratio of high-voltage-winding to low-voltage-winding turns
a_E	(I) 4	Stator/rotor phase voltage ratio
a_I	(I) 4	Stator/rotor phase current ratio $= q_1 a_E / q_2$
B	(sub)	Base or rated value
B		Magnetic flux density (webers/square meter or tesla)
B_{ave}		Average flux density of one pole of air-gap flux
B_{max}		Maximum or peak flux density
BR	(sub, I) 4	Under blocked-rotor conditions ($s = 1$)
b	(sub)	Of phase b
b	(I) 4	Asynchronous generator: Ratio of operating speed to rated speed
C		A constant
C	(D)	Total number of coils in the armature winding
C	(S)	Capability diagram: Proportionality constant relating phasor-diagram voltages to volt amperes
C_p	(D)	Number of coils in series in one armature current path
C_s	(D)	Number of armature coils shorted by brushes
c	(sub)	Of one coil. Of phase c

c_1, c_2	(I)	Constants related to variable-frequency speed control
D		Air-gap diameter
D_j	8	Generalized displacement at the jth mechanical output terminal
DMP	(I, 1ϕ) 4, 6	Developed mechanical power, watts
d	(D, S, sub)	Salient-pole machines: direct or pole axis
E		rms induced voltage
E_{ad}	(S) C	Armature reaction component of E_d
E_{ab}/E_{af}	(1ϕ) 6	Voltage induced in the auxiliary winding by the backward/forward rotating field
E_{aq}	(S) C	Armature reaction component of E_q
E_{ar}	(S) 2	Voltage model of the armature field: $E_{ar} = -jm\vec{F}_1$
E_c	(D, S) 2, 5	rms or d-c voltage induced in one armature coil
E_d	(D) 5	d-axis voltage induced in the armature
E_d	(S) C	Total d-axis voltage
E_f	(S) C	Voltage model of the field MMF: $E_f = -jm\vec{F}_2$
E_{f0}	(S) 2	Initial value of E_f
E_g	(D) 5	Voltage induced or generated in the armature
E_{ge}	2	Brushless dc motor: Armature induced voltage of the equivalent dc machine
E_{group}	(S) 2	Voltage induced in one phase group
E_i	(S) 2	Voltage behind the subtransient reactance. Assumed source voltage for fault calculations on loaded alternators
E_l	(S, T) 2, 3	rms leakage-reactance voltage
E_{mb}/E_{mf}	(1ϕ) 6	Main-winding voltages induced by backward/forward field components
E_q	(D) 5	Voltage induced in q axis of the armature
E_q	(S) C	Total q-axis voltage
E_R	(I) 4	Voltage induced in one rotor phase
E_R/E_S	2	Receiving-end/source-end voltages of a transmission line
E_1	(I) 4	Voltage induced in each stator phase by the air-gap field
E_1	(T) 3	rms voltage induced in the primary by the core flux, ϕ
E_n	(D, S) 2, 5	Voltage induced in the nth coil of the armature winding
E_2	(T) 3	rms voltage induced in the secondary by the core flux, ϕ
E_3	2	Third-harmonic component of an induced voltage
E_ϕ	(I, S) 2, 4	rms voltage induced in one stator phase by the air-gap flux
EL	8	Electrical loading; i.e., total number of ampere conductors in the armature winding under rated conditions
e		Instantaneous induced voltage: $e = \pm \, d\lambda/dt$
e_c	(D) 5	Instantaneous voltage induced in one coil
e_f	(D) 5	Voltage of self-induction of the field winding

e_{l1}/e_{l2}	(T) 3	Instantaneous voltage induced in the primary/secondary by the leakage flux of that winding. $e_{l1} = d\lambda_{l1}/dt$; $e_{l2} = d\lambda_{l2}/dt$
e_1	(I, T) 3, 4	Instantaneous voltage induced in the stator winding or primary by the core or air-gap flux
e_1'	(T) 3, 8	Total primary induced voltage, including leakage-flux voltage
e_2	(T) 3	Secondary voltage induced by core flux
e_2'	(T) 3, 8	Total secondary induced voltage, including e_{l2}
e_ϕ	(I, S) 2, 4	Instantaneous voltage induced in one stator phase by the air-gap flux
$\mathscr{F}$		Magnetomotive force (MMF), amperes
F		A particular value of $\mathscr{F}$; specifically the amplitude of a periodic $\mathscr{F}$ function
F_{Ad}/F_{Aq}	(D) 5	Component of armature MMF along the d/q axis
$\mathscr{F}_a, \mathscr{F}_b, \mathscr{F}_c$	(I, S) 2, 4	MMFs of the three stator phase windings
F_{a1}, F_{b1}, F_{c1}	(I, S) 2, 4	Amplitudes of the fundamental Fourier components of $\mathscr{F}_a, \mathscr{F}_b, \mathscr{F}_c$
F_a	(D) 5	Peak armature MMF
F_a	(1φ) 6	Peak MMF of the auxiliary winding
F_b	(1φ) 6	Amplitude of the backward-rotating MMF field
F_c	(T) 3	rms MMF of the common winding of an autotransformer
$\mathscr{F}_f$	(D) 5	MMF of the shunt-field winding of one pole
F_f	(1φ) 6	Amplitude of the forward-rotating MMF field
F_g	(D) 5	MMF applied to the air gap at some point
F_m	(1φ) 6	Peak MMF of the main winding
$\mathscr{F}_{pole}$	(D) 5	Total MMF of the shunt and series winding of one pole
F_{se}	(T) 3	rms MMF of the series winding of an autotransformer
FCL	(D) 5	Field copper loss
$\mathscr{F}_1$	(I, S) 2, 4	Stator MMF as a function of θ_1 and/or t
F_1	(I, S) 2, 4	Amplitude of the fundamental component of $\mathscr{F}_1$
$\vec{F}_1$	(I, S) 2, 4	Stator MMF, treated as a vector
$\vec{F}_{1b}$	(1φ Alt.) 2	Backward rotating vector component of stator MMF
$\mathscr{F}_{1b}$	(1φ Alt.) 2	Backward component of stator MMF as a function of θ_{1e} and t
$\mathscr{F}_{1f}$	(1φ Alt.) 2	Forward component of stator MMF as a function of θ_{1e} and t
F_1	(T) 3	MMF of the primary winding
$\mathscr{F}_2$	(I, S) 2, 4	Rotor MMF as a function of θ_2 and/or t
F_2	(T) 3	MMF of the secondary winding
$\vec{F}_2$	(I, S) 2, 4	Rotor MMF, treated as a vector
$\vec{F}_{1d}/\vec{F}_{1q}$	(S) C	d/q axis components of $\vec{F}_1$
$F_{1\phi m}$		Maximum instantaneous value of the fundamental component of the MMF of one phase group
f		Frequency, Hertz
f_B	(I) 4	Rated stator frequency
f_t	(I) 4	Blocked-rotor-test frequency

f_1	(I, S) 2, 4	Frequency of stator voltages/currents
f_2	(I) 4	Frequency of rotor voltages/currents
Fe	(sub)	Of the iron parts of a magnetic circuit
fl	(sub)	Full load; that is, rated load
G	(D) 5	Power gain of a metadyne
g		Radiolength of an air gap
g	(sub)	Of the air gap
H		Magnetic field intensity, amperes/meter
H_c	5	Coercive force of a magnetic material
H_{Fe}		H in the iron parts of a magnetic circuit
H_g		H in an air gap
H_{ci}	5	Intrinsic coercive force of a magnetic material
HV	(sub, T) 3	Of the high-voltage winding
h	(S) 2, C	Altitude of the Potier triangle
I		rms or dc current, amperes
I_a	(D) 5	Armature current, dc amperes
I_a	(1φ) 6	rms auxiliary-winding current
I_a, I_b, I_c		rms current in line or phase a, b, c
I_{ab}/I_{af}	(1φ) 6	Backward and forward symmetrical components of the auxiliary-winding current
I_{ae}	2	Brushless dc motor: Effective armature current of the equivalent dc machine
I_B		Base or rated current
I_{BR}	(I) 4	Line current during blocked-rotor test
I_c	(T) 3	Current in the common winding of an autotransformer
I_d	(D) 5	d-axis current in a metadyne rotor
I_d	(S) C	d-axis phasor component at the stator phase current, I_1
$\mathscr{I}_d$	(S) C	Scalar value of I_d whose sign indicates direction along the d axis
I_{dc}	(I) 4	dc current used in measuring stator winding resistance
I_{ex}	(I, T) 3, 4	rms value of the exciting component of the stator or primary current
I'_{ex}/I''_{ex}	(T) 3	Exciting current of step-up/step-down autotransformers
I_f	(D, S) 2, 5, 8	dc field current, amperes
I^*_f	(D) 5	A chosen value of shunt-field current, or the effective value: $I^*_f = I_f + (N_{se}/N_f)I_{se}$
I_{fI}	(S) 2	Field current corresponding to rated current on the short-circuit characteristic
I_{fs}	(S) 2	Field current corresponding to $\sqrt{3}|E_\phi|$ on the open-circuit characteristic curve
I_{fu}	(S) 2	Field current corresponding to $\sqrt{3}|E_\phi|$ on the air-gap line
I_{fV}	(S) 2	Field current corresponding to rated voltage on the open-circuit characteristic curve
I_{h+e}	(T) 3	rms value of that component of the primary current due to hysteresis and eddy-current core losses; that is, the core-loss current

I_I	2	Amplitude of CSI output current waves in a brushless dc machine system
I_L		Scalar line current in a balanced, three-phase system
I_L	(1ϕ) 6	Phasor line current
I_{L1}/I_{L2}	(T) 3	Primary/secondary line currents of a three-phase transformer or bank
I_m	(1ϕ) 6	rms main-winding current
I_{mb}/I_{mf}	(1ϕ) 6	Backward/forward symmetrical components of the main-winding current
I_q	(D) 5	q-axis current in a metadyne rotor
I_q	(S) C	q-axis phasor component of the stator phase current, I_1
I_R	(I) 4	rms current in one rotor phase circuit
I_{sc}	(S) 2	Current of the short-circuit characteristic; rms transient fault current
I_{se}	(D) 5	Current in the series field winding. $I_{se} = I_a$ in the absence of a diverter
i_{ss}	(S) 2	Steady-state value of the fault current
I_x	(S) 2	Value of short-circuit current found on the short-circuit characteristic at I_{fV}
I_ϕ	(I, T) 3, 4	Magnetizing component of the stator or primary current
$I_{\phi B}$		Rated current of one phase winding
$I_{\phi oc}/I_{\phi sc}$	(T) 3	Phase-winding current during open-circuit/short-circuit tests of three-phase transformers
$I_{\phi Y}/I_{\phi \Delta}$		Phase current in a Y/Δ connected three-phase circuit
I_1	(T) 3	rms phasor primary current
I_1	(I, S) 2, 3	rms phasor current in one stator phase winding
I_{1B}/I_{2B}	(T) 3	Rated value of I_1/I_2
I_{1m}	(S) 2	Stator phase current taken in such a direction in the winding that the power factor is positive in the motor mode
I_2	(I) 4	Load component of stator phase current: $I_2 = I_R/a_I$
I_2	(T) 3	rms secondary current. In three-phase transformers, the current in one secondary phase winding
i		Instantaneous current, amperes
i_a, i_b, i_c		Instantaneous phase currents in a three-phase system
i_c	(D) 5	Instantaneous coil current
i_{ex}	(T) 3	Instantaneous value of the core excitation current
i_f	(D) 5	Instantaneous field current
i_{h+e}	(T) 3	Instantaneous value of the core-loss component of i_{ex}
J		Current density, amperes/m^2
j		Operator that rotates a phasor 90° forward: $j^2 = -1$
K'	(D) 2	A coefficient which is a saturating function of field current such that $E_g = (K'I_f)\omega$ and $\tau_d = (K'I_f)I_a$. $K' = K_a \mathscr{P}_p N_f$
K''	2	Brushless dc motor: $K'' = \frac{1}{2}p N_{E1} N_f \mathscr{P}_p$ such that in the equivalent dc machine $E_{ge} = (K''I_f \cos \gamma)\omega$ and $\tau_d = (K'' I_f \cos \gamma) I_{ae}$
K_a	(D) 5	Armature voltage constant, speed in rad/s

K_a'	(D) 5	Armature voltage constant, speed in rev/min
$K_l \equiv L_l$	(T) 3	Constant of proportionality between winding current and leakage flux linkages, λ_l
k	(I) 4	Speed control; number of side-band poles in PAM windings
k	(T) 3	Coefficient of coupling in a transformer, such that $L_{12} = k\sqrt{L_{11}L_{22}}$
k_1	(T) 3	Primary coupling factor
k_2	(T) 3	Secondary coupling factor
k_d		Distribution factor of a winding
k_p		Pitch factor of a winding
k_s	(S) 2	Saturation factor, cylindrical-rotor theory
k_s		Skew factor of a winding
k_{sd}	(S) C	Salient-pole machines: saturation factor along the d axis
k_w		Winding factor: Product of pitch, distribution, and skew factors
L		Inductance, henrys
L_{11}	(T) 3, 8	Self-inductance of the primary winding
L_{12}	(T) 3, 8	Mutual inductance between windings 1 and 2
L_{22}	(T) 3, 8	Self-inductance of the secondary winding
L_{ff}	(D) 5	Self-inductance of the field winding
L_{l1}/L_{l2}	(T) 3	Leakage inductance of the primary/secondary winding
LV	(sub, T) 3	Of the low-voltage winding
ℓ		Magnetic circuits: Mean length of flux path
ℓ_{Fe}		Length of flux path in iron
l	(sub)	Due to leakage flux
lo	(sub)	Under load
M	(D) 5	Magnetization of a permanent-magnet material
M	5	Windings: an integer
ML	8	Magnetic Loading $= p \times$ flux per pole
m	(D) 5	The multiplicity of an armature winding. Simplex: $m = 1$; duplex; $m = 2$, etc.
m	(sub, 1ϕ) 6	Of the main winding
m	(I, S) 2, 4	The nonlinear factor relating circuit-model voltages to the machine MMFs they represent. (e.g., $E_f = -jm\vec{F}_2$). $m = \omega_e N_{E1} \mathscr{P}_p$ varies with saturation.
m', m_a'	(S) 2	$m' = N_f m$; $m_a' = N_f m_a$, where N_f is the effective turns per pole for the field winding
m_a	(S) 2	A value of m used in approximate calculations that assumes the induced phase voltage is constant and equal to rated phase terminal voltage
N		Number of turns in a winding or coil
N_a	(1ϕ) 6	Effective number of turns in the auxiliary winding
N_a, N_b, etc.	(S) 2	Numbers of turns in the coils of a cylindrical-rotor field winding
N_c	(D, S) 2, 5	Number of turns in a typical armature coil
N_c	(D) 5	Number of compensating-winding conductors in each pole face

N_c	(T) 3	Number of turns in the common winding of an autotransformer
N_{E1}	(I, S) 2, 4	Effective voltage turns per phase of the stator winding
N_f	(S) 2	Effective turns of the field winding, per pole, including winding factor for cylindrical rotors
N_f	(D) 5	Turns per pole of the shunt-field winding
N_g		Effective turns in one phase group $= (4/\pi)\sqrt{2}nN_ck_w$
N_{I1}	(I, S) 2, 4	Effective current turns of a three-phase stator winding such that $F_1 = N_{i1}I_1$. $N_{i1} = q_1N_g/2$
N_{I2}	(I) 4	Effective current turns of the rotor winding, such that $F_2 = N_{I2}I_R$
N_{se}	(D) 5	Turns per pole of the series-field winding
N_{se}	(T) 3	Turns in the series winding of an autotransformer
N_1/N_2	(T) 3	Number of turns in the primary/secondary winding
n		Mechanical speed in rev/min
n		Number of coils in a phase group
n_B		Rated speed in rev/min
n_s	(I, S) 2, 4	Synchronous speed, rev/min $= 120\, f_1/p$
nl	(sub)	No load
OCC	2, 5	Open-circuit characteristic
P		Average power, watts. Real component of the complex power, S
P_B	(D) 5	Base or rated armature power
P_{BD}	(D) 5	Brush-drop loss
P_{BR}	(I) 4	Total input power during blocked-rotor test
P_{Cu}	(T) 3	Copper loss
P_d	(D) 5	Developed or converted power
P_f	(D, S) 2, 5	Power input to field windings; excitation power
P_{fw}	(I, D) 4, 5	Sum of friction and windage losses
P_{fwc}	(I) 4	Sum of friction, windage and core losses
P_g	(I, S) 2, 4	Power crossing the air gap, stator to rotor, via the magnetic field
P_{gf}	(1ϕ) 6	P_g of the forward-rotating magnetic field
P_{gb}	(1ϕ) 6	P_g of the backward-rotating magnetic field
P_{h+e}	(I, T) 2, 3	Core loss: sum of hysteresis and eddy-current losses
P_{in}		Input power, watts
P_k		Losses independent of load. ("Constant")
P_k	(S) 2	A constant load level
P_{LL}	(I) 4	Stray load loss
P_{mag}	(D) 5	Magnetizing power
P_{max}	(S) 2	Power at which the machine loses synchronism
P_{mech}		Average mechanical power
P_{oc}	(T) 3	Input power during an open-circuit test
P_{out}		Output power
P_{rot}	(D, I) 4, 5	Rotational losses
P_{sc}	(T) 3	Input power during short-circuit test
P_ϕ	(S) 2	Power of one phase
$\mathscr{P}$		Effective permeance of a magnetic circuit, such that $\phi = \mathscr{P}\mathscr{F}$. Varies with saturation

$\mathcal{P}_d/\mathcal{P}_q$	(D, S) 2, 5, C	Permeance of the d/q-axis magnetic circuit
$\mathcal{P}_p$	(D, I, S) 2, 4, 5	Effective permeance of the magnetic circuit of one pole
$\mathcal{P}_{\ell 1}/\mathcal{P}_{\ell 2}$	(T) 3	Proportionality coefficient between effective leakage flux and the MMF of the primary/secondary
$\mathcal{P}_{12}$	(T) 3, 8	Permeance of magnetic circuit coupling the primary and secondary windings. Essentially, the permeance of the core
p		Instantaneous power, watts
p	(D, I, S, 1φ) 2, 4, 5, 6	Number of poles of the stator or rotor
p_a, p_b	(I) 4	Number of poles available in a PAM speed-control system
p_{mech}		Instantaneous mechanical power
p_1, p_m	(I) 4	Number of poles in the initial and modulating fields of a PAM speed-control system
p.f.		Power factor: cosine of the power-factor angle
Q		Reactive component of the complex power, S. $S = P + jQ$. $Q = \|S\| \sin \theta_\phi$
Q_j	8	Generalized force at the jth mechanical output terminal
q	(D, S)	Quadrature axis
q_1/q_2	(I, S) 2, 4	Number of phases in the stator/rotor winding
R		Resistance, ohms, or magnitude of a resultant MMF, amperes
R	(I) 4	Line to neutral load resistance per phase on an asynchronous generator
$\vec{R}$	(I, S) 2, 4	The MMF applied to the magnetic circuit of a machine, the resultant of the fundamental fourier components of the rotor and stator MMFs
R_a	(D) 5	Armature-circuit resistance including series field
R_B	(D) 5	Base resistance $= (V_B/I_B)$
R_c	(T) 3	Resistor representing core loss in the circuit model
R_d	(D) 5	Resistance of a diverter across a series field
R_{eq1}/R_{eq2}	(T) 3	Resistive components of Z_{eq1}/Z_{eq2}
R_{ex}	(D, I) 4, 5	Armature- or rotor-circuit resistance external to the machine
$\vec{R}_{oc}$	(S) 2	$\vec{R}$ on open circuit. $\vec{R}_{oc} \equiv \vec{F}_2$
R_L	(I) 4	Asynchronous generator: Resistive component of the equivalent series load per phase
R_R	2	Internal resistance of the rectifier feeding a CSI inverter in a brushless dc motor
R_R	(I) 4	Resistance of one rotor phase circuit
R_{rheo}	(D)	Field-rheostat resistance
R_{st}	(1φ)	Real component of the field impedance at starting $R_{st} = R_f$ when $s = 1$
R_T	(D)	Transfer resistance related to a given metadyne control field
R_1	(I)	Resistive component of Z_{Th}

$\mathcal{R}$		Magnetic reluctance. $\mathcal{P} = (1/\mathcal{R})$
r		Radius of the air gap
r_a	(D)	Resistance of the armature circuit between terminals
r_{ae}	2	Brushless dc machine: Effective armature resistance of the equivalent dc machine
r_f	(D)	Resistance of the shunt-field winding
r_{se}	(D)	Resistance of the series-field winding
r_1	(I, S)	Resistance of one phase of the stator winding
r_1	(T)	Resistance of the primary winding
r_2	(T)	Resistance of the secondary winding
r_2	(I) 4	Resistance of the rotor referred to one stator phase
r.f.		Reactive factor: sine of the power-factor angle
S		Complex power. $S = P + jQ$, volt amperes
S_B		Rated value of $|S|$ for a given device, volt amperes
S_1/S_2		Number of stator/rotor slots (or coils, in a double-layer winding)
SCC	(S) 2	Short-circuit characteristic
SCL	(I) 4	Stator copper loss, watts
SEL	8	Specific electrical loading: armature ampere conductors per meter
SML	8	Specific magnetic loading: average pole flux density in dc machines; rms air gap flux density in ac machines
s	(I) 4, 8	Slip: $s = (\omega_s - \omega)/\omega_s$
s_M	(I) 4	Slip at which maximum developed torque is developed
s_b	(1ϕ) 6	Slip of rotor with respect to the backward field; that is, slip of the backward motor
T		Period of a periodic function of time
T_a	(S) 2	Time constant of the decay of the d-c and double-frequency components of the fault current
T_d'	(S) 2	Transient time constant
T_d''	(S) 2	Subtransient time constant
t		Time, seconds
U	(D) 8	Slope of the armature MMF as a function of θ in a dc machine
V		rms or d-c voltage
V_a, V_{af}, V_{ab}	(1ϕ) 6	Auxiliary-winding terminal voltage and its forward and backward components
V_{ab}, V_{bc}, V_{ca}		Three phase: rms phasor voltage drops from lines a to b, b to c, c to a
V_{an}, V_{bn}, V_{cn}		Three phase: rms phasor voltage drops from lines to neutral
V_B		Base or rated voltage
V_{BD}	(D) 5	Brush-drop voltage. The sum of the drops at positive and negative brush sets
V_{BR}	(I) 4	Line voltage used in blocked-rotor test
V_{dc}	(I) 4	d-c voltage across two stator phases in measuring stator winding resistance
V_f	(D, S) 2, 5	d-c voltage at the field winding terminals

V_L		Line voltage or line to line voltage
V_m, V_{mf}, V_{mb}	(1ϕ) 6	Main-winding terminal voltage and its forward and backward symmetrical components
V_{nl}		Output voltage at zero load current
V_{nl}	(I) 4	Line voltage in no-load test
V_{oc}	(S) 2	Open-circuit voltage, line to line
V_{oc}	(T) 3	Voltage applied for open-circuit test
V_R	2, 4	Rectifier output voltage in a variable-frequency speed control or to a CSI inverter in a brushless dc motor
V_T	(D) 5	Armature terminal voltage
V_{Te}	2	Effective armature terminal voltage in a brushless dc motor
V_{Th}	(I) 4	Source voltage in Thevenin equivalent of stator circuit
V_ϕ		Three phase: rms phase voltage in either Y or Δ circuit
$V_{\phi B}$		Rated or base phase voltage
$V_{\phi oc}/V_{\phi sc}$	(T) 3	Three-phase transformers: phase voltage during open-circuit/short-circuit tests
V_1	(I, S) 2, 4	rms phasor terminal voltage of one phase of the stator winding
V_1	(T) 3	Primary terminal voltage
V_{1m}	(I) 4	$V_{1m} \equiv -V_1$. Phase terminal voltage in the motor mode
V'_{1m}	(I) 4	A reduced phase voltage approximately equal to V_{Th}. Used in Thevenin-reduced model and approximate circuit model
V_{1B}	(I, S) 2, 4	Rated stator terminal voltage of one phase winding
V_{1B}	(T) 3	Rated primary voltage
V_2	(T) 3	rms secondary terminal voltage
V_{2B}	(T) 3	Rated secondary terminal voltage
V_{2fl}/V_{2nl}	(T)	Secondary terminal voltage at full-load/no load
v		Instantaneous terminal voltage or voltage drop
v	7	Translational velocity of a linear motor
W_ϕ	5, 8	Energy stored in the magnetic field
W'_ϕ	8	System coenergy
w_{fld}	8	Magnetic field energy density, J/m³
X_{11}	(T) 3	Primary reactance $= \omega L_{11}$
X_{12}	(T) 3	Mutual reactance of a transformer $= \omega L_{12}$
X_{22}	(T) 3	Secondary reactance $= \omega L_{22}$
X_c	2	Fictitious reactance in CSI/synchronous machine model to accommodate $\gamma \neq 0$
X_c	(I) 4	Line to neutral shunt capacitive reactance of the load on an asynchronous generator
X_{eq1}/X_{eq2}	(T) 3	Reactive components of Z_{eq1}/Z_{eq2}; that is, the equivalent series reactances referred to the primary/secondary
X_L	(I) 4	Asynchronous generator: Capacitive reactance of the series load equivalent to R and X_c in parallel
X_m	(I, S, T) 2, 3, 4	The magnetizing reactance
X_{m1}/X_{m2}	(T) 3	Magnetizing reactances of the primary/secondary

X_{BR}	(I) 4	X'_{BR} corrected to stator frequency. $X_{BR} = x_1 + x_2$
X'_{BR}	(I) 4	Reactive component of the impedance Z_{BR}, measured at the terminals of one stator phase during a blocked-rotor test at frequency f_t
X_R	(I) 4	Leakage reactance of one rotor phase at rotor frequency
X_{RB}	(I) 4	X_R at blocked rotor; that is, at stator frequency
X_1	(I) 4	Reactive component of Z_{Th}
x		Steimmetz' exponent in calculating hysteresis loss
x_d	(S) 2	Direct-axis synchronous reactance. $x_d = X_m + x_1$
x'_d	(S) 2, C	Direct-axis transient reactance
x''_d	(S) 2	Direct-axis subtransient reactance
x_{da}	(S) 2	Approximate saturated synchronous reactance equal to V_{1B} divided by the short-circuit current corresponding to I_{fV}, the field current which gives $\sqrt{3}V_{1B}$ on the OCC
x_{du}	(S) 2, C	Unsaturated value of x_d
x_q	(S) C	Quadrature-axis synchronous reactance of salient-pole machines
x_{qu}	(S) C	Unsaturated value of x_q
x_1/x_2	(T) 3	Leakage reactance of the primary/secondary winding
x_1	(I, S) 2, 4	Leakage reactance of one phase of the stator winding, balanced currents assumed
x_2	(I) 4	Leakage reactance of the rotor as reflected in one stator phase at stator frequency
Y_c	(D) 5	Distance between coil lead connections to the commutator, expressed in segments
Y_r	(I) 4	Asynchronous generator: Rotor circuit admittance
Y_c	(I) 4	Asynchronous generator: Stator circuit admittance
Y_s	(D) 5	Armature coil pitch, expressed in slots
Z		Complex impedance, ohms
Z	(D) 5	Total slot conductors in an armature winding
Z_B		Base impedance. $Z_B = V_B/I_B$
Z_{BR}	(I) 4	Impedance looking into the terminals of one stator phase during a blocked-rotor test
Z_b	(1φ) 6	Backward field impedance
Z_c		Slot conductors per coil $= 2N_c$
Z_{eq1}/Z_{eq2}	(T) 3	Equivalent series impedance referred to the primary/secondary
Z_f	(I, 1φ) 4, 6	Forward field impedance
Z_L		Complex impedance of an electrical load
Z_{st}	(1φ) 6	Field impedance, Z_f, at $s = 1$
Z_{Th}	(I) 4	Impedance of the Thevenin-generator equivalent of one stator-phase circuit
Z_{12}	(1φ) 6	Mutual impedance between forward and backward meshes of the circuit model
z_1	(I, T, 1φ) 3, 4, 6	Stator or primary leakage impedance of one phase. $z_1 = r_1 + jx_1$
z_2	(I) 4	Rotor impedance referred to one stator phase: $z_2 = r_2/s + jx_2$
z_2	(T) 3	Secondary leakage impedance $= r_2 + jx_2$

z_{1a}	(1φ) 6	Sum of capacitor impedance (if any) and auxiliary-winding leakage impedance
z_{1m}	(1φ) 6	Main-winding leakage impedance
α	(D) 5	Arc subtended by a pole face
β	(D) 5	Angle of brush shift to reduce commutator sparking
β	(S)	Angle by which peak rotor MMF, $(\vec{F}_2)$ leads the peak stator MMF $(\vec{F}_1)$, electrical degrees
γ		Steinmetz' coefficient used on calculating hysteresis loss
γ	(S) 2	Slot pitch, usually in electrical degrees.
γ	(1;φ Alt.) 2	Angular rotation of $\vec{F}_{1f}$ from the axis of the stator winding
γ	2	Brushless dc motor: Angle between stator MMF and rotor q axis, equal to the angle between the dc voltages of the equivalent dc machine and E_f of the synchronous machine
Δ	(D) 8	Supposed shift in armature MMF wave, mechanical degrees, to permit calculation of dc machine torque by principle of virtual work
δ	(S) 2	The power angle of a synchronous machine, equal to the phase angle between the phase terminal voltage and E_f
δ	7	Hysteresis motor: Angle between the rotor field $\vec{F}_2$ and the resultant field, $\vec{R}$
δ	7	Reluctance motor: Angle between rotor d axis and resultant field $\vec{R}$
δ_{FR}	(S) 2	Angle between the rotor field vector $\vec{F}_2$ and the resultant of rotor and stator fields, $\vec{R}$
δ_{SR}	(S) 2	Angle between the stator field vector $\vec{F}_1$ and $\vec{R}$
η		Efficiency: P_{out}/P_{in}
θ_{BR}	(I) 4	Input power-factor angle during the blocked-rotor test
θ_L	(1φ) 6	Angle between line voltage and line current
θ_a/θ_m	(1φ) 6	Phase angle of auxiliary/main-winding current
θ_{nl}	(I) 4	Power-factor angle at no load
θ_0	(S) 2	Initial angle between $\vec{R}$ and the rotor d axis at the instant of a balanced, three-phase fault
θ_{oc}	(T) 3	Power-factor angle during an open-circuit test
θ_{sc}	(T) 3	Power-factor angle during a short-circuit test
θ_ϕ		Power-factor angle: Angle by which phase current leads or lags phase voltage
θ_ϕ'	(S) 2	Angle by which I_1 leads or lags E_ϕ, the flux-induced phase voltage
$\theta_{\phi BR}$	(I) 4	Full-voltage input power factor at blocked rotor
$\theta_{\phi m}$	(I, S) 2, 4	Power factor angle in the motor mode
θ_1	(S) 2	Space angle measured on the stator in mechanical degrees
θ_{1e}	(S) 2	θ_1 measured in electrical degrees
θ_2	(I) 4	Internal power factor angle of the rotor circuits
θ_2	(S) 2	Space angle measured on the rotor in mechanical degrees
θ_{2BR}	(I) 4	Angle between I_2 and V_{1m} at blocked rotor

θ_{2e}	(S)2	θ_2 measured in electrical degrees
λ		Number of flux linkages, weber turns
λ_c	(D, S) 2, 5	Flux linkages of a typical coil
λ_ℓ	(I, S, T) 2, 3, 4	Leakage flux linkages of a winding
$\lambda_{\ell 1}/\lambda_{\ell 2}$	(T) 3, 8	Leakage flux linkages of the primary/secondary
$\lambda_{m1}/\lambda_{m2}$	(T) 3	Flux linkages of the mutual flux, ϕ, with the primary/secondary winding
λ_1/λ_2	(T) 3	Total flux linkage of the primary/secondary winding
μ		Magnetic permeability
μ	(D) 5	Voltage gain associated with one control field of an amplidyne
μ_0		Permeability of free space; $4\pi \cdot 10^{-7}$ henry/meter in SI units
υ		Order of a field harmonic
ρ		Coil pitch; that is, angular distance between sides of a given coil, usually in electrical degrees
ρ_a, ρ_b, etc.		Pitch of coil a, b, etc.
ρ_p	(D) 5	Angular distance between axes of main stator poles
ρ_s	7	Surface resistivity of secondary of a linear motor
τ		Torque, newton meters or pound-feet
τ_d	(D, I, S) 2, 4, 5	Developed torque, calculated from converted power, losses neglected. $\tau_d = P_d/\omega$, $N - m$, $\tau_d = 7.04 P_d/n$, pound-feet. In an induction machine, $\tau_d = P_g/\omega_s$
τ_{fl}	(I) 4	Torque under rated conditions; that is, full load
τ_{max}	(S) 2	Maximum torque theoretically available without losing synchronism
τ_{max}	(I) 4	Breakdown or pull-out torque
τ_R	(S) 2	Reluctance torque of a salient-pole machine
τ_{st}	(I, 1ϕ) 4, 6	Starting torque, calculated for $s = 1$
ϕ	(D, I, S) 2, 4, 5, 8	Flux per pole, Wb
ϕ	(T) 3	Mutual or core flux
ϕ	(sub)	Of one phase of a three-phase system
$\overline{\phi}$		Effective flux $= \lambda/N$
ϕ_c		Instantaneous flux linking a coil; $\lambda_c = N_c\phi_c$
ϕ_c	(T) 3	Flux corresponding to the constant of integration when integrating Faraday's law
ϕ_d/ϕ_q	(D, S) 2, 5	d/q-axis fluxes
$\phi_{\ell 1}/\phi_{\ell 2}$	(T) 3	Effective leakage flux of the primary/secondary
ϕ_{max}	(T) 3	Peak core flux in the steady-state
ϕ_1/ϕ_2	(T) 3	Total effective flux linking the primary/secondary
ω		Angular velocity, rad/s. Usually the shaft speed of a machine
ω_B		Rated angular velocity
ω_e		Electrical angular velocity $= 2\pi f$
ω_{fl}		Speed at rated load
ω_{nl}		Speed at no-shaft load

INDEX

Air gap, 7, 17, 59, 61–64, 320, 377
Air gap line, 108, 110, 331
Air gap power, 257
Alternator, 1, 35, 42–43. *See also* Synchronous machines
  connected to infinite bus, 105
  isolated, 104
  permanent magnet, 123
Amortisseur winding, 94
Ampere's law, 49, 56, 360, 478
Ampere's right hand rule, 14
Amplidyne, 385–389
Angular velocity, 4
Apparent power, 506
Armature, 8, 11, 17
  copper loss, 349
  of dc machine, 11, 319, 325, 477
  induced voltage, 64–78, 82, 325, 328, 474
  paths, current, 325, 326, 328
  reaction, 50, 89, 331, 359–364
  of synchronous machine, 11, 17, 37–44, 50–59. *See also* Windings, polyphase
Asynchronous generator, 18, 239, 302–315
Asynchronous machines, 7. *See also* Induction machines
Automatic synchronizer, 107
Automatic voltage regulator, 33, 105, 377
Autotransformers, 227–231
  rating advantage, 229
Auxiliary winding, 398

Base current, 89, 193, 350, 352
Base impedance, 89, 193, 197, 354
Base speed, 330, 344
Base voltage, 89, 193
Basket weave motor, 441
Belt, 1, 7, 370
Belt factor, *see* Distribution factor
Blocked rotor (locked rotor), 253, 275, 279, 282, 284, 296
Blocked-rotor test, 279

Blondel two-reaction theory, 529–534
Breadth factor, *see* Distribution factor
Breakdown torque, 265, 267
Breathing field, 51
Brush drop, 20, 24
Brush drop loss, 20, 24, 350
Brushes, 24, 320
Brush holder, 320
Brushless dc machine, 129–141, 333
  characteristics, 137
  PM, 140
Brushless excitation, 44, 46, 124
Brush shift, 367, 410, 480
Brush shunt, 322

Capability curve, 141
Capacitor motors, 402
  calculations, 416
Capacitor-start motors, 403
Capacity of alternators, 125. *See also* Capability curve
Centrifugal switch, 402
Chopper drives, 348
Circle diagram, induction machine, 293–303
Code letters, NEMA, 284
Coenergy, 449, 467, 468
Coil pitch, 40, 48, 52, 56, 59, 68, 290
Coils, 8, 38, 306, 399
  coil sides, 38, 40, 454
  per group, 40, 41, 52, 57, 59
Commutating poles (interpoles), 12, 363–366
Commutation, 361–368
Commutator, 17, 129, 319, 320, 322, 333. *See also* Commutation
Commutator sparking, 364
Compensating windings, 362, 385, 386
Complex power, 101–103, 506, 513, 516
Compound dc machines, 335–336, 340–342, 368–370, 378–380
  cumulative, 335, 336, 340, 370, 379, 380
  differential, 335, 336, 340, 370, 380

Compound dc machines (*Continued*)
  flat-compound, 380
  long-shunt, 336, 378
  short-shunt, 336
  undercompounded, 379
Compounding curve, 100
Conductors, 38
Consequent poles, 288–289
Conservation of energy, 19, 453, 459
Constant losses, 26, 350
Cooling loss, 20, 24
Copper loss, 20, 28, 194, 334, 349
Core, 7, 37, 158, 160
  amorphous metal, 160
  armature, 319
  excitation of, 180
Core form, 160
Core losses, 20, 24, 28, 81, 92, 194
Counter emf (countervoltage), 5, 248, 323, 325
Counter torque, 6, 92, 325
Coupling, 7, 10
Coupling coefficient, 168
Critical resistance, 379
Current inrush, 185
Current ratio, 155, 172
Current source inverter, 130
Current transformers, 158, 232
Current turns per pole, effective, 54, 55, 243,
    474, 476
Cylindrical rotors, 40–45, 47
Cylindrical-rotor theory, 47

Damper winding, 93
d axis, *see* Direct axis
dc machines, 17, 319–394, 477–482
  air gap, 320, 478
    energy, 479
  amplidyne, 385
  armature, 319
    conductors, 328
    core, 319, 477
    inductance, 329
    MMF, 331, 359, 477–478
    reaction, 357–368
    windings, 319, 380–384, 477
      frog leg, 384
      lap, 325, 380–382
      multiplex, 380, 381
      parallel paths, 325, 326, 380–381
      pitches, 325, 383
      simplex, 380, 381, 383
      wave, 382
  armature constant, 329
  armature copper loss, 349
  base speed, 330, 344
  brush drop loss, 20, 24, 350
  brushes, 320, 477
  brush shift, 367, 480
    demagnetizing ampere turns, 367–368, 480
  circuit model, 329

commutation, 361–368
commutator, 319, 477
  insulation, 320
  segments, 320, 322, 323
compensating windings, 362–363, 385, 386
compound, 335–336, 340–342, 368–370
developed power, 331
developed torque, 331
diverter, 340
dynamic braking, 352
electrical loading, 481
  specific, 481
field coils, 320
field excitation, 320, 330, 334–336, 340–341
field poles, 320, 328, 477
  MMF, 330, 340, 477–478
  pole arc, 477–478
field-resistance line, 379
field rheostat, 343, 378
generated voltage, 325, 328
generator buildup, 377
  critical resistance, 379
generator characteristics, 379
generator operation, 325, 330, 377–380
interpoles, 12, 363–366
jogging, 357
long-shunt, 336
losses, 348
low inertia motors, 439
magnetic loading, 481
  specific, 481
magnetization curve, *see* Open-circuit
  characteristic
metadynes, 384
motor operation, 330
open-circuit characteristic, 330–331, 337–340
operating principle, 17, 322–325
permanent magnet, *see* PMDC
per-unit values, 348–351
plugging, 357
PMDC, 370–377
regeneration, 346
rotational losses, 350
runaway, 343
saturation curve, 330–331. *See also* Open-
  circuit characteristic
separately excited, 335
series, 336
short-shunt, 336
shunt, 336
speed control, 342
stabilized shunt motor, 330
starting, 352–359
stay-load loss, 351
stray power loss. *See* Rotational losses
terminal markings, NEMA, 336
torque motors, 442
turns ratio of field windings, 341
Deep-bar rotors, 272–274
Delta connection, 61, 73, 287, 508, 511

Demagnetizing ampere turns of magnets, 125
Direct axis, 47, 60, 128, 132–134, 385–388, 427, 529–536
Direct-axis synchronous reactance, 89. *See also* Synchronous reactance
Distribution factor, 58, 69
  harmonic, 74
Distribution transformer, 158
Diverter, 340
Dot convention, 167
Double-cage rotors, 272–274
Dynamic braking, 352

Eddy current loss, 20, 23, 194, 463
Eddy currents, 23, 160
Effective flux, 163
Efficiency, 18
  of dc machines, 348–352
  energy, 19
  of induction machines, 19, 256, 262
  maximum, 26
  power, 19
  of transformers, 25, 27, 194, 218–220
Electrical degrees, 42, 48, 62, 66, 68
Electrical loading, 472, 481
  specific, 472–473, 481
Electrical losses, 20
Electrical machines, 1, 7, 13, 17
Electromagnetic systems, 447–485
  coenergy, 449, 467, 468
  conservative, 452, 463
  doubly excited, 457
  force relationships, 454, 457, 463, 465
  linear, 449–460, 468
  magnetic field energy, 447, 459, 470
  non-linear, 460
  singly excited, 451, 461
  stored energy, 447–451
  torque, 451, 454, 456–457, 459, 461, 469, 471
    and design parameters, 472–482
End bell, 8
Energy efficiency, 19
Energy product, 372
English system of units, 4
Equalizers, 381
Exciter, 44
Exciter alternator, 36, 44, 46
Exciter rectifier, 46
Exciting current, 8
  of induction machines, 251, 252
  of transformers, 172, 180, 195

Faraday's law, 5, 8, 17, 21, 28, 66, 78, 129, 159, 161, 181, 239, 327, 444, 448
Field, 17, 38, 44, 334, 340–341
  copper loss, 334, 351
Field discharge resistor, 335
Field impedance, 261
  backward, 413
  forward, 261, 413

Field rheostat, 335, 350
Field winding, 8, 47–50, 320, 330, 334–336
  effective turns per pole, 50, 330
"Flashing the field", 377
Flashover, 361
Flux, 5, 14, 20, 59–63, 78, 163, 239
  transient, 181
Flux linkages, 5, 20, 65, 66, 79, 162–169
Flux per pole, 61, 83
Fourier Series, 47, 52
Fractional slot windings, 76
Friction and windage, 20, 24, 81, 92, 350

Gear, 7
Generalized forces and displacements, 464
Generator, 1, 5, 17, 38, 90–92, 104–107, 377–380
  buildup, 377
  generated voltage, 5, 64–78, 82
Governor, 104, 105
Group voltage, 68, 70

Hall-effect devices, 132
Harmonic order, 71
Harmonics, 71–78
  belt, 74
  slot, 75
  triplen, 73
High-slip motors, 276
"Hot-spot" temperature, 13
Hunting, 94
Hydro alternators, 44
Hysteresis, 20, 425
  loop, 20, 182
  loss, 23, 160, 194, 463
Hysteresis motor, 423

Ideal transformer, 171, 176
Induction machines, 239–318
  air-gap flux, 239, 240
  air-gap power, 256, 257–259
  approximations, 269
  asynchronous generator, 302–315. *See also* Generator mode
  blocked-rotor test, 279
  brake mode, 266. 267, 295
  breakdown torque, 265, 267
    slip for, 267
  circle diagram, 293–303
  circuit model, 247
    approximate, 294
  consequent poles, 288–289
  copper loss, 256, 259
  core losses, 256
  current turns ratio, 255
  dc test, 278
  developed mechanical power, 257, 258
  efficiency, 19, 257
  field impedance, 261
  generator mode, 18, 239, 302–315

Induction machines, generator mode (*Continued*)
  excitation, 307
 high slip, 272, 276
 leakage reactance, 249, 253, 256, 271, 273, 274, 281
 linear motors, 428
 locked-rotor current, 275. *See also* Blocked rotor
 magnetizing reactance, 251
 maximum torque. *See* Breakdown torque
 motor operation, 239
 no-load test, 277
 output power, 257
 performance calculations, 256, 260–264
 pole amplitude modulation (PAM), 289
 power factor, 298
 power flow, 256
 pull-out torque, 265, 268
 rotational loss, 256
 rotor, 14, 239–243, 271–276
  squirrel cage, 240–243, 271–276
   deep bar, 272–274
   double cage, 272, 274
  wound, 240, 241, 267
 rotor circuit, 239, 252
 rotor copper loss, 256, 259, 260
 rotor core, 240, 241
 rotor design, 271
  NEMA classes, 272–274
 rotor MMF, 244–245, 476
 rotor voltage, 239
 slip, 245
 speed, 18, 239
 speed control, 287
  by frequency, 293
 starting, 268, 284
  part-winding, 287
  Y–Δ, 287
 starting compensator, 285
 starting current, 240, 264, 285
 starting torque, 262–264
 stator, 18, 240, 242, 243
 stator circuit, 247
 stator MMF, 242, 243, 290
 stray-load loss, 256, 257
 synchronous speed, 242
 torque, 17, 258, 442, 476–477
  vs. speed, 264
  in synchronous watts, 258
 voltage turns ratio, 253
Infinite bus, 105
Instrument transformers, 231
Insulation systems, 7, 12
Interpoles (commutating poles), 12, 363–366
Inverter, 130, 131, 135–137, 333

Jogging, 357

Laminations, 23, 37, 38, 240, 241
Lap windings, 68, 325, 380–382

Leakage fields, 78, 163–168, 256
Leakage impedance, 81
Leakage inductance, 167
Leakage reactance, 80
 differential leakage, 78
 end leakage, 78
 of induction machines, 249, 253, 256, 271, 273, 274, 281
 Potier method, 119
 slot leakage, 78
 of synchronous machines, 78, 80
 of transformers, 169
Lenz's law, 23, 161
Linear motors, 6, 426
 end effects, 430
 tubular motor, 431
Line starting, 285
Line voltage, 507
Load current, 8
Locked rotor, 253, 275, 279, 282, 284, 296
Locked-rotor current, 275
Locked-rotor test, *see* Blocked-rotor test
Losses, 20, 24, 28, 81, 92
 brush-drop, 20
 constant, 26
 cooling, 24
 copper, 20, 28
 core, 20, 24, 28, 81, 92, 194, 256, 319, 425, 477
  eddy current, 20
  hysteresis, 20
 of dc machines, 348
 friction, 20, 81, 92
 of induction machines, 256, 257, 259
 stray-load, 20, 257
 ventilation, 24
 windage, 20, 24, 81, 92, 256, 349, 350

Machine speed, 7
Magnetic axes, 47, 54, 86, 132–134
Magnetic circuits, 491
 graphical methods, 496
 magnetization curve, 499
 problem types, 493
 superposition in, 500
Magnetic concepts, 489
Magnetic field:
 energy, 447, 459
  in linear systems, 449, 468
 energy density, 450, 470
 function of, in generator/motor, 5
Magnetic flux, *see* Flux
Magnetic loading, specific, 472–473
Magnetic polarity, 14
Magnetic poles, 11, 13, 16, 17, 44, 50, 52, 61
 of dc machines, 320, 323
 effective permeance of, 64, 83
 of embedded windings, 16, 48–50, 52
 of induction machines, 18, 243–247
 of polyphase windings, 17, 52

salient poles, 8, 13
  of synchronous machines, 11, 17, 44, 50, 52, 61
Magnetic units, 490
Magnetization curve, 499. *See also* Open-circuit characteristic
  of dc machines, 330–331, 337–340
  of induction machines, 307–310
  of transformers, 182
Magnetizing, 125
Magnetizing current, 8
  of induction machines, 251
  of transformers, 182, 184, 195
Magnetizing reactance, 89, 114
  of induction machines, 251, 313
  of synchronous machines, 89, 114
  of transformers, 167, 169, 184
Magnetizing winding, 320
Main winding, 398
Maximum efficiency, 26
Maximum power, 91, 141
Maximum power transfer, 267
Maximum torque:
  of induction machines, 265, 267
  of synchronous machines, 91, 98, 537–538
Mechanical degrees, 48
Mechanical losses, 20
Mechanical power, 3
Mechanical torque, 3
Mechanical work, 3
Metadynes, 384–390
Motor, definition, 1, 5
Motor–generator set (M–G set), 345
Mutual flux, 164, 166
Mutual inductance, 168

Nameplate, 27
NEMA:
  code letters, 284
  design classes, 272–276
  terminal markings, 289, 336
No-load test:
  of induction machines, 277
  of transformers, 220, 223

Open-circuit characteristic:
  of dc machines, 330–331, 337–340
  of synchronous machines, 108, 541–542
Open-circuit test, 220, 223
Open Y–open Δ connection, 213, 216

Pancake rotor, 440
Permanent-magnet alternators, 123
Permanent-magnet dc machines, *see* PMDC machines
Permanent magnet materials, 371–375
Permanent magnets, 370–377
Permeance per pole, effective, 64, 330, 472
  affected by saturation, 64, 330
Per-unit system:

in dc machines, 348–356
  in synchronous machines, 89
  in transformers, 193
Phase group, 40
Phase voltage, 67, 507
Phase winding 40–44, 51. *See also* Auxiliary winding
Pitch factor, 56, 66
  harmonic, 74
Plugging, 357
PMDC machines, 370–377, 481
  shear line, 374
  speed control, 342, 370
Pole-amplitude modulation, 289
Pole cores, 11, 321
Pole face, 320
Poles, *see* Magnetic poles
Pole shoes, 11, 320
Potential transformers, *see* Voltage transformers
Potier method, 119, 540, 542
  reactance, 121
  triangle, 119, 542
Power, 3
  apparent, 506
  electrical, 90
  mechanical, 3, 90
Power angle, 90, 536–539
Power efficiency, 19
Power factor, 19, 98–106, 506, 511
Power system, 155
Primary, 8
  of induction motor, 11
  of repulsion motor, 410
  of transformer, 11, 158
Prime mover, 104, 105
PSC motors, 404
Pulley, 7
Pull-out torque, 265, 267
Pulse modulation, 131

q-axis, *see* Quadrature axis
Quadrature axis, 47, 132, 385–389, 427, 529–540

Ratings, 27
Regeneration, 346
Regulation, 28
  speed, 28, 343
  in transformers, 29, 200
    by per unit, 203
  voltage, 28, 105, 200, 203, 379
Reluctance force or torque, 428, 432, 440, 456, 539
Repulsion-induction motor, 411
Repulsion motor, 410
Repulsion-start induction-run motor, 411
Resultant field, 469, 474
  in induction machine, 445, 448
  of synchronous machine, 59–66
  of three-phase winding, 25–59
  of transformers, 172

Rosenburg generator, 384
Rotational loss, 256, 350
Rotor, 7, 11
  basket weave, 441
  of dc machine, 319
  deep bar, 272–274
  direct axis, 47, 60, 128, 132–134, 385–388,
    427, 529–536
  double cage, 272, 274
  of induction machines, 14, 239–243, 271–276
  pancake, 440
  quadrature axis, 47, 128, 132–134, 385–388,
    427, 529–536
  squirrel cage, 240–243, 271–276
  of synchronous machine, 42–50
    MMF, 47
  wound, 240, 241, 267
Rotor speed, 7, 35
Runaway, 343
Run winding, 398

Salient-pole rotor, 11, 18, 44, 529–545
Salient poles, 8, 11, 13, 44
Salient-pole theory, 47, 529–545
Saturated synchronous reactance method, 115
  approximate, 121
Saturation curve, *see* Open-circuit characteristic
Saturation factor, 116, 535
Scott connection, 217
Secondary, 8
  of induction motor, 11
  of repulsion motor, 410
  of transformer, 11, 158
Selsyns, *see* Synchros
Series field, 336
  diverter, 340
Shaded-pole motor, 405
  speed control, 409
Shaft, 7
Shaft position sensor, 129, 132–134, 140, 334
Shaft speed, 7
Shear line, 374
Shell form, 160
Short-circuit characteristic, 108, 110, 541
Short-circuit ratio, 113
Short-circuit test, 222, 224
Short-circuit transients, 144
Shunt field, 335, 336
  discharge resistor, 335
  rheostat, 335
Single-phase alternators, 125–129
  reconnected three-phase, 125, 126
  third harmonic in, 129
Single-phase circuits, 505
Single-phase induction motors, 396
  auxiliary winding, 398
  capacitor-start motors, 403
  circuit model, 411
  PSC motors, 404
  rotating field, 399

  split-phase motor, 401
  starting switch, 402
  two-value capacitor motors, 404
Single-phase motors, 395
  induction motors, 396
  repulsion motors, 410
  shaded pole, 405
  universal motor, 395
Singly excited systems, 451
S.I. system of units, 3
Skew factor, 77
Skewing, 76, 77, 320
Slip, 245, 426
  for maximum torque, 267
Slip rings, 17
Slip test, 535
Slot conductors, 8, 13, 16, 38, 454
Slot harmonics, 75
Slot liner, 13
Slot pitch, 40, 68
Slots, 13, 14, 38, 40, 42, 44, 319, 454
Slots per pole, 68
Slot wedges, 13, 320
Solenoid, 451, 455
Space harmonics, 71
Speed control:
  by armature rheostat, 344
  by armature terminal voltage, 345
    chopper drive, 348
  of dc machinery, 342
  by frequency, 293
  of induction motors, 287
  of shaded-pole motors, 409
  shunt-field control, 343
  by terminal voltage, 345
  Ward–Leonard system, 345–347
Speed regulation, 28, 343
Spider, 240
Split-phase motor, 401
Stabilized shunt motor, 336
Stacking factor, 160, 494, 496
Starters, dc, 352–359
  constant-speed, 353
  counter-emf, 353
  time-delay, 535
  variable-speed, 353
Starting:
  dc motors, 344, 353–359
  induction motors, 268, 284
  part-winding, 287
  synchronous motors, 93
  Y-Δ, 287
Starting compensators, 285
Starting switch, 402
Starting torque, 262–264, 266
Stator, 7
  core, 7, 14, 37–40, 242
  of dc machine, 320, 321, 334–336, 362, 364–
    366, 386–387
  of induction machine, 18, 240, 242, 243

of synchronous machine, 36–44, 50–59

Stator winding, 7, 16, 17, 37–44, 240, 320, 334–336, 362, 364–366, 386–387

Steinmetz, Charles P., 165, 166, 171, 180, 247, 251, 255

Steinmetz formula, 23

Steinmetz model:
  of induction machine, 247–256
  of transformer, 165–198

Stepper motors, 431
  half-stepping, 433, 435
  hybrid type, 432, 435
  microstepping, 434
  permanent magnet type, 432
  slewing, 436
  step size, 434
  torque, 432, 438
  variable reluctance type, 432, 434

Stray-load losses, 20, 351

Subtransient reactance, 146

Switched reluctance motor, 438

Synchronizing, 106

Synchronous condenser, 92–93

Synchronous machine, 1, 7, 17, 35–153, 251, 334, 474–475, 529–545
  alternator, 1, 35, 36. *See also* Generator
  approximate x_d, 111
  armature, 7, 37–44, 50–59, 111. *See also* Windings, polyphase
    induced voltage, 64–78, 82
  armature MMF field, 50–59, 529
  armature reaction, 50, 89
  armature winding, *see* Armature
    double layer, 38
  Blondel two-reaction theory, 529
  brushless dc motor, 129
  brushless excitation, 44, 46, 124
  capability curve, 141
  characteristics, 135
  circuit model, 78–89
  coil voltage, 64–67
  compounding curve, 100
  construction, 36
  core losses, 81, 92
  cylindrical rotor, 12, 44
  cylindrical rotor theory, 47, 474, 529
  damper winding, 93–95
  direct axis synchronous reactance, 89. *See also* synchronous reactance
  d-q model, 529–534
  equivalent circuit, 82
  excitation calculation, 115–119, 121–123, 536, 540–544
  field windings, 11, 17, 47–50
  generator operation, 17, 38, 90–92, 104–107, 529–539
  hunting, 94
  hydroalternators, 44
  induced voltage, 64–78, 82, 474
  leakage impedance, 81
  leakage reactance, 80, 119–121
  losses, 81, 92
  magnetizing reactance, 89, 532
  maximum power, 91, 537–538
  maximum torque, 91, 98, 537–538
  MMF of armature (stator), 18, 50, 529
  motor operation, 18, 35, 92–102, 539–544
  open-circuit characteristic, 108, 541–542
  permanent magnet (PM), 36, 123
  per-unit system, 89
  phase voltage, 67
  phase winding, 36, 40–44, 51
  phasor diagram, 86–89, 531–537
  Potier method, 119–121, 535, 542
  Potier reactance, 121, 542
  Potier triangle, 119, 542
  power, 81, 90
  power angle, 90, 536–539
  power factor, 98–105
  for pumped storage, 35
  reconnections for single phase, 125
  resultant MMF, 59, 127
  rotor, 11, 42, 44
    magnetic axes, 47, 529–536
    MMF, 47
  salient-pole, 11, 18, 44, 529–545
    reluctance torque, 539
  saturation factor, 116, 535
  short-circuit characteristic, 108, 110, 541
  short-circuit ratio, 113
  short-circuit transients, 144
  single-phase, 125
    capacity, 125
    reconnected 3-phase, 125–127
    third harmonic in, 129
    third harmonic voltage, 139
  slip test, 535
  stator core, 37–40
  stator field, 50
  stator winding, *see* Armature
  subtransient 144, 146
  synchronizing, 106
  synchronous condenser, 92–93
  synchronous reactance, 89
    approximate, 108, 111
    direct-axis, 532
    measurement, 107
      approximate, 108, 111
      saturated, 115
      unsaturated, 113
    quadrature-axis, 532
    related to power and torque, 91
  synchronous speed, 17, 35
  terminal voltage, 71, 81, 101
  third harmonic voltage, 61, 72–73
  torque, 81, 90, 474–475
  transient reactance, 146
  transients, 144
  trapped flux, 145

Synchronous machine (*Continued*)
  V-curves, 99, 103, 106
  vector diagram, 81, 82–89
  voltage harmonics, 31, 71–78
  voltage induced, 64–78, 82, 474
  zero power factor characteristic, 541, 542
Synchronous motors, 18, 35, 92–102, 423, 427,
    428, 539
  effect of field current, 95–100
  hysteresis-reluctance type, 428
  hysteresis type, 423
  induction start, reluctance run, 428
  power factor, 97–100
  reluctance type, 427
  starting, 93
  torque, 81, 90, 474–475
  V-curves, 99, 103
Synchronous reactance, 89
  approximate, 108, 111, 112
  direct-axis, 89, 532
  measurement, 107, 534
  quadrature-axis, 532
  related to power and torque, 91, 536–539
  saturated, 115
  unsaturated, 113, 536
Synchronous speed, 17, 35, 242
Synchronous watts, 258
Synchros, 443
  control transformer, 445
  differential, 445

T-connection, 213, 216
Teeth, 320
Tertiary windings, 204, 208
Tesla, N., 239
Thevenin's theorem, 165, 264, 329
Three basic machines, 6
Three-phase circuits, 505
  Δ connection, 508, 511
  Y connection, 508, 514
Three-phase systems, 35, 239
  transformers for, 155, 158, 204–221
Torque, 3, 5, 17, 451, 454, 456–457, 459, 461,
    469, 471
  of dc machines, 17
  and design parameters, 472–482
  from field energy, 447–545
    in cylindrical air gaps, 469
  of hysteresis motors, 426
  of induction machines, 17
  of stepper motors, 432
  of synchronous machines, 17
  in synchronous watts, 258
Torque motors, 441
  ac types, 441
  dc types, 442
Transformers, 1, 155–237
  autotransformers, 227–231
    rating advantage, 229
  circuit model, 186

  approximate, 189, 192
  linear, 174
copper loss, 194
core, 158, 160
core excitation, 180
core form, 160
core loss, 194
core-loss current, 182, 195
coupling coefficient, 168
current, 158, 232
current inrush, 185
current ratio, 155, 172
distribution, 158
dot convention, 162
dry type, 159
eddy current loss, 194
eddy currents, 160,
efficiency, 194, 218–220
equivalent impedance, 190, 197
equivalent tee, 191
exciting current, 172, 180, 195
flux linkages, 162, 163–169
hysteresis loop, 182
hysteresis loss, 160, 194
ideal, 171, 176
instrument, 231
  accuracy class, 231
leakage inductances, 167
leakage reactance, 169
losses, 20, 194
magnetic fluxes, 159, 163
magnetizing current, 182, 184, 195
magnetizing reactance, 167, 169, 184
mutual inductance, 168
oil, 159, 160
open-circuit test, 220, 223
peak flux, 182
per-unit system, 193
potential, voltage, 158, 231
power, 157
in power systems, 155
primary, 158
primary voltage, 198
secondary, 158
shell form, 160
short-circuit test, 222, 224
single-phase, 158
stacking factor, 160
Steinmetz model, 165–198
  approximate, 189
step-down, 155
step-up, 155
tank, 159, 169
tertiary windings, 204, 208
three-phase, 155, 158, 204
  banks, 158, 204
  characteristics, 204
  core, 204
  efficiency of, 204
  equivalent Y–Y, 209–221

measurements on, 223
open Y–open Δ, 213, 216
Scott connection, 217
T-connected, 213, 216
third harmonic in, 208
tertiary windings, 204
transformers, 204
V–V, 213, 214
transient flux, 181, 185
turns ratio, 159
unit, 157
voltage, 158, 231
voltage ratio, 155, 171
voltage regulation, 200
by per unit, 203
volts per turn, 182
winding resistances, 165
Transient reactance, 146
Trapped flux, 145
Triplen harmonics, 73
Tubular motors, 431
Turboalternators, 11
Turns, of a coil, 38
Two-phase motors, 17–18, 398
Two-value capacitor motors, 404
Two-wattmeter method, 524

Unit transformer, 157
Universal motor, 395
Utilization voltage, 157

VARs, 100, 101, 506
V-curves, 99, 103, 106
Virtual work, principle, 447
Voltage dip, 285
Voltage transformers, 158, 230, 231
Voltage turns per phase, effective, 70, 249, 475
V–V connection, 213, 214

Ward–Leonard system, 345–347
Wattmeters, 524
Windage loss, 20, 24, 81, 92, 256, 349, 350
Winding factor, 69, 249
harmonic, 77

Windings, 8
bifilar, 436
for commutator machines, 380
basket weave, 441
equalization, 381
frog-leg, 384
lap, 325, 380–382
multiplex, 380, 381
printec circuit, 440
wave, 382
double layer, 38
functions of, 8
polyphase, 17, 36, 40, 67
coil pitch, 40, 48, 52, 56, 59, 68, 290
coils per group, 40, 41, 52, 57
coil voltage, 64–67
current patterns, 50–54
distribution factor, 69
effective current turns per pole, 54, 55, 243, 474
effective turns per phase group, 52
group voltage, 68, 70
leakage impedance, 81
leakage-induced voltage, 79
leakage reactance, 80
MMF of, 50–59
MMF per pole, 52
phase group, 40
phase voltage, 67
pitch factor, 56
rotating MMF, 50
single-phase connections, 126
skew factor, 77
slot harmonics, 75
slot pitch, 40, 52
slots per pole, 68
voltage harmonics, 71–78
voltage induced, 67, 474
voltage turns per phase, effective, 70, 475
winding factor, 69, 249
Wound rotor, 267

Y connection, 72, 240, 508, 514

Zero speed switch, 359

2250   1425
525   525